DATE DUE

JA 02 '07			
NO 08 '15			

Hinduism and Ecology

Publications of the Center for the Study of World Religions, Harvard Divinity School

General Editor: Lawrence E. Sullivan
Senior Editor: Kathryn Dodgson
Associate Editor: Eric Edstam

Religions of the World and Ecology

Series Editors: Mary Evelyn Tucker and John Grim

Cambridge, Massachusetts

Hinduism and Ecology

The Intersection of Earth, Sky, and Water

edited by

CHRISTOPHER KEY CHAPPLE

and

MARY EVELYN TUCKER

distributed by
Harvard University Press
for the
Center for the Study of World Religions,
Harvard Divinity School

Grateful acknowledgment is made for permission to reprint the following:

A revised version of Vinay Lal, "Gandhi and the Ecological Vision of Life: Thinking beyond Deep Ecology," *Environmental Ethics* 22, no. 2 (summer 2000): 149–68.

Library of Congress Cataloging-in-Publication Data

Hinduism and ecology : the intersection of earth, sky, and water / edited by Christopher Key Chapple and Mary Evelyn Tucker.
p. cm. — (Religions of the world and ecology)
Includes bibliographical references and index.
ISBN 0-945454-25-2 (hardcover : alk. paper)
ISBN 0-945454-26-0 (paperback : alk. paper)
1. Ecology—Religious aspects—Hinduism. 2. Nature—Religious aspects—Hinduism. 3. Natural History—India. 4. Hinduism—Doctrines. I. Chapple, Christopher. II. Tucker, Mary Evelyn. III. Series.

BL1215.N34 H56 2000
294.5'178362—dc21

00-059665

Acknowledgments

The series of conferences on religions of the world and ecology took place from 1996 through 1998, with supervision at the Harvard University Center for the Study of World Religions by Don Kunkel and Malgorzata Radziszewska-Hedderick and with the assistance of Janey Bosch, Naomi Wilshire, and Lilli Leggio. Narges Moshiri, also at the Center, was indispensable in helping to arrange the first two conferences. A series of volumes developing the themes explored at the conferences is being published by the Center and distributed by Harvard University Press under the editorial direction of Kathryn Dodgson and with the skilled assistance of Eric Edstam.

These efforts have been generously supported by major funding from the V. Kann Rasmussen Foundation. The conference organizers appreciate also the support of the following institutions and individuals: Aga Khan Trust for Culture, Association of Shinto Shrines, Nathan Cummings Foundation, Dharam Hinduja Indic Research Center at Columbia University, Germeshausen Foundation, Harvard Buddhist Studies Forum, Harvard Divinity School Center for the Study of Values in Public Life, Jain Academic Foundation of North America, Laurance Rockefeller, Sacharuna Foundation, Theological Education to Meet the Environmental Challenge, and Winslow Foundation. The conferences were originally made possible by the Center for Respect of Life and Environment of the Humane Society of the United States, which continues to be a principal cosponsor. Bucknell University, also a cosponsor, has provided support in the form of leave time from teaching for conference coordinators Mary Evelyn Tucker and John

Grim as well as the invaluable administrative assistance of Stephanie Snyder. Her thoughtful attention to critical details is legendary. President William Adams of Bucknell University and Vice-President for Academic Affairs Daniel Little have also granted travel funds for faculty and students to attend the conferences. Grateful acknowledgment is here made for the advice from key area specialists in shaping each conference and in editing the published volumes. Their generosity in time and talent has been indispensable at every step of the project. Throughout this process, the support, advice, and encouragement from Martin S. Kaplan has been invaluable.

Several others contributed to the shaping of this particular volume. Arvind Sharma served as co-convenor of the Hinduism and Ecology conference, which took place in October 1997. The volume editors wish also to acknowledge scholars who participated in the conference but whose contributions do not appear in this volume: Goswami Srivatsa, Ranchor Prime, Kamla Chowdhry, Anne Feldhaus, Frank Korom, Frank Clooney, Diana Eck, Kusumita Pedersen, Jack Hawley, David Eckel, and Julius Lipner. Appreciation also goes to Jharana Jhaveri, who presented and discussed her documentary on the Narmada Valley, *Kaise Jeebo Re!* At Loyola Marymount University, Hope Blacker helped immensely with editorial preparation of the manuscript and the bibliography. Jin Kahn, Clemens Philippi, Louise Dobbs, Michael Bennett, and Carol Turner assisted in various phases of the project. We are deeply grateful.

Contents

Preface

LAWRENCE E. SULLIVAN

Religion distinguishes the human species from all others, just as human presence on earth distinguishes the ecology of our planet from other places in the known universe. Religious life and the earth's ecology are inextricably linked, organically related.

Human belief and practice mark the earth. One can hardly think of a natural system that has not been considerably altered, for better or worse, by human culture. "Nor is this the work of the industrial centuries," observes Simon Schama. "It is coeval with the entirety of our social existence. And it is this irreversibly modified world, from the polar caps to the equatorial forests, that is all the nature we have" (*Landscape and Memory* [New York: Vintage Books, 1996], 7). In Schama's examination even landscapes that appear to be most free of human culture turn out, on closer inspection, to be its product.

Human beliefs about the nature of ecology are the distinctive contribution of our species to the ecology itself. Religious beliefs—especially those concerning the nature of powers that create and animate—become an effective part of ecological systems. They attract the power of will and channel the forces of labor toward purposive transformations. Religious rituals model relations with material life and transmit habits of practice and attitudes of mind to succeeding generations.

This is not simply to say that religious thoughts occasionally touch the world and leave traces that accumulate over time. The matter is the other way around. From the point of view of environmental studies, religious worldviews propel communities into the world with fundamental predispositions toward it because such religious worldviews are primordial, all-encompassing, and unique. They are *primordial* because they probe behind secondary appearances and stray thoughts to rivet human attention on realities of the first order: life at its source, creativity in its fullest manifestation, death and destruction at their origin, renewal and salvation in their germ. The revelation of first things is compelling and moves communities to take creative action. Primordial ideas are prime movers.

Religious worldviews are *all-encompassing* because they fully absorb the natural world within them. They provide human beings both a view of the whole and at the same time a penetrating image of their own ironic position as the beings in the cosmos who possess the capacity for symbolic thought: the part that contains the whole—or at least a picture of the whole—within itself. As all-encompassing, therefore, religious ideas do not just contend with other ideas as equals; they frame the mind-set within which all sorts of ideas commingle in a cosmology. For this reason, their role in ecology must be better understood.

Religious worldviews are *unique* because they draw the world of nature into a wholly other kind of universe, one that appears only in the religious imagination. From the point of view of environmental studies, the risk of such religious views, on the one hand, is of disinterest in or disregard for the natural world. On the other hand, only in the religious world can nature be compared and contrasted to other kinds of being—the supernatural world or forms of power not always fully manifest in nature. Only then can nature be revealed as distinctive, set in a new light startlingly different from its own. That is to say, only religious perspectives enable human beings to evaluate the world of nature in terms distinct from all else. In this same step toward intelligibility, the natural world is evaluated in terms consonant with human beings' own distinctive (religious and imaginative) nature in the world, thus grounding a self-conscious relationship and a role with limits and responsibilities.

In the struggle to sustain the earth's environment as viable for future generations, environmental studies has thus far left the role of religion unprobed. This contrasts starkly with the emphasis given, for example, the role of science and technology in threatening or sustaining the ecology. Ignorance of religion prevents environmental studies from achieving its goals, however, for though science and technology share many important features of human culture with religion, they leave unexplored essential wellsprings of human motivation and concern that shape the world as we know it. No understanding of the environment is adequate without a grasp of the religious life that constitutes the human societies which saturate the natural environment.

A great deal of what we know about the religions of the world is new knowledge. As is the case for geology and astronomy, so too for religious studies: many new discoveries about the nature and function of religion are, in fact, clearer understandings of events and processes that began to unfold long ago. Much of what we are learning now about the religions of the world was previously not known outside of a circle of adepts. From the ancient history of traditions and from the ongoing creativity of the world's contemporary religions we are opening a treasury of motives, disciplines, and awarenesses.

A geology of the religious spirit of humankind can well serve our need to relate fruitfully to the earth and its myriad life-forms. Changing our habits of consumption and patterns of distribution, reevaluating modes of production, and reestablishing a strong sense of solidarity with the matrix of material life—these achievements will arrive along with spiritual modulations that unveil attractive new images of well-being and prosperity, respecting the limits of life in a sustainable world while revering life at its sources. Remarkable religious views are presented in this series—from the nature mysticism of Bashō in Japan or Saint Francis in Italy to the ecstatic physiologies and embryologies of shamanic healers, Taoist meditators, and Vedic practitioners; from indigenous people's ritual responses to projects funded by the World Bank, to religiously grounded criticisms of hazardous waste sites, deforestation, and environmental racism.

The power to modify the world is both frightening and fascinating and has been subjected to reflection, particularly religious reflection, from time immemorial to the present day. We will understand ecology better when we understand the religions that form the rich soil of memory and practice, belief and relationships where life on earth is rooted. Knowledge of these views will help us reappraise our ways and reorient ourselves toward the sources and resources of life.

This volume is one in a series that addresses the critical gap in our contemporary understanding of religion and ecology. The series results from research conducted at the Harvard University Center for the Study of World Religions over a three-year period. I wish especially to acknowledge President Neil L. Rudenstine of Harvard University for his leadership in instituting the environmental initiative at Harvard and thank him for his warm encouragement and characteristic support of our program. Mary Evelyn Tucker and John Grim of Bucknell University coordinated the research, involving the direct participation of some six hundred scholars, religious leaders, and environmental specialists brought to Harvard from around the world during the period of research and inquiry. Professors Tucker and Grim have brought great vision and energy to this enormous project, as has their team of conference convenors. The commitment and advice of Martin S. Kaplan of Hale and Dorr have been of great value. Our goals have been achieved for this research and publication program because of the extraordinary dedication and talents of Center for the Study of World Religions staff members Don Kunkel, Malgorzata Radziszewska-Hedderick, Kathryn Dodgson, Janey Bosch, Naomi Wilshire, Lilli Leggio, and Eric Edstam and with the unstinting help of Stephanie Snyder of Bucknell. To these individuals, and to all the sponsors and participants whose efforts made this series possible, go deepest thanks and appreciation.

Series Foreword

MARY EVELYN TUCKER and JOHN GRIM

The Nature of the Environmental Crisis

Ours is a period when the human community is in search of new and sustaining relationships to the earth amidst an environmental crisis that threatens the very existence of all life-forms on the planet. While the particular causes and solutions of this crisis are being debated by scientists, economists, and policymakers, the facts of widespread destruction are causing alarm in many quarters. Indeed, from some perspectives the future of human life itself appears threatened. As Daniel Maguire has succinctly observed, "If current trends continue, we will not."[1] Thomas Berry, the former director of the Riverdale Center for Religious Research, has also raised the stark question, "Is the human a viable species on an endangered planet?"

From resource depletion and species extinction to pollution overload and toxic surplus, the planet is struggling against unprecedented assaults. This is aggravated by population explosion, industrial growth, technological manipulation, and military proliferation heretofore unknown by the human community. From many accounts the basic elements which sustain life—sufficient water, clean air, and arable land—are at risk. The challenges are formidable and well documented. The solutions, however, are more elusive and complex. Clearly, this crisis has economic, political, and social dimensions which require more detailed analysis than we can provide here. Suffice it to say, however, as did the *Global 2000 Report:* ". . .once such global environmental problems are in motion they are difficult to reverse. In fact few if any of the problems addressed in the *Global 2000*

Report are amenable to quick technological or policy fixes; rather, they are inextricably mixed with the world's most perplexing social and economic problems."[2]

Peter Raven, the director of the Missouri Botanical Garden, wrote in a paper titled "We Are Killing Our World" with a similar sense of urgency regarding the magnitude of the environmental crisis: "The world that provides our evolutionary and ecological context is in serious trouble, trouble of a kind that demands our urgent attention. By formulating adequate plans for dealing with these large-scale problems, we will be laying the foundation for peace and prosperity in the future; by ignoring them, drifting passively while attending to what may seem more urgent, personal priorities, we are courting disaster."

Rethinking Worldviews and Ethics

For many people an environmental crisis of this complexity and scope is not only the result of certain economic, political, and social factors. It is also a moral and spiritual crisis which, in order to be addressed, will require broader philosophical and religious understandings of ourselves as creatures of nature, embedded in life cycles and dependent on ecosystems. Religions, thus, need to be reexamined in light of the current environmental crisis. This is because religions help to shape our attitudes toward nature in both conscious and unconscious ways. Religions provide basic interpretive stories of who we are, what nature is, where we have come from, and where we are going. This comprises a worldview of a society. Religions also suggest how we should treat other humans and how we should relate to nature. These values make up the ethical orientation of a society. Religions thus generate worldviews and ethics which underlie fundamental attitudes and values of different cultures and societies. As the historian Lynn White observed, "What people do about their ecology depends on what they think about themselves in relation to things around them. Human ecology is deeply conditioned by beliefs about our nature and destiny—that is, by religion."[3]

In trying to reorient ourselves in relation to the earth, it has become apparent that we have lost our appreciation for the intricate nature of matter and materiality. Our feeling of alienation in the modern period has extended beyond the human community and its patterns of material exchanges to our interaction with nature itself. Especially in technologically sophisticated urban societies, we have become removed from the recognition of our dependence on nature. We no longer know who we are as earthlings; we no longer see the earth as sacred.

Thomas Berry suggests that we have become autistic in our interactions with the natural world. In other words, we are unable to value the life and beauty of nature because we are locked in our own egocentric perspectives and shortsighted needs. He suggests that we need a new cosmology, cultural coding, and motivating energy to overcome this deprivation.[4] He observes that the magnitude of destructive industrial processes is so great that we must initiate a radical rethinking of the myth of progress and of humanity's role in the evolutionary process. Indeed, he speaks of evolution as a new story of the universe, namely, as a vast cosmological perspective that will resituate human meaning and direction in the context of four and a half billion years of earth history.[5]

For Berry and for many others an important component of the current environmental crisis is spiritual and ethical. It is here that the religions of the world may have a role to play in cooperation with other individuals, institutions, and initiatives that have been engaged with environmental issues for a considerable period of time. Despite their lateness in addressing the crisis, religions are beginning to respond in remarkably creative ways. They are not only rethinking their theologies but are also reorienting their sustainable practices and long-term environmental commitments. In so doing, the very nature of religion and of ethics is being challenged and changed. This is true because the reexamination of other worldviews created by religious beliefs and practices may be critical to our recovery of sufficiently comprehensive cosmologies, broad conceptual frameworks, and effective environmental ethics for the twenty-first century.

While in the past none of the religions of the world have had to face an environmental crisis such as we are now confronting, they remain key instruments in shaping attitudes toward nature. The unintended consequences of the modern industrial drive for unlimited economic growth and resource development have led us to an impasse regarding the survival of many life-forms and appropriate management of varied ecosystems. The religious traditions may indeed be critical in helping to reimagine the viable conditions and long-range strategies for fostering mutually enhancing human-earth relations.[6]

Indeed, as E. N. Anderson has documented with impressive detail, "All traditional societies that have succeeded in managing resources well, over time, have done it in part through religious or ritual representation of resource management."[7]

It is in this context that a series of conferences and publications exploring the various religions of the world and their relation to ecology was initiated by the Center for the Study of World Religions at Harvard. Coordinated by Mary Evelyn Tucker and John Grim, the conferences involved some six hundred scholars, graduate students, religious leaders, and environmental activists over a period of three years. The collaborative nature of the project is intentional. Such collaboration maximizes the opportunity for dialogical reflection on this issue of enormous complexity and accentuates the diversity of local manifestations of ecologically sustainable alternatives.

This series is intended to serve as initial explorations of the emerging field of religion and ecology while pointing toward areas for further research. We are not unaware of the difficulties of engaging in such a task, yet we have been encouraged by the enthusiastic response to the conferences within the academic community, by the larger interest they have generated beyond academia, and by the probing examinations gathered in the volumes. We trust that this series and these volumes will be useful not only for scholars of religion but also for those shaping seminary education and institutional religious practices, as well as for those involved in public policy on environmental issues.

We see such conferences and publications as expanding the growing dialogue regarding the role of the world's religions as moral forces in stemming the environmental crisis. While, clearly, there are major methodological issues involved in utilizing traditional philosophical and religious ideas for contemporary concerns, there are also compelling reasons to support such efforts, however modest they may be. The world's religions in all their complexity and variety remain one of the principal resources for symbolic ideas, spiritual inspiration, and ethical principles. Indeed, despite their limitations, historically they have provided comprehensive cosmologies for interpretive direction, moral foundations for social cohesion, spiritual guidance for cultural expression, and ritual celebrations for meaningful life. In our search for more comprehensive ecological worldviews and more effective environmental ethics, it is inevitable that we will draw from the symbolic and conceptual resources of the religious traditions of the world. The effort to do this is not without precedent or problems, some of which will be signaled below. With this volume and with this series we hope the field of reflection and discussion regarding religion and ecology will begin to broaden, deepen, and complexify.

Qualifications and Goals

The Problems and Promise of Religions

These volumes, then, are built on the premise that the religions of the world may be instrumental in addressing the moral dilemmas created by the environmental crisis. At the same time we recognize the limitations of such efforts on the part of religions. We also acknowledge that the complexity of the problem requires interlocking approaches from such fields as science, economics, politics, health, and public policy. As the human community struggles to formulate different attitudes toward nature and to articulate broader conceptions of ethics embracing species and ecosystems, religions may thus be a necessary, though only contributing, part of this multidisciplinary approach.

It is becoming increasingly evident that abundant scientific knowledge of the crisis is available and numerous political and economic statements have been formulated. Yet we seem to lack the political, economic, and scientific leadership to make necessary changes. Moreover, what is still lacking is the religious commitment, moral imagination, and ethical engagement to transform the environmental crisis from an issue on paper to one of effective policy, from rhetoric in print to realism in action. Why, nearly fifty years after Fairfield Osborne's warning in *Our Plundered Planet* and more than thirty years since Rachel Carson's *Silent Spring,* are we still wondering, is it too late?[8]

It is important to ask where the religions have been on these issues and why they themselves have been so late in their involvement. Have issues of personal salvation superseded all others? Have divine-human relations been primary? Have anthropocentric ethics been all-consuming? Has the material world of nature been devalued by religion? Does the search for otherworldly rewards override commitment to this world? Did the religions simply surrender their natural theologies and concerns with exploring purpose in nature to positivistic scientific cosmologies? In beginning to address these questions, we still have not exhausted all the reasons for religions' lack of attention to the environmental crisis. The reasons may not be readily apparent, but clearly they require further exploration and explanation.

In discussing the involvement of religions in this issue, it is also appropriate to acknowledge the dark side of religion in both its institutional expressions and dogmatic forms. In addition to their oversight with regard to the environment, religions have been the source of enormous manipulation of power in fostering wars, in ignoring racial and social injustice, and in promoting unequal gender relations, to name only a few abuses. One does not want to underplay this shadow side or to claim too much for religions' potential for ethical persuasiveness. The problems are too vast and complex for unqualified optimism. Yet there is a growing consensus that religions may now have a significant role to play, just as in the past they have sustained individuals and cultures in the face of internal and external threats.

A final caveat is the inevitable gap that arises between theories and practices in religions. As has been noted, even societies with religious traditions which appear sympathetic to the environment have in the past often misused resources. While it is clear that religions may have some disjunction between the ideal and the real, this should not lessen our endeavor to identify resources from within the world's religions for a more ecologically sound cosmology and environmentally supportive ethics. This disjunction of theory and practice is present within all philosophies and religions and is frequently the source of disillusionment, skepticism, and cynicism. A more realistic observation might be made, however, that this disjunction should not automatically invalidate the complex worldviews and rich cosmologies embedded in traditional religions. Rather, it is our task to explore these conceptual resources so as to broaden and expand our own perspectives in challenging and fruitful ways.

In summary, we recognize that religions have elements which are both prophetic and transformative as well as conservative and constraining. These elements are continually in tension, a condition which creates the great variety of thought and interpretation within religious traditions. To recognize these various tensions and limits, however, is not to lessen the urgency of the overall goals of this project. Rather, it is to circumscribe our efforts with healthy skepticism, cautious optimism, and modest ambitions. It is to suggest that this is a beginning in a new field of study which will affect both religion and ecology. On the one hand, this process of reflection will inevitably change how religions conceive of their own roles, missions, and identities, for such reflections demand a new sense of the sacred as not divorced from the earth itself. On the other hand, environmental studies can recognize that religions have helped to shape attitudes toward nature. Thus, as religions themselves evolve they may be indispensable in fostering a more expansive appreciation for the complexity and beauty of the natural world. At the same time as religions foster awe and reverence for nature, they may provide the transforming energies for ethical practices to protect endangered ecosystems, threatened species, and diminishing resources.

Methodological Concerns

It is important to acknowledge that there are, inevitably, challenging methodological issues involved in such a project as we are undertaking in this emerging field of religion and ecology.[9] Some of the key interpretive challenges we face in this project concern issues of time, place, space, and positionality. With regard to time, it is necessary to recognize the vast historical complexity of each religious tradition, which cannot be easily condensed in these conferences or volumes. With respect to place, we need to signal the diverse cultural contexts in which these religions have developed. With regard to space, we recognize the varied frameworks of institutions and traditions in which these religions unfold. Finally, with respect to positionality, we acknowledge our own historical situatedness at the end of the twentieth century with distinctive contemporary concerns.

Not only is each religious tradition historically complex and culturally diverse, but its beliefs, scriptures, and institutions have themselves been subject to vast commentaries and revisions over time. Thus, we recognize the radical diversity that exists within and among religious traditions which cannot be encompassed in any single volume. We acknowledge also that distortions may arise as we examine earlier historical traditions in light of contemporary issues.

Nonetheless, the environmental ethics philosopher J. Baird Callicott has suggested that scholars and others "mine the conceptual resources" of the religious traditions as a means of creating a more inclusive global environmental ethics.[10] As Callicott himself notes, however, the notion of "mining" is problematic, for it conjures up images of exploitation which may cause apprehension among certain religious communities, especially those of indigenous peoples. Moreover, we cannot simply expect to borrow or adopt ideas and place them from one tradition directly into another. Even efforts to formulate global environmental ethics need to be sensitive to cultural particularity and diversity. We do not aim at creating a simple bricolage or bland fusion of perspectives. Rather, these conferences and volumes are an attempt to display before us a multiperspectival cross section of the symbolic richness regarding attitudes toward nature within the religions of the world. To do so will help to reveal certain commonalities among traditions, as well as limitations within traditions, as they begin to converge around this challenge presented by the environmental crisis.

We need to identify our concerns, then, as embedded in the constraints of our own perspectival limits at the same time as we seek common ground. In describing various attitudes toward nature historically, we are aiming at *critical understanding* of the complexity, contexts, and frameworks in which these religions articulate such views. In addition, we are striving for *empathetic appreciation* for the traditions without idealizing their ecological potential or ignoring their environmental oversights. Finally, we are aiming at the *creative revisioning* of mutually enhancing human-earth relations. This revisioning may be assisted by highlighting the multiperspectival attitudes toward nature which these traditions disclose. The prismatic effect of examining such attitudes and relationships may provide some necessary clarification and symbolic resources for reimagining our own situation and shared concerns at the end of the twentieth century. It will also be sharpened by identifying the multilayered symbol systems in world religions which have traditionally oriented humans in establishing relational resonances between the microcosm of the self and the macrocosm of the social and natural orders. In short, religious traditions may help to supply both creative resources of symbols, rituals, and texts as well as inspiring visions for reimagining ourselves as part of, not apart from, the natural world.

Aims

The methodological issues outlined above were implied in the overall goals of the conferences, which were described as follows:

1. To identify and evaluate the *distinctive ecological attitudes,* values, and practices of diverse religious traditions, making clear their links to intellectual, political, and other resources associated with these distinctive traditions.

2. To describe and analyze the *commonalities* that exist within and among religious traditions with respect to ecology.

3. To identify the *minimum common ground* on which to base constructive understanding, motivating discussion, and concerted action in diverse locations across the globe; and to highlight the specific religious resources that comprise such fertile ecological ground: within scripture, ritual, myth, symbol, cosmology, sacrament, and so on.

4. To articulate in clear and moving terms *a desirable mode of human presence with the earth;* in short, to highlight means of respecting and valuing nature, to note what has already been actualized, and to indicate how best to achieve what is desirable beyond these examples.

5. To outline the most significant areas, with regard to religion and ecology, in need of *further study;* to enumerate questions of highest priority within those areas and propose possible approaches to use in addressing them.

In this series, then, we do not intend to obliterate difference or ignore diversity. The aim is to celebrate plurality by raising to conscious awareness multiple perspectives regarding nature and human-earth relations as articulated in the religions of the world. The spectrum of cosmologies, myths, symbols, and rituals within the religious traditions will be instructive in resituating us within the rhythms and limits of nature.

We are not looking for a unified worldview or a single global ethic. We are, however, deeply sympathetic with the efforts toward formulating a global ethic made by individuals, such as the theologian Hans Küng or the environmental philosopher J. Baird Callicott, and groups, such as Global Education Associates and United Religions. A minimum content of environmental ethics needs to be seriously considered. We are, then, keenly interested in the contribution this series might make to discussions of environmental policy in national and international arenas. Important intersections may be made with work in the field of development ethics.[11] In addition, the findings of the conferences have bearing on the ethical formulation of the Earth Charter that is to be presented to the United Nations for adoption within the next few years. Thus, we are seeking both the grounds for common concern and the constructive conceptual basis for rethinking our current situation of estrangement from the earth. In so doing we will be able to reconceive a means of creating the basis not just for sustainable development, but also for sustainable life on the planet.

As scientist Brian Swimme has suggested, we are currently making macrophase changes to the life systems of the planet with microphase wisdom. Clearly, we need to expand and deepen the wisdom base for human intervention with nature and other humans. This is particularly true as issues of genetic alteration of natural processes are already available and in use. If religions have traditionally concentrated on divine-human and human-human relations, the challenge is that they now explore more fully divine-human-earth relations. Without such further exploration, adequate environmental ethics may not emerge in a comprehensive context.

Resources: Environmental Ethics Found in the World's Religions

For many people, when challenges such as the environmental crisis are raised in relation to religion in the contemporary world, there frequently arises a sense of loss or a nostalgia for earlier, seemingly less complicated eras when the constant questioning of religious beliefs and practices was not so apparent. This is, no doubt, something of a reified reading of history. There is, however, a decidedly anxious tone to the questioning and soul-searching that appears to haunt many contemporary religious groups as they seek to find their particular role in the midst of rapid technological change and dominant secular values.

One of the greatest challenges, however, to contemporary religions remains how to respond to the environmental crisis, which many believe has been perpetuated because of the enormous inroads made by unrestrained materialism, secularization, and industrialization in contemporary societies, especially those societies arising in or influenced by the modern West. Indeed, some suggest that the very division of religion from secular life may be a major cause of the crisis.

Others, such as the medieval historian Lynn White, have cited religion's negative role in the crisis. White has suggested that the emphasis in Judaism and Christianity on the transcendence of God above nature and the dominion of humans over nature has led to a devaluing of the natural world and a subsequent destruction of its resources for utilitarian ends.[12] While the particulars of this argument have been vehemently debated, it is increasingly clear that the environmental crisis and its perpetuation due to industrialization, secularization, and ethical indifference present a serious challenge to the world's religions. This is especially true because many of these religions have traditionally been concerned with the path of personal salvation, which frequently emphasized otherworldly goals and rejected this world as corrupting. Thus, as we have noted, how to adapt religious teachings to this task of revaluing nature so as to prevent its destruction marks a significant new phase in religious thought. Indeed, as Thomas Berry has so aptly pointed out, what is necessary is a comprehensive reevaluation of human-earth relations if the human is to continue as a viable species on an increasingly degraded planet. This will require, in addition to major economic and political changes, examining worldviews and ethics among the world's religions that differ from those that have captured the imagination of contemporary industrialized societies which regard nature primarily as a commodity to be utilized. It should be noted that when we are searching for effective resources for formulating environmental ethics, each of the religious traditions have both positive and negative features.

For the most part, the worldviews associated with the Western Abrahamic traditions of Judaism, Christianity, and Islam have created a dominantly human-focused morality. Because these worldviews are largely anthropocentric, nature is viewed as being of secondary importance. This is reinforced by a strong sense of the transcendence of God above nature. On the other hand, there are rich resources for rethinking views of nature in the covenantal tradition of the Hebrew Bible, in sacramental theology, in incarnational Christology, and in the vice-regency (*khalifa Allah*) concept of the Qur'an. The covenantal tradition draws on the legal agreements of biblical thought which are extended to all of creation. Sacramental theology in Christianity underscores the sacred dimension of material reality, especially for ritual purposes.[13] Incarnational Christology proposes that because God became flesh in the person of Christ, the entire natural order can be viewed as sacred. The concept of humans as vice-regents of Allah on earth suggests that humans have particular privileges, responsibilities, and obligations to creation.[14]

In Hinduism, although there is a significant emphasis on performing one's *dharma,* or duty, in the world, there is also a strong pull toward *mokśa,* or liberation, from the world of suffering, or *saṃsāra*. To heal this kind of suffering and alienation through spiritual discipline and meditation, one turns away from the world (*prakṛti*) to a timeless world of spirit (*puruṣa*). Yet at the same time there are numerous traditions in Hinduism which affirm particular rivers, mountains, or forests as sacred. Moreover, in the concept of *līlā,* the creative play of the gods, Hindu theology engages the world as a creative manifestation of the divine. This same tension between withdrawal from the world and affirmation of it is present in Buddhism. Certain Theravāda schools of Buddhism emphasize withdrawing in meditation from the transient world of suffering (*saṃsāra*) to seek release in *nirvāṇa*. On the other hand, later Mahāyāna schools of Buddhism, such as Hua-yen, underscore the remarkable interconnection of reality in such images as the jeweled net of Indra, where each jewel reflects all the others in the universe. Likewise, the Zen gardens in East Asia express the fullness of the Buddha-nature (*tathāgatagarbha*) in the natural world. In recent years, socially engaged Buddhism has been active in protecting the environment in both Asia and the United States.

The East Asian traditions of Confucianism and Taoism remain, in certain ways, some of the most life-affirming in the spectrum of world religions.[15] The seamless interconnection between the divine, human, and natural worlds that characterizes these traditions has been described as an anthropocosmic worldview.[16] There is no emphasis on radical transcendence as there is in the Western traditions. Rather, there is a cosmology of a continuity of creation stressing the dynamic movements of nature through the seasons and the agricultural cycles. This organic cosmology is grounded in the philosophy of *ch'i* (material force), which provides a basis for appreciating the profound interconnection of matter and spirit. To be in harmony with nature and with other humans while being attentive to the movements of the *Tao* (Way) is the aim of personal cultivation in both Confucianism and Taoism. It should be noted, however, that this positive worldview has not prevented environmental degradation (such as deforestation) in parts of East Asia in both the premodern and modern period.

In a similar vein, indigenous peoples, while having ecological cosmologies have, in some instances, caused damage to local environments through such practices as slash-and-burn agriculture. Nonetheless, most indigenous peoples have environmental ethics embedded in their worldviews. This is evident in the complex reciprocal obligations surrounding life-taking and resource-gathering which mark a community's relations with the local bioregion. The religious views at the basis of indigenous lifeways involve respect for the sources of food, clothing, and shelter that nature provides. Gratitude to the creator and to the spiritual forces in creation is at the heart of most indigenous traditions. The ritual calendars of many indigenous peoples are carefully coordinated with seasonal events such as the sound of returning birds, the blooming of certain plants, the movements of the sun, and the changes of the moon.

The difficulty at present is that for the most part we have developed in the world's religions certain ethical prohibitions regarding homicide and restraints concerning genocide and suicide, but none for biocide or geocide. We are clearly in need of exploring such comprehensive cosmological perspectives and communitarian environmental ethics as the most compelling context for motivating change regarding the destruction of the natural world.

Responses of Religions to the Environmental Crisis

How to chart possible paths toward mutually enhancing human-earth relations remains, thus, one of the greatest challenges to the world's religions. It is with some encouragement, however, that we note the growing calls for the world's religions to participate in these efforts toward a more sustainable planetary future. There have been various appeals from environmental groups and from scientists and parliamentarians for religious leaders to respond to the environmental crisis. For example, in 1990 the Joint Appeal in Religion and Science was released highlighting the urgency of collaboration around the issue of the destruction of the environment. In 1992 the Union of Concerned Scientists issued the statement "Warning to Humanity," signed by over 1,000 scientists from 70 countries, including 105 Nobel laureates, regarding the gravity of the environmental crisis. They specifically cited the need for a new ethic toward the earth.

Numerous national and international conferences have also been held on this subject and collaborative efforts have been established. Environmental groups such as World Wildlife Fund have sponsored interreligious meetings such as the one in Assisi in 1986. The Center for Respect of Life and Environment of the Humane Society of the United States has also held a series of conferences in Assisi on

Spirituality and Sustainability and has helped to organize one at the World Bank. The United Nations Environmental Programme in North America has established an Environmental Sabbath, each year distributing thousands of packets of materials for use in congregations throughout North America. Similarly, the National Religious Partnership on the Environment at the Cathedral of St. John the Divine in New York City has promoted dialogue, distributed materials, and created a remarkable alliance of the various Jewish and Christian denominations in the United States around the issue of the environment. The Parliament of World Religions held in 1993 in Chicago and attended by some 8,000 people from all over the globe issued a statement of Global Ethics of Cooperation of Religions on Human and Environmental Issues. International meetings on the environment have been organized. One example of these, the Global Forum of Spiritual and Parliamentary Leaders held in Oxford in 1988, Moscow in 1990, Rio in 1992, and Kyoto in 1993, included world religious leaders, such as the Dalai Lama, and diplomats and heads of state, such as Mikhail Gorbachev. Indeed, Gorbachev hosted the Moscow conference and attended the Kyoto conference to set up a Green Cross International for environmental emergencies.

Since the United Nations Conference on Environment and Development (the Earth Summit) held in Rio in 1992, there have been concerted efforts intended to lead toward the adoption of an *Earth Charter* by the year 2000. This *Earth Charter* initiative is under way with the leadership of the Earth Council and Green Cross International, with support from the government of the Netherlands. Maurice Strong, Mikhail Gorbachev, Steven Rockefeller, and other members of the Earth Charter Project have been instrumental in this process. At the March 1997 Rio + 5 Conference a benchmark draft of the *Earth Charter* was issued. The time is thus propitious for further investigation of the potential contributions of particular religions toward mitigating the environmental crisis, especially by developing more comprehensive environmental ethics for the earth community.

Expanding the Dialogue of Religion and Ecology

More than two decades ago Thomas Berry anticipated such an exploration when he called for "creating a new consciousness of the multiform religious traditions of humankind" as a means toward renewal of the human spirit in addressing the urgent problems of contemporary society.[17] Tu Weiming has written of the need to go "Beyond the Enlightenment Mentality" in exploring the spiritual resources of the global community to meet the challenge of the ecological crisis.[18] While this exploration has also been the intention of both the conferences and these volumes, other significant efforts have preceded our current endeavor.[19] Our discussion here highlights only the last decade.

In 1986 Eugene Hargrove edited a volume titled *Religion and Environmental Crisis.*[20] In 1991 Charlene Spretnak explored this topic in her book *States of Grace: The Recovery of Meaning in the Post-Modern Age.*[21] Her subtitle states her constructivist project clearly: "Reclaiming the Core Teachings and Practices of the Great Wisdom Traditions for the Well-Being of the Earth Community." In 1992 Steven Rockefeller and John Elder edited a book based on a conference at Middlebury College titled *Spirit and Nature: Why the Environment Is a Religious Issue.*[22] In the same year Peter Marshall published *Nature's Web: Rethinking Our Place on Earth,*[23] drawing on the resources of the world's traditions. An edited volume titled *Worldviews and Ecology,* compiled in 1993, contains articles reflecting on views of nature from the world's religions and from contemporary philosophies, such as process thought and deep ecology.[24] In this same vein, in 1994 J. Baird Callicott published *Earth's Insights,* which examines the intellectual resources of the world's religions for a more comprehensive global environmental ethics.[25] This expands on his 1989 volumes, *Nature in Asian Traditions of Thought* and *In Defense of the Land Ethic.*[26] In 1995 David Kinsley issued a book titled *Ecology and Religion: Ecological Spirituality in a Cross-Cultural Perspective,*[27] which draws on traditional religions and contemporary movements, such as deep ecology and ecospirituality. Seyyed Hossein Nasr wrote his comprehensive study *Religion and the Order of Nature* in 1996.[28] Several volumes of religious responses to a particular topic or theme have also been published. For example, J. Ronald Engel and Joan Gibb Engel compiled a monograph in 1990 titled *Ethics of Environment and Development: Global Challenge, International Response*[29] and in 1995 Harold Coward edited the volume *Population, Consumption and the Environment: Religious and Secular Responses.*[30] Roger Gottlieb edited a useful source book, *This Sacred Earth: Religion, Nature, Environment.*[31] Single volumes on the world's religions and ecology were published by the Worldwide Fund for Nature.[32]

The series Religions of the World and Ecology is thus intended to expand the discussion already under way in certain circles and to invite further collaboration on a topic of common concern—the fate of the earth as a religious responsibility. To broaden and deepen the reflective basis for mutual collaboration was an underlying aim of the conferences themselves. While some might see this as a diversion from pressing scientific or policy issues, it was with a sense of humility and yet conviction that we entered into the arena of reflection and debate on this issue. In the field of the study of world religions, we have seen this as a timely challenge for scholars of religion to respond as engaged intellectuals with deepening creative reflection. We hope that these volumes will be simply a beginning of further study of conceptual and symbolic resources, methodological concerns, and practical directions for meeting this environmental crisis.

Notes

1. He goes on to say, "And that is qualitatively and epochally true. If religion does not speak to [this], it is an obsolete distraction." Daniel Maguire, *The Moral Core of Judaism and Christianity: Reclaiming the Revolution* (Philadelphia: Fortress Press, 1993), 13.

2. Gerald Barney, *Global 2000 Report to the President of the United States* (Washington, D.C.: Supt. of Docs. U.S. Government Printing Office, 1980–1981), 40.

3. Lynn White, Jr., "The Historical Roots of Our Ecologic Crisis," *Science* 155 (March 1967):1204.

4. Thomas Berry, *The Dream of the Earth* (San Francisco: Sierra Club Books, 1988).

5. Brian Swimme and Thomas Berry, *The Universe Story* (San Francisco: Harper San Francisco, 1992).

6. At the same time we recognize the limits to such a project, especially because ideas and action, theory and practice do not always occur in conjunction.

7. E. N. Anderson, Ecologies of the Heart: Emotion, Belief, and the Environment (New York and Oxford: Oxford University Press, 1996), 166. He qualifies this statement by saying, "The key point is not religion per se, but the use of emotionally powerful symbols to sell particular moral codes and management systems" (166). He notes, however, in various case studies how ecological wisdom is embedded in myths, symbols, and cosmologies of traditional societies.

8. *Is It Too Late?* is also the title of a book by John Cobb, first published in 1972 by Bruce and reissued in 1995 by Environmental Ethics Books.

9. Because we cannot identify here all of the methodological issues that need to be addressed, we invite further discussion by other engaged scholars.

10. See J. Baird Callicott, *Earth's Insights: A Survey of Ecological Ethics from the Mediterranean Basin to the Australian Outback* (Berkeley: University of California Press, 1994).

11. See, for example, The Quality of Life, ed. Martha C. Nussbaum and Amartya Sen, WIDER Studies in Development Economics (Oxford: Oxford University Press, 1993).

12. White, "The Historical Roots of Our Ecologic Crisis," 1203–7.

13. Process theology, creation-centered spirituality, and ecotheology have done much to promote these kinds of holistic perspectives within Christianity.

14. These are resources already being explored by theologians and biblical scholars.

15. While this is true theoretically, it should be noted that, like all ideologies, these traditions have at times been used for purposes of political power and social control. Moreover, they have not been able to prevent certain kinds of environmental destruction, such as deforestation in China.

16. The term "anthropocosmic" has been used by Tu Weiming in *Centrality and Commonality* (Albany: State University of New York Press, 1989).

17. Thomas Berry, "Religious Studies and the Global Human Community," unpublished manuscript.

18. Tu Weiming, “Beyond the Enlightenment Mentality,” in *Worldviews and Ecology,* ed. Mary Evelyn Tucker and John Grim (Lewisburg, Pa.: Bucknell University Press, 1993; reissued, Maryknoll, N.Y.: Orbis Books, 1994).

19. This history has been described more fully by Roderick Nash in his chapter entitled “The Greening of Religion,” in The Rights of Nature: A History of Environmental Ethics (Madison: University of Wisconsin Press, 1989).

20. *Religion and Environmental Crisis,* ed. Eugene Hargrove (Athens: University of Georgia Press, 1986).

21. Charlene Spretnak, *States of Grace: The Recovery of Meaning in the Post-Modern Age* (San Francisco: Harper San Francisco, 1991).

22. *Spirit and Nature: Why the Environment Is a Religious Issue,* ed. Steven Rockefeller and John Elder (Boston: Beacon Press, 1992).

23. Peter Marshall, *Nature's Web: Rethinking Our Place on Earth* (Armonk, N.Y.: M. E. Sharpe, 1992).

24. *Worldviews and Ecology,* ed. Mary Evelyn Tucker and John Grim (Lewisburg, Pa.: Bucknell University Press, 1993; reissued, Maryknoll, N.Y.: Orbis Books, 1994).

25. Callicott, *Earth's Insights.*

26. Both are State University of New York Press publications.

27. David Kinsley, *Ecology and Religion: Ecological Spirituality in a Cross-Cultural Perspective* (Englewood Cliffs, N.J.: Prentice Hall, 1995).

28. Seyyed Hossein Nasr, *Religion and the Order of Nature* (Oxford: Oxford University Press, 1996).

29. *Ethics of Environment and Development: Global Challenge, International Response,* ed. J. Ronald Engel and Joan Gibb Engel (Tucson: University of Arizona Press, 1990).

30. *Population, Consumption, and the Environment: Religious and Secular Responses,* ed. Harold Coward (Albany: State University of New York Press, 1995).

31. This Sacred Earth: Religion, Nature, Environment, ed. Roger S. Gottlieb (New York and London: Routledge, 1996).

32. These include volumes on Hinduism, Buddhism, Judaism, Christianity, and Islam.

Introduction

CHRISTOPHER KEY CHAPPLE

India, the birthplace of Hinduism, boasts the world's largest environmental movement. Over 950 nongovernmental organizations dedicated to environmental causes can be found in India.[1] From the polluted cities to rural lands threatened by dams or deforestation, concerned persons are making their voices heard.

Environmentalism in India

In India, the environmental movement differs significantly from its counterparts in North America and Europe. Ramachandra Guha, for instance, suggests that a Western-style program of environmental preservation will not work in India, due to the immediate, pressing needs of local populations.[2] Madhav Gadgil and Guha suggest that a fissure has emerged in Indian society that divides the population into omnivores, ecosystem people, and ecological refugees.[3] The omnivores, following the development model of the West, absorb the raw material of India as fuel for the development of urban industrial centers. Benefiting from government support, these people tend to live in cities, seek advanced levels of education, have small families, and surround themselves with "modern amenities such as electricity and tap water, television and agrochemicals."[4] Most of these people are upper caste and constitute India's much-lauded burgeoning middle class, arguably the largest middle class in the world.

Ecosystem people, by contrast, are rural and largely uneducated. Ecosystem people tend to have large families because their children, not being in school, are able to produce much-needed income at a young age. The women of these communities spend much of their time fetching water and fuel. Because of their impoverishment, this population does not participate in any significant way in the industrial paradigm unless they become ecological refugees. These ecological refugees flee the hardscrabble life of the countryside and flock to the cities, where they generally become day laborers and servants. In most cases, this has not resulted in improved educational opportunities for their children but has created a seemingly permanent underclass in the shantytowns sprinkled throughout India's urban areas. Though the ecosystem people and the ecological refugees have the highest birth rates and contribute to India's population crunch, they consume the least amount of resources per capita. Because of India's failure to create a technological and educational infrastructure "to support the employment of every one of its citizens in the modern sector,"[5] the population of ecosystem people and refugees continues to increase.

The environmental movement in India has to respond to these competing constituencies. The urban masses want to enjoy modern material comforts. Rural villagers want access to arable land. Those villagers who fail to thrive in the countryside flee to the cities where they live often marginal lives as uneducated industrial or service workers. Government policies have not created a fully integrated modern society for all segments of the population. Furthermore, in the past the government has advocated massive development projects, such as the damming of the Narmada River, to support the consumerist urban life-style. This reflected little or no regard for the life-paths of traditional peoples whose existence within the ecosystem does not demand the excessive consumption of natural resources found in the cities. In many regards, the environmental movement in India pits the living past against the modernized present, with many traditional peoples asserting that "they will not tolerate the continued degradation of the environment that has resulted from India's forced march toward industrialization."[6]

According to Patrick Peritore, a political scientist who has typologized environmental activists worldwide, Indian ecological advocates fall into three typologies: "Greens," who emphasize bioregionalism and respect for traditional ways of knowing; "Ecodevelopers," who advocate responsible programs for economic growth; and "Managers," who "give priority to human needs and rational management of environmental processes."[7] All three see a need to develop a "Dharmic administrative model" that integrates traditional values with secularism and attempts to create a modern, ecologically responsible world. Peritore notes that "the Gandhian ethos provides the environmental movement with a coherent ethic, metaphysic, and method of struggle as well as strong legitimation on the national political scene."[8] However, he goes on to conclude:

> India's environmental movement has the advantages of Gandhian religion, strong links to native cultural ecomanagement practices, an excellent intellectual and political infrastructure, and multiple points of access to national and local government. But its sophistication and strength is dissipated by a corrupt and bureaucratically tangled government, by a declining economy, and by an ecological and population crisis that surpasses known techniques of environmental repair and management. The movement, far from being a vanguard, is fighting a rearguard action for cultural and ecological survival.[9]

While acknowledging the vibrancy of environmentalism in India, Peritore provides a grim assessment of the future prospects for environmentalism in India.

Modernization, Industrialization, and Pollution

Although Mahatma Gandhi's campaign at achieving self-rule (*svarāj*) drew deeply from the well of religious inspiration, Jawaharlal Nehru, who served as prime minister of India from 1947 to 1964, mounted a program of industrial development rooted in secularity.[10] He urged the newly freed Indians to pursue science and technology as the key to the modernization of India. Gerald Larson has commented that

> Just as Gandhi had successfully created a mass political movement based on a Neo-Hindu vision of universalism, firmness in the truth (*satyagraha*) and nonviolence (*ahiṃsā*) in pre-partition India, so Nehru successfully created a comparable mass political movement based on a translation, or perhaps better, a kind of 'demythologization,' of that same Neo-Hindu vision in terms of 'secularism,' 'socialism,' 'a mixed economy,' 'democracy,' and 'non-alignment' in post-partition India.[11]

In the building of India after independence, the resources of the state were devoted to supporting mass secular education (some minority communities also receive support for religious education) and increasing India's industrial base.

Nehru's drive for modernization, which received a boost from the liberalization of economic policies in the 1980s, has been accompanied by a large population increase. In 1951, India's population was 361 million. Today, it is approximately one billion. Consequently, several environmental problems have emerged. O. P. Dwivedi has identified seven major "side effects" of industrial development:

1. a brutal assault on the nation's limited common land
2. loss of forest cover due to shortage of firewood and fodder, harvesting of trees for commercial purposes, and illegal encroachment
3. minimal or no pollution abatement for heavy industries
4. industrial areas located too close to residential areas, as seen in the Bhopal Union Carbide Plant disaster
5. unplanned urbanization: in 1951 62 million Indians lived in cities; the number has increased to 217 million, over 30 per cent in slums
6. large scale projects such as dams and mines that have displaced over 14 million people; only 3.9 million have been adequately relocated
7. severe pesticide pollution, causing several hundred deaths each year.[12]

Additionally, according to the World Bank, more than forty thousand people die prematurely every year in India because of air pollution.[13] New Delhi's air is rated as among the most polluted in the world. Yet Anil Agarwal comments that "the Air Pollution Act does not provide for any government action to control it."[14]

Water pollution continues to be a huge problem as well, as noted in *India Today:*

> Each of India's 13 major river basins—making up 80 percent of the total surface area and home to nearly 85 percent of the population—is so polluted . . . that bacteria feeding on the water are the only things that have proliferated . . . river water laced with industrial toxins is irrigating farmland . . . and urban aquifers . . . are now filling with sewage.[15]

Dwivedi notes that only 27 percent of the urban population of India has even limited sewerage facilities and that "out of a total population of 846 million in 1991, only about 14 percent enjoyed adequate sanitation."[16]

One of the difficulties encountered by environmental activists stems from a lack of awareness on the part of the general population as well as the government regarding the severity of the ecological ravage being felt throughout India. Part of this is due to the rapid rate of growth. Anil Agarwal notes:

> In the period 1975–1995, during which the gross domestic product increased 2.5 times, the industrial pollution load increased four times and the vehicular pollution load increased by eight times. . . . In 1986, when the then Prime Minister Rajiv Gandhi had asked me to address his council of ministers on the environmental challenges facing the country, I had told this powerful group that rural environmental problems are more important than urban environmental problems. I had said this because the land and forest degradation affects the lives of hundreds of millions of poor people, especially poor rural women, extremely adversely. Delhi was still quite clean. I had no idea about the speed with which this capital city will turn into a hell-hole in less than ten years. And today every metro and small town is rapidly following suit. I now realize how stupid I was and what poor environmental leadership I had provided to the country's political leaders. I should have emphasized the importance of preventing the pollution disaster that was soon going to hit us. But I had no idea of the speed with which it would hit us.[17]

India today faces a level of pollution that once raged throughout North America but that has now been largely corrected through effective legislation and the compliance of government and industry. The level of air pollution in India resembles that of industrial steel towns, such as Pittsburgh, in the 1940s. The degradation of some rivers in India evokes memories of Ohio's Cuyahoga River, which burned in the 1960s.

In the United States and in Western Europe, public sentiment and concern for public health spawned a climate of environmental awareness that resulted in dramatic improvements in air and water quality. Lynn White, Jr., in a now-famous article, suggested that biblical attitudes toward the earth had encouraged overconsumption of natural resources and a callous attitude toward the realm of the nonhuman.[18]

Institutionalized religion was seen as an impediment to the development of ecological awareness. Lance Nelson and others have argued that aspects of Hindu tradition similarly downgrade the material world and can foster indifference toward the environment.[19] Anil Agarwal suggests, in this volume, that the insularity of the Hindu family works counter to the development of a healthy, community-minded ethic. This approach resembles the finger-pointing in the Judeo-Christian traditions, where blame has been placed on dominion theology, the notion that God created the world for human use and pleasure. According to White's analysis, this predisposed Jews and Christians to regard the environment in terms of its usefulness for human endeavor. This attitude might also be characterized as anthropocentrism, or putting the human person at the center of the cosmos. The Hindu tradition, with its emphasis on personal salvation (*mokṣa*) certainly is not exempt from this critique, though several authors in this volume seek to put forward an alternative, more ecofriendly view of Hinduism.

Although cultural values certainly help shape the worldview of a society, pollution is not necessarily a result of religious dogmas but an unfortunate, and probably unintentional, byproduct of the rise of technology, increased population, and the advent of manipulative consumerism in the modern era. Thomas Berry has argued that the technological trance, supported by advertising, has taken on mythic proportions and that it will take a new myth to undo the harm done to the planet by industrial pollution, habitat decimation, and a weakened sense of our place in nature.[20] The task of the series of conferences held at Harvard University's Center for the Study of World Religions has been to reconsider the world's religious traditions in light of this concern.

Hinduism and Environmentalism

In this volume, we will investigate the role of the Hindu religion in the development of ecological awareness in India. The word Hinduism carries many layers of meaning. In its original sense, as coined by the Persians a thousand years ago, it refers to the collective beliefs and practices of those people who live on the other side of the Indus River. This term was taken up by the British during the colonial era (1650 to 1947) and continues to remain in use during postcolonial times to refer to the religious practices of non-Muslim, non-Christian persons of Indian descent. Within Hinduism can be found many gods and goddesses; many competing belief systems, from atheistic materialism to profoundly emotional deistic devotion; various systems of prayer and meditation; and countless groupings and subgroupings of deities and of people.

The word ecology literally denotes a vast range of study, from the living habits of individual species to an overarching concern for the entire planetary system. Ecology, when interpreted in the Hindu context, cannot be separated from its place of origin, the Indian subcontinent, which is home to most of the world's Hindus. However, because of the diaspora of the Hindu community during both the British colonial period and in more recent times, Hindu views on ecological issues can be influential in such far-flung places as Guyana, Trinidad, Britain, the United States, eastern Africa, the Middle East, and Canada.

The various teachings of Hinduism have been learned and discussed by the Greeks, the Chinese, the Persians, the Arabs, the Europeans, and Americans over the course of twenty-three hundred years. Within India, there is no common agreement about what constitutes "Hinduism." Similarly, there can be no one definition applied by a non-Indian that can capture the essence of this dynamic, multifaceted tradition. On one extreme, through the prism of non-Indian, primarily missionary cultures, Hinduism has been stereotyped by Orientalists as caste-bound, retrogressive, and lethargic. On the other extreme, from the perspective of Theosophy and the sympathetic interpretations of Christopher Isherwood and others, Hinduism, particularly in its Vedantic form, is seen as a sublime unifying truth.

The anthropologist Agehananda Bharati postulated a threefold interpretation of Hinduism:

> a) "Village Hinduism" made up of grassroots, "little tradition" Hindu spirituality including shamanistic traditions of ecstatic experience but with some observance of all-India mainline Hindu practices
> b) literate or scripture-based "Sanskrit, Vedic Hinduism" of a "great tradition" variety, represented by Brahmin priests, pandits, itinerant ascetics or monastic practitioners and
> c) "Renaissance Hinduism" or Neo-Hinduism of what Bharati calls the urban alienate, a portion of the new urban middle class, [often followers] of Ramakrishna, Vivekananda, Satya Sai Baba and many others.[21]

Each of these perspectives will in some way be represented in this volume. Some chapters will investigate rituals associated with village life. Others will deal with the "great tradition" approach, which emphasizes text-based reflection. All the papers in some way seek to reexamine the Hindu tradition in light of the current ecological crisis that has thrust vast areas of India out of balance.

In earlier writings, I have noted that environmentalism in India has taken many forms: general information conveyed through the news media, direct action, as found in the Chipko and Narmada movements, and an emphasis on personal decision-making inspired by religious precepts.[22] Three primary varieties of religious expression influence this last component. These include tribal insights into ecosystems, Brahminical models that emphasize an intimacy between the human and the cosmos, and the renouncer orthopraxy of the Buddhists, Jainas, and Yogis that advocates nonviolence and minimization of possessions.[23] This collection of essays also examines these three approaches through an exploration of religious texts, folk metaphors and rituals, and Gandhian-inspired asceticism. Each of these avenues within the broad spectrum of Hindu faith can help contribute to and define the Hindu approach to environmentalism.

Hindu Scholars, Hindu Voices

In developing this volume, we attempted to incorporate as many voices as possible from the field of Hindu studies. This book includes essays by practicing Hindus. Some are of Indian descent living in India. Some are of Indian descent living overseas. Some are of non-Indian descent who have spent considerable time on the Indian subcontinent.

For the most part, the several essays by nonresident Indians reflect the perspectives of individuals who maintain close ties with the difficulties confronting contemporary India. Essays by scholars and environmental activists not of Indian descent who have studied and immersed themselves in various aspects of Hindu life and culture are included; in some aspects, they might be referred to as non-ethnic Hindus. And this volume also contains the voices of American academics who are experts in the field of Hindu studies but who retain their own Western-based culture as a primary orientation.

Using methods similar to that of constructive theology, which seeks to apply religious truths in contemporary contexts, various scholars examine nature themes from the Vedic and Upaniṣadic traditions, law books that recommend nature protection, and philosophical texts that advocate nonviolence. Using a more anthropological approach, other scholars in this volume examine the social realities of the environment in India today and in earlier periods in Indian history.

This book is divided into five sections. The first section examines how traditional concepts of nature from the Hindu tradition might inspire an ecofriendly attitude among modern Hindus. This section also takes a hard look at how traditional Hindu values might impede an environmentalist perspective. The first two essays, by O. P. Dwivedi and K. L. Seshagiri Rao, examine the Hindu notion of *dharma,* or cultural responsibility, in an ecological context and also discuss the concept of the five elements (*mahābhūta*), the building blocks of reality cited in Sāṃkhya philosophy that pervade Hindu discourse about the natural world. Laurie Patton discusses Vedic texts and warns against romanticizing the Vedic sacrificial tradition, which, in many ways, stands as the antithesis of some environmentalist values due to its ritual use of animals. Mary McGee studies the *dharmaśāstra* and *arthaśāstra* literature in light of nature protection, noting that forests, rivers, and other natural resources were to be protected by the king. T. S. Rukmani, in the fifth essay, discusses the role of nature in the Sanskrit literary tradition, with special reference to the story of Śakuntalā. Lance Nelson looks at the *Bhagavadgītā* through two prisms, noting that, on the one hand, ecological values can be developed from reading selected portions of the text, while from another perspective, the assertion that materiality is devalued seems to work counter to the idea of ascribing value to nature. Anil Agarwal, in the closing essay of this section, offers a self-criticism, questioning if the emphasis on self and family weakens Hinduism's capacity for responding to such a staggering social and ecological crisis.

The second section of the book looks to one of the founding fathers of modern India. Mahatma Gandhi mobilized India with his twin projects of nonviolence (*ahiṃsā*) and holding to truth (*satyāgraha*). Various sections of his voluminous memoirs advocate minimal consumption, self-reliance, simplicity, and sustainability—all clearly in accord with "green" values. Vinay Lal and Larry Shinn explore the viability of Gandhi's ethic in light of the contemporary problem of ecological destruction and suggest that his thinking might be readily adapted and embraced for this purpose.

Discussions of forests and groves comprise the third section of the book, beginning with David Lee's description of the forest biology contained in the *Rāmāyaṇa,* an epic text renowned for its sophisticated botanical details. The tension between the dark unknown forest and the safety of the city in the epic texts is explored in Philip Lutgendorf's essay. Moving more into the "little tradition" aspect of Hinduism, Frédérique Apffel-Marglin and Pramod Parajuli discuss the sacred grove tradition of Orissa. In the final chapter of this section, Ann Grodzins Gold provides a historical discussion of a Rajasthani king who made forest preservation a top priority of his rule.

The fourth section of the book examines three river systems of India: the Yamunā, the Gaṅgā, and the Narmada. The first two essays lament the degradation of the two major rivers of Uttar Pradesh, which have been fouled by industrial pollutants and human waste. David Haberman examines classical literary sources that underscore the sacrality of the Yamunā, and Kelly Alley discusses the reluctance of some Hindu religious leaders to provide leadership for the cleanup of the Gaṅgā. The last three essays in this section deal with a very different river system in western India. The Narmada River valley, in Maharashtra and Gujarat, remains largely undeveloped. The valley serves as the home of hundreds of thousands of tribal people. No major cities can be found along its course of more than four hundred miles. During the Nehru period, this river was slated for extensive damming to provide hydroelectric power and water for irrigation. In the process, at least one hundred thousand tribal people would have been displaced. Several villages have already been submerged as part of the preliminary phases of the project, thus ejecting thousands from their homes. Because of an extensive resistance campaign, the World Bank has withdrawn funding for the project. Chris Deegan explores the religious significance of the river; William Fisher discusses the political controversies; and Pratyusha Basu and Jael Silliman examine the role of women in the campaign for protection of the Narmada.

The final section of the book continues an exploration of the "little tradition" grassroots approaches to environmental protection. The first two essays, by Vijaya Nagarajan and Madhu Khanna, describe home-based rituals and "embedded ecology," a sensibility that arises from living within a particular landscape and biosystem, currently threatened by the spread of mass consumer, television culture. The book concludes with George James's essay on the Chipko movement and the environmentalism practiced by Sunderlal Bahuguna. James challenges the notion that the core theology of Hinduism allows for the degradation of the natural world.

Developing a Hindu Environmental Ethic

Throughout this volume, a tension can be detected, not unlike that generated by Lynn White, Jr., who laid blame for the West's environmental problems on the fundamental paradigm of exploitation espoused in the Bible. Like White, some of our authors, most notably Lance Nelson, Philip Lutgendorf, Laurie Patton, Kelly Alley, and Anil Agarwal, assert that Hindu philosophy, particularly as found in Vedanta and in select passages from the *Mahābhārata,* dismisses and perhaps denigrates the ontological status of the physical world. Simultaneously, renouncer tendencies place highest religious value on leaving behind the things of the world, again relegating the earth to a secondary status. Personal salvation, or *mokṣa,* as the primary religious value leaves little or no room for such worldly concerns as air quality or water quality. However, many other authors argue that "worldly" refers not to the five great elements (*mahābhūta*) but to karmic, ego-based concerns. Through ritual, meditation, and practices of yoga, one can leave behind the realm of pollution and embrace the purity of one's authentic being in a way quite harmonious with religious values. K. L. Seshagiri Rao and O. P. Dwivedi champion Hindu values as inseparable from environmental values; T. S. Rukmani and Mary McGee see explicitly ecofriendly values in traditional texts; David Lee, Ann Grodzins Gold, Frédérique Apffel-Marglin and Pramod Parajuli, Vijaya Nagarajan, Madhu Khanna, and George James see concrete evidence of a heightened consciousness of the earth in the day-to-day life of specific Hindus, both village and urban.

Toward the conclusion of the conference at Harvard University, Harry Blair of Bucknell University developed a useful scheme for organizing and defining Hindu approaches to ecology (see appendix 1). He outlined progressive stages, or categories, that indicate increasingly radical commitments to ecological harmony. The first category emphasizes use of natural resources. It promotes development and conquest of the natural order. The second category advocates utility, seeing nature to be in a reciprocal relationship with the human order. This approach emphasizes sustainability and social ecology. It clearly would give voice to persons seeking to support themselves in a simple manner, but it also would allow for some harvesting of natural resources. The third category, romance, sees ultimate reality manifesting itself in the natural world. It respects the divinity of nature and urges its adherents to practice deep ecology. The fourth category, according to Blair, emphasizes asceticism; I would suggest that it also emphasizes transcendentalism. This entails separation from nature through abstinence of various sorts. Following the tradition of the renunciant, or sadhu, this approach advocates withdrawing from the world. Though not practical for all persons and not at all supportive of "omnivore" culture, it nonetheless demonstrates an ecofriendly ethic.

In light of Gadgil and Guha's characterization mentioned earlier, the omnivores would fall into the use and utility groups. The ecosystem people would fall into the category of romance, while all committed environmentalists could be seen as practicing a form of asceticism. In Patrick Peritore's system, the greens would represent the phase of romance and asceticism, the ecodevelopers would fall into the category of utility, while the managers could be seen in terms of emphasizing use.

Various authors in this volume criticize the notion of use. Philip Lutgendorf notes that in much of the epic literature, the forest represents raw material to be exploited or destroyed. Laurie Patton reminds the reader that the early Vedic literary and ritual traditions conducted bloody sacrifices, using animals. Kelly Alley and David Haberman decry industrial pollution of India's rivers as an example of the philosophy of utility. Anil Agarwal suggests that the Hindu emphasis on the self and family allows for the use of people and objects outside one's own compound.

Frédérique Apffel-Marglin and Pramod Parajuli see practical utility in India's tradition of maintaining sacred groves. By allowing fields to stay fallow for a period of time and then bringing them back into cultivation, the landscape provides a utility without being subjected to undue harm. Ann Gold discusses the role of a twentieth-century raja in Rajasthan in assuring the balance of nature through strict land-use and tree preservaton policies. The *dharmaśāstra* and *arthaśāstra* injunctions explained by Mary McGee similarly reflect a utility approach to the use of natural resources.

In contrast, T. S. Rukmani's chapter on the portrayal of nature in traditional Sanskrit literature may be seen in terms of Blair's category of romance. In a somewhat similar romantic vein, David Lee extols the medicinal value of plants and the powerful beauty of the forest as portrayed in the *Rāmāyaṇa*. David Haberman lauds the once-pristine beauty of the Yamunā River.

Vinay Lal and Larry Shinn write about Mahatma Gandhi's commitment to nonviolence and abstemiousness. Gandhi may be seen as a prime example of Blair's description of asceticism. Lance Nelson similarly reads the *Bhagavadgītā* in terms of its emphasis on worldly transcendence through asceticism. O. P. Dwivedi and Seshagiri Rao, while presenting a romanticized view of the elements and love for nature, also emphasize an ascetic need for reducing consumption.

Lutgendorf's discussion of the forest provides an extended metaphor for the many potential expressions of a Hindu environmental ethic. When seen in terms of use, the forest is an exploitable resource, providing wood for fuel and structures, raw material for the building of cities. The utility of forest lands can be seen in the *jāṅgala,* which exists in reciprocity with human settlement and provides sustenance for domesticated animals, such as cows and goats, as well as a pastoral landscape for the unfoldment of the human drama. The forest can also be seen as a romantic paradise, a hiding place for lovers. The great ashrams, or ascetic retreats, have traditionally been located in the forests of India. These four interpretations of the forest illustrate various intersecting modalities of Hindu environmentalism. For the managers and ecodevelopers, the forest must be exploited and tapped. For the ecosystem people, the forest provides basic sustenance. For green activists, the forest must be protected from clearcutting and flooding, both to ensure its survival and for its own sake (the romantic view). For the ascetics, the forest must be maintained to provide a continued haven for spiritual retreats or ashrams.

Hinduism and Ecology: Future Prospects

What can be expected in the development of a Hindu-inspired environmentalism? This collection of essays suggests that several avenues can be pursued to lift up Hindu religious imagery and symbolism in the name of environmental protection. However, any visitor to India will see that it is just as likely that the same religious symbols might be used to promote the latest consumer product. As Vasudha Narayanan has noted, "A burgeoning middle class in India is now hungry for the consumer bon-bons of comfortable and luxurious living."[24] Nonetheless, despite the onslaught of advertising and industrialization, ecological consciousness is growing in India. Public urgency has caused wide public discussion of ecological degradation. Some visible changes have been made to improve the state of the environment. For instance, three-wheeled taxis have been banned from the New Delhi airports because they create a great deal of pollution. According to Narayanan:

> With the growing awareness of our ecological plight, Hindu communities are pressing into us the many dharmic texts and injunctions, using epics and Puranas as inspiration in planting gardens, and reviving customary lore on the medicinal importance of trees and plants. Women, through song and dance, communicate the assaults on women and nature.[25]

Narayanan describes a dance performed by Usha Vasanthkumar and Sudha Vijayaraghavan titled *Pancha Bootangal,* the Five Elements, that dramatically enacts humanity's greedy attack on the elements, resulting in ecological havoc.

Swami Agnivesh of the Arya Samaj order of monks works extensively with the poor and disenfranchised of India, seeking economic justice, protection of children, and universal education. Like Gandhi, he suggests that the village model provides for an ecologically sound life-style. He states that simplicity and contentment are needed to counter the juggernaut of industrial development and consumerism that leads to pollution and environmental degradation.[26] The swami's comments reflect a general distrust throughout India of modern life patterned on the Western paradigm. The rise to power of the Bharatiya Janata Party (BJP) during the 1990s came about in part due to a wariness of this cultural shift throughout India. As Thomas Blom Hansen has noted:

> The entire problematic of consumption of "western" Products—food, styles of dress, electronics gadgets, music—is among Hindu nationalists (and others) linked to the contamination, exposure, and corruption of the body . . . [A]ccess to electronic implements, motorized transport, and excessive watching of TV divert the attention away from healthier and more physically demanding pursuits.[27]

Nonetheless, the task of stopping or even slowing the spread of consumerism and industrialization seems quite impossible. It seems that the Nehruvian vision for India has prevailed and that such Gandhian notions as nonviolence (*ahiṃsā*) and nonpossession (*aparigraha*) are insufficient to bring about the changes needed to make India more environmentally conscious.

As Anil Agarwal has noted, this current urban ecological crisis took India by surprise. In his personal life, he has struggled with environmentally induced cancer. In 1994 he was diagnosed with lymphoma and has received extensive treatments in the United States for his condition. He eloquently writes:

> The elite of our nation [India] have failed to internalize the ecological principle that every poison we put out into our environment comes right back to us in our air, water and food. These poisons slowly seep into our bodies and take years to show up as cancer, as immune system disorders, or as hormonal or reproductive system disorders—affecting even the foetus . . . [T]he Indian people must not remain ignorant and nonchalant about the acute threats they face to their own health and to the health of their children.[28]

As the general population of India becomes more aware of the great harm and difficulty inflicted by industrial pollution and inappropriate use of resources, it will begin to awaken to the need for voluntary compliance with much-needed cleanup campaigns. Despite the good intentions of government agencies and the passage of various pieces of antipollution legislation, Indians have a long history of ignoring government regulations. Public advocacy, perhaps inspired by the memory of cleaner times, will eventually prompt the government and industries of India to be more attentive to the destructive nature of its current techno-industrial complex. The Hindu religion, with its vast storehouse of text, ritual, and spirituality, can help contribute both theoretical and practical responses to this crisis.

Notes

1. N. Patrick Peritore, "Environmental Attitudes of Indian Elites: Challenging Western Postmodernist Models," *Asian Survey* 33 (1993): 807.

2. Ramachandra Guha, "Radical American Environmentalism: A Third World Critique," in *Ethical Perspectives on Environmental Issues in India,* ed. George A. James (New Delhi: A.P.H. Publishing Corporation, 1999), 115–30. First published in *Environmental Ethics* 11, no. 1 (1989): 71–83.

3. Madhav Gadgil and Ramachandra Guha, *Ecology and Equity: The Use and Abuse of Nature in Contemporary India* (New Delhi: Penguin Books, 1995), 3–5.

4. Ibid., 180.

5. Ibid., 182.

6. Vikram K. Akula, "Grassroots Environmental Resistance in India," in *Ecological Resistance Movements: The Global Emergence of Radical and Popular Environmentalism*, ed. Bron Raymond Taylor (Albany: State University of New York Press. 1995), 127.

7. Peritore, "Environmental Attitudes of Indian Elites," 804.

8. Ibid., 817.

9. Ibid., 818.

10. Payal Sampat, "What Does India Want?" *World Watch* 11, no. 4 (1998): 31.

11. Gerald James Larson, *India's Agony over Religion* (Albany: State University of New York Press, 1995), 199.

12. O. P. Dwivedi, *India's Environmental Policies, Programmes, and Stewardship* (New York: St. Martin's Press, 1997), 22–23.

13. *India Today,* 16 December 1996, 39.

14. Anil Agarwal, electronic communication from Center for Science and Environment, 22 April 1999.

15 *India Today,* 15 January 1997, 121–23.

16. Dwivedi, *India's Environmental Policies,* 11.

17. Anil Agarwal, electronic communication, 3 February 1999.

18. Lynn White, Jr., "The Historical Roots of Our Ecologic Crisis," *Science* 155 (1967): 1203–7.

19. *Purifying the Earthly Body of God: Religion and Ecology in Hindu India,* ed. Lance Nelson (Albany: State University of New York Press, 1998), 61–81.

20. Thomas Berry, *The Dream of the Earth* (San Francisco: Sierra Club Books, 1988).

21. Larson, *India's Agony over Religion,* 20–21.

22. Christopher Key Chapple, *Nonviolence to Animals, Earth, and Self in Asian Traditions* (Albany: State University of New York Press, 1993).

23. Christopher Key Chapple, "Toward an Indigenous Indian Environmentalism," in *Purifying the Earthly Body of God,* ed. Nelson.

24. Vasudha Narayanan, "'One Tree Is Equal to Ten Sons': Hindu Responses to the Problems of Ecology, Population, and Consumption," *Journal of the American Academy of Religion* 65, no. 2 (1997): 321.

25. Ibid., 311.

26. Swami Agnivesh, personal conversation, 11 January 1999.

27. Thomas Blom Hansen, *The Saffron Wave: Democracy and Hindu Nationalism in Modern India* (Princeton: Princeton University Press, 1999), 233.

28. Anil Agarwal, "My Story Today, Your Story Tomorrow: An Environmentalist Searches for the Genesis of His Own Cancer," *Down To Earth* 5, no. 13 (30 November 1996): 30–37.

The Cultural Underpinnings: Traditional Hindu Concepts of Nature

Dharmic Ecology

O. P. DWIVEDI

The many environmental problems India faces may be summarized as follows: 1) continuous degradation, in varying degrees, of productive land (due to increased salinity and alkalinity, desertification, water-logging, and deforestation); 2) shortage of wood fuel and fodder for rural needs, which jeopardizes existing forests; 3) depletion of the forest cover, which in turn threatens the survival of indigenous biodiversity and affects wildlife habitat; 4) excessive and unwise use of pesticides and fertilizers and ill-advised agricultural practices (including mono-culture), which further stresses the fragile environment; and 5) poorly monitored and inadequately enforced environmental regulations for various natural resource extraction activities (such as mining, metal-lurgy, aggregate production, and other manufacturing industries).

A cursory glance at the extent of some of these major environmental problems reveals that all India's environmental issues are inter-connected and together constitute an increasingly deteriorating environment and rapid depletion of natural resources—whether the issue in question is health hazards caused by water and air pollution, such as the use of fuel wood or dung for cooking purposes; population pressure and urbanization straining the resources of local governments that have to provide various civic amenities, the dearth of which imperils the quality of life of people living in urban areas; myopic land management policies that are severely straining the ecosystem in order to meet people's requirements of food and other agricultural products; long-term degradation of land and increasing desertification which is jeopardizing people's futures; the use of hazardous chemicals to meet today's short-term needs without an eye to future costs; or the rapid shrinkage of natural resources such as grazing land and the lack of fodder, firewood, and timber due to deforestation. India faces yet another major challenge to its environmental well-being: continuing poverty coupled with growing population and the attendant side effects of enhanced industrial activities (including human settlement patterns and movements). The existence of poverty affects the meager natural resources that India wishes to protect and conserve: people are so desperate to improve their families' living conditions that they do not hesitate to damage the environment if the short-term gains ensure fulfillment of their family's daily basic needs.[1]

India's population of 844 million in 1991 is increasing at the rate of 2.11 percent annually, which means about 17 million people are added each year. Furthermore, India has probably the largest cattle population on the earth—about 500 million domesticated animals—with only 13 million hectares of grazing land. The multiplying population of both human beings and animals is putting tremendous pressure on India's environment. In the race for survival, both animals and human beings suffer. In addition, over 250 million children, women, and men suffer from malnutrition. The prospects for the future are alarming indeed.

Needless to say, India faces a double jeopardy in attempting to industrialize quickly while confronting poverty and a growing population. India's environmental problems are complex and the choices available are difficult. Vision and "environmentally sound" foresight based on a holistic approach to problem-solving are required and entail bringing the secular, socioeconomic, cultural, religious, and traditional domains together. In this essay, I will examine India's ecospirituality against the context of dharmic ecology from four interrelated perspectives. These are: 1) *Vāsudeva sarvam,* the Supreme Being resides in all beings; 2) *Vasudhaiva kuṭumbakam,* the family of Mother Earth; 3) *Sarva-bhūta-hitā,* the welfare of all beings; and 4) Dharmic ecology as a strategy toward putting into practice the Hindu concept of ecocare.

Vāsudeva Sarvam, the Supreme Being Resides in All Things

One of the main postulations of the *Bhagavadgītā* is that the Supreme Being resides in all. Chapter 7, verse 19, states:

> Only after taking many births is a wise person able to comprehend the basic philosophy of the creation; which is: whatever is, is Vāsudeva. If anyone understands this fundamental, such a person is indeed a Mahatma.[2]

Later, in chapter 13, verse 13, the Lord Kṛṣṇa says: "He resides in everywhere" (*sarvam āvṛtya tiṣṭhati*). As explained in the *Śrīmad Bhāgavata Mahāpurāṇa* (book 2, discourse 2, verse 41), "ether, air, fire, water, earth, planets, all creatures, directions, trees and plants, rivers, and seas, they all are organs of God's body; remembering this, a devotee respects all species." Further, the definition of a pundit is one who treats a cow, an elephant, a dog, and an outcaste with the same respect shown to the Brahman endowed with great learning and yet humble (*Gītā* 5.18). Thus, the basic concept is: seeing the presence of God in all, and treating the creation with respect without harming and exploiting others (*vāsudevaḥ sarvam iti*). The *Śrīmad Bhāgavata Mahāpurāṇa* (2.2.45) confirms this fundamental principle: a good devotee is the one who sees in all creation the presence of God (*sarva bhūteṣu yaḥ paśyed bhagvadbhāvamātmanaḥ*).

Such veneration, respect, and acceptance of the presence of God in nature is required of Hindus in order to maintain and protect the natural harmonious relationship between human beings and nature.[3] In the *Mahābhārata,* it is claimed that all living beings have soul, and God resides as their inner soul: *Sarvabhūtātmbhūtastho* (Mokṣadharma Parva, chapter 182, verse 20). It also means that all this universe and every object in it has been created as an abode of the Supreme God; it is meant for the benefit of all; individual species must therefore learn to enjoy its benefits by existing as part of the system, in close relationship with other species and without permitting any one species to encroach upon the others' rights. This stipulation is later endorsed in the *Mahābhārata,* where it is stated:

> The Father of all creatures, Lord God, made the sky. From sky he made water, and from water made fire (*agni*) and air (*vāyu*). From fire and air, the earth (*pṛthivī*) came into existence. Actually, mountains are his bones, earth is the flesh, sea is the blood, and sky is his abdomen. The sun and moon are his eyes. The upper part of the sky is his head, the earth is his feet, and directions are his hands. (*Mahābhārata,* Mokṣadharma Parva, 182.14–19)

Thus, at least for the Hindus of the ancient period, God and nature were one and the same. While Prajapati (the "Lord of Creatures" of the Ṛgveda) is the creator of the sky, the earth, the oceans, and all the species, he is also their protector and eventual destroyer. He is the only Lord of Creation. Human beings have no special privilege or authority over other creatures; on the other hand, they do have more obligations and duties.[4]

Hindu scriptures attest to the belief that the creation, maintenance, and annihilation of the cosmos is completely up to the Supreme Will. In the *Gītā,* Lord Kṛṣṇa says to Arjuna: "Of all that is material and all that is spiritual in this world, know for certain that I am both its origin and dissolution" (*Gītā* 10.8). "By my will it is manifested again and again and by my will it is annihilated at the end" (*Gītā* 9.8). Furthermore, the Lord says: "I am the origin, the end, existence, and the maintainer (of all)" (*Gītā* 10.32). Thus, for Hindus, God and *prakṛti* (nature) are interrelated.

Furthermore, the Hindu belief in the cycle of birth and rebirth, wherein a person may come back as an animal or a bird, means that Hindus are called to give other species not only respect, but reverence. This reverence finds expression in the doctrine of *ahiṃsā,* nonviolence (or non-injury) against other species and human beings alike. It should be noted that the doctrine of *ahiṃsā* presupposes the doctrines of *karma* and rebirth (*punarjanma*). The soul continues to take birth in different life-forms, such as birds, fish, animals, and humans. Based on this belief, there is a profound opposition in the Hindu religion (and in Buddhist and Jaina religions) to the institutionalized breeding and killing of animals, birds, and fish for human consumption. From the perspective of Hindu religion, the abuse and exploitation of nature for selfish gain is considered unjust and sacrilegious.

God Reincarnates in the Form of Animals and Humans

One of the central tenets of Hinduism is the doctrine of reincarnation, when the Supreme Being was himself incarnated in the forms of various species. The Lord says: "This form is the source and indestructible seed of multifarious incarnations within the universe, and from the particle and portion of this form, different living entities, like demigods, animals, human beings and others, are created" (*Śrīmad Bhāgavata*

Mahāpurāṇa 1.3.5). Among the various incarnations of God are a fish, a tortoise, a boar, and a dwarf. His fifth incarnation was as a man-lion. As Rāma, God was closely associated with monkeys, and, as Kṛṣṇa, he was surrounded by cattle. These are some examples where different species are accorded reverence.

Almost all the Hindu scriptures place a strong emphasis on the notion that God's grace cannot be received by killing animals or harming other creatures. That is why *not* eating meat is considered both appropriate conduct and one's *dharma*. For example, as mentioned in the *Viṣṇu Purāṇa:* "God, Keśava, is pleased with a person who does not harm or destroy other nonspeaking creatures or animals" (*Viṣṇu Purāṇa* 3.8.15). Further, the pain a human being causes other living beings to suffer will eventually be suffered by that same person, either in this life or in a later rebirth. It is through the transmigration of the soul that a link has been provided between the lowliest forms of life and human beings. In the *Manusmṛti,* the laws of Manu, a warning is given: "A person who kills an animal for meat will die of a violent death as many times as there are hairs of that killed animal" (*Manusmṛti* 5.38). The *Yājñavalkyasmṛti* warns those who kill domesticated and protected animals of hellfire (*ghora naraka*): "The wicked person who kills animals which are protected has to live in hellfire for the days equal to the number of hairs on the body of that animal" (*Yājñavalkyasmṛti,* Acaradhyayah, verse 180). And, the *Narasiṃha Purāṇa* states that a person who roasts a bird for eating will surely be a sinner:

> O Wicked person! what is the use of you taking a bath in sacred rivers, doing pilgrimage, worshiping, and performing *yajña*s if you roast a bird for your meals.[5]

The *Mahābhārata* describes an event in which the *ṛṣi*s and gods debated the merits of offering grain or the lamb (goat) as the sacrifice, *yajña*. The *ṛṣi*s insisted that, according to the Vedas, the sacrificial material ought to be the grain only, and thus no animal should be killed for the purpose of *yajña:*

> The *ṛṣi*s told the Gods that according to the Vedas and *śruti*s, the sacrificial offering should be of grain. The term "Aj" does not mean animal or goat as you denote, the term means only grain or seed. Thus, an animal must not be sacrificed. Further, that which sanctions the killing of any animal cannot be a true Dharma of a moral people.[6]

Thus, it would seem that the practice of grain sacrifice may have started from that time.

Ecological Unity in Hindu Mythological Diversity

How to protect and conserve the biological diversity is exemplified by the family and habitat of the god Śiva, his consort Pārvatī, and his two sons Kārttikeya and Gaṇeśa. His habitat is Mount Kailāsa, with snowy peaks representing the cosmic heavens. The nascent moon on his forehead denotes tranquillity; the constant stream of Gaṅgā's water from the interplaited lock of hair on his head indicates the purity and preeminence of water; Nandi, the bull, as his mount, represents livestock; serpents signify the presence of toxicity in nature; the lion used by his consort Pārvatī represents wildlife; the peacock, the mount of Kārttikeya, one of the most colorful birds, represents the avian species; and the mouse, the mount of Gaṇeśa, represents pests. Thus, various forms of animate and inanimate life are represented in the household and habitat of Lord Śiva. However, another important significance of the family of Lord Śiva is the harmonious relationship between natural enemies. In Lord Śiva's household, various natural enemies live in harmony with each other. The carnivorous lion's food is the vegetarian bull, the peacock is the enemy of the serpent, and the mouse is the serpent's food; nevertheless, all live together. Thus, when a devotee worships the family of Lord Śiva, he or she observes this coexistence and is influenced by what in contemporary times might be seen as analogous to the concept of ecological harmony and respect for biological diversity.

A story from Hindu scriptures illustrates this coexistence between good and evil. Cursed by the *mahārṣi* Dūrbhaṣa, the *deva*s (demigods) were deprived of their divine vigor and strength. Soon, they were defeated by the *asura*s (demons) and dislodged from their heavenly abode. They then prayed to Lord Viṣṇu to seek his intervention. Lord Viṣṇu suggested that the only course open to them lay in securing the Nectar of Immortality, which would make them invincible. In order to obtain this nectar, they had to put every herb in the cosmic sea of milk and churn the sea by using the mountain of Mandarachal as a pestle and the snake king, Vāsuki, as the noose. To churn the sea, they needed the assistance of the demons to hold one end of the rope. The demons agreed to assist on the condition that the nectar received from the churning would be shared equally. From their churning, the most sought after nectar was obtained, along with many precious stones, wealth, and worldly riches. But, along with these priceless and rare items, a most venomous poison was produced. If that poison was not immediately disposed of, *sarvanāśa* (total annihilation) of the entire universe could result; however, no one, god or demon, was willing to touch this most fatal toxic waste. Finally, all went to Lord Śiva for help in stopping the obliteration of the universe. Śiva agreed to take care of the toxic waste by drinking it, and thus the impending disaster was averted. Later, with the assistance of Lord Viṣṇu, the demigods were able to trick the demons out of their share of the nectar, so that only the gods became immortal. This story instructs us that the consequences of an activity can be both beneficial and disastrous. Both nectar and poison result from the same activity, and one cannot be acquired without the other. In contemporary terms, an ecological balance has to be maintained between the nectar of riches and the side effects (poisons) of technology and industrialization. As we live on the same planet, we suffer or benefit together.

Based on a metaphorical interpretation of the above stories, one can suggest the premise that every entity and living organism is part of one large extended family system (*kutumba*) presided over by the eternal Mother Earth, Devī Vasundharā. The development of humanity from creation until now has taken place nowhere else but on Earth. Our relationship with Earth, from birth to death, is like that of children and their mother. The mother, in this case Earth, not only bears her children but also is the main source of fulfillment of their unending desires. It is Earth which provides energy for the sustenance of all species. And, just as one ought not insult, exploit, or violate one's mother, but be kind and respectful to her, so should one behave toward Mother Earth. We are enjoined to take care of God's creation by engaging in ecostewardship.

***Vasudhaiva Kuṭumbakam,* the Family of Mother Earth**

In the *Atharva Veda,* an entire hymn, the Pṛthivī Sūkta, has been devoted to praise of Mother Earth. The hymn's sixty-three verses integrate many of the thoughts of Hindu seers concerning the concept of nature, the dependence of human beings on Earth, and the resultant respect required.[7] These verses are addressed to Devī Vasundharā, Mother Earth. Earth is seen as the abode of a family of all beings (humans and others alike). *Vasudhā* means "this earth," while *kuṭumba* means "extended family"—including human beings, animals, and all living beings. Every entity and organism is a part of one large extended family system presided over by the eternal Mother Earth. It is she who supports us with her abundant endowments and riches; it is she who nourishes us; it is she who provides us with a sustainable environment; and it is she who, when angered by the misdeeds of her children, punishes them with disasters:

> O Mother Earth! Sacred are thy hills, snowy mountains, and deep forests. Be kind to us and bestow upon us happiness. May you be fertile, arable, and nourisher of all. May you continue supporting people of all races and nations. May you protect us from your anger (natural disasters). And may no one exploit and subjugate your children.[8]

The hymn's composer, the Atharva *ṛṣi,* envisions Mother Earth in her pristine nature, not denuded of her natural cover. The Pṛthivī Sūkta also exemplifies the relevance of environmental sustenance, agriculture, and biodiversity to human beings.[9] The three main segments of our physical environment—water, air, and soil—are highlighted and their usefulness detailed. The various water resources, such as seas, rivers, and waterfalls, flow on Earth (*Atharva Veda* 12.3). Verse 2 of this hymn depicts the majesty of mountains and rivers, the beauty of small springs, and the invaluable tiny medicinal plants which save the lives of humans. But all these elements are treated equally by Mother Earth, who does not discriminate between the high and mighty (*uddhataḥ*), such as mountains and oceans, and the low (*prayataḥ*). For Mother Earth, all species are of equal value; she has not accorded any one species special authority over other species. Thus, humans do not have the authority to destroy the environment at the expense of others, including their own species, in nature.

The Pṛthivī Sūkta maintains that attributes of the earth (such as its firmness, purity, and fertility) are for everyone, and that no one group or nation has special authority over it. That is why the welfare of all and hatred toward none constitute the core values for which people on this planet ought to strive (verse 18). For example, there is a prayer for the preservation of the original fragrance of the earth (verses 23 and 25) so that its natural legacy is sustained for future generations. There is also a prayer which says that even when people dig the earth, either for agricultural purposes or for extracting minerals, they should do so in such a way that her vitals are not hurt and that no serious damage is done to her body and appearance (verse 35).

The Pṛthivī Sūkta enunciates the unity of all races and among all beliefs; further religious, linguistic, and cultural harmony is urged, with a prayer that Mother Earth bestow upon all the people living in any part of the world the same prosperity for which the *ṛṣi* Atharva of India has pleaded:

> Mother Earth, where people belong to different races, follow separate faiths and religions and speak numerous languages, cares for them in many ways. May that Mother Earth, like a Cosmic Cow, give us the thousandfold prosperity without any hesitation, without being outraged by our destructive actions.[10]

In verse 63 of this hymn, a prayer is offered to Mother Earth and her blessings are sought for all:

> O, our Mother Earth! May we possess the intellect and wisdom which enable us to speak in concord with heavenly beings, may we continue to enjoy your blessing of hidden riches, glory, and realization of material and spiritual well-beings.[11]

In summary, it can be said that the Pṛthivī Sūkta, whose sixty-three verses have been dedicated to Mother Earth, is the foremost ancient spiritual text from India, enjoining all human beings to protect, preserve, and care for the environment. This is beautifully illustrated in verse 16, which says that it is up to us, the progeny of Mother Earth, to live in peace and harmony with all others:

> O Mother Earth! You are the world for us and we are your children; let us speak in one accord, let us come together so that we live in peace and harmony, and let us be cordial and gracious in our relationship with other human beings.[12]

These sentiments denote the deep bond between the earth and human beings and exemplify the true relationship between the earth and all living beings, as well as between humans and other forms of life. The Sūkta guides us to behave in an appropriate manner toward nature and defines our duty toward the environment.

Sarva Bhūta Hitā, the Welfare of All Beings

How is the welfare of all beings related to the Hindu view of life? As mentioned earlier, the prerequisite for understanding this concept in Hindu thinking is to accept the view that Brahman (the Supreme Being) is the ultimate source and cause of the universal common good, not only of humans but of all beings in creation. That common good, for Hindus, is the concept of *sarva-bhūta-hitā,* the highest ethical standard that Hindus ought to apply, according to their *dharma*. Further, the Hindu tradition requires that a common good (such as protection of the environment, welfare of the poor and needy, or the well-being of other living beings) takes precedence over a private good (including individual material and personal well-being). Under such a system, a dharmic citizen should act for *sarva-hitā:* enhancing the common good of all together. In other words, the term "common good" can be related to the concept of *sarva-kalyāṇkarī-karma,* which denotes a deed resulting in the common good of all. This also relates to the concept of "caring for others," meaning a deed (*karma*), which results in universal welfare based on mutual cooperation and respect. To reiterate, the question of *sarva-bhūta-hitā* care and taking care of the others or serving others, is entirely intertwined with the concept of dharmic ecology. It is an obligation that human beings owe, not only to each other, but also to all of nature and the entire cosmos.

The requisite duties and mode of conduct to perform *sarva-bhūta-hitā,* and to protect and sustain the common good, has been provided in Hindu scriptures. For example, this advice is given by Lord Kṛṣṇa in the *Gītā:*

> One ought to understand what is duty, and what is forbidden in the commands laid down by the scriptures [*śāstra*s]. Knowing such rules and regulations, one should behave as ordained by scriptures.[13]

Lord Kṛṣṇa later says:

> O Partha! that understanding by which one knows what ought to be done and what ought not to be done, what is to be feared and what is not, what is obligatory and what is permitted, leads to the righteous path [*Sattvika Pravṛtti*].[14]

That righteous path in the Hindu religion is called *dharma*.

Dharma is one of the most intractable and unyielding terms in Hindu,

Buddhist, and Jaina religions and philosophies. Instead of discussing the root and theological or philosophical definitions of the term, the author of the *Mahābhārata* goes directly to the common usage of the term:

> Dharma exists for the general welfare (*abhyudaya*) of all living beings; hence, that by which the *welfare* of all living creatures IS sustained, that for sure is *Dharma*.[15]

Thus, *dharma* can be considered an ethos, a set of duties, that holds the social and moral fabric together by maintaining order in society, building individual and group character, and giving rise to harmony and understanding in our relationships with all of God's creation.

Duty toward humanity and God's creation is an integral part of Hindu ecology and *dharma*. While all other species conduct themselves according to the *dharma* of their kind, only human beings, because of free will, think that they are very powerful and act in an adharmic manner. Such acts are to be avoided, as *ṛṣi* Markandeya says in the *Mahābhārata* during a conversation with the Pandavas:

> O king, all creatures act according to the laws of their specific species as laid down by the Creator. Therefore, none should act unrighteously (*adharma*), thinking, "It is I who is powerful."[16]

Dharma requires that one consider the entire universe an extended family, with all living beings in this universe members of the same household. This is also known as the concept of *vasudhaiv kuṭumbakam,* discussed above. Only by considering the entire universe as a part of one's extended family can one develop the necessary maturity and respect for all other living beings. The welfare and caring of all (*sarva-kalyānkarī-hitā*) is realized through the golden thread of spiritual understanding and cooperation.

Dharma can help us master our baser characteristics, such as our greed and our exploitation, abuse, mistreatment, and defilement of nature. Before we can hope to change the exploitative tendencies of society, it is absolutely essential that we discipline our own inner thoughts. This is where the role of *dharma* comes into play. It is important to appreciate that the concept of *dharma* can be used in any culture (although the term has become synonymous with the practices and rituals of Hinduism, Buddhism, and Jainism). *Dharma* in its pure form can be the mechanism that creates respect for nature, since it transcends institutional structures, bureaucratic impediments, and rituals associated with organized religions; it enables people to center their values upon the notion that there is a cosmic ordinance and a natural or divine law that must be maintained. *Dharma* thereby provides a code of conduct as well as a vision. It serves both as a model and an operative strategy for the transformation of human character. And, if the goal of transformation is to achieve an environmentally conscious and sustainable world, then *dharma*'s precept that reward after death can be attained through actions in this world may provide the incentive for humanity to seek peace with nature.[17] The manifestation of *dharma* necessitates, however, the acceptance of the concept of *karma*.

Karma *and the Environment*

The term *karma* comes from the root *kṛ,* meaning "to do," and thus has a general connotation of "action," but in its broadest sense it applies also to the effects of such actions. People often confuse the law of *karma* with the law of destiny. This misunderstanding needs to be rectified because an appropriate understanding is essential to appreciate the use of this precept in Hindu and Buddhist ways of life. A brief definition of the law of *karma* is that each act, wilfully performed, leaves a consequence in its wake. These consequences, also called *karma-phala* (fruits, or effects, of action), will always be with us, although their impact may not be felt immediately. Thus, the law related to *karma* tells us that every action performed creates its own chain of reactions and events, some of which are immediately visible, while others take time to surface. Environmental pollution is but one example of the *karma* of those people who thought that they could continue polluting the environment without realizing the consequences of their actions for future generations. For example, those who buried the toxic waste in Love Canal (near Niagara Falls, in New York State) thought that by concealing their actions, the problem would go away; instead it surfaced a few years later. Every action creates its own reaction. What is important to know is that a right action, that is a dharmic action, generates beneficial results, while an adharmic action results in harmful effects. It is not always easy to foresee the consequences of one's actions, but one should be ready either to overcome obstacles that arise or suffer the repercussions of one's actions.

Once *karma* has started, it continues without a break; even though a person may be dead, his or her *karma* survives in the form of a memory and carries over into the next life. This is stated in the *Mahābhārata:*

> An action, which has been committed by a human-being in this life, follows him again and again (whether he wishes it or not).[18]

Furthermore, it is stated in the *Brahmavaivarta Purāṇa* that whatever action, good or bad, is knowingly performed by a person, he or she must face its consequences.[19] Sometimes karmic justice is indecipherable to individuals, when an individual's family and later descendants are forced to pay for the past crimes and mistakes of that person. Bhīṣma acknowledges this to King Yudhiṣṭhira in the *Mahābhārata:*

> O king, although a particular person may not be seen suffering the results of his evil actions, yet his children and grandchildren as well as great-grandchildren will have to suffer them.[20]

Such suffering may continue to visit humanity unless we realize that the destruction we are inflicting on our natural surroundings will result in dire consequences for current and future generations. Only if we recognize and act on this philosophy of life will people start paying due respect to nature and taking active responsibility for the care of the environment. Understanding this is the key to the point being made here, and it draws on the concept that people living in any part of the world are our brothers and sisters. All of our actions are interrelated with and interconnected to what eventually happens in this world. Although we may not face the consequences individually, someone is going to be burdened by or benefit from our actions. It is in this context that the concepts of *dharma* and *karma* become meaningful.

Once our *dharma* and *karma* to the environment are appropriately understood, their precepts recognized, and their relevance to environmental protection and conservation accepted, then a common strategy for ecospirituality and stewardship can be developed. Such a strategy will depend much upon how different people together 1) perceive a common future for society; 2) act both individually and as a group toward that end; and 3) realize that each individual has a moral obligation to support his society's goal, since his acts will have repercussions on the future of society and on his own destiny.

Toward a Dharmic Ecology and Environmental Stewardship

Since the late 1980s, there has been a steadily growing awareness among the people of India about the ecological challenges facing their society. There has also been an impressive growth of regulatory and administrative institutions to deal with the problems of pollution and environmental conservation, at both national and state levels. At the same time, the mounting pressures of population, expanding urbanization, and growing poverty have led to the ecologically unsustainable exploitation of natural resources that is threatening the fragile balance in India. The nation would do well to draw on its ecospirituality and rich cultural heritage as well as on traditional conservation practices, such as those preserved by the Bishnois.

A Hindu Strategy for Dharmic Ecology?

The effectiveness of any religion in protecting the environment depends upon how much faith its believers have in its precepts and injunctions. Its value also depends upon how those precepts are transmitted and adapted in everyday social interactions. In the case of the Hindu religion, some of its precepts became ingrained in the daily life and social institutions of a certain segment of the population. Two such examples illustrate this point.

The Bishnois, Defenders of the Environment: The Bishnois, a small community in the state of Rajasthan, practice environmental conservation as a part of their daily religious duty. They believe that cutting a tree or killing an animal or bird is sacrilege. Their religion, an offshoot of Hinduism, was founded by Guru Maharaj Jambaji, who was born in 1451 C.E. in the Marwar area. When he was young, he witnessed how, during a severe drought, people cut down trees to feed animals; when the drought continued, nothing was left to feed the animals, which resulted in their deaths. Jambaji thought that if trees were protected, animal life would be sustained, and his community would survive. So he formulated twenty-nine injunctions. Principal among them was a ban on the cutting of any green tree and killing of any animal or bird. His community accepted these injunctions. Over time, their geographic area developed into a lush dense forest with substantial trees. About three hundred years later, when the king of Jodhpur wanted to build a new palace, he sent his soldiers to the Bishnoi area to secure timber for it. Villagers protested, and when soldiers would not pay any attention to the protest, the Bishnois, led by a woman, encircled the trees in order to protect them with their bodies. The soldiers had orders to bring timber, so they began killing the villagers. As the soldiers continued, more and more of the Bishnois came forward to honor the religious injunction of their guru. Finally, when the king heard about this human sacrifice, he ordered his soldiers back to Jodhpur and gave the Bishnois state protection for their beliefs. Even today, the Bishnoi community continues to protect trees and animals in their area with the same zeal. Their dedication became the inspiration for the Chipko movement of 1973.

The Chipko Movement: In March 1973, in the town of Gopeshwar in Chamoli district, Uttar Pradesh, villagers formed a human chain and encircled earmarked trees to keep them from being felled for a nearby factory producing sports equipment. The same situation later occurred in another village, when forest contractors wanted to cut trees under license from the Government Department of Forests. Again, in 1974, women from the village of Reṇi, near Joshimath in the Himalayas, protested logging by hugging trees and forced the contractors to leave. Since then, the Chipko Andolan (movement) has continued to grow from a grassroots ecodevelopment movement.[21]

The genesis of the Chipko movement has its background not only in religious belief, but also in ecological or economic concerns. Villagers have noted how industrial and commercial demands have denuded their forests, how they cannot sustain their livelihood in a deforested area, and how floods continually play havoc with their small agricultural communities. Women, specifically, have seen how men tend not to mind destroying nature in order to get money, while they themselves have to walk miles in search of firewood and fodder or other suitable grazing. In a sense, the Chipko movement is a feminist movement to protect nature from the greed of men. In the Himalayan areas, the pivot of the family is the woman. It is the woman who worries most about nature and its conservation in order that its resources are available for her family's sustenance. On the other hand, men often go away to distant places in search of jobs, leaving the women, children, and elders behind.

These two examples are illustrative of the practical impact of Hinduism on environmental conservation, or dharmic ecology in action. The Bishnoi and Chipko experiences demonstrate that when appeals to secular norms fail, one can draw on cultural and religious sources for environmental conservation.

There is no doubt that Hindu religion and culture, in ancient and medieval times, has provided a system of moral guidelines for environmental preservation and conservation. Environmental ethics, as propounded by the ancient Hindu scriptures and seers, was practiced not only by common persons, but by rulers and kings. They observed these fundamentals, sometimes as religious duties, often as rules of administration or obligations for law and order, but always as principles properly entwined with the Hindu way of life. That way of life did enable Hindus as well as other religious groups residing in India to use natural resources but to have no divine powers of control and dominion over nature and its elements. The Hindu belief that so long as Mother Earth is able to sustain magnificent mountains, lush forests, streams and rivers, and related endowments, she will be able to nourish all, particularly the human race and its progeny. This is expressed in the following prayer to the goddess Durgā:

> So long as the earth is able to maintain mountains, forests, trees, etc., until then the human race and its progeny will be able to survive.[22]

If such has been the tradition, philosophy, and ideology of Hinduism, what then are the reasons behind the present environmental crisis facing India? As we have seen, Hindu ethical beliefs and religious values do influence people's behavior toward others, including our relationship with all creatures and plant life. If, for some reason, those noble values become displaced by other beliefs that are either thrust upon the society or transplanted from another culture through invasion, then the faith of the masses in the earlier cultural tradition is shaken. As appropriate answers and leadership are not provided by the religious leaders and priests, the masses become ritualistic, caste-ridden, and inward-looking. However, besides the influence of alien cultures and values, what has really damaged India's environment are the forces of materialism, consumerism, individualism, and corporate greed, the blind race to industrialize the nation immediately after achieving independence, and the capriciousness and corruption among forest contractors and ineffective enforcement by forest officials.[23] All these acted against the maintenance, or resurgence, of respect for nature in India. Under such circumstances, religious values that had acted as sanctions against environmental destruction were sidelined as insidious forces worked to inhibit the transmission of ancient values that encouraged respect and due regard for God's creation.

How can those ancient values and wisdom be transmitted into practice? Can there be a practical dharmic ecology? It is not sufficient to examine and extol the ancient wisdom of Hindu seers, to dwell on the Vedic heritage, and then simply hope that a self-correcting process for environmental problems will set in. What is more important is how to put into practice that ecocare vision and make it relevant to modern times. I would propose the following strategy. The Hindu religion and its followers should become effective advocates and practitioners of the concept of ecocare and dharmic ecology, rather than staying on the sidelines. Hindu religious leaders should: 1) take the initiative and help secular institutions by providing timely and appropriate advice to encourage greater integration of ecocare heritage into educational curricula; 2) strengthen the capability of secular institutions to meet their goals of sustainable development and environmental conservation; 3) promote the concept of *sarva-bhūta-hite ratāḥ* (to serve all beings equally); 4) take the lead in promoting the concept of *vasudhaiv kuṭumbakam,* the family of Mother Earth, and the obligation of humanity to accept a world of material limits; 5) protect and restore places of ecological, cultural, aesthetic, and spiritual significance; and 6) build partnerships across social, economic, political, and environmental sectors, including dialogue with other religions and spiritual traditions.[24] The choice before the Hindu religion (as well as before all other religions) is either to care for the environment or be a silent participant in the destruction of planetary resources. Partnership with secular institutions must be forged and cooperation fostered at local, regional, national, and international levels. An environmental and sustainable development strategy, based on the lines suggested above, could offer a way of bridging the gap and making the essential link between secular, scientific, and spiritual forces.

An environmental stewardship that draws upon the Hindu concept of *dharma* and *karma* to the environment can provide new ways of valuing and acting. It can promote policies for sustainable development and introduce environmental protection initiatives. *Dharma,* if globally manifested, will provide the values necessary for an environmentally caring world and will not advance economic growth at the cost of greed, poverty, inequality, and environmental degradation. There is an urgent need to instill in all people a respect for nature and to strengthen decision-making processes in favor of environmental protection. This must be the focal point for a new global consciousness in an environmentally caring world.

In summary, a new universal consciousness must be developed that believes in at least two dictums: "what we sow is what we reap" and "everything is connected to everything else"; and our inherent *dharma,* or obligation, is toward the environment. These two concepts are intertwined with a third: *sarva-bhūta-hite ratāḥ,* serve all beings equally. The Hindu religion, like other religions and spiritual traditions, has the capacity to move the individual toward the divine because of its belief in divinity in nature; thus, it is imperative that such an inherent capacity is strengthened to its ultimate end. To achieve this, we need a new paradigm of thought, a dharmic ecology, perhaps. By developing such a paradigm, drawing upon the concept of *dharma* and *karma,* and based on the notions of *vasudeva sarvam, vasudhaiv kuṭumbakam,* and *sarva-bhūta-hitā,* we may be able not only to sustain the present generation, but also to leave a healthy legacy for future generations. Can the Hindu religion take up this challenge and act upon it as an integral part of its conscious strategy and vision for a sustainable future?

Notes

1. O. P. Dwivedi, *India's Environmental Policies, Programmes and Stewardship* (London: Macmillan, 1997), 21.

2. Bahūnāṃ janmanām ante jñanvān māṃ prapadyate,
Vāsudevaḥ sarvam iti sa mahātmā sudurlabhaḥ.
(*Bhagavadgītā* 7.19)
All translations are my own.

3. O. P. Dwivedi, "Vedic Heritage for Environmental Stewardship," *Worldviews: Environment, Culture, and Religion* 1, no. 1 (April 1997): 25–36.

4. O. P. Dwivedi, B. N. Tiwari, and R. N. Tripathi, "Hindu Concept of Ecology and the Environmental Crisis," *Indian Journal of Public Administration* 30, no. 4 (January-March 1984): 33–67. We first presented this paper at the annual conference of the Canadian Asian Studies Association, University of British Columbia, Vancouver, 4 June 1983. When it appeared in the *Indian Journal of Public Administration,* it may have been the first such research paper to be published.

5. Pakṣi dagdhaḥ sudurbudhhe pāpātman sāmprataṃ vṛthā,
Vṛthā snānam vṛthā tirthaṃ vṛthā japtaṃ vṛthā hutaṃ.
(*Narasiṃha Purāṇa,* 13.44)

6. Bijaiyarjñesu yaṣṭvyamiti vai vadikī śruti
Aj sanjñani bijani cchāgan no hantumarhatha
Naiṣa dharmaḥ satām devā yatra vadhyeta vai pasuḥ
Idaṃ kṛtayugaṃ śreṣṭhaṃ kathaṃ vadhyeta vai pasuḥ.
(*Mahābhārata,* Santiparva, Moksadharma, 337.4–5)

7. O. P. Dwivedi, *Vasudhaiv Kutumbakam: A Commentary on Atharvediya Prithivi Sukta,* 2d ed. (Jaipur: Institute for Research and Advanced Studies, 1998).

8. Giryaste parvatā hima vanto raṇyaṃ te pṛthivī syonamastụ
Babhrum kṛuṣhṇaṃ rohīṇīm vishvarūpāṃ dhruvaṃ bhūmiṃ pṛithivī mindraguptāṃ
Ajītohato akṣhatoadhyathāṃ pṛithivīmaham.
(*Atharva Veda,* Kanda 12, hymn 1, verse 11)

9. Dwivedi, *Vasudhaiv Kutumbakam.*

10. Janaṃ bibhratī bahudhā vivācasaṃ nāndharmāṇaṃ pṛithivī yathaukasaṃ
Sahastraṃ dhārāḥ draviṇasya me duhāṃ dhruveva dhenurana pasphurantī.
(*Atharva Veda,* Kanda 12, hymn 1, verse 45)

11. Bhūme mātarni dhehi bhadrayā supratisthitam
Saṃvidana divā kave shṛiyā mā dhehi bhūtyām.
(*Atharva Veda,* Kanda 12, hymn 1, verse 63)

12. Tā naḥ prajāḥ saṃ duhatāṃ samagrā vacho madhu pṛithivi dhehi mahyaṃ.
(*Atharva Veda,* Kanda 12, hymn 1, verse 16)

13. Tasmāc chāstraṃ pramāṇaṃ te kāryākāryavyavasthitau
Jñatvā shāstra vidhānoktaṃ karma kartum ihārhasi.
(*Gītā* 16.24)

14. Pravṛttiṃ ca nivṛttim ca kāryākārye bhayābhaye
Bandhaṃ mokṣaṃ ca yā vetti buddhiḥ sā Pārtha sāttvikī.
(*Gītā* 18.30)

15. Prabhavārthaya bhūtānām dharma pravacanaṃ kṛtaṃ
Yaḥ syāt prabhav saṁyuktaṃ sa dharma iti niścayaḥ.
(*Mahābhārata,* Shanti Parva, chapter 109, verse 10)

16. Sarvāni bhūtāni Narendra paśya tathā yathāvad vihitaṃ vidhātra
Svayonitaḥ karma sadā caranti neśe balasyeti cared dharmaṃ.
(*Mahābhārata,* Vanaparva, chapter 25, verse 16)

17. O. P. Dwivedi, "Our Karma and Dharma to the Environment," in *Environmental Stewardship: History, Theory, and Practice,* ed. Mary Ann Beavis, 59–74 (Winnipeg: Institute of Urban Studies, University of Winnipeg, 1994).

18. Yesāṃ ye yāni karmāṇi prak sṛiṣṭyam pratipedire
Tāny eva pratipādyante sṛigyamānāḥ punaḥ punaḥ.
(*Mahābhārata,* Shanti Parva, chapter 232, verse 16)

19. Na Bhuktam Kṣīyate Karma Kalpa koṭiṣtairapi
Avaśyamaiva bhoktavyam kṛtam karma śubha ā subhaṃ.
(*Brahmavaivarta Purāṇa,* Prakṛti. 37.16)

20. Pāpaṃ karma kṛtam kiṃcid yadi tasmin na dṛśyate
Nṛpate tasya putreṣu pautreṣu api ca naptriṣu.
(*Mahābhārata,* Shanti Parva, chapter 139, verse 22)

21. Chandi Prasad Bhatt, "Chipko Movement: The Hug That Saves," in *The Hindu Survey of the Environment, 1991* (Madras: The Hindu, National Press, 1991), 17.

22. Yāvadbhūmaṇḍalaṃ dhatte sāsailvavana kānanaṃ,
Tāvat tiṣṭhati medinyaṃ śāntātiḥ putra-pautrikī.
(*Durgā Saptaśati,* Devi kavacham, verse 54)

23. Ashish Kothari, "Forest Bill: Old Wine in a New Bottle," in *The Hindu Survey of the Environment, 1995* (Madras: The Hindu, National Press, 1995), 51–54.

24. Dwivedi, *India's Environmental Policies, Programmes and Stewardship.*

The Five Great Elements (*Pañcamahābhūta*): An Ecological Perspective

K. L. SESHAGIRI RAO

Introduction

The environmental problems of today are immense because they affect the entire planet, the whole of humanity, and all life-forms. The environmental ravage looms over all countries of all continents and threatens the survival of humankind. The problems cannot be solved on a local or partial basis; they have to be tackled globally. The warming of the global climate, for example, will put the ecosystems of all countries at great risk. Only a cosmocentric, global consciousness can help. It is, therefore, useful to explore what our ancient texts say about environmental protection and the kind of relationship they recommend between humans and their natural environment.

Unfortunately, for many, awareness of the ecological crisis is close to nonexistent. They do not realize that they are damaging the environment and ultimately themselves. They are cutting the very branches on which they are sitting; they do not realize their responsibility to keep these branches healthy. Human ignorance, shortsightedness, and egoism have given rise to an environmental catastrophe. A very important task, therefore, is to arouse people's public consciousness and environmental activism. This should be done on the basis of ancient and modern knowledge.

In this paper I shall explore the ecological vision that the Hindu scripture and traditions advocate. I believe that Hindus can benefit from these insights and revive traditional sustainable practices, many of which they have forgotten, to their own peril. I submit that these traditional practices also have a wider relevance. In any case, all of us have a duty to muster greater knowledge and available resources to prevent further damage to the life-promoting and life-sustaining characteristics of the planet.

Dharma

A fundamental feature of the Hindu tradition is that there is no dividing line between the sacred and secular. There is no area of life that is alien to spiritual influence. Secular and sacred concerns are inextricably interwoven. *Dharma* deals with moral as well as physical order. It espouses an integral view of life and the world by providing coherence to the diverse activities of life. One's duty to one's family, society, humanity, the natural elements, and God are all part of *dharma.*

Dharma is a comprehensive term, incorporating duty, morality, law, and justice. It also refers to the structure of reality. It is a cosmic law, the law of life and development. It is *sanātana,* eternal. It is a support of the whole universe. The purpose of *dharma* is to maintain and conserve the society and the world. *Dharma* operates in accordance with time, space, and conditions, yet maintains a balance, lest society stray from eternal principles.

The moral code of conduct prescribed by the *dharmaśāstra*s counteracts the individual's tendency to pursue exclusively selfish interests. *Dharma* requires from each member of society a way of life consistent with the general welfare of humankind. Lord Kṛṣṇa teaches in the *Bhagavadgītā* that only those actions are worthy and valuable that contribute to the welfare of the whole world, with all living beings in it. The prosperity and happiness of people are dependent on an ordered society, on nature, and on the cosmos as a whole. Just as the health and happiness of one person depend on all the sense organs working cooperatively, the world organism only functions smoothly when the great elements perform their work together and well. According to the *Bhagavadgītā* (3.10–13):

> The lord of all creatures, having created beings and deities declared thus: by sacrificial action you will attain prosperity; this is for you the desired milch cow, with this you cherish gods (cosmic powers) and the

> gods will cherish you. By mutual care and concern, there will reign general prosperity for all. The deities, being satisfied by the performance of *yajña,* supply the needs of humans. But one who enjoys these gifts without offering them in return is certainly a thief.

The concept of *dharma* provides a model of cosmic and social equilibrium; it represents the principle of preservation. *Dharma* sustains and is to be sustained in turn through human efforts; following *dharma* leads to worldly development and spiritual fulfillment (*mokṣa*). Prosperity (*abhyudaya*) is the result of proper care and management of natural and human resources. Even divine incarnation works to restore cosmic *dharma*. Śrī Kṛṣṇa, by personal example, encourages an attitude of caring and commitment toward the earth and her creatures.

Natural World

Prakṛti, cosmic matter, is the matrix of the entire material creation. The multiplicity and variety of things that we experience in ordinary life is traced to this single substance. According to the philosophy of Sāṃkhya in Indian thought, primordial matter is a bundle of energy. In this bundle there are positive, negative, and neutral forces, which are in a state of constant flux, each one trying to dominate the others. These are known as the three *guṇa*s (thread or quality) of *prakṛti* (matter), which is of great value to conscious beings. *Prakṛti* is rich and bountiful. It is powerful and fierce. It is also benevolent, tender, and delicate. It shies away from exploitation and abuse. It needs to be looked after with affection and adoration. It is compared to a delicate dancer who serves us but feels hurt if exploited. Nature serves us but reacts when abused. What nature wants is a judicious use of its resources for progress and prosperity, not indiscriminate exploitation. Initially, it cautions us to rectify our wrongs; but if we do not heed nature's warnings, it reacts vigorously.

The five great elements that constitute our environment are evolutes of *prakṛti;* they are distinctive elaborations of primal energy. Every element has its own life, form, and location, but all are interconnected and interdependent. They are the conditions and blessings of our individual and collective life. A sequential ordering of these elements, according to their decreasing subtlety—namely, space, air, fire, water, and earth (*ākāśa, vāyu, tejas, ap, pṛthvī*)—is stated in the *Nārāyaṇa Upaniṣad.* The Upaniṣads explain the doctrine of the five elements, in a holistic way, in relation to Brahman, the ultimate reality. The world cannot be understood wholly within itself. Brahman is the cause of the universe. The *Taittirīya Upaniṣad* says: "From Brahman arises *ākāśa,* from *ākāśa* arises *vāyu,* from *vāyu* arises *tejas,* from *tejas* arises *ap,* and from *ap* arises *pṛthvī.*" After creation, the transcendental Brahman enters into the universe as its life and consciousness. This is the basis of the affinity of humans with the natural world. In the *Maitrāyaṇī Upaniṣad* it is stated that the three-quartered Brahman has its roots above, and its branches below; the branches are earth, water, fire, air, and space.

The individual's body is a microcosm of the great cosmic body, which is the Supreme Being's mode of self-expression. The five subtle elements (*tanmātra*s) are imperceptible; they are associated with the five senses, the gateways of human knowledge. Nose is related to earth, tongue to water, ice to light or fire, skin to wind, and ears to space. Gross elements are the manifestation of the subtle ones. Each element has a unique attribute: space is qualified by sound; air possesses both sound and touch; fire consists of sound, touch, and color; water, in addition to the other three, has taste; and earth has odor as well as the first four attributes. Nature and the environment are not outside us; they are not alien and hostile to us. They are an inseparable part of our existence.

Divine Source

Nature owes its significance to a transcendent principle whose mind is reflected in it; it reveals a supreme intelligence. The physical world, both animate and inanimate, is a manifestation of the ultimate reality. This world is God's world and all the things thereof belong to him. The five elements are the propelling forces of God. God is independent of the five elements, but the five elements are not independent of God. Waves belong to the ocean; the ocean does not belong to the waves. According to the *Bhāgavata Purāṇa,* the existence of God is to be realized in two stages: one is concrete and the other abstract. The abstract is the all-pervading spirit and the concrete is the physical world, visible and changing. The five elements are the ingredients of the world and also the support of life on the earth. The *Maitrī Upaniṣad* states: “The three-quarter Brahman has its roots above, . . . its branches are *ākāśa, vāyu, tejas, ap,* and *pṛthvī.*”

The *Bhagavadgītā* (7.45) says that primordial nature is eightfold: the five elements plus mind, intelligence, and ego-sense, which constitute the lower nature of the Divine. The higher nature of the Divine relates to living beings, which uphold the world. Life has a higher value than things or property. There is not a thing in the world that is apart from the Divine, however. Kṛṣṇa says in the *Bhagavadgītā* (8.39): “Whatever is the seed of all beings, I am that, O Arjuna. There is no being either animate or inanimate which can exist without me.” In the *Puruṣasūkta* of the *Ṛgveda* (10.90), the world order is compared to a cosmic being, a living organism in which each living being is related to the life of the whole. “The moon was created from his mind, the sun from his eyes, *Indra* and fire from his mouth, the wind from his breath, the atmosphere issued forth from his navel, the heaven from his head, the earth from his feet, the quarters from his ears.”

Wholesome Environment

The environment has far-reaching effects on every creature. The life of our world depends upon the purity and balance of the five elements that surround us. All things are the result of the combination and permutation of these elements. Even our bodies, which at death disintegrate and dissolve into nature, are made up of these five elements. These elements are not spiritually vibrant all the time and in all places; but they can be made so. They have to be kept pure and in balance. If we consume polluted elements, our bodies become subject to disease and distortions. Distortions of the elements put all life-forms in danger; hence, the elements should be protected. To live in nature and in accordance with nature is to create an ideal environment free from pollution. Love of nature is the secret to protection of the environment.

There is a Purāṇic story that the gods and demons churned the milky ocean to obtain nectar *(amṛta).* They used Mount Mandara as the churning rod and chief of the serpents Vāsuki as the rope. When they began to churn, many things other than nectar emerged, among them the deadly poison known as *hālāhala,* before which everyone fled in fright. The poison began making people giddy. Soon it became clear that it would destroy the whole world. The people sought refuge with the ever-auspicious Lord Śiva. Śiva took all the poison in his palm and swallowed it in his overflowing compassion; the poison could not harm him, but left a blue stain on his throat. Hence, he became known as *Neelakanṭha*.

Industrialization has churned the five elements excessively. In the name of development, nature has been badly mauled. Development has, no doubt, given rise to some immediate profits and material gains for some persons; but the world has become overburdened and perturbed. Development has also given rise to terrible poison in the shape of environmental pollution and long-term damage, resulting in the disequilibrium of the five elements. We need to follow the example of Śiva and strive to remove the toxic components and make the five elements whole and wholesome once again.

Natural elements nourish life; they promote, heal, and rehabilitate life. Humans need to conserve these elements wisely, and the elements will in turn confer upon humans vitality, strength, and splendor. All forms of life are an integral part of nature. All species need to be appreciated. My father, who lived most of his life in a village in South India, used to say that we should not underestimate even the termites. They eat up wood and dead trees and turn them into precious soil. All species serve creation and work for a divine purpose. There are birds, fishes, and animals of many kinds. The human species, other species, and the world of divinities are all interdependent. We are not isolated. We are participants in a large and grandly meaningful whole.

Purification of the Elements

Whenever a worship service is performed in the Hindu tradition, it is preceded by the purification of the elements, *bhūtaśuddhi*, as a necessary preliminary. The five elements are purified within the body and without. Purification comes before sanctification. It establishes harmony between macrocosm and microcosm. The offerings that are made to the deity represent the best of each of the five elements. Fragrance offered represents the essence of the earth, flowers represent the sky element, incense the wind element, water represents the water element, and light the fire element.

Actions leave both their good and bad effects on the mind, the subtle body. The mind, with its character, endures the death of the gross body. The quality of actions performed in the present life conditions the future life. The present life itself is the result of past actions. Thus, one's psychological endowment and fortunate or unfortunate circumstances are the consequences of past deeds. Hindus speak of three bodies of a person: gross, subtle, and causal. The gross body is made up of the five elements. It is also nourished by these very elements, by what we eat, drink, breathe, and so on. It is impelled by the subtle body of the soul, which is constituted by the threefold energies of *prakṛti*. Karma is stored in the subtle body. When the gross body collapses, the soul, the subtle body, survives and is born again. This is reincarnation. Hindu scriptures speak not only of reincarnation but also of transmigration, that is, regression into subhuman births—which, incidentally, is another argument in favor of vegetarianism.

Nature and Life

Humans grow in the lap of nature and receive sustenance from her. The basic needs of human beings are met by her gifts. These natural resources are not given for the selfish exploitation of one group or nation or generation by another, but are to be shared by all creation. Human life, animal life, and plant life are all interlinked. Yet humans have been abusing other forms of life; they have shown themselves to be shortsighted and selfish. The very industrial achievements of humans frequently undermine the natural environment. The crisis today is not only what humankind is doing to the environment, but also what the environment is doing to humankind. Humans are faced with air pollution, water contamination, diminished wildlife, the hazards of radiation, and polluted vegetation. Their health and prosperity are undermined. They are deprived of direct contact with nature in overgrown cities. They are led to artificial living, crime, and chaos. Rabindranath Tagore writes in *Sādhana: The Realisation of Life:*

> When by physical and mental barriers we violently detach ourselves from the inexhaustible life of nature; when we become merely man, but not man in the Universe, we create bewildering problems and, having shut off the sources of their solution, we try all kinds of artificial methods each of which brings its own crop of interminable

> difficulties. When man leaves his resting place in the universal nature, when he walks on the single rope of humanity, it means either a dance or fall for him; he has ceaselessly to strain every nerve and muscle to keep his balance at each step, and then in the intervals of his weariness, he fulminates against Providence and feels a secret pride and satisfaction in thinking that he has been unfairly dealt with by the whole scheme of things. (London: Macmillan, 1915, 9)

Space

Ākāṣa is the generic term for space, the subtlest and most pervasive of the five basic elements. It is practically omnipresent. It is not void or emptiness; rather, it fills any void or vacuum. It manifests itself as gravitational force. It represents openness, brightness, expansiveness, and the fullness of blooming capacity.

At Cidambaram in Tamilnadu in South India, space is worshiped as a manifestation of the Divine. It is said that God Śiva lives there in the form of *ākāṣa,* or sky. The priest lifts a screen and allows the devotees to have a peep at the roof. It is said to be the most sacred spot in Cidambaram. *Mahākāśa* is infinite space. *Antarikṣa* is the intermediary space between heaven *(svarga)* and earth *(bhū). Loka* is the perceptible space that fills space. *Loka* sustains people and *loka* has to be sustained by cosmic powers. Due to industrialization, space has become congested in urban areas. There is too much noise and pollution levels are dangerously high in factories and metropolitan areas. Industrialization is causing physical and psychological ailments. Overcrowding frequently leads to unrighteousness and psychological pollution.

Air

Air is an essential part of our natural environment. The atmosphere is a positive blanket of gases surrounding our planet, saving it from the hostile radiation from outer space and thus sustaining life on the earth. We all inhale and exhale; we are all connected through air. If any one of us pollutes the air, we harm all of us, just as the secondary smoke of smokers is dangerous to all the rest. Unwholesome air is responsible for diseases and epidemics. If we save the air from pollution, we increase our longevity. The air over many cities around the globe is becoming increasingly dirty. Asthmatics and elderly people with respiratory problems need clean air. They suffer in a polluted atmosphere. And yet, noxious gases are released into the atmosphere. We have disrupted the equilibrium of the carbon cycle by our heavy use of automobiles (400 million in the United States alone), the nitrogen cycle through repeated usage of chemical fertilizers and pesticides, and the sulfur cycle by burning crude oil and coal. We are destroying the protective ozone layer by the heavy use of chlorofluorocarbons. At the same time, we are cutting down trees that are needed to absorb the ever-increasing quantities of carbon dioxide released into the atmosphere by our industrial activities. Every minute of every day, we are losing fifty acres of our rain forests.

Nitrogen makes up 78 percent of the air we breathe, while oxygen makes up 21 percent. Nitrogen is a necessary part of the protein substance in the body-building material of humans, animals, and plants. Human bodies cannot use pure nitrogen; we acquire it when taking in proteins obtained by eating certain plants or plant-eating animals. The need for plant proteins requires that we protect the world's vegetation.

Fire

Fire is considered divine. Sun, lightning, and fire are forms of this element. The lustrous dawn (Uṣas) and the glorious orb of the sun (Sūrya) are its expressions. Sunlight is life giving and life supporting under certain conditions. For example, if the temperature is too high or too low, life cannot be sustained. The friendly aspect of the sun is personified as Mitra. He is the stimulator of life (Pūṣaṇ). He shines in every direction (*vivasvat*). Fire is a messenger to the gods. No impure thing should be thrown in the fire. Sacrificial smoke purifies sky (space) and air. The seer prays in the *Ṛgveda* (1.1): "Oh, Agni, like a father to his son, be thou easy of access to us; be with us for our welfare."

The divine sun has been worshiped in India since ancient times. There are sun temples in Kashmir (Martanda), Gujarat (Modhera), and Orissa (Konarak). The *Gāyatrī Mantra* is addressed to the sun god. It is believed that if we recite it regularly, our intellect becomes sharp. Fire possesses the power to destroy various types of poisonous elements and diseases. It can also purify the environment.

The cooperation of different elements produces benefits for living beings. While our lives are sustained by the chemical energy that we obtain from the food we eat and the oxygen we breathe, the energy actually comes from one source, sunlight. Similarly, when the sun heats air, it rises into the sky, laden with moisture. When the warm air gets to a certain height in the atmospheric space, it cools. Moisture then changes into small drops of water, which form clouds, and then it rains. And the rains help the vegetation to grow.

Water

Water is a life-sustaining, life-preserving and life-purifying element. It is one of the most common of all substances. Water constitutes about two-thirds of the earth's surface. Of this, 0.1 percent accounts for pure ground water—from all the lakes, streams, rivers, and rainfall—which is unevenly distributed throughout the world. Water not only provides an environment for life, it is a part of life itself: no water, no life. All creatures depend on water. Without water, there can be no agriculture, no fruits or vegetables. Water is sacred. Bathing and its symbolic substitute, *ācamana* (sipping water), are essential in religious rites.

Today, water has become scarce due to the profound damage to the global water system. Vast ocean shores are littered with the debris of our industrial culture. Millions of tons of untreated sewage are released into the seas. Rivers, too, are polluted by industrial and organic waste. For example, as a byproduct of making paper and paper products, mercury and other toxic chemicals are dumped into streams and rivers, resulting in long-term, irreversible, and harmful effects. Household rubbish and human effluents are released into waters. Acid rain has damaged lakes and forests.

The importance of the purity of drinking water, the cleanliness of river water, and the protection of forests and mountains are discussed in the Ayurvedic texts. Contaminated waters are dangerous to the health of people in both villages and cities. In this connection, the early life of Kṛṣṇa—the eighth incarnation of Lord Viṣṇu—who lived among the cowherds, is inspiring. He is concerned about the life-sustaining quality of the waters of the Yamunā and the life-promoting nature of Mount Govardhana. Lord Kṛṣṇa plays with the common, simple, and pure cowherds and endears himself not only to his contemporaries and humanity at large, but also to cows, other animals, and the natural world.

In the Purāṇic story, Kāliya, the serpent king, polluted the waters of the river Yamunā by emitting poison, causing harm to the people and the cattle living nearby. Kṛṣṇa jumped into the water and onto the serpent. Kāliya attacked him with a lifted hood. Kṛṣṇa grew in size and danced on its hood, and the serpent could not bear the weight of the steps, each of which brought blood from its mouth. Finally, the serpent prayed that it might be pardoned. Kṛṣṇa spared it and commanded it never again to pollute the Yamunā waters, to leave the river immediately, and to go to the deep of the ocean. Today, the many CEOs of chemical and fertilizer companies responsible for the pollution of rivers and lakes deserve to be so disciplined.

Flourishing civilizations historically developed on riverbanks. It is at these centers that trade and commerce developed and connected the world with the ideas and ideals of philosophy, religion, and science. The rivers and their confluence became places of pilgrimage in ancient cultures. Ceremonies of initiation and death were and still are performed at these places. It is because of their life-giving and life-fulfilling properties that the appellation "mother" is bestowed on such rivers as Mother Gaṅgā. The *Gautama Dharmasūtra* says: "One must not splash water with his feet nor enter water when one is full of dirt; one must not discharge into water blood, excreta, urine, spit, and semen."

Earth

We are part of the earth and the earth is part of us. Earth and we are part of the same whole. We are dependent on and participate in earth's life. Earth is wonderful; it is beautiful. Earth shelters a vast number of species, including the humans. Hindus look upon the earth as God's body. This perception explains the Hindu morning prayer, which is said daily when the first footstep is taken on the earth: "Forgive me even for treading on you."

The earth is a living organism. Hindus regard the earth as a mother deserving our reverence. She feeds us, provides us with shelter, and gives material for clothes. Earth's resources are for the entire humanity. Without her gifts, human survival is impossible. Mother Earth is called Dharaṇī, as she sustains and supports all her children. Whatever befalls Earth also befalls the children of the earth. There is a mutuality of interests between humans and Mother Earth. We cannot strip the earth and remain unaffected. If, as children, we do not take care of her, we diminish her ability to take care of us. Because of soil pollution, the earth looses its fertility and has adverse effects on agricultural production. Now, the earth is on the brink of disaster: the future health of our planet is under serious threat. How to bring redress to her? The *Atharva Veda* (12.1.64) says:

> Bearer of all things, hoard of treasures rare, sustaining Mother, Earth the golden breasted. . . . Impart to us those vitalizing forces that come O earth, from deep within your body. Whatever I dig up of you, O earth May you of that have quick replenishment.

If we are to rejuvenate the earth, we need to reawaken our love for her.

The green earth is suitable for the life of the animal and human worlds. Trees with wide leaves absorb moisture and give it to the earth. They prevent soil erosion. The vegetable and animal worlds are linked in the web of life. Organic farming is good for human health and good for the earth. The earth provides us with abundant foodstuffs. A vegetarian diet provides the required substances for a healthy life. It feeds more people and is less wasteful. *Ahiṃsā* and vegetarianism are embedded in the Hindu ideals of life. These practices may be summarized: go vegetarian; it makes you live longer and better; it is also good for the economy and the global environment.

The principle of cow protection symbolizes human responsibility to the subhuman world. It indicates reverence for all forms of life. Humans are related to and dependent on the whole creation. The cow serves humans thoroughout its life and even after death. The milk of the cow runs in our blood. Its contributions to the welfare of the family and society are countless. Hindus seek daily blessings for the welfare of the cows. When cows are cared for, the world, on all levels, finds happiness and peace.

The population explosion is causing natural resources to diminish at a terrific pace. By the year 2000, the world population will pass six billion and India's population one billion. The strain the world's population imposes on land, water, and other natural resources must be seriously considered. The rate of population growth is not in keeping with the availability of food. There is a blatant imbalance here. There is a limit to how much burden the earth can bear. Hindus, hopefully, will see the writing on the wall. Family planning, which they call the "family welfare program," is to be vigorously pursued. There is no religious prohibition against it.

Mental pollution causes physical and environmental ravages. Disharmony, conflict, sickness, and degradation of the natural elements are the results of greed, lust, and anger. Ignorance and selfishness cause evil deeds. All actions start in the mind. The mental environment has to be kept clean and focused. *Yajña* is the sacrificing of one's ego and the burning of impurities. When this is accomplished, one's environment and society are purified. A wiser thought, a cleaner vision, and a greater kindness appears. Honesty, love, and selfless service emerge.

Need and Greed

The bounty of nature is a source of wealth and provides for the needs of all creatures on the earth. Animals by instinct adhere to natural requirements. Humans, however, have a tendency to indulge excessively and accumulate things beyond their needs; we also waste valuable resources. We squander this bounty through the multiplying of our superfluous desires and wants. This dichotomy between basic needs and excessive desires is crucial to the understanding of our ecological problem. The wealth of the world is based on the five elements; all peoples must share this gift among themselves and with other creatures. But consumerism and advertising mislead people into thinking that they need more and more. They obscure basic human needs. This leads to a "violent" society, since exploitation provides its basic motivation. By plundering and destroying nature, technology is ironically bringing humans into a position of conflict with nature. Ruthless exploitation of nature results in disaster. Exploding consumption in industrialized countries and population explosion in developing countries are ruining the resources, health, and balance of the natural elements. Unless we place limits on our extravagant consumption and throwaway mentality, we will face natural shortages and worse. Only if we stop wasting our wealth and energy can the peoples of the world be happy and contented.

The global ecological crisis is due to the quest for superfluous worldly goods. It is the result of a materialist worldview. The persistent and demonic passion for an ever-increasing standard of living is leading us to a global disaster. If technological and material development overlooks the needs of the spirit, is this really advancement? We have lost our perspective on what constitutes wealth. Our goal should be an optimally viable economic and social organization in relation to the entire planet and the good life of all humankind. Growth must be sustainable on a worldwide basis. It is unfortunate that we have traded a good life for a goods life. It is foolish to think that merely by multiplying one's wants, one is achieving happiness. The more our desires are satisfied, the greater our desires grow. The *Manusmṛti* (4.2) declares: "Happiness is rooted in contentment; its opposite is rooted in misery."

To minimize consumption means to minimize harm to the environment. Even the wealthy should live a simple life and so prevent the rapid reduction of the earth's resources. Heavy industrialization inflicts violence on the environment. Bhopal, Chernobyl, and Love Canal are tragic examples. To harm the environment is to harm ourselves. Our life-style should become less and less violent. The time has come to reach for an optimum standard of living, rich in spiritual qualities and values. This means a minimization of wants for society as a whole—in other words, simple living and high thinking.

Conclusion

The earth is our common home; humanity is our family. We have to manage our household properly. Accordingly, we need to plan and work for short-term and long-term goals simultaneously. We need to adhere to environmentally friendly procedures and learn to utilize all resources efficiently. The art of balancing and accommodating the interests of the present as well as of future generations is to be practiced. Conservation relates to a care for the future. We need to promote life-styles that will improve and protect for posterity the life-sustaining characteristics of our earth, water, fire, air, and sky. Do we have any examples to follow? I invite you to look at the life of a tree, as described by an anonymous author. We may be inspired to draw from the tree's example some ecological lessons:

> It gives its leaves to animals; it shares its flowers with bees. It gives its fruits to peoples. It provides shelter to birds and insects. It takes carbon dioxide and gives oxygen to the living world. Finally, it sacrifices itself to be used as construction material or fuel. It lives and dies for the service of others. However, it never asks anything in return. Wouldn't it be nice if we could appreciate and express our good will for it.

Let me conclude with a parable from the *Bṛhadāraṇyaka Upaniṣad* (5.2.1–3). Demons, humans, and gods, the progenies of the Creator (the Father), performed penance and pleased the Father; and they asked Him for a boon. The Father responded to their request by repeating one syllable: "da, da, da." The gods could not comprehend the meaning of the pronouncement. Father again repeated the syllable, "da," and asked, "Could you comprehend?" The gods replied, "Yes, Father; 'da' means control of the senses (*damana*); we are too much involved in seeking pleasures. Lest we be carried away by sensory indulgence and forget our duties, you have given us this boon for our good. We need to exercise self-restraint." Next was the turn of the humans. The Father pronounced the syllable, "da," and asked, "What did you understand?" Humans replied, "We are greedy and covetous; so you have commanded us to give gifts and make charitable donations (*dāna*) so that our greedy and grabbing nature be controlled." Lastly, when the demons asked for a boon, he repeated the same syllable, "da," and asked, "What did you comprehend?" The demons replied, "We are very cruel; so 'da' signifies that we must be kind." The point of the above parable is to show, in short, that living a life of *dharma* is to live a full life, that self-control, charity, and kindness link the individual to the entire society and the universe; they lead to life affirmation and world affirmation.

The five elements were blessings for millions of years. Since the time of industrialization, they have become commodities. Sunderlal Bahuguna, a well-known follower of Mahatma Gandhi and an activist in the field of ecology, says that proponents of modern development believe in "converting nature into cash." They view nature as a "commodity." Our luxuries have been gained at great cost to the environment and to other life-forms. Before we make our ecological plans for the twenty-first century, we must retrace our steps. We need to decentralize and develop small-scale industries. We need to develop beneficial, appropriate, and nonpolluting technologies, such as solar energy, nuclear fusion, wind power, wave power, and geothermal energy. The sources of these powers are the five great elements; and they are beneficial and benevolent.

Nature Romanticism and Sacrifice in Ṛgvedic Interpretation

LAURIE L. PATTON

"Do you adore the green grass, with its terror beneath?"
Mary Oliver, "Peonies"

Sacralizing Nature

What does it mean to sacralize nature? How and why do we use ancient texts in the service of an idea of nature or ecology? As Norman Crowe puts it in his most recent theory of the environment, *Nature and the Idea of a Man-Made World,*[1] to claim something about ecology is, in part, to claim a certain kind of unity: to invoke nature is to invoke primordial balance, holism, biological economy, understanding of interrelationship between things. In the study of religion this description is called a cosmology, a story about the origins of the universe, which describes just that interrelationship and balance between the world of nature and the world of humans. To put it succinctly, cosmology is usually in the service of some form of ecology. In the recent study of Indian ecology, the focus has been similar, stressing cosmological texts and ritual practices which, as one recent Indian publication claimed, "sacralize nature, so that man [sic] can live as an integral part of the universe."[2]

Both ecologists and Indologists have resorted to ancient texts to describe Indian ideas about the cosmological workings of nature and to recover an understanding of balance and harmony within that world.

As we shall see below, both appeal to a kind of Romantic ideal of "nature in the abstract," which is in fact only part of the picture suggested by the ancient Indian milieu itself. A more realistic glance at that textual tradition is the object of this article; our readings will teach us lessons about how and why we use ancient texts and what they in turn can teach us about human relationship to the natural world.

A word is in order about the process of Romanticization that I am arguing against here. I am defining Romanticism as the idea that there is an integral and harmonious connection between the human being and the natural world. A voluminous literature is developing about the roots of the Romantic movement, the idea of nature, and present-day ecological philosophy.[3] A lesser, but nonetheless important, literature is developing on the Romanticization of India, particularly in its relationship to its status as a culture in touch with nature and the earth. The literature, taken as a whole, is ambivalent; on the one hand, the last two decades have witnessed a massive postcolonial critique of European Romanticization of India as a place of natural mystery and beauty, a beauty which is essentially female and which should be plundered (raped) and "developed" out of its primitive state.[4] On the other hand, the last decade has also witnessed a reassertion of these same values in Hindu culture as positive ones, to be celebrated as unique and ancient and to be used as resources for future ecological movements.[5]

Herein lies the philosophical question about Nature Romanticism and India. What if the texts that are used to support both of these arguments (both for and against the Romancing of India) are in fact the same texts? What if the "seeds" of a nonviolent, harmonious, relationship to nature also contain explicit (and transcendentally sanctioned) scenes of destruction as well? Is it incumbent upon us as ecologically minded readers of texts to acknowledge the presence of both attitudes? I want to argue here that it is indeed incumbent upon us to do so. One participant in the Harvard conference on Hinduism and ecology, Harry Blair, made a very eloquent plea on behalf of Romanticization —for without that process of affirmation of harmony between humans and nature, the respect and veneration which could determine our constructive behavior toward nature would be lost. While I take his point, I also think that the argument could be made that any reading that "leaves something out" of the ancient assessment of nature can lead to a dishonest practice of reading, which ultimately does not help the ecological cause. For if, in fact, balanced harmony can be read into the *Ṛgveda,* so too can the destruction of nature. And such destruction can be seen as equally sanctioned by the Vedic texts. Any citation of ancient texts lands ecologically minded hermeneuts on a very sticky wicket, and, because of their benign but quite evident selectivity, leaves them open to being less, and not more, persuasive.

An important word of caution is necessary here. Although the texts I will be interpreting in this article are sacrificial texts which do involve a level of violence to nature, it is not my intention to define sacrifice as inherently violent. As Frédérique Apffel-Marglin has pointed out,[6] there are ways in which the sacrificial paradigm could be used in a nonviolent way to talk about ecological balance and harmony. A definition of sacrifice does not necessarily entail violence. But the sacrificial worldview can inform our ecological reading of ancient India.

Even more importantly, it is the early Indian case, beginning with the sacrificial practices and metaphors of the Vedas and coming to full flower in the Upaniṣads, that teaches us that "everything is eating everything else."[7] Although space does not permit a detailed discussion of a vast scholarly literature, some general review of ancient Indian sacrificial practices might be useful here. As many scholars imply, the sacrifice (*yajña*) of the *Ṛgveda* was probably a basic offering of hospitality, without the extreme elaboration and detail that the later philosophical and performative texts, the Brāhmaṇas and the Sūtras, describe. Yet, even in these earliest of times, the sacrifice played a crucial role in Vedic religion: it is the religious act par excellence. One might describe its earliest form as an act of homage to the divinities of the Vedic world, consisting of a gift of oblation, called *havis,* made in order to obtain certain benefits, such as prosperity, health, long life, abundance of cattle, male offspring, and, in the later rites, even more elaborate requests.

Before we get to the specifics of such sacrifice in the Ṛgvedic hymns, I want to add that, in the view of the Aryans, the sacrifice also played a major role in the survival of the Aryan group and its ability to conquer enemies, both natural and human, along the way. In a later Vedic work, the *Śatapatha Brāhmaṇa,* there is mention of the story of the sacrificial fire, which came out of the mouth of Videgha Māthava and then rolled eastward to the boundary river of Videha in eastern India, purifying as it rolled along. Thus, the image and practice of sacrifice also was integrally bound up with the development of Vedic society.

Not surprisingly, what is offered in Vedic sacrifice is what one likes to eat oneself: milk in its various forms of curd, warmed and clarified butter, barley and rice, especially in the form of cakes. In the later ritual literature there are elaborate prescriptions of the preparation of the cakes of the various gods. These edible offerings are also accompanied by scents, seats, and clothing, especially in the domestic rituals. The animal sacrifices (*paśu*) should be contrasted with the vegetable sacrifices (*iṣṭi*). The animal victim was usually the goat, but could also have been another animal, such as a bull, a donkey, or a dog, one that is used particularly in the sacrifice to the Rakṣas. The Aśvamedha, or sacrifice of the horse, was one of the most elaborate of Vedic ceremonies, involving rituals lasting up to a year and the political appropriation of land, depending upon where the horse wandered.

In addition to the vegetable and animal offerings, the Vedic offering par excellence is the *soma,* the intoxicating drink. It promotes an ecstatic state of mind, and is believed to confer immortality. In all of the later ritual texts, and certainly throughout the entire ninth book of the *Ṛgveda,* the purchase of the plant and the various sacrificial operations connected with it are described in detail: the purchase of it, the transportation in carts, the procession of carrying it, the drawing of the water, the pressing of the stems that are swelled by the water, and the straining of the juice into the decanter.

The sponsor of the sacrifice and the officiant priest usually partake of the remains of the dishes offered. While there is mention of the remains in the *Ṛgveda,* later in the ritual texts of the Brāhmaṇas the question of the why's and wherefore's of eating the remains becomes one of great speculative importance. In the later ritual, the goddess Iḍā is called to descend to earth and enter that part of the offering, also called *iḍā,* in order to convey to the sacrificer the quality of the victim: this has been called by Hermann Oldenberg an absorption of strength or healing, by Henri Hubert and Marcel Mauss a transubstantiation. It has also been analyzed as a communion or as a ceremony of alliance.

Perhaps the most essential part of the ritual is the sacrificial fire: the fire acts as an intermediary between the sacrificer and the gods, the carrier of food between the worlds. Fire was also from its very inception an agent of purification, which protected against evil spirits. A frequent practice is circumambulation about an object, an animal victim, a consecrated instrument, and especially the fire. The movement takes place in the direction the sun circles, with the right side presented to the honored object, thus the name *pradakṣiṇa*.

In sum, as many Vedic hymns and later ritual texts discussed above indicate, sacrifice of an animal into the fire was part of the ecological balance in the ancient Vedic world; the killing and distribution of the animal was part of a larger understanding of human harmony with natural forces. Jan Heesterman has constructed a kind of cyclical "potlatch" scenario, wherein animal sacrifice is countered by animal sacrifice from a competing lineage. In the human realm, the stakes involved death to the loser, and, on a higher scale, nothing less than the disruption of cosmological forces. While some scholars might disagree with his more violent interpretation of the ancient Indian world, all do agree that cosmological functions and sacrificial functions are radically interdependent upon each other. The gods are given food and return it through their natural bounty; thus, the ecology of sacrificial food production and consumption is the central, guiding metaphor for the survival of earthly and celestial worlds.

What does this brief description of animal sacrifice in ancient India teach us? To be sure, it is an ecological commonplace to claim that, as inherent processes in nature, decay and violence are necessary for nourishment, and that creativity requires a movement between life and death. However, in our ecological readings of the ancient world we very rarely incorporate these insights and cautions from ancient sacrificial processes into our ecological sensibilities of the twentieth and twenty-first centuries.

Given that this is the case, our responsibility is to a holistic reading that realistically takes in both harmonious and disharmonious, creative and destructive accounts of ecology.[8] If we read ancient sources in this way, we take lessons from the *Ṛgveda* other than those of constant natural harmony—but these are nonetheless important lessons for us to learn. While as ecological activists we may not want to condone the practice of animal sacrifice today, the Vedic scenario can teach us far more than beatific contemplation of nature. It can teach us what it means to redistribute the natural world in the process of destruction; what it means to hasten the processes of life and death, growth and decay, on behalf of a larger religious perspective; and how ideas of harmony with nature and sacrifice can frequently go hand in hand.

The Sources

Let us turn to the texts to see how ecology and sacrifice play themselves out in particular exegetical situations. The most appealing of these texts are the cosmologies found in the *Ṛgveda*. They are poetic and can read (as my students often tell me) like an enigmatic version of Walt Whitman's *Leaves of Grass,*[9] filled with natural imagery of waters being released (*RV* 1.32); cows careening down well-trodden paths (*RV* 3.31); the sun rising like a wife coming to meet her husband (*RV* 1.92) or racing across the sky like a chariot driver (*RV* 1.115); healing plants, which emerge gently and magically from the earth's soil (*RV* 10.97); elements emerging from the human body back into the wind, sun, sky, and water (*RV* 10.16); spirits of the forest who show gentleness and a lack of fear (*RV* 10.146). This ecological appeal brings us back to the heart of the issue: how and why do we use ancient texts in the service of a harmonious idea of nature?

In Indological writing since the nineteenth century, this appeal rests on a basic premise: that interaction with, and invocation of, the world of nature in cosmological terms goes hand in hand with mystical vision and/or beatific contemplation. Max Müller's understanding of sun worship, the initial human experience of the awesomeness of nature, and thereby of God, was based on the hymn to Sūrya, or sun hymn, *RV* 1.115.[10] Willard Johnson, and Abel Bergaigne before him, saw the goddess of speech as the essence of *ṛta,* the orderer of the cosmos, and thereby a kind of goddess in the natural idiom (*RV* 10.125, 10.71, etc.).[11] The basic understanding of balance and harmony is traced by A. B. Keith to the threefold cosmological stride of Viṣṇu, who not only orders the world, but embodies all creatures within himself (*RV* 7.99, 7.100, 1.154).[12]

Let us be even more specific: the Ṛgvedic hymns were seen as concepts and poetic images of contemplation, which we ourselves can also contemplate. They are extractable from the context of contemplation. Indeed, as Karl Figlio points out in his influential article "Knowing, Loving, and Hating Nature,"[13] when nature becomes an extractable "idea," the world of nature is born. We can contemplate it just as the ancients did and we can use it just as the ancients did, because it exists apart from us and, as such, is manipulatable.

This notion of extracting and contemplating is the convergence of Romanticism and science, and, I would argue, the convergence between Indological and some ecological writing about the Veda. In other words, we *as scholars* do intellectually with texts what we *as environmentalists* advocate *against* doing to the world of nature itself: we plunder and use. I want to argue that it is far more instructive to behave in the same environmental way with Vedic texts as we would like to do with the earth itself. If we refrain (as much as is humanly possible) from extraction, we learn a great deal more about ancient modes and forms of interaction with the world of nature than we do if we plunder the image.

In what follows I conduct a detailed reading of the three images most frequently used in the service of the idea of nature: the sun, the idea of eloquence as nature goddess, and the balance of Viṣṇu's stride. I want to place them in their first ecological context: the world of the sacrifice,[14] and even more specifically, the world of their *viniyoga,* or application. To be even more ecological in our reading principles: instead of theorizing, in aggregate, about general Vedic ideas of the sun, let us look at particular nature hymns in situ, or, to put it more ecologically, in their "host environment."

Ṛgveda 1.115

Our discussion might well begin with *RV* 1.115, Kutsa's hymn to Sūrya. It invokes awe and wonder and glorious pastoral imagery, and we can see how tempting it would have been to extrapolate from the hymn, as Müller did, into experiential contemplation of the natural world, and thus, of the universe.

> 1. The glorious group of rays has arisen; the eye of Mitra, Varuṇa, and Agni; the sun is the self of the moving and fixed; and has filled the heaven, earth, and the atmosphere.
>
> 2. The sun follows the godly and shining dawn, as a man [follows] a woman, [dawn is] where god-like men span the ages, that which is auspicious for the sake of good.
>
> 3. The auspicious, swift horses of the sun, with beautiful limbs, crossing the roads, well deserving of praise, honored by us, have ascended to the dome of the sky, and quickly circle around earth and heaven.
>
> 4. Such is the divinity and majesty of the sun that, when he has set, he withdraws that which has been spread in the middle of things. When he has unyoked his racehorses from his chariot, then night spreads the veiling darkness over all.

> 5. The sun, in the sight of Mitra and Varuṇa, displays his form in the middle of the heavens, and his rays (horses) extend his eternal shining power, and bring on the blackness of night.
>
> 6. O gods, this day deliver us from severe wrongdoing, and may Mitra, Varuṇa Āditi, ocean, earth and heaven, be favorable to this [mantra].[15]

One can see why Müller and others favored this hymn: dawn is personified; the sun has his racehorses; the sun is also the "self" of all things and therefore the one who has the power to give and withdraw light. In verse two, there is a lovely wordplay on *yugāni,* or "ages," which could also mean "yokes" for ploughs; thus, the sacrificers who invoke the sun and "span the ages" are also agriculturalists who work the plough. The analogy is continued in verse four, where the sun can also cause darkness to be spread over "the middle of things," that is, the cultivator with the plough must stop in the middle of cultivating his field. The hymn is, in one sense, profoundly ecological, presenting levels of human and cosmic activity in delicate and poetic interrelationship.

What of the interpretive frame that Indologists and ecologists have given it? Max Müller quotes this hymn in several of his works on the origin of religious ideas.[16] The sun is the beginning of humankind's understanding that there is a power beyond itself, and humans stand trembling before that power in awestruck wonder, stammering out the first few syllables of proto-Indo-European. In a later vein, Willard Johnson also views this hymn as a kind of description of a contemplative act, filled with aesthetic power (*samvega*).[17] In his book published for the Centre of Advanced Study in Sanskrit, *Ancient Indian Theories of the Earth,* S. R. N. Murthy also describes the sun as one of the many divinities who were "natural powers personified"; the Vedic Aryans worshiped the sun as a kind of geothermal field, and their philosophy was dynamic. According to Murthy, they did not make the mistake of modern science, in seeing these forms and fields as without life.[18] Murthy is right to emphasize the dynamic nature of the sun, and its clear relationship to fire and water, in Vedic texts. Yet, the claim that Vedic natural forces can be interpreted as kinds of geothermal fields is to extract and contemplate in a way similar to those who used the Ṛgvedic sun as the "origin" of religion.

Our first question, then, as historians, is to inquire as to how dynamic these forces actually were in their own environments—most particularly, the ritual environment, where one would expect just such texts to be used. At first glance, we can see hymn 1.115 used in two well-known Ṛgvedic ritual texts, the *Āśvalāyaṇa Śrauta Sūtra* and the *Śāṅkhāyana Śrauta Sūtra*. Both texts claim that *RV* 1.115, the hymn to Sūrya, is used in the Āśvinaśāstra—the offering to the healing deities the Aśvins, at the first *soma*-pressing, just before sunrise. (The *soma*-pressing is the physical crushing of a sacred plant into a sacred drink—a process which is a basic building block of Vedic sacrificial procedure.) Once again we can see the balance between human and cosmic activity reflected here, and thence the appeal of this hymn to ecologists. As is basic to the Agniṣṭoma rite, the *soma* drink is offered in a kind of bridge between the worlds of human and natural activity.

However, there is another, lesser-known, ritual usage, one which suggests a much different scenario: *Āśvalāyaṇa Śrauta Sūtra* 3.8 also states that verses one through three are recited as the "inviting" verses for the god Sūrya, when an animal is about to be sacrificed to him. Verse one, describing the rising of the sun, invites one to the *vāpa,* or roasting of the omentum, or abdominal membrane, of a cow on a spit of *udumbura* wood and the spreading of clarified butter over and under the omentum. Verse two, describing the sun following the maiden dawn, invites one to the *puroḍāśa,* or grain offering of barley cakes, which typically follows the offering of the limbs or the omentum of the animal. Verse three, describing the limbs of the steed, invites one to the *havis,* or main offering, usually, in the case of a *paśu* sacrifice, the limbs of the animal, which follows the offering of the omentum. Verses five and six of *RV* 1.115, describing the onset of darkness, are sung not as prefatory verses, as the earlier ones are, but when the *vāpa* and *puroḍāśa* offerings are made. They are sung as the animal is actually sacrificed.

What is revealed when the hymn is put into this context? First, we learn that, in the Vedic time and context, the praise of nature is intimately associated with killing animals for food, and that eating animals, both symbolically and literally, is a function of ecology. Second, we see that each poetic image of the movement of the sun is linked, metonymically, with the movement and actions of the sacrificer engaged in the immolation of a sacrificial victim. Verses one through three illustrate this fact. And, at verse four, at the actual offering of the victim, there is a kind of break in thematic imagery—in this case, the introduction of darkness, which interrupts a task left unfinished.

Ṛgveda 7.99

Let us move to another case study. *Ṛgveda* 7.99, a hymn to Viṣṇu, is a companion hymn to *RV* 7.100, which depicts the three wide strides of Viṣṇu mentioned above. Both together are frequently invoked as a kind of ecological hymn representing preserving and sustaining action and the balance of the earth in the three strides of the god.

> 1. Widening with a body beyond measure, Viṣṇu, men don't understand your greatness; we know your two worlds from the earth, but you, divine Viṣṇu, know the highest.
>
> 2. No being that exists, or that has been born, divine Viṣṇu, has reached the utmost limit of your magnitude, which you have held up in wide and beautiful heaven; you have sustained the eastern horizon of the earth.
>
> 3. Heaven and earth, abounding with cattle, yielding rich fodder, you are ready to be liberal to man. You, Viṣṇu, have helped up these two, heaven and earth, and secured the earth around with mountains.
>
> 4. You two, Indra and Viṣṇu, have made the expansive world for the sake of sacrifice, giving birth to the sun, the dawn, and the fire; you leaders have confounded the machinations of the *dāsa* Vṛṣaśipra in the battles of armies.
>
> 5. Indra and Viṣṇu, you have destroyed the ninety-nine strong cities of Śambara, you have killed in one moment the hundred thousand heroines of the demon Varchin.
>
> 6. This ample praise magnifies you both, you are mighty, striding wide, and imbued with power; to you two, Viṣṇu and Indra, I offer praise at sacrifices, grant us food in battles.
>
> 7. I offer Viṣṇu the oblation placed before you with the exclamation *vaṣaṭ,* be pleased, Śipiviṣṭa, with my offering. May my hymns of praise magnify you; value us always with blessings.[19]

As before, we can see the ecological import of this hymn in its first three cosmological verses. Viṣṇu holds up the highest heaven, the meeting point in the eastern horizon, and gives birth to the sun, the dawn, and fire. He secures the entire earth with mountains as a kind of ecological carpenter. Alfred Hillebrandt agrees with the earliest of Indologists in dubbing Viṣṇu a solar god who is combined with a vegetation demon.[20] Jaan Puhvel speaks of Viṣṇu's deeds as part of the concept of "the meadow of the otherworld" in Indo-European tradition.[21] Moreover, the hymn has its less controversial ritual uses; according to the *Āśvalāyaṇa Śrauta Sūtra,* in one part of the Agniṣṭoma sacrifice, which is focused on recitation, the Ukthya ceremony, *RV* 7.99 is used as a special invitation (*acchavaka śastra*).

However, once again there is a sacrificial side to the *viniyoga,* or application, of this hymn. The same verses, one through three, are used in exactly the same pattern as the previous hymn, dedicated to Sūrya. They are recited when a victim is immolated for Viṣṇu (itself a rather startling idea given our own understanding and reception of Vaiṣṇavism in classical Hinduism). Verse one is the invitation to the *vāpa,* or cooking of the omentum, when the animal is killed, and describes the wide expanse of Viṣṇu. Verse two, describing the vastness of his magnitude, accompanies the barley-cake offering. And verse three, describing the giving of cattle, accompanies the offering of the limbs of the cattle.

Even more interestingly, *RV* 7.99 reveals another pattern similar to *RV* 1.115, the hymn to Sūrya. There is a "break" in verse four and onwards: the acts of battle and war, like the introduction of the interrupting darkness in the previous hymn, *RV* 1.115, occur at verse four. Verses four and five are recited at the actual offerings, following the pattern of the hymn previously discussed. Thus, the cosmological dimensions (vv. 1–3) involve the invitation to the animal sacrifice, and the mention of slaying and battle (vv. 4–5) accompanies the actual offering of the animal.

Ṛgveda 10.125

Let us turn to one final well-known Vedic hymn, cited many times by ecological and Indological sources alike as a kind of beautiful description of "nature" in the Vedas. *Ṛgveda* 10.125 is a hymn on the creation of the world by the goddess of speech. It, too, is an excellent lens onto the uses of cosmological descriptions in ritual. The hymn consists of eight verses; while speech herself is not named, she is named in the commentarial literature as the "seer" of the hymn—literally, the sage who sees the hymn in a vision and then utters it out loud.

> 1. I move with the Rudras, the Vasus, with the Ādityas and the All-Gods. I carry both Mitra and Varuṇa, both Indra and Agni, and both of the Aśvins.
>
> 2. I carry the swelling soma, and Tvaṣṭṛ and Pūṣan. I give wealth on the diligent sacrificer who presses the soma and offers the oblation.

3. I am the queen, the gatherer of riches, the artful one who is first among those worthy of the sacrifice. The gods divided me up into many parts; I abide in many places and enter into many forms.

4. The one who eats food, who really sees, who breathes, who hears what is said, does so through me. Though they do not realize it, they dwell in me.

5. Listen, you whom they have heard: I tell you something which should be heeded. I declare, by myself, that which gives joy to gods and men. The one whom I love I make formidable, I make him a sage, a wise man, a brahmin.

6. I stretch the bow for Rudra so that his arrow will strike down the enemy of prayer. I incite the contest among the people. I have pervaded sky and earth.

7. I gave birth to the father on the head of this [world]. My womb is in the waters, within the ocean. From there I spread through all creatures and touch the sky with the crown of my head.

8. I am the one who breathes forth like the wind, embracing all creatures. Beyond the sky, beyond this earth, so vast am I in my greatness.[22]

Many Vedic texts cite *RV* 10.125 as one to be recited at the beginnings of things: the *Kauṣītaki Brāhmaṇa* (57.31, 139.15) requires it to be recited at the beginning of the Upanayana ceremony, the investiture of the sacred thread, and the Utsarjana ceremony, the beginning of Vedic study. The hymn is about the relationship between eloquence and the beginning of the world; thus any ritual of "entrance" into the rites of the twice-born, particularly Vedic study, would make this hymn an apt one to use.

And, *RV* 10.125 has attracted some of the more Romantic interpretations of Indologists and some environmentalists. Indologists, while textually thorough, have nonetheless tended to view the goddess of speech in an "inspiring muse" kind of role, either abstract or concrete; however, they hardly ever acknowledge her as the recipient of an animal sacrifice.[23] S. R. N. Murthy sees the goddess of speech as one example of the "geospheric" theories found in the Ṛgvedic understandings of the earth, akin to Sarasvatī, the river goddess and nurturer of the soil.[24] Many other ecological writers see her as the Vedic equivalent of the "great mother" in early Indian thought.[25]

Yet, if we look at the ritual application of this hymn, we see a pattern exactly like that of the two earlier hymns. As *Śāṅkhāyana Śrauta Sūtra* 6.11.11 prescribes, the cosmological description of the first three verses are meant to be an invitation to offer the slaughtered animal to Vāc (again, a rather odd-sounding phrase, given our own predispositions to see the goddess of eloquence as rather more mellow). Verse one, describing the activities of all the gods, is used as the invitation to offer the omentum of the cow; verse two, describing the particular wealth-giving action of the goddess, accompanies the invitation to offer the barley cake; and verse three, describing the division of the goddess into various parts, calls for the offering of the limbs of the animal. Once again we see a metonymic relationship between the verses of the hymn and the action of the sacrificer. And here, just as in the two earlier hymns, the later verses describe the tumultuous earthly action after the cosmological beginnings; as verse six says, the goddess stretches the bow for Rudra so that his arrow will strike down the hater of prayer. She also incites the contest, which could mean either war or the *brahmodya,* or verbal contest of the sacrificers. These later verses may also be used for the actual offering of the animal, as in the earlier cases.

Conclusions

How, then, can we best sum up this brief reading of some of the favorite "nature hymns" of the *Ṛgveda* and their uses for present-day environmental theory? First and foremost, if we read the texts carefully, with a view to their own "environment," it is clear that an understanding of sacrifice followed close on the heels of an appreciation of the natural world and its ecology. We can see this in two different ways. First, in all of the hymns, *RV* 1.115, 7.99, and 10.125, the celebration of the balance and beauty of the creation of the universe in the first three verses is immediately followed by a mention of interruption or breakage, whether it be the onset of darkness, or, more dramatically, of slaying, or of harm done to those who are outside the system.

Second, we can see the integration of sacrifice in the fact that this very intriguing pattern in thematic structure of cosmological hymns reflects the ritual usage found in the Ṛgvedic ritual texts of animal sacrifice. The first three verses of cosmogony accompany the invitation to offer the slaughtered animal, and the following verses of dramatic, interruptive deeds on earth accompany the actual offering. (This correspondence between cosmogony and the detailed specifics of the ritual action of animal sacrifice has not, as far as I know, been much remarked upon by scholars and therefore may be of some historical use outside of the concerns of this volume.)

Thus, what we learn from the Vedas about the environment is not the contemplation of the beauty of nature or the experience of the sun, as Bergaigne and Müller would have it. Nor is it that the Vedas are hymns which describe the action of various geological agencies in their pristine form identifying themselves with nature and with harmony, as Murthy might have it.[26] While these themes are present, the celebration of beauty accompanies the offering of part of that beauty (cows, representing light, as the primary animal sacrifice) up for slaughter. The balance of the universe, of such natural elegances as the sun, the weaving together of heaven and earth, and the power of speech to sustain the earth, all depend on the integration of cosmology and killing.

For the purposes of larger discussion, I want to infer three principles of "reading for the environment" from this small case study:

1) When we go to ancient texts such as the Vedas, we must be textually environmentalist in our approach: we must take historical milieus, and, therefore, ritual interpretations of texts involving nature, as seriously as we do our own environmental concerns.

2) Second, and following on this point, we can be less romantic (extracting and contemplating) in our hermeneutics of retrieval and emphasize instead our agreements and disagreements with the ancient world. We may in fact disagree with what we see in the ritual interpretation of inspiring Vedic cosmologies, but we cannot *refuse* to see it.

3) Third, we can use ancient texts not to "help" or "hinder" our discussion of the environment, but to relativize our own concerns about it. I want to be very clear here: I am not advocating that the excellent ecological theories built by Hindu philosophers and theologians be undone by a few historical flourishes. Rather, I want to argue that, as Hindu philosophers concerned with the environment, we can ask ourselves more sophisticated questions, ones that are informed by careful textual reading and by history. (Frank Korom, in his multilayered interpretation of the sacred cow in India, argues for the necessity of a textual as well as an ethnographic perspective in the development of an ecological ethic.)[27]

We can ask ourselves, why do we need this authoritative text, and how do its constructions of the natural world teach us something about our own constructions? In the Vedic case, since the ancient Aryans did not think it possible, in fact, to have cosmological balance without the inevitable accompaniment of sacrifice, are our own assumptions similar? Those involved in the present-day issues of the Narmada Dam and deforestation in the lower Himalayas might ask the same questions. If, in fact, we are scholars committed to conservation of natural resources, how might we be more attentive to that factor about which the Veda teaches us so well, that sacrifice has its own powerful aesthetic and, in many systems of thought, is not different from nature's beauty?

In addition, we can also ask ourselves about how we view the present practice of animal sacrifice in many parts of contemporary India. Here, too, as discussed in the Vedic case, the practice has ecological meanings which tend to be overshadowed in controversies about violence and nonviolence to animals. James Preston, B. B. Lal, and Alf Hiltebeitel, among many others, have recently discussed animal sacrifice as a kind of balancing of the cosmos.[28] In discussing the controversy surrounding sacrifice at the Chandi Temple to the goddess in Cuttack, South India, Preston writes:

> A ritual sacrifice is never meant as a violent act against a particular animal. The ceremonies that surround the event are intended instead to establish relationship between man and god. The animal's life is offered to the deity as an act of love, thanksgiving, and conciliation. It would be a mistake to equate everyday slaughter for food, with this higher dimension of the ritual sacrifice. Thus, the meaning of sacrifice may be missed by those who find it repulsive or brutal. It is neither "primitive" nor "barbaric," but rather an expression of special significance for devotees who strive to establish a sense of harmony with the cosmic order.[29]

As many local citizens expressed to Preston during his fieldwork, animal sacrifice is seen as human beings willing to play their part in the divine drama, an admission of the tumultuous forces of the universe as part of reality.

These principles of textual reading also imply that we develop an ecology of memory: it is only in the memory of what we have lost—the sacrificial price we have paid—that we can truly understand the nature of our greed, and understand the difficult Gandhian distinction between need and greed. If we do not remember what it was like for us (or our ancestors) to have killed, then we will not know the nature of our ethical choice of restraining from killing. For these reasons, we might keep the sacrificial parts of the ancient Hindu tradition before us. The maturity of our ecological thinking depends on our ability to be fully honest about the traditions we embrace.

One particularly salient Vedic procedure gives us imagery more instructive than all the pastoral images of harmony put together. This procedure usually follows those in which the cosmological hymns discussed above would have been sung. As the *Āśvalāyaṇa Śrauta Sūtra* states in describing the model animal sacrifice, the sacrifice should end by remembering the particular animal victims. After the stick holding the heart of the animal is driven into the ground, the priests then go out through the *tīrtha* passage, or crossing, *without passing through the fire or the animal parts*. They then return into the sacrificial arena, with *explicit instructions to avert their eyes* while passing the heart spit. They build three fires, while still avoiding all physical and visual contact with the central symbol of killing, and, after offering prayers (*RV* 1.23.23), they conclude the animal sacrifice.

The text does not say why the priests are explicitly instructed to look away from the center of the animal immolation. But it does ask us, as ecological readers of ancient India, to contemplate the degree to which a system inexplicably requires looking away from the heart of destruction. The passage also asks us to consider the degree to which a system places psychological and physical distance between the actors and the acted upon in any given ecological system. Its starkness inquires of us: how is a man, enjoying the fruits of his divinely sanctioned sacrifice on a spit of *udumbura* wood, unable to look, even as he enjoys the fruits and the mantras about harmony and balance still ring in his ears? In what ways are we like him, moving in and out of our own postindustrial sacrificial arenas, averting our eyes because it is deemed right, respectful, and customary to do so, while the smoke from the newly burnt food is rising only feet away?

Notes

I would like to acknowledge the help of Christopher Chapple, Mary Evelyn Tucker, Julius Lipner, Jack Hawley, David Haberman, Philip Lutgendorf, Frédérique Apffel-Marglin, Kusumita Pederson, Diana Eck, Tara Doyle, Maggie Kulyk, Joyce Flueckiger, Frank Korom, and Shalom Goldman in the preparation of this article. It was first delivered at the conference on Hinduism and ecology at the Center for the Study of World Religions at Harvard University in October 1997.

1. See especially Norman Crowe's chapter, "Unity and the Idea of Harmony," in *Nature and the Idea of a Man-Made World: An Investigation into the Evolutionary Roots of Form and Order in the Built Environment* (Cambridge, Mass.: MIT Press, 1995), 108ff.

2. From the promotional material for *Prakṛti: The Integral Vision,* ed. Kapila Vatsyayan (New Delhi: Indira Gandhi National Centre for the Arts; D. K. Printworld, 1995).

3. See, among many others, Jonathan Bate, *Romantic Ecology: Wordsworth and the Environmental Tradition* (London and New York: Routledge, 1991); Samuel R. Levin, *Metaphoric Worlds: Conceptions of a Romantic Nature* (New Haven: Yale University Press, 1988); Bernard Rosenthal, *City of Nature: Journeys to Nature in the Age of American Romanticism* (Newark: University of Delaware Press, 1980); Paul Henri Thiry, Baron d'Holbach, *The System of Nature* (New York: Garland Publishers, 1984); H. R. Rookmaaker, *Towards a Romantic Conception of Nature: Coleridge's Poetry up to 1803: A Study in the History of Ideas* (Amsterdam and Philadelphia: J. Benjamins, 1984); Desmond King-Hele's *Erasmus, Darwin, and the Romantic Poets* (Basingstoke, Hampshire: Macmillan, 1986); and Christopher Thacker, *The Wildness Pleases: The Origins of Romanticism* (London: Croom Helm; New York: St. Martin's Press, 1983).

4. See A. Leslie Willson, *A Mythical Image: The Ideal of India in German Romanticism* (Durham, N.C.: Duke University Press, 1964); John Drew, *India and the Romantic Imagination* (Delhi and New York: Oxford University Press, 1987); Ronald Inden, *Imagining India* (Oxford: Basil Blackwell, 1990); Bernard S. Cohn, *An Anthropologist among the Historians and Other Essays* (Delhi and New York: Oxford University Press, 1991); Partha Chatterjee, *The Nation and Its Fragments: Colonial and Postcolonial Histories* (Princeton, N.J.: Princeton University Press, 1993); *Orientalism and the Postcolonial Predicament: Perspectives on South Asia,* ed. Carol A. Breckenridge and Peter Van Der Veer (Philadelphia: University of Pennsylvania Press, 1993); *After Colonialism: Imperial Histories and Postcolonial Displacements,* ed. Gyan Prakash (Princeton, N.J.: Princeton University Press, 1995); and Henry Schwarz, *Writing Cultural History in Colonial and Postcolonial India* (Philadelphia: University of Pennsylvania Press, 1997).

5. While the literature on India and the environment is voluminous, the texts which address the use of ancient Indian texts are worth citing here: O. P. Dwivedi and B. N. Tiwari, *Environmental Crisis and Hindu Religion* (New Delhi: Gitanjali Publishing House, 1987); *The Hindu Survey of the Environment* (Madras: Kasturi and Sons, 1992–); Ranchor Prime, *Hinduism and Ecology: Seeds of Truth* (London: Cassell, 1992); Marta Vannucci, *Ecological Readings in the Veda: Matter, Energy, Life* (New

Delhi: D. K. Printworld, 1994); Banwari, *Pancavati: India's Approach to Environment,* trans. Asha Vohra (Delhi: Shri Vinayaka Publications, 1992); *Nonviolence to Animals, Earth, and Self in Asian Traditions,* ed. Christopher Chapple (Albany: State University of New York Press, 1993); *An Ecology of the Spirit: Religious Reflection and Environmental Consciousness,* ed. Michael Barnes (Lanham, Md.: University Press of America, 1994); and Christopher Chapple, "Toward an Indigenous Indian Environmentalism," in *Purifying the Earthly Body of God: Religion and Ecology in Hindu India,* ed. Lance E. Nelson (Albany: State University of New York Press, 1998).

6. Frédérique Apffel-Marglin, oral communication, October 1997.

7. I am grateful to conversations with Mary Evelyn Tucker in refining my theoretical frame for this material.

8. An excellent overview of the nonviolent possibilities of Asian literature is given in Chapple, *Nonviolence to Animals, Earth, and Self in Asian Traditions;* in Hindu literature more specifically, see Chapple's "Toward an Indigenous Indian Environmentalism."

9. For an intriguing reading of Whitman's influence on contemporary culture, see John Berryman, "On Whitman" in Alfred Kazin, *The Open Form: Essays for Our Time,* 2d. ed. (New York: Harcourt, Brace, and World, 1965)

10. See Max Müller, *Lectures on the Origin and Growth of Religion* (New York: AMS Press, 1976), 268ff; and *Lectures on the Science of Religion,* 73ff.

11. Willard Johnson, *Poetry and Speculation of the Rg Veda* (Berkeley and Los Angeles: University of California Press, 1980), 93–94. Abel Bergaigne, *Religion Védique, d'après les hymnes du Rig-Veda.* Abel Bergaigne's Vedic Religion, trans. V. G. Paranjpe (Delhi: Motilal Banarsidass, 1978), 314.

12. Arthur Berriedale Keith, *The Religion and Philosophy of the Veda and Upanishads,* 2 vols. (Cambridge, Mass.: Harvard University Press, 1925).

13. Karl Figlio, "Knowing, Loving, and Hating Nature," in *FutureNatural: Nature, Science, Culture,* ed. George Robertson et al. (New York and London: Routledge, 1996), 72–85.

14. Frédérique Apffel-Marglin, oral communication, October 1997.

15. 1. citráṃ devā́nām úd agād ánīkaṃ cákṣur mitrásya váruṇasya agnéḥ/
ā́prā dyā́vāpṛthivī́ antárikṣaṃ sū́rya ātmā́ jágatas tasthúṣaś ca//
2. sū́ryo devī́m uṣásaṃ rócamānām máryo ná yóṣām abhí eti paśćā́t/
yátrā náro devayánto yugā́ni vitanvaté práti bhadrā́ya bhadrám//
3. bhadrā́ áśvā harítaḥ sū́riyasya citrā́ étagvā anumā́diyāsaḥ/
namasyánto divá ā́ pṛṣṭhám asthuḥ pári dyā́vāpṛthivī́ yanti sadyáḥ//
4. tát sū́ryasya devatváṃ tán mahitvám madhyā́ kártor vitataṃ sáṃ jabhāra/
yadéd áyukta harítaḥ sadhásthād ā́d rā́trī vā́sas tanute simásmai//
5. tán mitrásya váruṇasyābhicákṣe sū́ryo rūpáṃ kṛṇute dyór upásthe/
anantám anyád rúśad asya pā́jaḥ kṛṣṇám anyád dharítaḥ sám bharanti//
6. adyā́ devā úditā sū́riyasya nír áṃhasaḥ pipṛtā́ nír avadyā́t/
tán no mitró váruṇo māmahantām áditiḥ síndhuḥ pṛthivī́ utá dyaúḥ//

16. Müller, *Origin and Growth,* 297.

17. Johnson, *Poetry and Speculation,* 55.

18. S. R. N. Murthy, *Ancient Indian Theories of the Earth* (Pune: Centre of Advanced Study in Sanskrit, University of Poona, 1992), 14.

19. 1. paró mā́trayā tanúvā vṛdhāna ná te mahitvám ánu aśnuvanti/
ubhé te vidma rájasī pṛthivyā́ víṣṇo deva tvám paramásya vitse//
2. ná te viṣṇo jā́yamāno ná jātó déva mahimnáḥ páram ántam āpa/
úd astabhnā nā́kam ṛṣvám bṛhántaṃ dādhártha prā́cīṃ kakúbham pṛthivyā́ḥ//
3. írāvatī dhenumátī hí bhūtáṃ sūyavasínī mánuṣe daśasyā́/
ví astabhnā ródasī viṣṇav eté dādhártha pṛthivī́m abhíto mayū́khaiḥ//
4. urúṃ yajñā́ya cakrathur ulokáṃ janáyantā sū́ryam uṣā́sam agním/
dā́sasya cid vṛṣaśiprásya māyā́ jaghnáthur narā pṛtanā́jiyeṣu//
5. indrāviṣṇū dṛṃhitā́ḥ śámbarasya náva púro navatíṃ ca śnathiṣṭam/
śatáṃ varcínaḥ sahásraṃ ca sākáṃ hathó aprati ásurasya vīrā́n//
6. iyám manīṣā́ bṛhatī́ bṛhántā urukramā́ tavásā vardháyantī/
raré vāṃ stómam vidátheṣu viṣṇo pínvatam íṣo vṛjáneṣu indra//
7. váṣaṭ te viṣṇav āsá ā́ kṛṇomi tán me juṣasva śipiviṣṭa havyám/
várdhantu tvā suṣṭutáyo gíro me yūyám pāta suastíbhiḥ sádā naḥ//

20. Alfred Hillebrandt. *Vedic Mythology,* vol. 2, trans. Sreeramula Rajeswara Sarma (Delhi: Motilal Banarsidass, 1990), 201–5.

21. Jaan Puhvel, "Meadow of the Otherworld in Indo-European Tradition," *Zeitschrift für vergleichende Sprachwissenschaft* 83 (1969): 64–69.

22. 1. ahám rudrébhir vásubhiś carāmi ahám ādityaír utá viśvádevaiḥ/
ahám mitrā́váruṇobhā́ bibharmi ahám indrāgnī́ ahám aśvínobhā́//
2. aháṃ sómam āhanásam bibharmi aháṃ tváṣṭāram utá pūṣáṇam bhágam/
aháṃ dadhāmi dráviṇaṃ havíṣmate suprāvíye yájamānāya sunvaté//
3. aháṃ rā́ṣṭrī saṃgámanī vásūnāṃ cikitúṣī prathamā́ yajñíyānām/
tā́m mā devā́ ví adadhuḥ purutrā́ bhū́riṣṭhātrām bhū́ri āveśáyantīm//
4. máyā só ánnam atti yó vipáśyati yáḥ prā́ṇiti yá īṃ śṛṇóti uktám/
amantávo mā́ṃ tá úpa kṣiyanti śrudhí śruta śraddhiváṃ te vadāmi//
5. ahám evá svayám idáṃ vadāmi júṣṭaṃ devébhir utá mā́nuṣebhiḥ/
yáṃ kāmáye táṃ-tam ugráṃ kṛṇomi tám brahmā́ṇaṃ tám ṛ́ṣiṃ táṃ sumedhā́m//
6. aháṃ rudrā́ya dhánur ā́ tanomi brahmadvíṣe śárave hántavā́ u/
aháṃ jánāya samádaṃ kṛṇomi aháṃ dyā́vāpṛthivī́ ā́ viveśa//
7. aháṃ suve pitáram asya mūrdhán máma yónir apsú antáḥ samudré/
táto ví tiṣṭhe bhúvanā́nu víśvā utā́mū́ṃ dyā́ṃ varṣmáṇópa spṛśāmi//
8. ahám evá vā́ta iva prá vāmi ārábhamāṇā bhúvanāni víśvā/
paró divā́ pará enā́ pṛthivyā́ etā́vatī mahinā́ sám babhūva//

23. In addition to Johnson's discussion of this hymn in *Poetry and Speculation,* see Fritz Staal's allusion to it in his article, "RgVeda 10:71 on the Origin of Language," in *Revelation in Indian Thought: A Festschrift in Honour of Professor T. R. V. Murti,* ed. Harold Coward and Krishna Sivaraman (Emeryville, Calif.: Dharma Publishing), 3–14; and my own treatment of it in passing, "Vac: Myth or Philosophy?" in *Myth and Philosophy,* ed. Frank E. Reynolds and David Tracy (Albany: State University of New York Press, 1990), 183–214. For earlier treatments see W. Norman Brown, "The Creative Role of the Goddess Vāc in the Ṛg Veda," in *Pratidanam: Indian, Iranian, and Indo-European Studies Presented to Franciscus Bernardus Jacobus Kuiper on His Sixtieth Birthday,* ed. J. C. Heesterman, G. H. Schokker, and V. I. Subramoniam (The Hague: Mouton, 1968), 393–97; B. Essers, *Vac: Het Woord als godsgestalte en als godgeleerdheid in de Veda* (Assen: Van Gorcum, 1952); Louis

Renou, "Les Pouvoirs de la Parole dans le Rg Veda," in *Etudes Védiques et Paninéènnes,* 1 (Paris: E. De Boccard, 1955): 1–27; C. A. Scharbau, *Die Idee der Schopfung in der vedischen Literatur* (Stuttgart: W. Kohlhammer, 1932), 123.

24. Murthy, *Ancient Indian Theories of the Earth,* 22.

25. See, in particular, *God, Humanity, and Mother Nature,* ed. Gilbert E. M. Ogutu (Nairobi: Masaki Publishers, 1992), among many other examples; Pupul Jayakar, *The Earth Mother,* rev. ed. (Delhi: Penguin Books, 1989); Carol Bolon, *Forms of the Goddess Lajja Gauri in Indian Art* (University Park: Pennsylvania State University Press, 1992).

26. Murthy, *Ancient Indian Theories,* 9.

27. Frank Korom, unpublished manuscript, 19.

28. James Preston, *The Cult of the Goddess: Social and Religious Change in a Hindu Temple* (New Delhi: Vikas Publishing House, 1980), 69. B. B. Lal, "Hindu Perspectives on the Use of Animals in Science," in *Animal Sacrifices: Religious Perspectives on the Use of Animals in Science,* ed. Tom Regan (Philadelphia: Temple University Press, 1986). Alf Hiltebeitel, *The Cult of Draupadi,* vol. 2 (Chicago: University of Chicago Press, 1991).

29. Preston, *Cult of the Goddess,* 69.

State Responsibility for Environmental Management: Perspectives from Hindu Texts on Polity

MARY McGEE

He is a true king who is able to acquire resources, to increase them, to guard them and to distribute them.[1]

Tirukkuṟaḷ 39.5

What do ancient Sanskrit texts on polity have to do with environmental management in India today? This essay provides an overview of environmental concerns found in classical Indian texts on kingship and state responsibility and points out ways that these texts can contribute not only to an environmental history of India and South Asia but also to the continued cultivation of responsible stewardship of the environment by the state. That the Indian king or state has long provided the leadership and been responsible for maintaining the earth's resources is evident not only in Sanskrit texts which elaborate on the duties of kings, but also in portraits of more recent rulers in princely India. For example, Vansh Pradip Singh, the twentieth-century ruler of Sawar in North India, is portrayed by Ann Grodzins Gold in her essay in this volume as caring deeply (even excessively) about the trees that flourished within his royal jurisdiction; she quotes several of his subjects who remember him as saying, "if you cut the smallest branch of a tree it is just as if you cut my finger."[2] Concerned with understanding Singh's environmentalism, Gold asks: Why did Vansh

Pradip Singh care about his trees? Why did he proclaim his own physical identification with them? And how did his stance impact on the common good?[3] Her analysis leads her to investigate Singh as a model environmentalist, who understood his political responsibility to maintain his kingdom as a moral responsibility that extended well beyond the public welfare of his human subjects to the protection of flora and fauna as well. After his death, she notes, everything in this jurisdiction changed: the politics, the economy, and most visibly, the landscape. Rapidly, the area became deforested and woodlands vanished. Was Singh's care for the trees merely the idiosyncratic doting of a nature-loving ruler with a special concern for the trees that shaded thoroughfares and enhanced the beauty of the Sawar estate? Or was his protection of the natural environment part of his *dharma,* his moral and social responsibility as a king?

Vansh Pradip Singh in several ways provides an interesting case study by which we may examine what are otherwise largely theoretical policies on kingly use and management of the natural environment.[4] He is certainly not the first ruler to extend his kingly *dharma* to the environment; "Asoka, Akbar and Shivaji were said to be ardent and enthusiastic lovers of trees, who took keen interest in planting shade trees and groves on highways."[5] One of the pillar edicts of Aśoka says:

> I have ordered banyan trees to be planted along the roads to give shade to men and animals. I have ordered mango groves to be planted. I have ordered wells to be dug every half-kos [about a half-mile], and I have ordered rest houses built. I have had many watering stations built for the convenience of men. These are trifling comforts. For the people have received various facilities from previous kings as well as from me. But I have done what I have done primarily in order that the people may follow the path of Dharma with faith and devotion.[6]

While such public improvements may be acts of pious liberality (*purta*), as in the case of Aśoka, rather than acts of governmental rule (*śāsana*), there is no question that the king had the right and responsibility to develop the land and manage its resources in his pursuit of the economic and social welfare of the kingdom (or in pursuit of his own personal piety). To what extent a king's commercially driven concerns—such as highway improvements, excavating wells and irrigation ditches, management of forests and mines—were ecologically informed is more difficult to ascertain from ancient texts, though it is clear that the earth was given in trust to a king for his protection and development. Romila Thapar has argued that Aśoka's enhancements of the environment and public facilities were driven not by piety or commercial venture but by his deep commitment to *dharma* (*dhamma*), which he understood as social responsibility; and as a king, he had a responsibility to set an example and lead the way in cultivating attitudes and practices of social responsibility.[7]

The protection and enhancement of the environment was indeed within the scope of dharmic duties of kings, but who assumes that duty and responsibility today in secular, democratic India? Do the principles of stewardship found in ancient texts on polity (*nītiśāstra*), kingly duties (*rājadharma*), and socioreligious law (*dharmaśāstra*) find counterparts and analogues in contemporary treatises of Indian government? This question I will consider in my conclusion, as my main purpose herein is to examine the environmental stewardship of kings within classical Indian texts. This essay sketches out broad features of the theoretical ideals that governed how rulers regulated and used their natural resources, balancing their material concern for revenue with their responsibility to protect the earth.

Classical Paradigms of Kings as Protectors of the Earth

Vansh Pradip Singh's protectionism toward the trees and wild pigs that populated his kingdom, as described by Ann Gold in this volume, fits quite neatly within the paradigm of an ideal king according to classical Hindu treatises on kingship, right down to the metaphorical identification with trees. In the *Śukranīti* (5.12–13),[8] one of the later authoritative treatises on public administration, the entire workings of the state are summed up in an arboreal analogy:

> The king is the root of the tree of state, the ministry is its trunk;
> the military leaders are branches, the army are the leaves of the tree,
> and the subjects are its flowers; prosperity of the land its fruit,
> and the lands its final seed.

According to the *Śukranīti,* the *mūla,* or root, is the most important part of the plant and the tree withers away when the root decays, just as a kingdom withers if a king loses his power, or *śakti* (cf. *Mahābhārata*

1.1.65–66, *Matsya Purāṇa* 219.34; Varāhamihira 47.1). According to classical Sanskrit texts on *nītiśāstra* and *rājadharma,*[9] the king had a moral responsibility to protect and maintain the people and resources of his kingdom. In other words, the common good of the kingdom—its social, economic, political, and ecological welfare—was rooted in the *dharma* of the king.[10]

In the *Nītivakyāmṛta,* another late *nītiśāstra,* the opening line reads: "Salutations to the State, the tree of *dharma* (moral and social order), *artha* (wealth), and *kāma* (pleasure)."[11] Just as Vansh Pradip Singh was identified as *darbār,* "the court,"[12] the king in classical texts is identified with the state, his kingdom and its lands.[13] Therefore when I refer to the "state" in this essay, it carries with it multiple layers of meaning and references to the king, the kingdom, and the political machine of the kingdom. As the root of the tree (the tree being identified with the state), the king was responsible for the protection and preservation of the lands. Indeed, in the coronation oaths found in the *Yajurveda Saṃhitā* (9.22) and prescribed in the *Śānti Parva* of the *Mahābhārata* (59.106.107), the key commitment made by the king-elect is to nurture and protect (*pālayiṣyāmy aham*) all that belongs to the earth (*bhauma*). Upon his coronation, the king took an oath (*pratijñā*) to uphold and regulate the state. Part of this oath states:

> To you [the state is given] for agriculture,
> for well-being, for prosperity, for development.

In the *Yajurveda Saṃhitā,* from which this oath derives, the twenty-second mantra in chapter nine begins with salutations to Mother Earth (*pṛthvī mātṛ*), and it is implied by the verses that follow that the king is entrusted with the well-being and prosperity of Mother Earth. The land is not a gift to the king but rather a trust which is made sacred by this oath and the accompanying rites. A king who failed to keep this oath or broke it (*asatyapratijñā*) was, in theory, expected to forfeit the title to the throne (though few kings did). Kingship was a trust, and a king was expected to be responsible for the protection, prosperity, and growth of the country. The *Mahābhārata* (*Anuśāsana Parva* 61.32.5) warns that

> A king who does not protect his subjects [aṇd] robs them of their wealth
> is an incarnation of evil (*kālī*) and must be killed by the people (*prajā*).

This verse from *Manusmṛti* (8.307) carries a similar warning:

> If a king who does not protect collects taxes on crops, land taxes, tolls and duties, daily gifts, and fines, he soon goes to hell.

A toned-down version is found in the *Śukranīti* (2.173):

> The kingdom where wealth is amassed by forsaking morality and by oppressing the people is destroyed.

The *Śukranīti* enumerates the eight tasks of a king, among which are the protection of his subjects (*prajāyāh paripalānam*) and the extraction of wealth from the land (*bhūmer ūpārjanam bhūyaḥ*) (1.123–124; cf. *Agni Purāṇa* 239:44–45). The king had a duty to attend to the public welfare, and this included agrarian reforms, the maintenance of forests and pastures, and management of lands in cases of natural disaster. The king was entrusted with the well-being, prosperity, regulation, and growth of land. This relationship with the land led to many epithets for kings, all associated with the earth (*bhū*); among them are *bhūpati* (lord of the earth), *bhūbharaṇa* (maintainer of the earth), *bhūpālana* (protector of the earth), *bhūbhojana* (enjoyer of the earth), *bhūdhana* (one whose property is the earth), *bhūpa* (earth-protector), *bhūpāla* (earth-guardian), *bhūbhartṛ* (husband of the earth), *bhūbhuj* (earth-possessor/enjoyer), and *bhūvallabha* (earth's favorite). In the *Arthaśāstra,* the earliest extant *nītiśāstra* we have, the earth is *artha,* both wealth and the means to wealth, and the king is *arthapati,* the Lord of Wealth.

This royal concern for the maintenance and prosperity of the earth is amply evident in the *Arthaśāstra* of Kauṭilya (fourth century B.C.E.–fourth century C.E.), the most well-known *nītiśāstra,* which deals with public administration and the duties of rulers.[14] Specifically, according to its opening verse, this manual for kings and councilors is concerned with the king's responsibility to acquire and protect the earth (*pṛthvī*). The premise of this treatise is that the livelihood of humans depends on material wealth (*artha*) and that land (*bhūmi*) provides that wealth (15.1.2). In other words, the earth is a source of wealth and enjoyment for humankind, and the king has the responsibility to protect and maintain this source of wealth, as well as to distribute its bounty. The economy of the kingdom and its subjects depended on the prosperity of the land; thus, it behooved a king to guard and ensure that prosperity. The king's relationship to the earth, as the manager of her resources, is captured in this *śloka* from the *Paraśurāmapratāpa,* an unpublished work discovered by P. V. Kane:[15]

> Brahman arranged that the king was (to be) the owner of all wealth and specially (wealth) that is inside the earth. The protector (*bharta*) is he, the overlord of gods.

The care of the earth and its resources was seen as an economic value, a political value, and a moral value. According to the *Arthaśāstra* (3.1.41), a king who carries out his duties responsibly and justly, and who protects his subjects, is rewarded with heaven. Thus, his duty to protect also had a religious value attached to it.

The Environment of Kauṭilya's Country and the Management of Natural Resources

The kingdom and duties of statecraft described in Kauṭilya's *Arthaśāstra* are ideals that reflect certain political ideas and characteristics of the culture in which they were formulated, namely, the Mauryan rule of ancient India.[16] As a treatise on politics which sets out principles for effective government by a king, the *Arthaśāstra* has often been compared to Plato's *Republic;* however, as D. D. Kosambi has observed, while the Greek treatise makes excellent reading and good thinking, the *Arthaśāstra*'s teachings were more practical to implement.[17] According to Madhav Gadgil and Ramachandra Guha, the *Arthaśāstra* provides a good example of environmental management in India during "the 'Age of Empires' (500 BC–AD 300)." During this period, note Gadgil and Guha, there was a shift in environmental management as empires such as the Mauryans, Kushanas, Chalukyas, and Sangam Cholas increased "external trade and warfare," augmented their lands and agricultural production, and expanded their systems of irrigation.[18]

The natural features present in an ideal kingdom as discussed by Kauṭilya—mountains, valleys, plains, deserts, forests, lakes, and rivers—were not in actuality present in every kingdom under the Mauryans, yet a treatise of this type had to cover all contingencies. The *Arthaśāstra* ideal presents a system that reflects ideas of ecological and economic balance current at that time.[19] The late D. D. Kosambi describes the landscape during the formative period of the

Arthaśāstra as one still dominated by Aryan pastoral tribes, which were "gradually falling apart under the internal pressure of private holdings in land. The overwhelming primeval forest was yet to be cleared, the land it covered was naturally without private owners. The Kautalyan state appears so fantastic today because it was the main land-clearing agency, by far the greatest landowner, the principal owner of heavy industry, and even the greatest producer of commodities."[20]

The dependency on the earth and her natural resources was so important to the economic well-being and defense of kingdoms that kings appointed superintendents to oversee the care and management of the kingdom's resources and to promote the study of the environment. Kauṭilya's *Arthaśāstra* lists some thirty superintendents (*adhyakṣa*s) who managed different departments of government. The public works ministries of Kauṭilya oversaw such concerns as the working of mines; digging irrigation ditches; maintaining preserves and grazing grounds; the upkeep of land and water routes essential for trade; the construction of dams and bridges; planting of fruit trees and medicinal plants; and the protection of disabled and infirm animals.

Management of the environment extended to controlling resources, making sure that every opportunity to use a natural resource for economic gain or defense was taken advantage of, and fixing the distribution and prices of products resulting from natural resources (fruit, salt, metals, medicinal herbs). In the Kautilyan state, village accountants (*gopa*) were entrusted with preparing a census not only of human resources but of natural resources as well (*Arthaśāstra* 2.24.3–5; hereafter AS). Accounts were kept of cultivated and uncultivated land, vegetable and fruit gardens, forests and pastures, irrigation works and drinking water distribution centers, roads, temples, and pilgrimage sites. An account was also to be prepared of all two- and four-legged animals in the region. The state conducted a census of all reservoirs and followed a careful plan of constructing new reservoirs and repairing old ones (AS 2.35.2). Private construction of irrigation facilities was encouraged by the state, but if owners neglected the maintenance of these resources, they lost the privileges and revenues associated with them. These censuses were used to assess land revenues, but also aided in the planning and carrying out of comprehensive land policy that would increase the wealth of the kingdom and its inhabitants. Most sections in the *Arthaśāstra* concerning the protection, maintenance, and use of natural resources are grouped around one or more of the following concerns, which I briefly illustrate below: revenue, defense, and pleasure.

Natural Resources as a Source of Revenue

In Kauṭilya's ideal government, the protection, as well as exploitation, of natural resources was assigned to specialists who oversaw the complete processing cycle from raw material to commercial product.[21] I single out three such managers here for brief attention where particular responsibilities required specialized knowledge of the environment.[22] These three are the superintendents of agriculture (*sītādhyakṣa*), of forest produce (*kūpyādhyakṣa*), and of mines (*ākarādhyakṣa*). The concerns of these managers in ancient India find parallels in the portfolios of contemporary government officials, namely, India's ministers of agriculture, environment and forests, and steel and mines.[23] The contemporary ministers, however, have much more to contend with than did their ancient counterparts, in that they are confronted daily by rapidly changing technology and scientific developments, environmental watchdog agencies, international trade regulations, and demands from a much more influential private sector.

According to the *Arthaśāstra,* crown lands (*sītā*),[24] whether cultivated by paid labor or leased out to others, were the prime concern of the superintendent of agriculture, who was expected to be knowledgeable in the science of cultivation, water management, and plant care (AS 2.24.1). He was responsible for overseeing the collection of seeds, roots, and tubers; preparing seeds and land for planting; fertilization appropriate to the plant, soil, and season; and harvesting and threshing. The revenue from the produce of crown lands was a significant part of the king's treasury. If a field lay dormant too long it was reclaimed by the state and given to those who would work the land. Fines were levied for jeopardizing agricultural production, including instances of laborers showing up late for work or without the proper tools for the task of sowing and reaping (2.24.4). Agriculture was serious business and those who did not cultivate arable lands leased to them were charged a fine to reimburse the state for the loss of production (2.1.9). In its discussion of cultivated lands (2.24), the *Arthaśāstra* provides detailed information about farming, crop rotation, and irrigation relevant not only to India's environmental history but also to contemporary environmentalists. For example, the *Arthaśāstra* enumerates various forms of cultivation and organic farming (e.g., AS 2.24.24–25), which may be relevant to those seeking alternatives to chemical fertilizers and pesticides that have polluted our environment in recent times. Also of significance are some of the irrigation techniques detailed in the *Arthaśāstra* specifically designed to suit particular environments within India, such as the *parabandhi* system of rotational distribution, which is still implemented in parts of India today.[25]

Another great source of revenue for the state came from resources found within the forests.[26] A man appointed to the position of superintendent of forest produce, according to Kauṭilya, needed not only good management skills, but also the specialized knowledge of a botanist and an arborist. He supervised the collection of forest products (*kūpya*)[27] and oversaw the production and sale of goods from these raw materials. Familiarity with plant life or botany was required of the superintendent of the forest, for he needed to identify and classify plants and trees according to their different uses. Products from the forests provided trade for revenue, as well as food and medicine. It is clear from the *nītiśastra*s that the sciences of medicine, agriculture, horticulture, and sylviculture were greatly developed in ancient India. Botany was perhaps the science in most demand, and the *Arthaśāstra* and related *śāstra*s single out for attention *vṛkṣāyurveda,* the knowledge of the long life of trees. *Bheṣajavidyā,* the knowledge of medicine, is also discussed in this context, since most medicinal substances came from plants. Kauṭilya also speaks of *gulmavṛkṣāyurveda,* the practice of the applied botanist, who specialized in the art of seed collection and selection, planting, manuring, crop rotation, cultivation in different meteorological conditions, plant health and treatment, plant classification and identification, landscaping, and so on. The *Arthaśāstra* names and classifies numerous plants, trees, vines, creepers, soil types, and reptiles found in the forests, and is especially attentive to the medicinal properties of different plants. These botanical details provide us with important information on the biodiversity and varied terrain of early India as well as about applied biological knowledge of the day.[28]

In addition to agriculture and forest produce, mines provided a significant source of revenue for the Mauryan Empire, and the chief superintendent of mines in a Kautilyan state was expected to have expertise in geology, with a specialty in metal ores; he also had to be familiar with techniques of smelting and the identification of gems (AS 2.12.1). He supervised the work of the superintendent of metals (*lohādhyakṣa*); the superintendent of gem mines (*khanyadhyakṣa*); the master minter (*lakṣanādhyakṣa*); the examiner of coins (*rūpadarśaka*); and the salt commissioner (*lavaṇādhyakṣa*) (AS 2.12). Among the responsibilities of the chief superintendent of mines were the identification of mine locations and discerning the quality of the ore. Those mines that were easily worked and exploited were excavated by his crew, whereas those that proved more challenging were leased out to private individuals. All mining had to be authorized by his office, which controlled and centralized the sale and purchase of metals. The manager of gem mines (*khanyadhyakṣa*), working under the chief superintendent, saw to the harvesting of mother-of-pearl, quartz, mica, coral, pearls, sapphire, beryl, diamonds, and other gems, whereas the manager of metals (*lohādhyakṣa*) oversaw the recovery of copper, tin, and iron as well as the manufacturing of copper alloys such as brass, bell metal, and bronze. A manager of precious metals and jewelry (*suvarṇādhyakṣa*) was primarily responsible for the purification and processing of silver and gold, and supervised the controller of goldsmiths and silversmiths and the operation of the crown workshop. The salt commissioner (*lavaṇādhyakṣa*) coordinated the leasing of salt mines and the collection of the state's salt, and regulated the price and sale of salt as well as taxes on imported salt. The *Arthaśāstra* provides our most significant evidence on mining in early India in its discussion of mineral exploration, various mineral ores and methods for their purification, and processes for softening metals.

While each of these three areas of revenue—cultivated lands, forests, mines—depended on the exploitation of natural resources, the *Arthaśāstra* shows evidence of what ecologist Madhav Gadgil has called "ecological prudence." By ecological prudence Gadgil means "the exercise of restraint in the exploitation of natural resources such that the yields realised from any resource are substantially increased in the long-run even though the restraint implies foregoing some benefit at the present."[29] The king and his superintendents had to find a balance between exploitation and conservation of the natural resources on which the state's well-being depended. Responsible and knowledgeable management of resources required technical expertise and included efforts to sustain rather than exhaust the resource base.

If a resource base was exhausted—for example, the overcultivation of farm land—measures were taken to let the land lay fallow for a time and be used as pasture land, thereby allowing the land to replenish itself (with the help of organic fertilizer provided by grazing cattle).

Natural Resources as Sources of Defense

Kauṭilya, while emphasizing the economic importance of mines for building the State's treasury, also points to the significance of mined metals for the development of weapons used in defense. Much of the *Arthaśāstra* is concerned with defense strategies for protecting the kingdom from enemies. The natural landscape was evaluated for both its strategic advantages and its disadvantages for establishing fortresses and fighting battles. Kauṭilya discusses the use of rivers, mountains, deserts, marshlands, forests, and thickets as natural fortifications against enemies and advises that each quadrant of the kingdom have some natural fortress established and manned near the borders of other territories (AS 2.3; 7.12). Natural boundaries—mountainous regions, tree lines, rivers, and forests—were to be variously used as protective barriers as well as to mark political boundaries.[30] For example, forests populated by large predatory animals could be a deterrent to many an enemy or interloper. In peace time, trees and forests were also used as buffer zones between settlements and to mark boundaries of lands variously set aside for grazing, agriculture, or town development (see AS 2.1.3; 3.9.10–14).

Natural Resources as Sources of Pleasure

Shade trees, flowering plants, and various animals, in particular deer, are prized by the *Arthaśāstra* for the enjoyment they bring. I am reminded once again of the great pleasure Vansh Pradip Singh derived from his trees and wild pigs, as described by Ann Gold's informants (see her essay in this volume). The *Arthaśāstra* also attends to the king's pleasure and describes a pleasure forest especially designed for the king's enjoyment. Such a forest was to be surrounded by a moat and have a single entrance, which would secure the forest for the king's protection but also keep the chosen animals within the enclosure. This forested pleasure park was to be populated by trees bearing sweet fruits, thornless plants, soothing ponds, tamed deer, a family of elephants, and wild animals that had been defanged and declawed (AS 2.2.3). Several of these pleasure parks, some turned into hunting forests for the sport of kings and their colonialist cronies, have in contemporary times become part of India's network of protected nature preserves.[31] Vansh Pradip Singh also fostered a pleasure forest, according to some Rajputs familiar with the royal family who told Ann Gold stories of the *darbār* bringing animals to Sawar from Binai: "So Vansh Pradip Singh made this jungle very good, and he created a Forest Protection Department [which didn't exist previously]."[32] Just like the king of the *Arthaśāstra,* Vansh Pradip Singh also "dug wells and step-wells and planted gardens, and he planted the shade trees by the side of the road"; these trees were there to enhance the beauty, shade, and hospitality of the village, and the *darbār* took great pleasure in this.[33] The *Arthaśāstra* too values the beauty of flowers, advocates the cultivation of gardens, and appreciates the natural solace of the forest as a habitat conducive to meditation and learning. After all, the earth was to be enjoyed.

State Domination and Protection of Natural Resources

The state not only oversaw the maintenance of natural resources but also dominated and exploited them for purposes of pleasure, defense, and revenue. As is quite evident from the texts, custodial concern for the environment was largely driven by economic needs and to some extent by concerns for defense. Laurie Patton, in her essay in this volume, has warned us of the tendency to over-romanticize Hindu attitudes toward the earth, especially as we search ancient texts for philosophical insights that can be used to promote a greater reverence for the earth and a renewed sense of environmental stewardship. In mining such texts for wisdom we cannot ignore the coal that imbeds the diamond or the polluted waters from which springs the fragrant lotus. Even Ann Gold's portrait of tree-loving, environmental enthusiast Vansh Pradip Singh is sullied by his zealousness to protect his trees, which was often at the expense of some other ecological unit. Gold provides an account of the *darbār* demanding a "wholesale slaughter of a species of lizard in order to protect trees," a protective attitude toward wildlife which she says was "neither indiscriminate nor necessarily ecological."[34]

There is no hiding the exploitation of the environment in the *Arthaśāstra,* where laborers slave to clear fields, till land, mine forests, and build dams for the benefit of the royal treasury. The duties of the king as articulated in the *Arthaśāstra* and other *rājadharma* sources involve the domination and exploitation of natural resources. Yet the king was also required to protect these resources. The state and its superintendents of natural resources were not oblivious to the results of this exploitation and often undertook measures to address the devastation of the physical environment. For example, proclamations and injunctions are found in the *Arthaśāstra* concerning such matters as the use of trees to mark boundaries; planting saplings of shade trees along public roads; laying out gardens for public use; special protection of aged and diseased animals; the establishment of parks or enclosures for the protection of animals;[35] the exemption from taxes of lands and fields overrun by enemy troops or oppressed by foresters, disease, and famines; and rotation of pasture land and of crops so as not to exhaust the natural resources.

The administration of fines also provides us with insight into the state's relationship to and responsibility for the environment (and public welfare), while also providing us with possible clues to practices of this period that were considered harmful to the common good as well as to the environment. Fines were to be levied for such things as injuring, torturing, or killing animals; catching deer; cutting, injuring or felling plants and trees, especially those in city parks, sanctuaries, holy places, or cremation grounds, or which provide fruit, flower, or shade; the destruction of waterways; damage to dams and reservoirs; failure to maintain irrigation systems; and failure to cultivate arable land.

Public health was also an important concern and fines were to be imposed for violations of any of the following injunctions, all concerned with pollution of the public domain (AS 2.36.26–33): no one shall throw dirt on the streets or let mud and water collect there; no one shall pass urine or feces in a holy place, a water reservoir, a temple, or a royal property, unless it is unavoidable for reasons such as illness, medication, or fear; no one shall throw out dead bodies of animals or human beings inside the city. As P. Leelakrishnan notes in his study *Law and Environment,* "environmental pollution was controlled rigidly in ancient times." He points in particular to the sanctions imposed on those who brought harm to the environment, adding that knowledge of these sanctions motivated people in their duty to protect the environment.[36] This is in contrast to India's constitutionally mandated environmental protection directives which contain either no sanctions or nonjudicable sanctions, a point to which I shall return.

The *Arthaśāstra* not only provides for the protection of natural resources from overuse by humans, but it also is concerned with protecting humans from the calamities of nature. It counsels the state on how to cope with natural disasters derived from fires, floods, disease, famine, drought, rats, locusts, snakes, and wild animals (see AS 4.3). Among the responsibilities of the state is to keep on hand in the royal store provisions for emergencies resulting from natural disasters, such as famines and floods.

The *Arthaśāstra*'s concern for the protection of plants and animals derives more from their economic value than their religious value.[37] However, Kauṭilya does assign a higher value to sacred trees, doubling fines when trees of special religious value are injured (AS 3.19.30; cf. *Yājñavalkya* 2.230–31). Other texts that deal with the *dharma* of kings, such as the *Śānti Parva* of the *Mahābhārata* and relevant sections of *Manu* and *Yājñavalkya,* give more attention than does Kauṭilya to the religious value of the earth and provide more philosophical insights on Hindu understandings of the sentience of plants and animals. For example, in the *Mahābhārata* (*Śānti Parva* 184.1–7), we find a very graphic description of the life of trees and how they feel pain and pleasure. In this context, the penalty for injury of a tree is a religious one, as someone has caused harm to a sentient being, and an appropriate penance (*prāyaścitta*) is assigned. The *Yājñavalkyasmṛti* (2.276) advises that

> If a man has cut a tree, a thicket, a creeper or a shrub, he should recite one hundred vedic verses. If a man is guilty of recklessly cutting a medicinal plant, he should subsist on milk for one day and follow a cow.

According to Yājñavalkya's guidelines, the late Vansh Pradip Singh should have undergone an expiation consisting of a three-day fast on milk for the death he brought upon the lizards who were falsely blamed for harming his trees.[38] But the *Arthaśāstra* is a more secular-oriented text, and the punishments (*daṇḍa*) it prescribes are due to civil or criminal offenses, not religious ones, whereas *dharmaśāstra* texts, such as the *smṛtis* of Yājñavalkya and Manu, incorporate sections that deal with violations of civil, criminal, *and* religious law.[39] For example, *Yājñavalkya* 2.242 mentions that a minor sin is incurred by the superintendent of mines, as his work involves harm to the earth and the creatures within it, something that is unavoidable given his job description. Here we have a succinct example of how one's occupational duty (*dharma*) can come into conflict with the moral and religious order (also *dharma*) as perceived within Hinduism. A king faces a similar conflict, as he is entrusted with the well-being of the earth as well as his subjects, for the welfare of his subjects often entails detrimental exploitation of the earth's resources. None of the *dharmaśāstra* or *arthaśāstra* texts advise that a miner, a king, or anyone else should abandon his or her occupational responsibilities (*varṇadharma*) on the grounds that some sin may be incurred by the mere fulfillment of the basic job requirements; but the texts do, as in this example from Yājñavalkya, take care to remind people of the implications (or side effects) of their *dharma* within the larger framework of the universe and warn them about the misuse of their skills.

The Politics of the Environment

In the opening verse of its final book (15.1), the *Arthaśāstra* reminds us that politics is concerned with methods for attaining and protecting the earth on which the livelihood of human beings depend. The word earth (*pṛthvī*) used in this verse is especially significant as it refers both to the source of the income as well as to the human society supported by it. The Kauṭilyan king, concerned with the security and prosperity of the state, aimed to protect life and property and to further the material aspect of the lives of his citizens. The state controlled land, labor, industry, capital, trade, and prices. The state was expected to be benevolent and protective of families, both socially and economically. It is clear from many classical Indian texts on the duties of kings that a king had a responsibility to reap the benefits of nature's resources for the sake of his kingdom and its citizens;[40] yet it is also clear that the successful and ideal king had to carefully manage these resources and respect the earth.

While a good part of the *Arthaśāstra* attends to the attainment of land through tactics often described as Machiavellian, the actual management and maintenance of resources, as detailed in Book Three of the *Arthaśāstra,* reveals both a deep knowledge and an appreciation of the earth's resources. While *artha* (material gain) certainly drives the state's excavation of these resources, it is tempered by the expert knowledge of specialists who demonstrate familiarity with the fragility as well as resilience of the ecosystem.

According to the *Arthaśāstra,* the king was entrusted with the protection and maintenance of land. It was his responsibility to maintain the land, to augment it, to develop and use its resources toward the health, welfare, and happiness of his kingdom. A king's reputation and his kingdom's livelihood and economy depended on agriculture to such an extent that farmers and fields were afforded special protection during times of war.[41] Concern for productivity of the land was so important to the welfare of a kingdom that fines were levied against any action that might contribute to a slowdown in productivity, whether it was a laborer who came late to the fields, a farmer who failed to cultivate arable lands leased to him, or damage to dams or irrigation systems.

The *Arthaśāstra* is clearly a text concerned with material gain and with the political and economic value of land and natural resources. Certainly this has a contemporary ring. In reading Kauṭilya's *Arthaśāstra* we are reminded that exploitation of the land is not a new phenomenon, and neither are the efforts of humans to live in harmony with nature, even as they reap the fruits of nature's bounty. The *Arthaśāstra* emphasizes the need for specialized knowledge about the life of plants and animals and for responsible stewardship, for even in ancient India men and women knew how dependent they were on these resources not only for their own survival, but also for their pleasure and material gain. But being aware of our dependency on nature is not enough, as Ramachandra Guha and David Arnold, among others, have pointed out; we must also take responsibility for the havoc human needs (and whims) have introduced into the ecosystem:

> the [ecological] relationship is a reciprocal one, for man more than any other living organism also alters the landscape, fells trees, erodes soils, dams streams, kills off unwelcome plants and predatory animals, installing favoured species in their stead. The awareness of man's dependence upon nature has a long ancestry; but a sense of man as the maker and unmaker of nature has only more recently dawned upon us,

and with it an awesome sense of our own capacity for mischief and mayhem.[42]

Ecology does not just concern the study and preservation of plants, animals, and microscopic organisms; it is very much concerned with human interaction with the physical environment. The *Arthaśāstra* itself has much to offer us as a resource for information on the study, cultivation, and preservation of plants and animals,[43] as well as mineral resources, in ancient India. Yet it also provides us with an ecological portrait of the time, revealing not only how humans managed the physical environment but why and toward what end.[44] This essay provides a brief overview of ecological awareness and the uses of natural resources as evinced in traditional Sanskrit treatises on polity and the duties of kings. We can conclude from a study of the *Arthaśāstra* that the state's dependence on agriculture and natural resources led the state to develop policies and methods aimed at soil conservation, protection of certain species, land management, and forest preservation. Presumably such methods sustained rather than devastated the ecosystem, but for how long and to what extent is a question for environmental scientists. My aim in writing this essay has been to suggest ways in which the *Arthaśāstra,* and other texts like it, can contribute to an environmental history of South Asia.[45] But of what value are such ancient Hindu texts on polity to environmental scientists? What possible interest could classical Hindu texts such as *arthaśāstra*s and *dharmaśāstra*s have for today's ecologists? Can the information in these texts inform our ecological crisis today? How can the indigenous knowledge in these texts be used toward contemporary environmental policies and concerns?

Classical Sources as Aids to Contemporary Environmentalism

The study of ancient Hindu texts has usually been the domain of historians of religion and Indologists, but increasingly such texts are providing significant fragments of information of interest to scientists. For example, David Lee, a biologist, has looked to editions of the *Rāmāyaṇa* for evidence about species of flora in diverse regions of India; and researchers at the Centre for Science and the Environment, a Delhi-based nongovernmental organization, have reviewed "the Arthaśāstra for the latest State of India Environment Report which focuses a lot on management of water systems."[46] While environmental information in these texts may be limited, classical Indic texts are clearly of some value to both environmental historians and scientists, and perhaps even to policy makers.

Contributions to Historical Knowledge about Ecosystems

David Lee, a biologist writing in this volume, has reminded us that ecologists study the dynamism and interaction of organisms in the ecosystem and care about the preservation of biological diversity. They are particularly interested in how an ecosystem responds to destruction and change in its environment. His concerns helped me to better understand the potential value of a text like the *Arthaśāstra* to contemporary environmental biologists, as these texts and others like them provide us with resourceful information that tells us something about a structured relationship between humans and the environment, the biodiversity present in a particular region and period, and the human and technological management of that diversity.

Texts like the *Arthaśāstra* provide us with lists of flora and fauna indigenous to the area at a particular time. Such catalogs may provide critical evidence about the biological diversity that was present in earlier times, but they also provide practical knowledge that can be used to reintroduce diversity into a particular environment. Frank Korom, in comments at the Harvard conference on Hinduism and ecology, raised questions about the trustworthiness of texts in providing accurate knowledge about regional biodiversity, since many texts are copied verbatim from generation to generation and region to region—this is an important caution to be aware of if classical texts are to be used as a reference for reintroducing particular flora and fauna into different regions of India.[47] At the same time, I should point out that the details in the *Arthaśāstra* go beyond mere copious lists and include attention to climate, rainfall, typologies of plants, and soil analysis, and these may be relevant particularly to scientists like David Lee. Historical information about the environment and the management of resources in these texts, and particularly in the *Arthaśāstra,* could also be drawn on to complement elementary and secondary school science education in India. Various passages could be used to acquaint students with India's rich environmental history; create greater awareness of biodiversity; teach about the value of plants and plant care; introduce students to methods of organic farming; inspire a desire to work to preserve diversity; and enhance respect for the environment and our dependency on it.

Paradigms for Government Protection and Management of the Environment

The *arthaśāstra*s and *dharmaśāstra*s are not texts that are rich in metaphor, story, or poetry. However, they are full of prescriptions for proper behavior, outlining duties and responsibilities of the different sectors of society aimed at maintaining social, moral, and ecological order. With its focus on the economic and political order of society, the *Arthaśāstra* offers us a model of the state's responsibility to protect and maintain the earth as well as of the state working with private enterprise to responsibly develop land and excavate resources.[48] But do we have any evidence that these models were put into practice?

In spelling out the responsibilities and workings of the state, the *Arthaśāstra* clearly portrays the custodial duties of the king in relation to the earth and its resources. Is this the same stewardship that Ann Gold speaks of when she writes of Vansh Pradip Singh's "*zimmedārī*"? Gold points out that Singh's care for trees and pigs was part of his moral responsibility, his *zimmedārī*. "The *darbār* had embodied *zimmedārī* [responsible authority]—in his rootedness, his vigilance, his personal care."[49] This sentiment of responsible authority is also present in the ancient Hindu texts on economic polity, kingship, and socioreligious law, in which kings are chastised for letting greed or personal welfare sway them from their moral responsibility to protect the well-being of the people. Singh fits well into the profile of the righteous king, even with his foibles.[50] The *darbār* was concerned with protecting the forest and envisioned the forest as an extension of his body and the trees as members of his court. I support Gold's interpretation, which views this attitude and the practices of Vansh Pradip Singh as an extension of the traditional and paramount duty (*dharma*) of kings to protect. Protection of the kingdom, its people, and its resources was not just a political duty but a moral responsibility, since the common good depended on the king's protection as well as his enforcement of law and *dharma*.

The maintenance of natural resources was also traditionally rooted in the king's *dharma*, as we have seen in our brief review of pertinent sections of the *Arthaśāstra*. But with the loss of power to kings under the British and in postcolonial times, who is to assume that responsibility? Both Ann Gold and Atluri Murali observe in their respective case studies that under colonial rule, with its market-oriented economy, new pressures were brought to bear on India's natural resources. "[T]he entire ecosystem was transformed under the influence of the colonial model of 'private property' in land, water and other natural resources."[51] Vansh Pradip Singh's passion for trees flourished during the twilight of colonial rule, as he ruled the principality of Sawar from 1917–1947. Ann Gold's sense is that the *darbār*'s campaign to protect and cultivate trees was not just a personal indulgence but was rooted in some cultural value or historical precedent associated with kings. If we look to Murali's study of forest practices and local communities in Andhra between 1600 and 1922, we see a similar pattern of environmental oversight by kings in which an effort is made to sustain an ecological balance between forests, agricultural lands, and pastoral regions. Murali's evidence on the state's responsibilities for the environment—drawn from historical documents, travelogues, poetry, oral legends, village chronicles, and *stalapurāṇa*s—follows even more closely than does the rule of Vansh Pradip Singh the principles and practices set forth in the *Arthaśāstra*. The state's involvement under Krishnadeva Raya, a sixteenth-century king of the Vijayanagara Empire, in the development of irrigation systems, grazing lands, agricultural production, and the maintenance of forests provides an exemplary study of the *Arthaśāstra* principles at work. The clearing of forests, construction of village settlements, and the conversion of wasteland to farmland, which are described by Murali as state policies that were implemented in the Deccan in precolonial times, have direct correspondence to state responsibilities articulated in the *Arthaśāstra*.

The Economics of Environmental Policies

There is a continuity of values and responsibilities apparent from the environmental policies prescribed in the *Arthaśāstra* to those observed in the respective kingdoms of Krishnadeva Raya and Vansh Pradip

Singh. Responsible management of resources stands out as a key duty of kings who want to achieve a healthy economy and happy kingdom. Yet, while kings may have a moral responsibility to maintain the environment, they also need economic resources as well as political authority. With the land reforms introduced by the British, observes Gold, the princes lost much of their power and authority. There was "no longer an incentive for the court to protect the jungle" but rather there was "considerable incentive to exploit."[52] While there may still be hereditary maharajas in India today, there no longer are princely states or royal purses, so who is going to assume the economic responsibility that must accompany the moral responsibility toward maintaining forests and biodiversity? Who is to oversee the competing needs of the people, and who has the authority to make judicious decisions about the management of resources? What is to drive those decisions: demands for more electricity to the corporations that bring capital into India or the needs of the farmer dependent on reservoirs and irrigation systems for the success of his crops? And what about sustaining the health of the land in the face of these varied economic and agricultural demands?

The *Arthaśāstra* has often been presented as a text that advocates the ends rather than the means, sacrificing ethics to political and economic gain. While the *Arthaśāstra* certainly is more aggressive and perhaps at times Machiavellian in its politics as compared to the *rājadharma* sections in the *dharmaśāstra*s, especially when it comes to land acquisition, there are no deceptive strategies prescribed for the management of land resources. Agricultural and excavation strategies are aimed at working responsibly with the earth to develop her resources so that she will yield abundantly. Abuse of the land—through overdevelopment, overworking of arable fields, deforestation—can lead to loss of revenue, which jeopardizes the welfare of the state and its citizens. Other than a respect for the powers of nature and a desire to benefit from and share her resources, it is hard to derive systematic environmental ethics or policies from the *Arthaśāstra*. Interaction with the environment, as depicted in the *Arthaśāstra,* is driven by practical concerns, human needs, and human desires. Like the *darbār* of Sawar, the environmental protectionism of Kauṭilya's king served a particular aim. Vansh Pradip Singh's practice of tree conservation was driven by his enjoyment of trees, and any apparent conservation methods advised in the *Arthaśāstra* are prescribed for the purpose of economic gain. In other words, while both Vansh Pradip Singh and the *Arthaśāstra* advocate conservation strategies and demonstrate awareness of the impact of human interaction with the environment, neither of them consciously formulates their ecological actions based on an environmental ethics. What we observe is that their environmental strategies are, to a large extent, a by-product of a king's *dharma*.

If responsible management of resources traditionally depended on a king's fulfillment of his duty, what happens to the stewardship of the environment when kings are stripped of their authority and power? Whose responsibility is it now, asks Ann Gold, since the princely states of India are a thing of the past? Who is to assume this responsibility? Who has the power and the resources to maintain the land? Does this responsibility now fall to the central government in India, to individual states, or to the private sector? Do the ancient theoretical models have any correspondense with more recent government policies and practices?[53] Is there any continuity between the princely states and the modern states when it comes to public welfare and state responsibility? Who assumes the responsibility for the maintenance and protection of lands, fields, forests, dams, and reservoirs now that there are no longer kings for whom such responsibility was their duty? Who is to coordinate the use of natural resources and oversee their replenishment? Does the modern state have a political responsibility to maintain the health and wealth of this diversity? Presumably these concerns fall to the Government of India, which has cabinet-level positions for ministers who hold portfolios in such areas as the environment and forests, steel and mines, and agriculture, among others.

State Responsibility for the Environment: Past and Present

The kings of ancient India took an oath to protect the earth and its subjects, and to develop the earth so that its fruits were abundant enough to satisfy the needs and desires of those who depended on the king for protection and order. This oath bound them personally and morally to the fulfillment of their duty; to break the oath incurred more than just the wrath of the people and amounted to a sin often accompanied by a sentence to a period in hell. The secular government of India in recent times has made promises as well to its citizens; for example, it has promised resettlement packages to tribals and villagers in the regions of the Sardar Sarovar and Bargi dam development projects along the Narmada River.[54] The government's representatives have proclaimed the great bounty that will flow to everyone as a result of these dam projects. As citizens in ancient India were expected to trust their king to protect them, so the villagers of Bijasen and tribals communities in the Narmada valley have trusted government promises. But studies have documented the devastation to the land and to the livelihood of families who have been uprooted by these dam projects. What is to be gained from this great expense? Where is the common good? What is the government's responsibility in these situations? How is an ecological balance to be maintained and guarded, which, while utilizing the natural resources to benefit humankind, also sustains those resources and the homelands of Indian families?

The *Arthaśāstra* reflects a period in Indian history when there was a move toward a universal government for all of society, breaking from the model of competing tribes.[55] The Magadha Kingdom under the Mauryans aimed to be that universal government, and the *Arthaśāstra,* formulated as a manual of polity within that milieu, offers us early formulations of ideas about state responsibility. In ancient Indian texts such as the *Arthaśāstra,* we find relevant information about biological diversity as well as models for sustaining natural resources developed as part of the state's responsibility to guard and increase nature's bounty.[56] Yet, given that the state's immediate goal was to acquire and distribute this bounty to benefit the common good, we should not be surprised to find within these same texts evidence that points to patterns of manipulation and attitudes of exploitation of both human and natural resources. These seemingly contradictory attitudes of protection and exploitation of the natural environment have contemporary counterparts in current debates about the use of the environment; such debates bring out competing humanistic goals and values, often pitting major internationally financed development projects (big fish) against small villages and family farms (little fish). The decisions are not easy ones, and often result in sacrificing the economic well-being of one family or one village to benefit what is expected to be the increased economic welfare of a larger, more extensive community. It is the old theory of *mātsya-nyāya,* big fish devouring little fish (AS 1.4), which the *Arthaśāstra* hopes to control with puissant administration.[57] But as we are learning in this age of ecological consciousness and environmental studies, the little fish must survive as well; they are a necessary part of the rich web of biodiversity on which we all depend.[58] The government often has tough decisions to make, as did the king in times past. The key duty for both is protection and maintenance of the state, its residents, its resources, its boundaries, its big fish and little fish.

Certainly one of the most significant shifts we must consider in our examination of state responsibility for the environment in India's history is the shift in forms of governance from multiple monarchies to centralized socialist democracy.[59] No longer is environmental stewardship solely the dharmic responsibility of kings and their ministers; within a democracy this becomes a more broadly shared responsibility. Indeed, through the Forty-second Amendment to the Constitution of India passed in 1976, environmental stewardship, along with nine other specified responsibilities,[60] became a duty of every Indian citizen. The relevant section of this amendment is contained in Part IVA, Article 51A, clause "g":[61]

> It shall be the duty of every citizen of India . . . to protect and improve the natural environment including forests, lakes, rivers and wild life, and have compassion for living creatures.

The addition of a section on "Fundamental Duties" of citizens to the Indian Constitution (Part IVA) was designed to complement the Constitution's original sections on "Fundamental Rights" (Part III), on the rights of individual citizens, and "Directive Principles" (Part IV), on the duties of the state.

In addition to the new section on fundamental duties, the Forty-second Amendment also introduced a new directive within Part IV concerning the state's responsibility toward the environment;[62] the language of this directive, specifically Article 48A, is very similar to that of Fundamental Duty (g):

> The State shall endeavor to protect and improve the environment and to safeguard the forests and wildlife of the country.

Note that both clauses use the language "protect and improve," which emphasizes that environmentalism is not merely about protection of the ecosystem, but must include efforts to ameliorate the environment, which translates into social activism, educational programs, scientific research, policy making, and implementation and enforcement of relevant laws. Together these two amended sections place a shared responsibility for the care of the environment upon the shoulders of both the government and individual citizens.

It is significant that neither the duties of the state, as articulated under Directive Principles, nor the duties of citizens enumerated in the 1976 amendment are enforceable in the courts. That is, there is no provision to ensure that these individual or state duties are carried out, nor is there any punishment prescribed for their violation.[63] Technically, the directives and duties as spelled out in the Constitution are nonjusticiable.

The Directive Principles of State Policy, Part IV, Articles 36–51, of India's Constitution, are one of the most instructive and important parts of the Constitution, as they were designed to inspire the ideals of a democratic welfare state.[64] They were also meant to avert conflict between the rights of individuals set out in the section preceding (Part III: Fundamental Rights). These Directive Principles are not laws in and of themselves but are designed to inform law and policy making. Article 37 of the Directive Principles states: "It shall be the duty of the State to apply these principles in making laws." Framed as mandates, the Directive Principles carry a social if not moral imperative rather than a legal imperative.[65] The addition in 1976 of a section on fundamental duties of individuals, significantly incorporated into Part IV rather than Part III of the Constitution, is meant to provide a citizen's counterpart to the Directive Principles by emphasizing that individuals, in addition to the state, also have moral obligations to contribute to and uphold the integrity of the Indian nation. The amendment articulating citizens' duties, coming some thirty years after the constitutional formulation of citizens' rights, also came with the stated realization that there can be no rights without duties, a popular refrain of Mahatma Gandhi. By including a section on the duties of citizens alongside the constitutional treatment of the state's duties, the amenders were stressing not only the need for individuals to share in the ideals and goals of the state, but also their fundamental obligation to uphold, pursue, and work for those goals. Implicitly at work in this articulation, I believe, is the fundamental notion of *dharma* (social and moral order; social responsibility) so important to understanding ideas of interdependence, morality, law, duty, and work ethic embedded in Indian culture. The state cannot fulfill these constitutional goals without the help of the people, but what is to impel individuals to comply with these nonjusticiable, constitutionally framed duties, and specifically the duty to protect and improve the environment?

As I have previously pointed out, in the times of the *Arthaśāstra* the state resorted to fines and various other punishments to enforce rules governing everything from the pollution of the environment to the failure to work arable lands. Respect for certain flora and fauna was also fostered by religious sentiments and teachings, specifically the principle of *ahiṃsā,* nonviolence to living creatures; violation of different life-forms constituted a form of sin (*pāpa*), which could in some cases be mitigated by observing prescribed penance (*prāyaścitta*), but which, due to the theory of *karmavipāka,* very often had an impact on one's next life and environment. Since the welfare of society (not to mention an individual's future rewards and next life) depended on each person fulfilling his or her *dharma,* violations of both individual duties (*svadharma*) and common duties (*sādhāraṇadharma*) had prescribed punishments, which we find articulated in the *arthaśāstra*s and *dharmaśāstra*s.[66] Yet, the duties of the state and of individuals as articulated in the Constitution of India, while projecting moral responsibilities and a national ideal, are not enforceable, as is the section on individual rights in the Constitution.[67] This would cause one to question how the state can fulfill the directive to protect and improve the environment.

Despite the fact that neither the Directive Principles nor the Fundamental Duties can be enforced in the courts, acts and laws derived from them may. Justice E. K. Venkataramiah of the Supreme Court of India, in his book *Citizenship: Rights and Duties,* explains:

> Clause (g) of Article 51-A of the Constitution is concerned with the duty of protection of the natural environment. This is one of the duties that has received the most intensive and extensive legislature and executive attention during the last few years. The far-reaching Environment (Protection) Act 1986 is the most apt example. The range of its provisions and the far-reaching (almost radical) measures that can be undertaken through statutory rules and orders that can be issued under the Act show that the duty to 'protect and improve' the natural environment has been spelt out quite elaborately in our law. It now remains for citizens to comply with its provisions.[68]

Yet, it is not just citizens who need to comply with the laws and rules generated by the Environmental Protection Act (the legal instrument of Articles 43A and 51A); the central and state governments must also comply by enforcing the principles and laws to protect and improve the environment. One reason citizens may feel compelled to observe their duty to protect the environment is that their own welfare could be greatly enhanced by improved environmental conditions, but often they have to compel the government to comply in order to reap that benefit. A case in point is *L. K. Koolwal versus State of Rajasthan*. In 1988, L. K. Koolwal, on behalf of the citizens of Jaipur, filed a petition before the court to "compel municipal authorities to provide adequate sanitation." He based his petition on Article 51A of the Constitution, arguing that if a citizen has a "constitutional duty to protect the environment (Article 51A), the citizen must also be entitled to enlist the court's aid in enforcing that duty against recalcitrant state agencies." The judge in the case accepted the writ petition and directed "the Municipality to remove the dirt, filth etc. within a period of six months and clean the entire Jaipur city."[69]

While much legislation has been enacted to "protect and improve" the environment at both national and state levels since the incorporation of Articles 48A and 51A into the Constitution,[70] it remains to be seen how the state can instill and foster "compassion for living creatures," a duty, according to Article 51A (g), to be required of every Indian citizen. As India is a secular socialist democracy, the quality of compassion (*dayā*) expected of every citizen must be understood as a humanistic value, though it has deep roots in the dominant ethico-religious teachings and ideologies of India.[71] Compassion demands sensitivity and awareness best cultivated through example and education. Like most of the fundamental duties prescribed in Article 51A, protection and care of the environment and its host of life-forms demands public awareness and public education. As Kashyap observes in his book on constitutional law:

> In the ultimate analysis, the only way to bring about adherence to fundamental duties is through public opinion and education in citizenship values and duties, and building adequate awareness and a congenial climate wherein every citizen feels proud and bound to perform his constitutional duties to the nation and pay his debt to society.[72]

That the care and protection of the environment is framed within sections of the Constitution that reflect more the moral norms and social ideals, rather than legal norms, of the established political order suggests to me that the environmentalism of the contemporary

Indian government, like that of Kauṭilya's king, is grounded in a moral ecology. That is, environmental laws and ecological activism grow out of or are compelled by a moral imperative rather than by a scientific or legal imperative. Both Kauṭilya's king and the Indian government have their scientific experts to guide them, but their motivation to act upon the advice of these experts as well as their authority to legislate fines and penalties against offenders of the environment is driven by a sense of morality which seeks to promote the common good. That morality, both in Kauṭilya's time and in contemporary times, is fed largely by practical concerns, which recognize human dependence on earth's resources for our material welfare, and to a lesser extent by religious values embedded in Indian culture, such as respect for the life-breath and sentience which courses through multiple life-forms.

The king in ancient India was expected to use his powers to protect the earth, which was entrusted to him for agriculture, well-being, prosperity, and development (*Mahābhārata* 59.106.107; *Yajurveda* 9.22). With the Environmental (Protection) Act of 1986, the Indian government recognized its "power" to protect and improve the quality of the environment;[73] its responsibility to employ this power is mandated in Article 48A of the Constitution. Kauṭilya's *Arthaśāstra* and the Indian Constitution articulate ideals and goals to be pursued by the king and the government, respectively. The pursuit of those ideals demands action in the form of judicious leadership, legislation, and sanctions aimed at the goal of the common good. The "common good" is also an ideal, but now more than ever before in history, the ideal of the common good entails more than just the health and prosperity of human beings and extends to include the health and welfare of all of earth's resources in recognition of our common interdependence. As the *Śukranīti* astutely observes (1.359-60):

> Wealth and life are preserved by men for enjoyment. But what avails a man to have wealth and life who has not protected the land?

Notes

The presentations of Ann Grodzins Gold, an anthropologist, Philip Lutgendorf, a scholar of Hindi literature, and David Lee, a biologist, at the 1997 Conference on Hinduism and Ecology held at Harvard University, significantly influenced the shape and content of this essay; each of them has an essay in this volume based on their presentations. My textual study of environmental consciousness in the *arthaśāstra*s and *dharmaśāstra*s benefited greatly from the unprecedented collaboration among scholars from the sciences, social sciences, and humanities, along with environmental activists and policy makers, present at the Harvard conference. It became clear at this conference that many new insights might be gained from such collaborations as we work together to formulate policies, ethics, and strategies that will provide for a more careful management of the earth and her shared resources. My own research for this project, as well as the Harvard Conference on Hinduism and Ecology, was subsidized in part by Columbia University's Dharam Hinduja Indic Research Center. The focus of this essay also benefited from subsequent communications with Ann Gold and editorial input from Gary A. Tubb, who keeps me honest.

1. This expectation of a king, which broadly addresses a king's responsibility for the management of the natural environment and its resources, is reiterated in many classical Indian texts, this verse being taken from the ancient Tamil classic known as the *Tirukkuṟaḷ;* H. A. Popley, *The Sacred Kural* (Caclutta: Y.M.C.A. Publishing House, 1958). It is not a great leap from this verse to the mandate found in Chapter II.3.1. of India's Environment (Protection) Act of 1986, which states: "Subject to the provision of this Act, the Central Government shall have the power to take all such measures as it deems necessary or expedient for the purpose of protecting and improving the quality of the environment and preventing, controlling and abating environmental pollution" (Environment Protection Act 1986, reproduced in *Law and Environment,* ed. P. Leelakrishnan [Lucknow: Eastern Book Company, 1992], 285).

2. Ann Grodzins Gold, "'If You Cut One Branch You Cut My Finger': Court, Forest, and Environmental Ethics in Rajasthan," in this volume, 325.

3. Ibid., 320.

4. Without the intention of doing so, Ann Gold's study of V. P. Singh, as well as Atluri Murali's detailed study of the Vijayanagara king, Krishnadeva Raya, in his article entitled "Whose Trees? Forest Practices and Local Communities in Andhra, 1600–1922," provide excellent examples of arthaśāstric principles at work in connection with environmental stewardship. It becomes evident from reading Gold's and Murali's respective case studies, alongside relevant sections of the *Arthaśāstra,* that there is a continuity of theory and practice in India when it comes to environmental stewardship. I have made more extensive use of Gold's case study in my essay since readers of this volume may immediately turn to it to read more about Vansh Pradip Singh and his environmentalism, but Murali's study, published in David Arnold and Ramachandra Guha's edited volume, *Nature, Culture, Imperialism: Essays on the Environmental History of South Asia* (Delhi: Oxford University Press, 1995), provides even more evidence of correspondence between the historical conservationism of rulers (in this case Krishnadeva Raya) and *Arthaśāstra* theories on environmental responsibilities of kings.

5. P. Leelakrishnan, "Forest Conservation: Dawn of Awareness," in his *Law and Environment* (Lucknow: Eastern Book Company, 1992), 52.

6. N. A. Nikam and Richard McKeon, eds. and trans., *The Edicts of Asoka* (reprint; Chicago: University of Chicago Press, 1978), 64–65.

7. Romila Thapar, *A History of India,* vol. 1 (reprint; Harmondsworth: Penguin Books, 1972), 86.

8. The dates of the *Nītisāra* of Śukrācārya, also known as the *Śukranīti or Śukranītisāra*, are greatly disputed. Lallanji Gopal considers it a forged work from the nineteenth century and thinks that Dr. Gustav Oppert, who translated the text for the Madras government in 1882, was duped into thinking it was an ancient text of polity; Lallanji Gopal, *The Śukranīti: A Nineteenth Century Text* (Varanasi: Bharati Prakashan, 1978). Others who have studied the text and intertextual evidence say that a *nītiśāstra* by one Śukra is referred to in texts as early as the fourth century C.E. Some scholars believe it was compiled between 900 and 1200 C.E., and others say it is a seventeenth-century text. For further discussion on the date of this work, see Benoy Kumar Sarkar, *The Positive Background of Hindu Sociology: Introduction to Hindu Positivism* (Delhi: Motilal Banarsidass, 1985); and Vandana Nagar's summary of the disputed dates in *Kingship in the Sukra-Niti* (Delhi: Pushpa Prakashan, 1985), 7–9. Quotations from the *Śukranīti* in this essay are my translations from the *Śukranītisāraḥ* (Bahalagarha, Sonipata: Rsidevi Rupalala Kapura Dharmartha Trasta, 1997).

9. *Nītiśāstra*, *rājadharma* or *rājanīti*, *daṇḍanīti,* and *arthaśāstra* are traditional Sanskrit terms used to refer to the study of politics and to classical treatments of the duties of kings, which usually include attention to such matters as economics, politics, defense, foreign policy, and civil and criminal law. *Nīti* means rules, guidance, direction, and implies a moral orientation; *nītiśāstra* refers to the teachings that provided direction to the king in administering the law righteously. The term *rājadharma,* meaning the duties (*dharma*) of kings (*rāja*), is used more widely in the *dharmaśāstra* texts and epics, since monarchy was the normative form of government; these texts incorporate more references to the divine origin of kings than do the *nītiśāstra*s. *Daṇḍanīti* means the rule (*nīti*) of force (*daṇḍa*) and is used in reference to the political duties of kings who were expected to enforce laws through the use of a strong rod (*daṇḍa*). *Arthaśāstra* is a term that also refers to the science of politics and was chosen as the title for Kauṭilya's work, the most well-known *nītiśāstra*. Technically the title means the science or teaching (*śāstra*) of wealth or gain (*artha*). For Kauṭilya, *artha* mainly meant land. Kauṭilya's *Arthaśāstra* (abbreviated as AS) is the main textual source for this essay, though other texts on *nīti* and *rājadharma* have been consulted. Other key texts that deal with the science of government include Kāmandaka's *Nītisāra,* a.k.a. *Kāmandakīyanītisāra* (300–800 C.E.?); the *Nītivākyāmṛtam* of Somadevasūri; and the *Śukranīti* of Śukrācārya (see n. 8). Significant treatments of *rājadharma* are also found in the *Śānti Parva* of the *Mahābhārata;* in key sections of the *Rāmāyaṇa* (especially the *Ayodhyā* and *Yuddha kāṇḍa*s); in *Manusmṛti* (chaps. 7–9) and *Yājñavalkyasmṛti* (1.305–67); and in several *purāṇa*s, e.g., *Matsya* chaps. 225–29, *Agni* chaps. 218–42, and *Garuḍa* chaps. 108–15. Among the relevant *nibandha*s treating *nīti* are the *Nītimayūkha* of Nīlakaṇṭha, the *Nītiratnākara* of Caṇḍeśvara, the *Nītiprākaśa* of Vaisampāyana, the *Rājadharmakāṇḍa* of Lakṣmīdhara, and the *Rājanītiprākaśa* of Mitramiśra. Quotations from the *Arthaśāstra* in this essay are my

translations from the Sanskrit text published along with R. P. Kangle's translation (*The Kauṭilya Arthaśāstra,* Sanskrit with English translation by R. P. Kangle, 2d ed., 2 vols. [Bombay: University of Bombay, 1969]). Verses from the *Mahābhārata* have been translated from the critical edition of the *Mahābhārata* (Pune: Bhandarkar Oriental Research Institute, 1971–76). Selections from *Yājñavalkyasmṛti* and the *Mitākṣara* are my translations based on the *Yājñavalkyasmṛti* with the commentary of *Mitākṣara* of Vijñaneśvara (Delhi: Nag Publishers, 1985). Verses from *Manusmṛti* are taken from the translation by Wendy Doniger with Brian K. Smith, *The Laws of Manu* (New York: Penguin Books, 1991). References to the *Matsya* and *Agni purāṇa*s, the *Yajurveda Saṃhitā,* and the *Bṛhatsaṃhitā* of Varāhamihira are from the following editions: *Matsyamahāpurāṇam*, ed. Kumar Pushpendra (reprint, Delhi: Meharacanda Lachamanadasa, 1984); *Agnipūraṇam* (Pune: Ananda Ashram, 1987); *Yajurveda Saṃhitā,* Sanskrit text with English translation of Ralph T. H. Griffith, ed. and enlarged by Surendra Pratap (Delhi: Nag Publishers, 1990); and Varāhamihira, *Bṛhatsaṃhitā,* 2d ed. (Varanasi: Sampurnananda Samskrta Visvavidyalayasya, 1996).

10. The welfare of the state and its citizens was so dependent on the king's fulfillment of his duties and responsibilities that some classical texts conclude that *rājadharma* is the root or best of all *dharma*s (see *Mahābhārata, Śānti Parva* 56.3; 63.25; 141.9–10). Indeed, the king had a responsibility to make sure those under his rule fulfilled their own respective *dharma*s (cf. AS 1.3; *Matsya* 215.63, *Viṣṇusmṛti* 3.2–3; *Mahābhārata, Śānti Parva* 57.15, 77.33); *Viṣṇusmṛti with the Commentary of Keśavavaijantī of Nandapaṇḍita,* ed. V. Krishnamacharya (Madras: Adyar Library Research Center, 1964).

11. The *Nītivākyāmṛtam* of Somadevasūri, dated around the tenth century C.E., is largely indebted to the *Arthaśāstra* of Kauṭilya; Somadevasūri, *Nītivākyāmṛtam* (Varanasi: Sri Mahavira Jaina Granthamala, 1976). The term I have translated in this verse as "tree" is actually "*phala,*" usually translated as "fruit" or "reward." I take my translation cue from the commentators who gloss *phala* as *vṛkṣa,* tree, in this context.

12. Gold, "'If You Cut One Branch You Cut My Finger,'" 321.

13. *rājā rājyamiti prakṛtisamkṣepaḥ. Arthaśāstra* 8.2.

14. The *Arthaśāstra* of Kauṭilya is traditionally attributed to one Kauṭalya/Kauṭilya, alias Cāṇakya (sometimes also identified as Viṣṇugupta), believed to have been the minister of Candragupta, the first Mauryan emperor (ruled 321–297 B.C.E.), whose dates are contemporary with Alexander the Great; thus, late fourth century B.C.E. is the earliest date assigned to the *Arthaśāstra.* However, linguistic studies such as Thomas Trautmann, *Kauṭilya and the Arthaśāstra: A Statistical Investigation of the Authorship and Evolution of the Text* (Leiden: E. J. Brill, 1971), and Hartmut Scharfe, *Investigations in Kautalya's Manual of Political Scienc*e, Untersuchungen zur Staatrechtslehre des Kautalya, 2d rev. ed. (Wiesbaden: Harrassowitz, 1993), are convincing in dating the text later, somewhere in the early Common Era. It is likely that the work existed by the third century C.E. Trautmann's examination finds multiple dates and authors for different parts of the *Arthaśāstra,* and he observes that "[i]n its impersonal and abstract way it sums up ancient Indian beliefs about the state with an authority which no individual creation could possess" (187). He concludes that the *Arthaśāstra* was "compiled by one person, but has no one creator" (186).

15. P. V. Kane, *History of Dharmaśāstra,* vol. 3, 2d ed. (Poona: Bhandarkar

Oriental Research Institute, 1973), 196.

16. Kauṭilya's *Arthaśāstra* does not provide a historical construction of the Magadha Empire under the Mauryans, but rather a portrait of an ideal Hindu kingdom which draws on principles of knowledge perhaps current under the Mauryans about how an ideal king should rule. The author was certainly influenced by the values and traditions of his period, but his overall aim was to present theories and principles of statecraft.

17. D. D. Kosambi, *The Culture and Civilization of Ancient India in Historical Outline* (London: Routledge and Kegan Paul, 1965), 141.

18. Madhav Gadgil and Ramachandra Guha, *This Fissured Land: An Ecological History of India* (Berkeley and Los Angeles: University of California Press, 1992), 85–87.

19. Francis Zimmerman has noted that "[i]n the *Arthaśāstra,* a geographical web of interconnections determines how royal power should be exerted over the space in question, a space conceived right from the start to be ecologically complete, with its double interplay of expansion in the direction of distant lands and enclosure around the established nuclei of civilization. Its immense central plain is bounded in the distance by mountains, dense forests, or marshes, and within the plain there is the jungle (wasteland), also distant from the village centers"; Francis Zimmerman, *The Jungle and the Aroma of Meats: An Ecological Theme in Hindu Medicine* (Berkeley and Los Angeles: University of California Press, 1987), 50.

20. Kosambi, *The Culture and Civilization of Ancient India,* 143.

21. The main sources of revenue for the state, according to the *Arthaśāstra* (6.24.1–10), derived from the following seven areas (via fees, taxes, fines, shares, etc.): 1) *durga* (the fort; activities and transactions within the fortified city related to such things as customs duty, fines, passport fees, liquor, gambling, prostitutes, animal slaughter, artisans, markets, temples, etc.); 2) *rāṣṭra* (includes income from the rural areas outside the city, such as agricultural produce from the crown lands, shares of produce from private lands, fees from ports, ferries, pastures, etc.); 3) *khani* (income from mines); 4) *setu* (irrigation works, including flower gardens, fruit orchards, wet and root crops, and vegetable gardens); 5) *vana* (forest produce [*kūpya*], elephants, horses, and other cherished animals that could be traded, gifted, or used for defense); 6) *vraja* (herds such as cow, buffalo, goat, sheep, camel, and mules); and 7) *vaṇikpatha* (income derived from tariffs and trade routes).

22. An examination of the departments devoted to game-keeping and slaughterhouses (*sūnādhyakṣa*), crown herds (*go-adhyakṣa*), the elephant corps (*hastyadhyakṣa*), the cavalry corps (*aśvādhyakṣa*), and the elephant forester (*nāgavanādhyakṣa*) provides additional evidence of the *Arthaśāstra*'s wide knowledge about the health and habits of animals. However, I have limited the examples in this essay to natural resources in the environment other than animals, as a discussion of the maintenance and protection of the animal kingdom as evinced in these texts would make this essay too unwieldy. Some of the responsibilities of the superintendent of pasture lands (*vivītādhyakṣa*) are also relevant here, for he was especially concerned with grass lands, orchards, and gardens, but his main concern was the health and safety of cattle that grazed the pastures, so for similar reasons I have not elaborated on the responsibilities of this superintendent. Fields not suitable for agriculture were designated pasture lands and maintained by the *vivītādhyakṣa*. Care was taken to rotate grazing

lands so that pastures would not be exhausted of their resources, and different pasture lands were used in different seasons to suit the health of the cattle.

23. According to its web site, the Ministry of Environment and Forests is a "nodal agency in the administrative structure of the Central Government, for the planning, promotion, and co-ordination of environmental and forestry programmes. The principal activities undertaken by Ministry of Environment & Forests, consist of conservation & survey of flora, fauna, forests and Wildlife, prevention & control of pollution, afforestation & regeneration of degraded areas and protection of environment. . . ." See <http://www.nic.in/envfor/mef/mef.html>.

24. A distinction was made between crown lands (*sītā*) and those of independent farmers; this distinction is reinforced in several places in the *Arthaśāstra* (see 5.2.8–11; 2.15.2-3; 2.6.3). Kosambi observes that "[t]he state was not the sole land-clearing agency. Any group could move out into the jungle on their own, usually organized as a guild (*śreṇi*) for land clearance with temporary or permanent occupation" (*The Culture and Civilization of Ancient India,* 151). See also Scharfe's discussion about disputes and misunderstandings about whether the king was the sole proprietor of the land, in which he points out that "peasants had possession of their land and on occasion sold it"; *Investigations in Kautalya's Manual,* 241, 240–48. While the king is the protector of the land, he is not necessarily the owner of all the lands in his kingdom; those lands which he was at liberty to lease, give away, and reap full profit from were known in the *Arthaśāstra* as *sītā,* herein translated as "crown lands." Yet, there is clear evidence in the *Arthaśāstra* that there were also lands owned by individual families, from which the king did reap a share of the profits. Romila Thapar's discussion of different classification of land and various types of land tenure during this period is also relevant on this point; see her book *From Lineage to State: Social Formations in the Mid-First Millennium in the Ganga Valley* (Delhi: Oxford University Press, 1990), 104–7, 123.

25. See Nirmala Sengupta's essay "Irrigation: Traditional vs. Modern," *Economic and Political Weekly* 20, no. 45-47 (1985): 1919–38, in which he examines the merits of traditional irrigation systems for today's world with specific reference to India.

26. Kosambi, *The Culture and Civilization of Ancient India,* 146–47, observes that the "*Arthaśāstra* must appear strange and unreal to those who visualise the Indian countryside in its later form." Districts or *janapada*s "were separated by considerable forests . . . [which supplied] . . . fuel, timber, hay, game, edible produce, and grazing." The *Arthaśāstra* (2.2) recognizes four different types of forests: 1) untamed forests; 2) productive forests of economical value; 3) recreational forests (used for hunting, etc.); and 4) elephant forests (sources for wild elephants, which were captured and tamed for use in the king's defensive troops). Each of these forests has a separate superintendent. The *Arthaśāstra* also recognizes the value of forests as places of refuge, and the king is advised to designate protected sections of the forest for study and teaching, and for meditation. My discussion here is only concerned with the superintendent of productive forests.

27. Forest products (*kūpya*) included hardwood, reeds, creepers, fibrous plants, leaves, flowers, medicinal plants, and poisons variously derived from plants, insects, and reptiles.

28. For information on India's current biodiversity, see <http://www.wcmc.org.uk/igcmc/main.html>.

29. Madhav Gadgil, "Toward an Ecological History of India," *Economic and Political Weekly* 20, no. 45-47 (1985): 1909–13.

30. Murali's case study of the Vijayanagara king, Krishnadeva Raya, shows a similar use of nature for defensive and protective strategies. See Murali, "Whose Trees?"

31. While nature preserves and national parks are generally assumed to be good ways of promoting environmentalism and preserving biological diversity, some studies of Indian parks and sanctuaries have shown how imposed limits, controls, and laws within some preserves are detrimental to local economies and, in some cases, can even thwart some kinds of biological diversity which often benefit from (limited) human intervention. See Gadgil and Guha, *This Fissured Land,* 233ff., as well as relevant discussion in their more recent book, *Ecology and Equity: The Use and Abuse of Nature in Contemporary India* (London: Routledge, 1995), especially 91–94, 148–75.

32. Gold, "'If You Cut One Branch You Cut My Finger,'" 324.

33. Ibid., 325–26.

34. Ibid., 326, 327. The slaughter of lizards to preserve trees, while in this case the result of misinformation, does raise questions for the environmental ethicist and drafters of environmental policy. Holmes Rolston III relates a case in which the U.S. Fish and Wildlife Service "planned to shoot two thousand feral goats to save three endangered plant species" on San Clemente Island. "After a protest, some goats were trapped and relocated . . . [but] thousands were killed. In this instance, the survival of plant species was counted more than the lives of individual mammals; a few plants counted more than many goats"; Holmes Rolston III, "Environmental Ethics: Values in and Duties to the Natural World," in *Applied Ethics: A Reader,* ed. Earl R. Winkler and Jerrold R. Combs (Cambridge, Mass.: Blackwell Publishers, 1993), 280. Rolston cites several examples of such decision making in his article and analyzes the logic of these decisions as he develops an environmental ethic based on the "intrinsic value" in all life.

35. Animals afforded special protection included birds, deer, calves, bulls, cows, animals that did not prey on other animals, and certain marine life that resembled elephants, horses, or humans. Birds and animals that had a religious value also received special protection by the state.

36. Leelakrishnan, *Law and Environment,* 3, 22.

37. Leelakrishnan, in his book *Law and Environment,* states that "[a] hard look at history makes it increasingly clear that politics and religion fundamentally differed in their approach towards preservation of forest. It is commonly believed that deforestation with a view to earning revenue started only when the Britishers wanted wood from the forest for building ships and railway sleepers. This is only partly correct when relationship between the State and the forest at different stages of Indian history is examined. Holders of political power in the past had adopted always a materialistic approach towards forests considering it as a source of revenue to the State and the people. Religion viewed forests in a different manner; preservation was for the sake of maintaining ecological balance" (53). While I agree with Leelakrishnan that the state primarily saw and used forests for their economic value, I have pointed to examples of ecological prudence by the state which aimed at maintaining

an ecological balance. It is to be expected that the state and "religion" may view and use resources differently toward their different aims, but it is clear in the *Arthaśāstra* that the state recognized the religious value of certain flora, fauna, and environments (e.g., sacred groves), and took that into consideration when determining fines and implementing laws of protection.

38. According to the account related to Ann Gold, the *darbār* ordered the Bhils to be summoned to kill all the lizards between Ghatiyali and Sawar, after he was falsely advised that it was the lizards who were eating his trees, rather than the goats of the local goatherd. As Gold observes, no one really cared about the lizards. See Gold, "'If You Cut One Branch You Cut My Finger,'" 326–27.

39. Reverence for trees pervades Indian literature. That reverence, which led the authors of *dharmaśāstras* and *nītiśāstras* to provide laws for the protection of plants and trees, is derived from several perspectives: 1) a recognition of plants as sentient beings with consciousness, which therefore should be protected from harm (advocacy of *ahiṃsā*); 2) a designation of certain trees as sacred objects, due to their religious value, identification with certain deities, or resident *bhūtas*; 3) a respect for the economic value of trees; 4) a legal interpretation that recognized certain trees as personal property. *Manu* 11.143 (cf. *Yājñavalkya* 2.276) says that "For cutting fruit trees, shrubs, vines, creepers or flowering plants, a thousand Vedic verses should be chanted." A penance (*prāyaścitta*) was prescribed in proportion to the sin, and the seriousness of the sin had to do with the value of the tree. Thus, useful fruit-bearing trees such as breadfruit and mango garnered heavier penalties, and plants yielding blossoms like the jasmine creeper were also ranked high for penalties (see *Manu* 11.144; cf. *Viṣṇusmṛti* 50.48). The cutting of a tree for the purposes of firewood was considered a minor sin (*upapātaka*) (cf. *Manu* 11.64; *Yājñavalkya* 3.24; *Viṣṇusmṛti* 37.24). The *Mitākṣarā* (640), commenting on Yājñavalkya, explains: "if the cutting was for the purpose of *pañcayajña* there is no violation of the rule." This rule is derived in analogy with the killing of animals which are only allowed to be killed for religious or sacrificial purposes. In other words, for religious purposes, trees and plants could be cut or injured (and animals sacrificed). When the tree or plant that is violated is someone's private property, according to civil law articulated in the *dharmaśāstras*, the culprit had to pay damages equivalent to the loss of profit the owner suffered due to the injury (*Manu* 8.285); if such a tree also had a religious value, then presumably the individual doing the damage would have to perform a *prāyaścitta* as well as pay damages. For further discussion on religious and civil penalties for injuring trees, see: *Manu* 8.285; 11.42–46, 64; *Yājñavalkya* 2.230–31; 3.276–77; *Viṣṇusmṛti* 50.48–50.

40. A term commonly used in the *Arthaśāstra* to refer to the king's subjects is *prajā*, which is of significant interest as it means progeny. The king's duty to protect is much like that of a father's responsibility to protect his family. For some relevant discussion on the paternal relationship toward the earth, see Vasudha Narayanan, "'One Tree is Equal to Ten Sons': Hindu Responses to the Problems of Ecology, Population, and Consumption," *Journal of the American Academy of Religion* 65, no. 2 (1997): 291–332.

41. Despite the necessity and value of agriculture attested to in these texts, Brahmins and rulers are advised in *Manu* and elsewhere to "try hard to avoid

farming, which generally causes violence and is dependent on others. Some people think, 'Farming is a virtuous trade,' but as a livelihood it is despised by good people, for the wooden [plough] with the iron mouth injures the earth and the creatures that live in the earth" (*Manu* 10.83–84). H. A. Popley, in commenting on verse 104.1 in the *Tirukkuṟaḷ,* an ancient Tamil treatise on proper conduct (a third of which is devoted to the conduct of kings), says that Tiruvaḷḷuvar, the author of the *Kuraḷ,* "disputes the Brahmanical theory of Manu that the farmer is one of the lower castes. Professor Chakravarti comments: 'The agriculturalist is the linch-pin of the whole social chariot'"; *The Sacred Kural,* trans. H. A. Popley (Calcutta: Y.M.C.A. Publishing House, 1958), 136.

42. Arnold and Guha, *Nature, Culture, Imperialism,* 3.

43. The *Arthaśāstra* contains many references to fish and fisheries, to rearing cows, buffaloes, goats, and sheep, and pays special attention to the care of horses and elephants (with a chapter devoted to each), which were particularly valuable to the defense of the state.

44. While the *Arthaśāstra* displays great awareness of the diversity and riches of the natural environment, with its emphasis on human management of natural resources, it is neither a scientific treatise nor an ethical treatise, and thus we must be cautious in attributing to its author(s) a developed environmental ethics or even particular environmental policies.

45. See David Arnold and Ramachandra Guha's nuanced discussion of what constitutes and contributes to environmental history, in their introduction to *Nature, Culture, Imperialism,* 1–4. My essay does not actually take on this task of doing an environmental history through a case study of the *Arthaśāstra,* but rather suggests ways that this source could be used in developing such a history. Madhav Gadgil consulted the *Arthaśāstra,* among other sources, for his essay "Toward an Ecological History of India." In particular he observes that the *Arthaśāstra* "emphasises the need to bring new lands under cultivation by settling hunting-gathering tribes who generate little surplus for the state" (1913).

46. Personal Communication from Anil Agrawal, Director, Centre for Science and Environment, New Delhi, October 1997.

47. Akhil Gupta, *Postcolonial Developments: Agriculture in the Making of Modern India* (Durham: Duke University Press, 1998), in his discussion of "indigenous" knowledge concerning agronomy, raises a similar question, although his critique is slightly different. What, he asks, is the relationship between "a classificatory system adumbrated in a few 'great' books with the activities of practitioners across the centuries?" He explains that the "problem I have in reading the traces of a great textual tradition in present practices is that no evidence is ever offered of the mechanisms that translated those texts into practices and then preserved the entire system or elements of it intact over centuries" (158–59). I take his point, but I am not arguing that these methods of agronomy found in the *Arthaśāstra* or other classical texts have been continuously practiced in India, although some other scholars have made this point (for example, Ajay Mitra Shastri in his study of the *Bṛhatsaṃhitā,* an important classical work of Varāhamihira, notes that some of the arbori-horticultural practices recorded by Varāhamihira are "more or less followed by Indian peasants even to this day"; *Ancient Indian Heritage: Varāhamihira's India,* vol. 2, *Economy, Astrology, Fine Arts, and Literature* (New Delhi: Aryan Books International, 1996), 272.

Rather, my point is that we find specific methods of planting, harvesting, irrigating, etc., described in these texts, and it may be worthwhile to take a look at them to see if they are more ecofriendly and as efficient as current methods. As Nirmala Sengupta shows in "Irrigation: Traditional vs. Modern," there are certain techniques of agriculture that have been developed to fit the particular soil, climate, and seasonal changes of different regions of India, and some of these technologies (several still in use today) are described in ancient texts such as the *Arthaśāstra*. It is also likely that those who worked the land continued to refine and adapt such techniques over the centuries as technology and environmental conditions changed, in which case these texts would provide us with some information for developing an environmental history of India, including changes in farming and mining techniques, as well as in flora, fauna, crops, seasons, water resources, etc. Gupta's rigorous scrutiny of the terms "indigenous" and "indigenousness" in chapter three of *Postcolonial Developments* provides an important critique and significant caution to our scholarly and activist enterprises, which involve searching out "indigenous" attitudes toward or methods of environmentalism.

48. The state encouraged private enterprise to cultivate wastelands, to work difficult mines, to develop and maintain irrigation systems; it did this by supplying aid (grains and cattle) and purchasing their produce. The state benefited in two ways from this partnership: it made the territory within the kingdom more productive, and the state took a percentage of the crop or earnings from the results of this enterprise. By private enterprise I mean individual groups or families who took on the costs of the labor, as compared with those mines and fields that were worked by laborers paid by the state.

49. Gold, "'If You Cut One Branch You Cut My Finger,'" 323.

50. In the *Arthaśāstra,* revenues gathered by the state, while used to protect resources and sustain the people through times of war, famine, and drought, were also used to build pleasure parks for the king and to maintain his hunting forests, among other things. Gold mentions several instances of the *darbār*'s self-indulgence in "'If You Cut One Branch You Cut My Finger,'" in this volume.

51. Murali, "Whose Trees?" 120.

52. Gold, "'If You Cut One Branch You Cut My Finger,'" 322–23.

53. As I use the term "state" in this essay to refer to "the king, the kingdom, and the political machine of the kingdom" of the monarchal states of pre-independence India, I similarly use "government" to refer to the operations and offices of post-independence, democratic India.

54. For more discussion on the damming of the Narmada and its opposition, see *The Dam and the Nation: Displacement and Resettlement in the Narmada Valley*, ed. Jean Drèze, Meera Samson, and Satyajit Singh (Delhi: Oxford University Press, 1997); also Ali Kazimi's *Narmada: A Valley Rises*, Anand Patwardhan and Simantini Dhuru's *A Narmada Diary* (New York: First Run/Icarus Films, 1995), and Anurag Singh and Jharana Jhaveri's *Kaise Jeebo Re! (How Do I Survive, My Friend!)*, three documentary films related to Narmada dam projects and their effects on the environment and population within these areas. There are also several relevant web-sites related to these concerns; one such starting place is <http://www/irn.org/programs/narmada/> sponsored by the International Rivers Network.

55. See D. D. Kosambi's and Romila Thapar's discussions of this transition in their respective studies, *The Culture and Civilization of Ancient India in Historical Outline* and *From Lineage to State: Social Formations in the Mid-First Millennium B.C. in the Ganga Valley.*

56. While the *Arthaśāstra* does not provide us with systematic environmental policies or principles for an environmental ethic, it does provide guidelines for policy making in its theory of *mantra* that may be prudent for us to review here. Such guidelines are certainly applicable to contemporary policy decisions concerning the economic benefits of environmental management and development, though ancient textual evidence suggests that they were used largely in private counsel between kings and ministers to assess strategies of war. The *Arthaśāstra* theory of *mantra* refers to lines of policy derived after consultation with various ministers (*mantrī*). In most treatises on *nīti, mantra* consists of five elements to be considered before making a policy decision (cf. AS 1.15; *Agni Purāṇa* 241.4). I have formulated a question out of each of these five technical points concerning policy development: 1) What are the benefits which will result from the course of action? (*kāryasiddhi*); 2) Does the state have the means to carry out this course of action? (*karmaṇām ārambhopāya*); 3) Does the state have the necessary personnel and materials for this undertaking? (*puruṣadravyasampat*); 4) Has the proper time and place for this undertaking been discerned or scrutinized? (*deśakālavibhāga*); 5) Is the state prepared to handle any calamities that may result from this undertaking? (*vinipātapratīkāra*).

57. However, unlike the *dharmaśāstra*s, which teach that all human goals (*puruṣārtha*s) should be tempered by *dharma,* or moral responsibility, the *Arthaśāstra* of Kauṭilya privileges *artha* over the pursuit of *dharma* and *kāma* (pleasure), arguing that economic well-being is critical to the pursuit of pleasure and the fulfillment of one's duties. This position has led many to conclude that Kauṭilya sacrifices ethical concerns to material gain. But I think that the principles of Kauṭilya should be considered alongside other treatises on *nīti* and *rājadharma,* which support the idea that a king's *dharma* to protect is contingent on the successful pursuit of *artha,* which provides the wherewithal by which a king might fulfill his duty. The pursuit of *dharma* and *artha* need not be viewed as competing values, if we understand that the *artha* pursued by the king is in service to his *dharma*. The *Arthaśāstra* (8.2.12) itself counsels that a king who mocks the teachings of the *dharma-* and *arthaśāstra*s and rules without justice will ruin his kingdom.

58. Holmes Rolston observes that "[e]ven before the rise of ecology, biologists began to conclude that the combative survival of the fittest distorts the truth. The more perceptive model is coaction in adapted fit. Predator and prey, parasite and host, grazer and grazed, are contending forces in dynamic process in which the well-being of each is bound up with the other—coordinated as much as heart and liver are coordinated organically" ("Environmental Ethics," 286). This model is quite similar to Hindu notions of the natural as well as social world, which emphasize interdependence, natural duties, and inherent natures.

59. Whereas I have used "state" in most of this essay to refer to the Kauṭilyan State, using "government" to distinguish the centralized government of post-independence India, as I come to the close of my essay, I ask my readers' indulgence as I now make yet another shift. "State" (*rājya* in Hindi) is the technical term em-

ployed in the Constitution, and thus in derivative texts and commentaries, to refer to the governing body and operations of modern India. I think it will be clear from the context of this concluding section when I am using state to refer to contemporary governance; but it is also my intention to imply by this dual usage of "state" that certain ideas cohere to both notions of state and indeed that, when it comes to environmental attitudes, practices, and responsibilities, there are some continuities between the ideals of the Kauṭilyan State and those of the modern State of India as articulated in her Constitution. It is of particular interest that the term used for state in the Hindi version of the Constitution is "*rājya,*" a term from the Sanskrit meaning kingdom, country, empire or government. In the Indian Constitution the term "State," or "*rājya,*" is defined in Article 12 as including the government and parliament of India, the government and legislature of each state, and all local authorities within the territory of India or under the control of the Government of India. Compare this with Kauṭilya's definition of "state" (AS 6.1.1), which has seven constituent parts, namely, territory (*janapada*), king (*svāmī*), minister (*amātya*), fort (*durga*), army (*daṇḍa*), treasury (*kośa*), and allies (*mitra*).

60. Part IVA: Fundamental Duties (Hindi: *mūla kartavya*) was inserted by the Constitution (Forty-second Amendment) Act, 1976, s. 11. This section by no means constituted the main or most important amendment within the 1976 Act, and was only one of several sections of the Forty-second Amendment to be ratified. However, it surprises me that it has received very little attention by commentators and scholars. Durga Das Basu, a retired judge and former Dean of the Faculty of Law at Calcutta University, in her annotations to this section of the Constitution, notes that the Fundamental Duties were added "in accordance with the recommendations of the Swaran Singh Committee. It will bring our Constitution in line with Art. 29(1) of the Universal Declaration of Human Rights and the Constitutions of Japan, China, U.S.S.R."; *Constitutional Law of India,* 3d ed. (New Delhi: Prentice-Hall of India Pvt. Ltd., 1983), 115. When committees were at work on the original Constitution of India, K. M. Munshi submitted to the Subcommittee on Fundamental Rights a document entitled "Citizens, Their Fundamental Rights and Duties," in which he differentiated between "rights" and "duties." Similarly, the Indian National Congress Committee submitted a list of "fundamental rights and duties" to this same committee. Some of their suggestions, such as number 33, "It is the duty of every citizen to safeguard and protect public property and to prevent its misuse . . .," while not originally incorporated, seem to have been taken up in the 1976 amendment on "Fundamental Duties." (See the appendices in K. C. Markandan, *Directive Principles of State Policy in the Indian Constitution* [Jalandhar: ABS Publications, 1987], for the memos submitted by Munshi and the Congress Committee, specifically pp. 446, 490, 493.) It is of particular interest to me that this new text was not added to Part II of the Constitution, "Citizenship," or to Part III, "Fundamental Rights" of citizens, but rather was added under Part IV, "Directive Principles of State Policy." However, it should be noted that "Fundamental Duties" was given its own discrete section, Part IVA, to signify its importance. Ten clauses in this section prescribe the duties of Indian citizens. Most of the duties are aimed at upholding and respecting "the sovereignty, unity and integrity of India," clause (c). For example, clause (a) prescribes respect for the national flag and national anthem; clause (e) promotes the "spirit of common brotherhood" within a

diverse country and calls attention to "practices derogatory to the dignity of women"; clause (i) is concerned with "safeguarding public property"; and clause (j) counsels all citizens to strive for excellence in all spheres of activity. See the Constitution of India, Part IV, article 51A.

61. Protection of the environment was not included in earlier drafts of the amendment's section on duties—which initially set out eight fundamental duties—prepared by the Congress Committee on Amendments appointed by then Congress president D. K. Boraoah in 1976.

62. The Constitution already contained language in Article 39 directing that the state should guide policy concerning the ownership and control of material resources so that they are distributed to serve the common good of the community. Article 48 directs the state to oversee the organization of agriculture and the prohibition of cow slaughter.

63. According to Mangal Chandra Jain Kagzi, professor of law at the University of Rajasthan, proponents of a separate section of the Constitution devoted to fundamental duties of citizens "further recommended that any violation of such duties should be deemed an offence punishable by law"; *The Constitution of India,* vol. 2, 4th rev. ed. (New Delhi: Metropolitan Book Co., 1984), 1001.

64. In formulating these Directive Principles, the framers of the Constitution were influenced by the Irish Constitution of 1937, the 1789 French Declaration of the Rights of Man and of the Citizen, and the Declaration of Independence of the United States.

65. Justice K. S. Hedge has explained that the purpose of the Directive Principles "is to fix certain social and economic goals for immediate attainment by bringing about a non-violent social revolution. Through such a social revolution the Constitution seeks to fulfill the basic needs of the common man and to change the structure of our society. It aims at making the Indian masses free in the positive sense"; quoted in Subhash C. Kashyap, *Our Constitution: An Introduction to India's Constitution and Constitutional Law* (New Delhi: National Book Trust, 1994), 127.

66. *Svadharma* refers to the duties and responsibilities of specific classes (*varṇa*) and stages of life (*āśrama*), which are set out more fully in the *dharmaśāstra*s and summarized in chapter 3, section 1 of the *Arthaśāstra,* as it was the duty of the king to make sure that all beings fulfill their special duties (AS 1.3.16). Duties common to all, according to AS 1.3.13, include nonviolence (*ahiṃsā*), truthfulness (*satya*), probity (*śauca*), freedom from malice (*anasūya*), compassion (*ānṛśaṃsya*), and forbearance (*kṣamā*). *Manu* 6.91–92 says that men of the top three classes must fulfill ten common duties throughout their life, namely, patience, forgiveness, self-control, not stealing, purification, control of senses, wisdom, learning, truth, and lack of anger.

67. In his book *The Constitution of India,* Kagzi recognizes the moral basis of these duties but questions their legality: "these duties only set-forth what would be the desired expectations from the citizens. . . . These are vaguely and mildly prescribed standards of behaviourism, public morality, social morality and decency and political allegiance to the Constitution and the laws. They impose demands of loyalty to the establishing institutions of national life and the new social order. They must deprive the citizens of their natural right of dissent to the established socio-political order. They would seem to deny them the right of the disobedience (*Satyagraha*) and

non-cooperation against the Constitution and the Constitutional State. The question whether these fundamental duties are legal duties must be answered in the negative. These duties have no legal forms or force. They are so many oughts, and moral duties. They are not corelative to any legal rights. . . . They must be deemed to be imperfect and must be likened to the pious obligations with which law can hardly be concerned" (1002–3).

68. E. K. Venkataramiah, *Citizenship: Rights and Duties* (Bangalore: Naga Publishers, 1988), 62. In his book on India's Constitution and constitutional law, *Our Constitution,* Subhash Kashyap makes a similar point: "Article 51A (g) regarding protection of the environment has particularly come up before the courts. Oral orders were issued by the Supreme Court for stopping quarrying operations in some areas in the State of Uttar Pradesh. Also, orders were issued in respect of declaring certain disputed areas as reserved forests" (136).

69. Armin Rosencranz, Shyam Divan, and Martha L. Noble, *Environmental Law and Policy in India: Cases, Materials, and Statutes* (Bombay: Tripathi Pvt. Ltd., 1991), 243–45.

70. Environmental legislation did not all of a sudden come about as a result of the relevant amendments in 1976. As Suresh Jain and Vimla Jain note, "[p]lans and programmes in fields of soil conservation, public health, forest and wildlife protection, industrial hygiene etc. have been in existence in India for many decades" (*Environmental Laws in India,* ed. Suresh Jain and Vimla Jain (Indore: Lawyers Home, 1984), 601. In fact, such plans have been in existence even longer, as I have shown in this essay, drawing out relevant portions from the *Arthaśāstra* on the use and care of the environment. An example of the value and relevance of the *Arthaśāstra* even for today's environmentalist is the inclusion of selected passages from the AS in the section on "Environmental Policy in India" in Rosencranz, Divan, and Noble, *Environmental Law and Policy in India,* a book supported by the Ford Foundation in India and designed to meet the needs of public interest lawyers dealing with environmental issues. The second chapter on environmental policy begins with excerpts from Kauṭilya's *Arthaśāstra* and concludes with excerpts from the 1989–90 Annual Report for the Ministry of Environment and Forests, Government of India. In modern times, according to Jain and Jain, "the first formal recognition of the need for integrated environmental planning was made when the Government of India constituted the National Committee on Environmental Planning and Coordination (NCEPC) in 1972" (*Environmental Laws in India,* 601) in preparation for the 1972 United Nations Conference on the Human Environment from which came the Stockholm Declaration. Jain and Jain omit from their study legislation prior to independence, such as the Indian Forest Act of 1927, which generated much opposition over the years. This and other pre-independence Acts extended to parts or most of British India; other relevant Acts prior to independence include the 1857 Oriental Gas Company Act and the 1905 Bengal Smoke Nuisance Act, both aimed at reducing pollution in Calcutta; the 1853 Shore Nuisance Act dealing with coastal and water pollution in the Bombay region; the Cattle Trespass Act of 1871; the Elephant Preservation Act of 1879; Wild Birds and Animal Protection Act of 1912 (which regulated hunting seasons); the Indian Fisheries Act of 1897; the 1912 Bombay Smoke Nuisance Act to curtail smoke in the Bombay area; and the Hailey National Park Act of 1936 (one of the first acts aimed

specifically at the protection of wildlife and their habitats). Among the legislation generated and legislated since 1976 are the following: The Water (Prevention & Control of Pollution) Cess Act, 1977 (with Rules passed in 1978); The Forest (Conservation) Act, 1980 (Rules in 1981); The Air (Prevention & Control of Pollution) Act, 1981 (with Rules passed in 1982); and the Ganga Action Plan of 1985. The most significant, far-reaching, and stringent act to date is the Environment (Protection) Act of 1986, instigated in large part by the Bhopal disaster of 1984, which claimed some three thousand lives. Legislation predating the 1976 amendment was less comprehensive, less integrated, and less ecologically informed; these acts were more concerned with rights and privileges of humans, addressing pollutants effecting the lives and livelihoods of individuals, and were less concerned with (and less aware of) the need to protect the ecosystem. These included such acts as the Factories Act, 1948 (which included provisions for controlling environmental pollution); the Mines Act, 1952; the River Boards Act, 1956; the Inter-State Water Dispute Act, 1956; the Merchant Shipping Act, 1958; the Atomic Energy Act, 1962; the Insecticides Act, 1968; the Wildlife (Protection) Act, 1972; and the Water (Prevention and Control of Pollution) Act, 1974. It was only after the Stockholm Declaration in 1972 that we begin to see legislation focusing specifically on protection of the environment (rather than regulating the environment, controlling pollutants, etc.). The Stockholm Declaration brought a new environmental awareness to all legislative bodies around the world and subsequent Indian legislation reflects the impact of the Stockholm accord on Indian environmental activism, policy making, and legislation. Under Article 246 of the Constitution, jurisdiction over different matters, including the environment, is allocated to one of the following categories: 1) the Union (central government); 2) individual state governments; 3) concurrent (joint administration). Water (including irrigation), land (including agriculture), and fisheries come under the jurisdiction of individual states; forests, wildlife, and birds, originally the concern of the states, became a joint concern of the states and the union with the Forty-second Amendment; mines, minerals, oilfields, interstate rivers, and major ports come under the jurisdiction of the Union. In addition to legislation and laws (Acts and Rules), many government agencies (of various types such as research, policy, enforcement) have been created by the central government (as well as by individual state governments) since 1976, which in turn have generated many environmental policy statements and plans. The key oversight is provided by the Ministry of Environment and Forests (which originated as a Department of Environment in 1980); two of the key agencies it has developed are the National Wasteland Development Board and the Central Ganga Authority.

71. Book Two of the *Tirukkuṟaḷ* deals with the pursuit of *artha* and is largely addressed to the conduct of kings. Within this section we find several verses promoting the practice of compassion in the pursuit of wealth and success, such as those translated by P. S. Sundaram (*The Kural* [New Delhi: Penguin Books, 1990], 78): "It is compassion, the most gracious of virtues, which moves the world" (2.58.1); "Compassion is human; lacking it men are a burden on earth" (2.58.2); "The world is his who does his job with sympathy" (2.58.8).

72. Kashyap, *Our Constitution,* 134.

73. See n. 1 above, which includes a quote from this Act.

Literary Foundations for an Ecological Aesthetic: *Dharma*, Ayurveda, the Arts, and *Abhijñānaśākuntalam*

T. S. RUKMANI

This chapter deals mainly with different literary traditions in Sanskrit that have an ecological outlook and are rooted in the broad paradigms of *dharma*. First, the discussion will cover the many-sided nature of the *dharma* concept and then it will look at the way *dharma* is linked with nature and the healing traditions in Ayurveda. I will point out the close connections that Yoga philosophy has with nature and then look at the use of animals in the moral fables of the *Pañcatantra*. I will then briefly touch upon the importance of forests as retreats in the philosophical history of ancient India and, in doing so, explore a couple of examples from the *Araṇyakāṇḍa* (chapter on forests) of Vālmīki's *Rāmāyaṇa* and some examples from the Purāṇic literature. Some brief comments on classical dances and Sanskrit poetry will be followed by a detailed examination of Kālidāsa's *Abhijñānaśākuntalam* (hereafter Śakuntalā). I explore this Sanskrit literature in the belief that each one of these themes opens up new ways to negotiate the current ecological crisis in the world.

I will start with a quotation from Vandana Shiva's book *Staying Alive* that pinpoints the crisis facing the modern world:

> In December 1987, two prizes were awarded in Stockholm: the Nobel Prize for economics was given to Robert Solow of MIT for his theory of growth based on the dispensability of nature. In Solow's words: "The

> world can, in effect, get along without natural resources, so exhaustion is just an event, not a catastrophe" [*Economic and Political Weekly,* Vol. XXII, No. 45, Nov. 7, 1987]. At the same time, the Alternative Nobel Prize (the popular name for the Right Livelihood Award), instituted 'for vision and work contributing to making life more whole, healing our planet and uplifting humanity', honoured the women of the Chipko movement who, as leaders and activists, had put the life of the forests above their own and, with their actions, had stated that nature is indispensable to survival.[1]

The above quotation juxtaposes two cultural viewpoints which have come to be associated with a Cartesian, dualistic perspective and a holistic view. These dominant cultural patterns, as is well known, have shaped human behavior and have impinged on ecological issues as well. For instance, Hinduism works its values around the individual, not as opposed to, but as related to everything in the cosmos, including in its orbit the sun, the moon, the planets, and the stars. Ideally, in Hinduism, life is regulated by the notion of *dharma,* a concept which is overriding and which governs all aspects of life and which is basically a friendly, ecological approach to nature.

While there is no One God in Hinduism to whom obedience is mandatory, as in the Abrahamic religions, the moral and physical order that is intuitively realized in the heightened vision of the wise *ṛṣi*s is spoken of as *ṛta/dharma*. Since *ṛta* and *dharma* deal with the totality of moral and physical order, it is but logical to take within their embrace the entire universe and all its inhabitants. Thus, though rooted in a religious milieu, *dharma* has never been confined to the sphere of ritual and religious actions alone, nor only to religious literature.

Dharma

Dharma appeals to the conscience for its different interpretations and therefore has great flexibility in being understood and practiced in accordance with the necessities of different situations.[2] Lessons in *dharma* were handed down over generations through the art of storytelling: through the enactment of stories from the epics (the *Rāmāyaṇa* and the *Mahābhārata*) and the *purāṇa*s, as well as through dance and drama by professional and folk theater groups. Both religious and secular literature promoted the regulating principle of *dharma* in its many forms. This widespread presence of the *dharma* paradigm sensitizes the reader to its arresting presence in the culture as a whole.[3]

We can divide Sanskrit literature broadly into that which is secular and that which is religious, depending on the meaning put on the word *dharma* in each of these contexts. In a religious milieu there would be an overemphasis on the transitoriness of worldly existence. Thus, *dharma* would advocate withdrawal from overinvolvement with the world (*saṃsāra*) and it would emphasize what is called *nivṛtti dharma*. *Nivṛtti* stands for this withdrawal underlying the notion of "less of the world" in preparation for ultimate liberation. Secular literature, on the other hand, has dealt largely with *pravṛttidharma,* or activity in the world in accordance with *dharma*.

Forests and Movements

Reverence for trees and, by extension, for all that is in nature is advocated in Hinduism. In this context, it is interesting to observe that movements in the West, such as Greenpeace International, largely draw their support from that section of society which, due to reflection based on education, is able to respond to the long-term defects of a Cartesian developmental model. There is also support for such movements from feminists for a different reason altogether, as they are fighting the dominant patriarchal paradigm of thought in society. Movements like the Chipko in India, on the other hand, have been supported by so-called illiterate villagers, mainly women rooted in *dharma,* who somehow could relate to the destruction of the forests as the killing of their own kith and kin. In fact, the very word "Chipko," meaning "clinging" or "hugging" (in this case clinging to trees in order to save them from loggers), has been used to name a movement that predates the twentieth century, a fact not generally understood. The premise behind the movement goes back to the Bishnoi (Vishnoi) community in a village in Rajasthan, to some time in the fifteenth century of the common era, when Jambhoji of Bikaner, prompted by his love for nature, prohibited the killing of wild animals and the felling of trees. His followers are the Bishnois, who have continued this tradition all these years. It is said that in 1753, three hundred and sixty-three men, women, and children sacrificed their lives protecting their trees by hugging them. Amrita Devi, the woman who led them in this heroic struggle, is immortalized and figures prominently in the legendary folklore of the region.[4]

Nature and the Healing Tradition of Ayurveda

The science of healing in ancient India was called Ayurveda (knowledge of long life). This medical system is rooted in the theory of "natural balance," a balance within and with the laws of nature. Ayurveda views all diseases of the body as caused by an imbalance of the natural humors *vāta* (wind), *pitta* (bile), and *kapha* (phlegm) in the body and traces even mental illness to the activities of the *guṇa*s (qualities) of *rajas* (passion, energy) and *tamas* (darkness, lethargy).[5] A holistic approach to health aims at restoring the balance of the humors through therapies "based on spirituo-ritualistic devices as well as the relevant medicinal prescriptions."[6] Mental health is also assured by "spiritual and scriptural knowledge, patience, memory and meditation."[7] One clearly sees the emphasis on the interactive nature of physical and mental well-being in what is known as "health" in Ayurveda. "Life (*āyus*) is a productive and dynamic aggregate of sense organs, mind, body and self . . . ," says the *Caraka Saṃhitā*.[8]

The activities of the three humors result in the production of *rasa* (nutrient fluid), which has to be preserved with great care. *Rasāyana* (rejuvenation therapy, rejuvenate health care) and *vājīkaraṇa* (revitalization therapy) depend on the consumption of proper food, while living a life in accordance with *dharma* will ensure a long, healthy life. "The principle idea underlying these two therapies is the presupposition that appropriate diet (*āhāra*), when supported by a prescribed code of moral behavior (*vihāra*) will activate the rejuvenatory force (*vāja*)."[9] This rejuvenation takes place continuously through the consumption of proper food that maintains the balance between the three humors.

When the balance is upset, help is sought in natural products that are capable of supplying the necessary ingredients to reset the equilibrium between the three humors. Thus, deficiency in the body is corrected through the foods eaten and specially commissioned to restore the balance of the humors. The three humors are, in turn, affected by the seasonal changes of the yearly solar and monthly lunar movements. Nature is thus not just part of an ecological outlook but is embedded in an intimate relationship with the general well-being, both physical and mental, of humans. The same truth is described in the following verses from the *Jñānabhaiṣajyamañjarī* (hereafter JBHM) of Gumani, combining it with herbal prescriptions for many ailments:[10]

> Delusions are removed when the mind has devotion (to God) along with faith, just as a decoction of *śuṇṭhī* (dry ginger) mixed with honey gets rid of fever due to excessive *vāta*.
>
> Discernment that discriminates between Truth and Falsehood alone is well known for the destruction of worldly misery, just as a decoction of *Parpaṭa* (*Fumaria Vaillantii*) can cure fever caused by excessive *pitta*.
>
> Mental equanimity combined with contentment pacifies worldly misery always, just as a decoction of *vibhītika* (*Terminalia belerica*) mixed with *pippalītaṇḍula* (*Piper longum*) removes fever caused by excessive *kapha*.
>
> Knowledge combined with detachment destroys at once the threefold (worldly) misery, just as the decoction of *guḍuci* (*Tinospora cordifolia*) mixed with *Kaṇṭakārī* (*Solanum indicum*) cures the fever caused by the excess of all the three *doṣas* (*vāta, pitta,* and *kapha*).
>
> If sense control along with detachment cannot get rid of worldly bondage then even *harītaki* (terminalia chebula) mixed with honey cannot cure *viṣamajvara* (irregular chronic fever).
>
> The knots of (bondage of) the heart cannot be broken without knowledge based on the meaning of the *mahāvākyas*. (So also) debilitating (chronic) fever cannot be cured without the foam of cow's milk.

The JBHM and other works on Ayurveda mention many plants and herbs used in the treatment of various ailments. Among them are *kuṭaja* (*Holarhena antidysentrica*) and *indraja* (the roots of *kuṭaja*) for diarrhea; *agniśikhā* (*Plumbaga zeylanica*) and *nāgaramoṭha* (*Cyperus rotundus*) for piles; *trikaṭu* (ginger, *Piper longum,* and black pepper) for indigestion; *yavānikā* (*Hyocymus niger*) for removing worms from the stomach, and so on.[11] The healing properties of *haridrā* (turmeric), *śatāvarī* (*Asparagus recemosus*), *niṃbā* (*Azadirachta indica*), *vāsa, dhānyaka* (coriander seed), and *kesara* (*Mesua forrea*) are well known.[12] Rooted in nature, Ayurveda has very few adverse effects and thus is becoming quite popular worldwide as an alternative medical system.

Ayurveda, in turn, crosses boundaries and draws on a philosophical notion of the balance of the three *guṇas*—*sattva, rajas,* and *tamas*—to ensure physical and mental good health. The Sāṅkhya and Yoga philosophical systems also emphasize the equilibrium of the three *guṇa*s in their metaphysical thought. Patañjali's *Yogasūtra,* though discussing an esoteric science, brings the whole of nature into its discourse. The *siddhi*s, or supernatural powers that come to a *yogī* on practice of *saṃyama* (*dhāraṇā, dhyāna,* and *samādhi*),[13] are illustrated using different examples from nature itself. Thus, *saṃyama* is advocated with reference to animals, such as an elephant, and toward the sun, the moon, the polestar, and space.[14]

Haṭhayoga, which specializes in various poses, has a large number of postures modeled on animals and other living beings. Some of these are the cat pose (*mārjārāsana*), the peacock pose (*mayūrāsana*), the cock pose (*kukkuṭāsana*), the tree pose (*vṛkṣāsana*), the lotus pose (*padmāsana*), and the lion pose (*siṃhāsana*). These *āsana*s acknowledge the uniqueness of each living thing and accord each being a status of equality with humans; these amazing *āsana*s could only have been created through close observation of, attention to, and respect for nature as a whole.

Animals in the *Pañcatantra*

The *Pañcatantra* is in some ways a political treatise written for the instruction of princes. The instruction is carried out by animals and the teaching is all about good or wise conduct (*nīti*). Management of life in the lived world, with all its risks, requires tremendous skill and ingenuity. The story of the monkey and the crocodile, for instance, is a lesson about all the greed seen in the world and the skills needed to get out of difficult situations. The story is of a monkey, called Raktamukha, who lived in a tree always laden with sweet, juicy jumbu fruits. Once a crocodile, called Karālamukha, visited that place and Raktamukha entertained him as a guest (*atithi*), offering him the fruits of the tree. The two became great friends and met daily under the tree. Karālamukha would also carry back some of the sweet fruits to his wife. One day, eating the nectar-like fruits, the wife said, "The heart of the person who eats these nectar-sweet fruits constantly must indeed be nectar itself. If you want me to continue as your wife, get me the heart of that monkey; if we eat that heart, we shall live happily forever."[15] In spite of all his persuasions against these evil thoughts, Karālamukha's wife insisted on eating the monkey's heart, failing which, she threatened to commit suicide. Karālamukha then hatched a plot very reluctantly, and told Raktamukha that his wife had scolded him for not returning the hospitality of his friend by bringing him home and entertaining him. He assured the monkey that his home was on a sandbank, and therefore it would not be a problem for him as he would carry the monkey there on his back. The monkey agreed and climbed onto the back of the crocodile. Once in the deep waters, the crocodile decided to divulge the secret plan as he knew that the monkey was now his prisoner. When Karālamukha told him that his wife wanted to eat Raktamukha's heart, the monkey did some quick thinking and said, "O friend, if you had only told me earlier that my friend's wife needs my heart, I would have brought it along with me. I normally keep it safely hidden in a cavity of the tree. So you have brought me without my heart to no purpose."[16] Hearing this the crocodile was delighted and, saying, "Please give me your heart so that my wife will not commit suicide," he carried the monkey back to the tree. Once safely back in the tree, the monkey ridiculed the crocodile for his foolishness and broke off his friendship with Karālamukha.

This story is found in the chapter called "Loss of what one has" (*Labdhapraṇāśam*) due to greed and foolishness. Viṣṇuśarman narrates these stories of worldly wisdom using animals and birds as varied as lions, bulls, jackals, pigeons, snakes, tigers, cranes, crows, cats, and many more. Humans and others share the world equally in the *Pañcatantra,* and they are all governed by the same natural laws. We are reminded of Kālidāsa's *Raghuvaṃśam,* in which the reason king Dilīpa is without an heir is attributed to his showing disrespect to Surabhi, the divine cow; to compensate for that lapse, Dilīpa and his queen, Sudakṣiṇā, look after Nandinī, Surabhi's daughter, at great risk to their lives and are then forgiven for their earlier mistake.[17]

Philosophical Forests (*Āraṇyaka*s)

Forests, fauna, and flora figure in all aspects of life for a Hindu. Nature (*prakṛti, pṛthvī*) as a whole and forests in particular are interwoven into the lives of Vedic sages as spaces wherein much of the thinking and teaching was conducted. We have the *Āraṇyaka*s of the Vedic period, books which were composed in the forests, as the word *āraṇyaka* (belonging to the forest) seems to suggest. These *araṇya*s were peaceful resorts where sages like Yājñavalkya lived and maintained their *āśrama*s, both for personal contemplation and for the instruction of students (*brahmacārī*s).[18] Some daring philosophical ideas were perhaps debated in these surroundings before they culminated in the holistic vision of the Advaita Vedānta of the Upaniṣads.

> Thus the Āraṇyaka represents the forest portion of the Brāḥmaṇa and points to the development of forest life as an institution by itself in the social life of the community. It is to the forest life and to the solitary little sylvan seats of learning that . . . we must chiefly ascribe the depth of speculation, the complete absorption in mystic devotion by which the Hindus are so eminently distinguished, and, accordingly, we find the Āraṇyakas bear this character impressed upon them in a most marked degree.[19]

It would be plausible to believe that, but for the forests, Hindu thought would have been deprived of the unique contribution to world philosophy in the form of Advaita.[20]

To continue with Vedic thought, we find the same reverence for nature in all of its extant literature. Nevertheless, it finds a special place in the *Pṛthvīsūkta* of the *Atharva Veda,* which, in its sixty-three verses, emphasizes the mutuality of interests between humans and Mother Earth.[21] Later we find that both the epics, the *Rāmāyaṇa* and the *Mahābhārata,* have a large part of their action set in natural surroundings. They include chapters entitled the *Araṇyakāṇḍa* and *Vanaparva,* respectively. I provide here a couple of examples from the *Rāmāyaṇa* that illustrate the effortless way in which nature seems to blend itself into the consciousness of the people.

The *Araṇyakāṇḍa* of the *Rāmāyaṇa*

In the *Araṇyakāṇḍa,* Sītā confronts Rāvaṇa, using many examples from nature to contrast him to Rāma. Thus, she says:

> The same disparity exists between Śrī Rāma and yourself as does between a lion and a jackal in the forest, between a brook and a sea, between nectar and *sauvīraka* (a sour drink), between gold and base metal, between sandal paste and mud and between an elephant and a cat in the jungle. The same disparity exists between you and Śrī Rāma as does between a crow and Garuḍa, between a diver bird and a peacock and between a vulture and a swan.[22]

Rāma describes with great tenderness the beauty of Pañcavaṭī, where he decides to stay.[23] The *parṇaśāla,* or hermitage, which is constructed at Pañcavaṭī brings out the close relationship with Mother Earth. Thus, it says:

> Lakṣmaṇa erected a lovely and excellent abode with a wall of mud supported on good pillars and roofed with long bamboo sticks, looking very attractive and charming; thatching it with boughs of *śamī* tree, it was fastened with strong cords and was also well covered with blades of *kuśa* and flowers of *kāśa* grass and reeds and the floor was well levelled.[24]

The *purāṇa*s also abound in examples pointing out the importance of trees and comparing the planting of trees to the merit of begetting progeny.[25] According to the *Bhaviṣya Purāṇa,* providing fruits, flowers, leaves, and the shade of the trees will also lead to liberation.[26]

Nature in the Arts

If we now turn our attention to the domain of the arts—music, dance, and drama—which are all found in Bharata's *Nāṭyaśāstra,* we realize the holistic approach that Bharata adopts in his work. For Bharata, "All the cosmos is a stage"—taking liberties with Shakespeare—and he deals with the seven islands that were known at the time and with all living creatures in the world.[27] The Sanskrit plays, unlike plays of the Greek, treat the whole universe as their canvas and action takes place in all the three realms of the sky (*antarikṣa*), the nether regions (*pātāla*), and the earth (*pṛthvī*). Bharata was conscious of the sacredness of Mother Earth and the violence done to her during any entertainment. He therefore laid down rules for propitiation to the earth before the commencement of a play. He also cursed with dire consequences, anyone who did not worship the sacred space before starting a play.[28]

The dances that developed, based on the hand gestures of the *Nāṭyaśāstra* (*Hastābhinaya*), thus have the entire universe as their sphere. These gestures (*hasta*s or *mudra*s) became so sophisticated in classical dance that in the style known as Kathakali, there are almost five hundred *mudra*s pressed into service for depicting life as a whole, including various flora and fauna. When we look at the meters of the poetical works, we realize that many of them are constructed on the movement of birds and animals. Thus, *gajavilasita* (sport of an elephant), *vṛṣabhaceṣṭita* (action of a bull), *śārdūlavikrīḍita* (sport of a tiger), *aśvalatita* (horse's grace), *bhujaṅgavijṛmbhita* (opening like a snake), *mattamayūrī* (intoxicated peahen), *vasantatilaka* (gem of spring), *kusumitalatāvellita* (tree encircled with creeper blooming with flowers) *śaśikalā* (disk of the moon), *vāsantī* (belonging to spring), and many others share, in some way, the natural domain which they try to evoke. One can argue that it is this sensitivity to nature that led the composers of these meters to name them in this manner.

Literature and Nature

Having taken this sojourn into territories that overlap literature, let us try and define what it is that one understands by the term literature. Literature by its very definition has to do with imaginative writing and must have an emotional appeal in order to survive as good literature. Only literature that addresses values dear to the heart of the audience it speaks to can survive through the ages. Thus, all works of good literature "are acts of writing that call forth acts of reading" to use Jacques Derrida's phraseology.[29] This includes all literature that conveys some dominant meaning to the receiver. It is possible to argue that literature should not be read on the assumption that it conveys a dominant meaning. On the other hand, it is possible to imagine that there is a dominant meaning in every literary piece, but the way in which each reader internalizes that meaning can be different. In such a view, it is left to the reader to conclude what interpretation to put on the meaning of the work. Since one cannot overlook cultural specifics and differing ways of perceiving things, it is also reasonable to assume that literary works represent different worldviews and also convey the values held dear by that group of people. Therefore, for the purposes of this paper, which deals with Sanskrit literature, I shall hold that most Sanskrit literature conveys the dominant worldview of *dharma,* which is *ecologically friendly*. It is also important to define the kind of human being who can fit into the ecologically friendly universe. It is the kind of individual whom we might call an "eco-person" and whom the Hindus call a dharmic individual: one who can live in harmony with nature as a whole. The stress here is on the inner transformation of an individual, a transformation which allows the sharing of the planet and its resources in a sense of mutuality. E. F. Schumacher and others have stressed the need for this inner change, and Ken Jones "places special emphasis on the belief that, in the final analysis, no solution to environmental problems is viable which does not recognize the need for an inner transformation to take us beyond our current ego-centered ideologies."[30]

One talks about the extinction of many rare species of flora and fauna due to environmental degradation, but one never thinks of that rare species of humans who learned to live in harmony (with others) and the whole nature. This involves a sensitivity to the feelings of others and to nature. Today, we are addressing the problem of ecology from the wrong end; for instance, we talk about preserving the ozone layer, not because that is the right thing to do, but because the thinning of the ozone layer will lead to all kinds of cancer; thus, the emphasis is again on the human being. The Cartesian duality is still present in this discourse, and nature as an object is being manipulated within an ecological framework to accommodate the needs of humans. If the threat to the health of humans disappears, then, in this kind of thinking, humans will revert back to their earlier habits of living. One is aware of the subject/object dichotomy again creeping in through the back door. Therefore, it is imperative to recall the dharmic human being sought to be created by Sanskrit literature. Unless humans learn, or relearn, how to live in a community with consideration for all that is in nature, they will not understand the true meaning of ecological balance. The *Daśarūpaka* mentions that an ideal hero in a Sanskrit play must be a *tyāgī* (one who can sacrifice self-interest for the larger good), a *śāstracakṣuḥ* (one who is well versed in the *śāstra*s), and a *dhārmikaḥ* (one who follows the path of *dharma*).[31]

When reading Sanskrit literature, one needs to be aware of the double bind of literature and philosophy interwoven together in these writings. One is reminded of Derrida's view of the need "to grasp together the literature/philosophy couple."[32] Thus, I shall be crossing boundaries in reading literary texts which, though being singular, also serve as carriers of general philosophical notions within which are included the ecological virtues as well.

We can now say that literature has a function to play in shaping the mind. This shaping of the mind can work in two ways: first, it can create an ontic idea of what it is to be a human being in general; and second, it can distinguish an individual as belonging to a particular group (religion, culture, ethnic group, and so on). In a global society, there would ideally be only one kind of literature that fulfills the general function of creating an ideal individual having the "new ecological postmodernist view," which rediscovers a reality "that transcends cultural constructions" and has resonances with the *Advaita Vedānta Brahman* of the Upaniṣads and with the Buddhist Nāgārjuna's concept of *śūnyatā* (a state of spiritual insight).[33] In such a view, then, all literature, which includes Sanskrit literature, has a universal message to convey, which need not be culture specific. This is a view that is gaining support among postmodernist writers like Charlene Spretnak and others. "While post-modernist critics gleefully strip away layers of cultural conceptualization, they arrive at positions and offer perspectives which are in some respects central to the great wisdom traditions of the past, most especially those of the East."[34] This is still not a mainstream opinion and it has to wait, like all new ideas, to touch large groups of people. For the purposes of this paper, we can say that all literature is culture specific and thus addresses a special audience which can resonate with nuances of the literature being read. As Gayatri Spivak argues, in the context of India, the successful reader or listener is one who "learns to identify implicitly with the value system figured forth by literature."[35]

Thus, literature performs its task of transmitting values within an accepted worldview of the "implied reader," to use Spivak's words. There is, then, a symbiotic relationship between the worldview, the literature that grows and becomes popular within that worldview, and the "implied reader" that the literature addresses. Literature, besides creating an ontic idea of oneself, also has an epistemic value as the reader/listener is informed of certain ways in which to know oneself. Thus, when we examine a cross section of popular, secular Sanskrit writings, we are able to realize that the literature promotes and molds the thinking in keeping with the dharmic notion that we discussed earlier.

Daṇḍin, the famous Sanskrit literary critic of the sixth century C.E.,[36] has laid down some guidelines for Sanskrit literary works in his well-known work, *Kāvyādarśa*. According to Daṇḍin, a great literary work, called a *mahākāvya,* must necessarily describe nature in detail in order for it to qualify as a *mahākāvya*.[37] The *Kāvyādarśa* also laid down the rule that a good literary work had to deal with the acknowledged four goals of life,[38] and it further stipulates that the hero must be one having generous qualities.[39]

Kālidāsa's Śakuntalā

Returning to the purpose of literature as promoting an ontic and epistemic view of human beings, I would like to look at one work in detail, the Śakuntalā of Kālidāsa.[40] I would like to believe that the ontic view that Sanskrit literature emphasizes is that of an ultimate entity that keeps migrating into the world and out of it in accordance with the results of its *karma,*[41] until such time as it gains liberation.[42] As only a life lived in accordance with *dharma* can liberate an individual from the bondage of *saṃsāra,* it is important not to transgress the ethical and moral dimensions of *dharma*. The epistemic dimension that flows from such a worldview is to know of oneself as existing not only for one's sake, but understanding oneself as being just one among many others in the "whole." It emphasizes again and again the dharmic concept of interpersonal relationships in one's own family and in the wider society,[43] and teaches the individual to look upon herself or himself as just a small speck in the ocean of existence—but as a speck that has the obligation and responsibility to behave in such a manner that behooves one's existence as a conscious, self-aware individual. In such a view, even when the goal is economic gain (*artha*), or sensual gratification (*kāma*), it can only be within the code of *dharma*. The Śakuntalā depicts all these various dimensions of the Hindu worldview.

The Śakuntalā, like all Sanskrit literary works in general, begins with a benedictory verse.[44] The benedictory verse itself acknowledges the fact that, for a work to be completed without obstacles, there is the need for the individual to view oneself as only an instrument dependent on something outside of oneself to help in the task of completing the work undertaken.

Duṣyanta, the king, is the hero of the play, and the story is about his meeting Śakuntalā in the forest abode (*āśrama*)[45] of Kaṇva, falling in love with her, and marrying her. Back in his kingdom, he forgets her, due to a curse, and eventually is united with her when the curse is removed.

Kālidāsa begins the play with Duṣyanta hunting in the forest.[46] In the course of his hunt, he strays into the *āśrama;* but, the moment he realizes he has entered the *āśrama* area, he stops the hunt and refrains from killing any animal in the area.[47] One of the prime duties of the king is to maintain the peaceful atmosphere of the forests where sages resided. Using modern terminology, one could argue that large areas in the forests were protected reservations where killing was forbidden.[48] Thus, Duṣyanta refrains from hunting in the *āśrama* region. Though boundaries were probably not strictly drawn between *āśrama* and non-*āśrama* areas, it is possible to argue that in a climate of mutual respect and trust, such boundaries were not necessary. The fact that sages resided in the forests must have been common knowledge then, as it is even today in India. The allusions in the play[49] to noise pollution and lack of peace in the city confirm the conjecture that the forests were protected territory, for humans to practice their spiritual paths and for the care of flora and fauna.

Great pains are taken to depict the "ecoperson" in the actions of the king and Śakuntalā. Even in the most difficult situations and in moments of great passion, the characters are anxious not to cross the boundaries of dharmic behavior out of consideration for the feelings of others.

When Duṣyanta and Śakuntalā meet for the first time, their unfolding love is described, keeping in mind the sensibilities of the other characters on the stage and of the audience watching the play.[50] Similarly, when Duṣyanta is placed in the difficult situation of protecting the interests of the forest dwellers (*ṛṣi*s) and respecting the wishes of his mother, he tries to tend to both.[51] An ecoperson is one who is conscious of the interrelationship of one human with all the other humans and with all that inhabits the universe. For instance, Duṣyanta, while speculating on the reasons for the sages from Kaṇva's *āśrama* to be visiting his palace, wonders if the animals roaming in the sacred forests have been hurt or whether the growth of trees in the area has been stunted due to his adharmic deeds.[52] That the actions of one person may in some way affect others or nature as a whole is embedded in this view.[53] This, then, makes us understand why *dharma* was placed at the head of the *puruṣārtha*s, and why there is the insistence that *dharma* regulate the *puruṣārtha*s—*artha and kāma*. The *Mahābhārata* also discusses the distinction between dharmic *artha* and adharmic *artha*.[54] It is interesting to observe that, in Hindu mythology, *artha* is the son of *dharma* and *buddhi,* while *kāma* is the son of *dharma* and *śraddhā*.[55] Without this regulating principle of *dharma,* within which *artha* and *kāma* are enclosed, we would logically end up within a consumerist society having *artha* and *kāma* as the be all and end all of existence.

Having married Śakuntalā in the forest, Duṣyanta goes back to his palace, promising to send for her soon. Meanwhile Śakuntalā, who is pregnant, is lost in thoughts of the king and neglects to attend to the hospitality of the sage Durvāsas and is cursed by him. The curse was that the person Śakuntalā was thinking about would forget her because of her neglect of her dharmic duty. This was softened somewhat by adding, on the plea of her friends, that the concerned person would regain his memory upon seeing any token given to Śakuntalā, which in this case was a royal ring with the king's name inscribed on it. While the purpose of a play (*nāṭaka*) is the entertainment of all people[56] by creating different moods (*rasa*s),[57] the underlying aim, of instruction toward wisdom, is never forgotten.[58] Was it wrong for Śakuntalā to be lost in love and not attend to the sage Durvāsas? Does the curse mean that, no matter what the circumstances, one must always be the ecoperson, aware of *dharma?* This seems to be the message conveyed. Here, as Śakuntalā's one error is her unconscious disrespect for the sage, it is significant to note that the reunion of Śakuntalā and Duṣyanta takes place later in another *āśrama,* the heavenly *āśrama* abode of the sage Mārīca, a place where *dharma* reigns supreme.[59]

To continue with the story of the play: Śakuntalā is sent by the sage Kaṇva to the palace of Duṣyanta, but on the way the token ring, which would enable the king to recognize her, is lost in a pool while Śakuntalā is bathing. So Duṣyanta refuses to accept her, because of the curse, and Śakuntalā is carried away by her mother, Menaka, the heavenly nymph, to the *āśrama* abode of Mārīca. After Śakuntalā had been carried away, the ring is discovered in the stomach of a fish and is returned to the king, who then recalls all that had happened. An opportunity arises for him to go to the heavenly regions to help Indra in a battle. On his way back, in Mārīca's *āśrama,* he is reunited with Śakuntalā and their three-year-old son Bharata.[60]

Rabindranath Tagore, in his critique of the play Śakuntalā, finds the earlier love between the king and Śakuntalā tainted with physical passion; it therefore had to be purified by both the characters going through moral elevation (*tapasyā*); once they achieve this, they are granted a higher union in the hermitage of Mārīca.

A holistic vision runs through Kālidāsa's entire play. Here, nature is not just a backdrop to the characters and, as such, serving the purpose of humans. Nature is intertwined into the plot and helps the progress of the play. Kālidāsa and Shakespeare have been compared as great dramatists from two different cultures, though separated from each other by more than one thousand years, Kālidāsa being the earlier. William Jones was the first Indologist and scholar to compare the two. While Shakespeare excelled in depicting human nature and its frailties,[61] nature never takes center stage in his plays and is always a backdrop to serve human ends. Kālidāsa, on the other hand, in keeping with the *ṛta/dharma* worldview, intertwines all that exists in the world to help in the movement of the play. Nature in its totality blends into the action of the different acts, and if we were to remove nature from the scenes, the play would falter and cease to be.

The sentiment of love (*śṛṅgāra rasa*), which is the dominant mood of the play Śakuntalā, is interwoven into all aspects of nature. For instance, in the first act, Priyaṃvadā, Śakuntalā's friend, asks Śakuntalā to stand next to the *kesara* tree, for the tree, she says, now looks married; Śakuntalā is compared to a beautiful creeper that now has her mate in the strong *kesara* tree. This kind of interplay between humans, flora, and fauna comes naturally and does not in any way seem contrived. The audience also has no problem identifying with this approach as they translate such practices into their daily lives.[62]

> There is not an object in heaven and earth that a Hindu is not prepared to worship—sun, moon and stars; rocks, stocks, and stones; trees, shrubs and grass; seas, pools and rivers; his own implements of trade; the animals he finds most useful; the noxious reptiles he fears; . . . each and all come in for a share of divine honour or a tribute of more or less adoration.[63]

In keeping with the view that "All the universe is a stage" (again paraphrasing Shakespeare), the scenes in the play are set in all the worlds. Śakuntalā is introduced as a child of a heavenly nymph and a sage, and the birds (*śakunta*) are mentioned as having looked after her until she was found by the sage Kaṇva.[64] The sixth and seventh acts deal with the heavenly and atmospheric regions where, in the peaceful surroundings of Mārīca's *āśrama,* Śakuntalā is reunited with Duṣyanta.[65] Thus, the practice of moving between all worlds (known and imagined), in thought and spirit, is reiterated in the play. When the jester Vidūṣaka describes himself using the metaphor of a reed bent by the strong current of the river (that is, Duṣyanta), not only are his and Duṣyanta's characters etched in our minds, as is the close relationship that they have, but the interconnection between parts of nature is conveyed.[66]

Natural herbs, lotus stalks, and lotus leaves are used for relief from the heat, and the oil of the *ingudī* tree is used for healing wounds. The moon's rays are said to help in the general well-being of humans. Five of the seven acts are set in natural surroundings, and even the sixth act takes place in the gardens within the city.

The fourth act is considered the best because of the tender farewells exchanged all around. The anguish caused by the separation of loved ones is compared to the fading of *kumuda* flowers (lilies) once the moon has set.[67] The deities of the forest take part in the decoration of Śakuntalā by providing the ornaments she needs before she leaves for the palace.[68] When all the residents of Kaṇva's *āśrama* prepare to bid farewell to Śakuntalā, Kaṇva himself addresses the plants, asking them to permit Śakuntalā to leave, and he interprets the cuckoo's call as an answer to that request.[69] The lakes fill with lotuses, the route is lined with thick shade trees, and the path Śakuntalā is to take on her way to the palace is strewn with the pollen of lotuses;[70] thus does nature truly come alive in this act. We find the flora and fauna in the *āśrama* sharing in the sadness at parting from their dear Śakuntalā. The deer spit out the grass they are eating, the peacocks give up dancing, and the creepers weep by shedding tears in the form of leaves.[71] Śakuntalā fondly bids farewell to the doe heavy with child and asks Kaṇva to let her know when it delivers safely; her foster child, the little deer, pulls at her garment, refusing to let her go.[72]

Thus, in many ways, this fourth act is a picture of ecological ethics. The loss of Śakuntalā's ring while she bathes in the river and the fish swallowing it[73] can be contrasted to Iago contriving the stealing of Desdemona's handkerchief in Shakespeare's *Othello*. Shakespeare uses the cunning of human nature to serve his end, but Kālidāsa uses natural acts like a bath in the river and a fish swallowing the ring to fulfill the same purpose. Duṣyanta, when confronted with Śakuntalā's convincing stories, compares the untutored cleverness of the female species to the way the cuckoos cleverly place their eggs in the nests of crows.[74]

I have used the Śakuntalā as a model of Sanskrit literature. One could say that, while the style and choice of subject differ from author to author, the same worldview informs other literary works of this genre. Thus, if we look at the play *Uttararāmacaritam* by Bhavabhūti (seventh century C.E.), we find the same ontic and epistemic values being promoted: Rāma, as a dharmic king, goes to the extent of proclaiming that in order to fulfill his *rājadharma* he would give up all personal pleasures, and even Jānakī herself.[75] The *Uttararāmacaritam* is replete with natural metaphors and allusions and the entire play is set in natural surroundings. Bhāsa's plays (second century B.C.E.) also depict the same worldview. In the *Svapnavāsavadattam,* the very first scene opens with an emphasis on the *vānaprastha āśrama* practiced by both men and women, thus drawing attention to the "restraint on consumerism" referred to earlier. One finds the same values being advocated in almost all Sanskrit works, irrespective of the dates they were written.[76]

Conclusion

As I come to the end of my survey of different Sanskrit writings that reflect an ecological outlook, the question remains as to the relevance of these ideas in modern times. One is painfully aware today of pollution, large-scale corruption, and a general devaluation of the spiritual values encoded in *dharma*. The phenomenon of globalization and the explosion of media technology do not make it any easier to retain these ancient values. It is, however, reassuring to note that, in spite of the hurdles posed by modernization, there is in India a serious concern for sustainable development by effective management of natural resources. Newspaper articles and books, written in both English and the regional languages, highlight environmental ethics.[77] Attention has also been given to "geoinformation" on natural resources in order to plan "the strategy of sustainable development." The following report in a leading newspaper sheds some light on the direction that ecodevelopment is taking:

> Geoinformatic technology provides an effective tool for integration of information on natural resources and ancillary plan for sustainable development taking into account social, cultural, and economic needs of the people. . . . Under IMSD [Integrated Mission for Sustainable Development] sustainable development of land and water resources [sic] are generated on watershed bases, by integrating natural resources information generated from sattelite [sic] data with collateral/conventional information and socio-economic inputs.[78]

Another example is the international conference, entitled "Geoinformations: Beyond 2000," sponsored by the Indian Institute of Remote

Sensing (IIRS), Dehra Dun, and the International Institute for Aerospace Survey and Earth Sciences, the Netherlands, and held in March 1999 at the IIRS.[79]

In addition to the Integrated Mission for Sustainable Development, a number of other institutions, such as the Centre for Science and Environment, the Centre for Environment Education, and the Centre for the Study of Developing Societies, have spread across India, using both literary and other means to educate people about environmental issues.

Much of the inspiration for these institutes, for their ideas and direction, has come from the cultural roots of the land and from the literature that expresses this culture so eloquently. It is in this context that writings in Sanskrit can play a constructive role in molding the thinking of those engaged in the developmentalist process.

Notes

1. Vandana Shiva, *Staying Alive: Women, Ecology, and Development* (London: Zed Books, 1988), 218.

2. According to a story in *India Today*, 15 March 1997, Thimakka and her husband, a childless couple, were inspired to plant three hundred saplings near Bangalore and rear them as their children. Planting saplings is a dharmic duty and is still practiced as a religious duty.

Temples like the famous Tirupati Devasthānam offer saplings as *prasāda* (a kind of food distributed to devotees after it has been offered and blessed by the deity) instead of *laddus* (sweets), which again is in the same tradition. This new custom of offering saplings has come into vogue recently, side by side with the earlier custom.

3. *Dharma* is an all-encompassing concept and is flexible enough to include a wide range of behavior. There are many books on *dharma;* for an erudite discussion on the multifaceted dimensions of *dharma*, consult P. V. Kane's multivolume work *The History of Dharmaśāstra*, 2d ed., rev., 5 vols. (Poona: Bhandarkar Oriental Research Institute, 1968–1977).

4. This information came in personal communication from a friend called Vishnoi who belongs to this community and takes pride in it. The present-day Narmada valley movement, led by Medha Patkar and others, also draws its strength from a large number of villagers who are supposedly illiterate.

5. *Vāyuḥ pittam kaphaścoktaḥ śārīro doṣasaṅgrahaḥ*
mānasaḥ punaruddiṣṭo rajaśca tama eva ca.

Jñānabhaiṣajyamañjarī, ed. R. K. Sharma (Delhi: Nag Publishers, 1998), xviii (hereafter JBHM). All quotations from the JBHM are from this edition.

6. Ibid., xix.

7. Ibid.

8. See Shrinivas Tilak, *Religion and Aging in the Indian Tradition* (Albany: State University of New York Press, 1989), 64.

9. Ibid., 117.

10. *Bhaktirvibhāvyamānā vidhunoti śraddhayānvitā moham*
Anilajvaramiva śuṇṭhī kvathitā madhudhārayopetā.
Sadasadviveka ekaḥ prakṛṣṭabhavatāpavāraṇaḥ prathitaḥ
Pittajvarāpahārī kevala iva parpaṭakvāthaḥ.
Śamayati śamaḥ sametaḥ santuṣṭyā santatam bhavakleśam
Kvāthaḥ kalidrumotthaḥ kaphajvaram sopakulya iva.
Vidyā viraktiyuktā tritāpatāpam kṣaṇādapākurute
Amṛtānvitā bṛhatyā jvaramiva doṣatrayodbhūtam.
Uparatirasaṅgagarbhā yadi hi samutsārayenna saṃsāram
Viṣamajvaram na jarayettarhi śivā kṣaudrasaṃpannā.
Na mahāvākyapadārthajñānādanyacchinatti hṛdgranthim
Na ca gopayojaphenādaparam jīrṇajvaram hanti. JBHM 2–7.

11. Ibid., 8, 9, 11, and 12.

12. Ibid., introduction.

13. *Saṃyama* is an intensive internal yogic practice combining attention (*dhāraṇā*), meditation (*dhyāna*), and identity or oneness with the object of medita-

tion (*samādhi*). This is intended to transfer the properties of the object of *saṃyama* onto the *yogī*. See *Yogasūtra* 3.4 (hereafter YS).

14. YS 3.24, 26–28.

15. *yah sadaivāmṛtaprāyāṇīdṛśāni phalāni bhakṣayati tasya hṛdayam amṛtamayam bhaviṣyati. Tadyati mayā bhāryayā te prayojanam tatastasya hṛdayam mahyam prayaccha. Yena tadbhakṣayitvā jarāmaraṇarahitā tvayā saha bhogānbhunajmi. Pañcatantra of Viṣṇuśarman,* ed. M. R. Kale (Delhi: Motilal Banarsidass, 1965), 190.

16. *bhadra yadyevam tatkim tvayā mama tatraiva na vyāhṛtam yena svahṛdayam jaṃbūkoṭare sadaiva mayā suguptam kṛtam . . . tvayāham śūnyahṛdayo'tra kasmādānītaḥ*. Ibid., 193.

17. Kālidāsa, *Raghuvaṃśam* 1.76, 79.

18. Cf. *Bṛhadāraṇyaka Upaniṣad* and *Chāndogya Upaniṣad.* See also R. K. Mookerji, *Ancient Indian Education* (Delhi: Motilal Banarsidass, 1989), and W. Cenkner, *A Tradition of Teachers* (Delhi: Motilal Banarsidass, 1983).

19. Mookerji, *Ancient Indian Education,* 151.

20. M. Hiriyanna, *Outlines of Indian Philosophy* (London: G. Allen and Unwin, 1951).

21. *Atharva Veda* 12.

22. *Yadantaram siṃhasṛgālayorvane yadantaram syandanikāsamudrayoḥ*
surāgryasauvīrakayoryadantaram tadantaram dāśarathestavaiva ca
yadantaram kāñcanasīsalohayoryadantaram candanavāripaṅkayoḥ
yadantaram hastibiḍālayorvane tadantaram dāśarathestavaiva ca
yadantaram vāyasavainateyayoryadantaram madgumayūrayorapi
yadantaram haṃsakagṛdhrayorvane tadantaram dāśarathestavaiva ca.

Vālmīki *Rāmāyaṇa* 3.47.45–47. (The translation is from Kalyana Kalpataru, *The Vālmīki-Rāmāyaṇa,* 4, no. 12 (January-December 1963).

23. Ibid., 3.15.10–19.

24. *parṇaśālām suvipulām tatra saṃghātamṛttikām*
sustaṃbhām maskarairdīrghaiḥ kṛtavamśām suśobhanām
śamīśākhābhirāstīrya dṛḍhapāśāvapāśitām
kuśakāśaśaraiḥ parṇaiḥ suparicchāditām tathā
samīkṛtatalām ramyām cakāra sumahābalaḥ
nivāsam rāghavasyārthe prekṣaṇīyamanuttamām.

Ibid., 3.15.20–23. (The translation is from Kalyana Kalpataru, *The Vālmīki-Rāmāyaṇa,* 4, no. 12 (January-December 1963).

25. *aputrasya hi putratvam pādapā iha kurvate*
yatnenāpi ca viprendra aśvatthāropaṇam kuru.
The trees are projeny of those who have no projeny
Therefore try to plant *aśvattha* even with great effort.
Bhaviṣya Purāṇa, Madhyamaparva 1.10.37.

26. *Vāpīkūpataḍāgāśca udyānapravahāstathā*
punaḥ punaśca saṃskāryaḥ labhate mauktikam phalam.
(One who repeatedly constructs lakes, wells, ponds and gardens will attain the fruit of liberation.) Ibid., 1.9.24–25. See Kala Acharya, "Ecology in Purāṇas," in *Purāṇa-Itihāsa-Vimarśaḥ,* ed. R. I. Nanavati (Delhi: Bharatiya Vidya Prakashan, 1998).

27. Bharata, *Nāṭyaśāstra,* 1.1.17.

28. Ibid., 2.42; 3.14–100.

29. Jacques Derrida, *Acts of Literature,* ed. Derek Attridge (London: Routledge, 1992), 2.

30. Ken Jones, in J. J. Clarke, *Oriental Enlightenment: The Encounter between Asian and Western Thought* (London: Routledge, 1997), 175.

31. The three qualities that characterize the Vedic view are: detachment, which leads to sacrificing the sense of ego and self-interest; sound learning, which alone can result in this sense of sacrifice; and the implicit knowledge that only one who is ecologically related to everything in the universe will be able to live a dharmic life.

32. Derrida, *Acts of Literature,* 13.

33. Clarke, *Oriental Enlightenment,* 221.

34. Ibid., 220.

35. Gayatri Spivak, "The Burden of English in Orientalism," in *Orientalism and the Postcolonial Predicament: Perspectives on South Asia,* ed. Carol A. Breckenridge and Peter van der Veer (Philadelphia: University of Pennsylvania Press, 1993).

36. Like most authors of Sanskrit works, Daṇḍin's date is not known, but he lived before the date of the Aihole inscription (634 C.E.). Literature is generally translated as *kāvya* in Sanskrit, and it stands for the three categories of prose, poetry, and drama. *Kāvyādarśa,* 1.11. It is important to know that there is a long tradition of aesthetics and literary criticism in Sanskrit.

37. *Kāvyādarśa,* 1.14 and 16:

> Here is stated the qualities of a *mahākāvya* connected by *sargas* (sections)
> With homage to divinity and statement of contents.
> Adorned with description of cities, oceans, mountains, seasons, moonrise, sunrise
> Sports in gardens, pools, drinking scenes, festivals and enjoyment of love.

38. Collectively called the *puruṣārthas*, they are *dharma, artha, kāma,* and *mokṣa.* In general, we can translate them as ethical behavior in all situations, economic pursuit within the code of *dharma,* gratification of sensual pleasure within the code of *dharma,* and freedom from the trammels of worldly bondage by transcendence through spiritual insight. These goals are linked singly and severally with the four social institutions: the student's stage of life (*brahmacarya*), the householder's stage (*gṛhastha*), a preliminary training in renunciation by withdrawing into the forest space (*vānaprastha*), and finally, total renunciation in quest of spiritual realization (*sannyāsa*).

39. The plot from either the epics or the *purāṇas*

> Or some equally good source must be
> Describe *puruṣārthas* fourfold
> With hero loftyminded, wise. *Kāvyādarśa,* 1.15.

40. Goethe, the German poet, who read Śakuntalā in translation, was struck by the nature-embedded personality of its heroine Śakuntalā; his appreciation of the character, in translation, reads as follows:

> Would'st thou the young year's blossoms and the fruits of its decline
> And all by which the soul is charmed, enraptured, feasted, fed,
> Would'st thou the Earth and Heaven itself in one name combine
> I name thee O Śakuntalā! And all at once is said.

Quoted in *Tagore's Śakuntala: It's Inner Meaning* (Bengali); English rendering by Jadunath Sarkar (n.p.: Das Gupta and Binyon, n.d.).

Kālidāsa is supposed to have lived any time between the first century B.C.E. and the fifth century C.E.

41. The *karma* theory in conception has a moral dimension, by making each person responsible for the results of her/his action, either in this life or in future lives to come. Thus, its corollary is the transmigration of the self (*ātman*) through many lives until liberation is attained.

42. Most philosophical systems in India, whether they believe in a permanent entity or not, subscribe to the notion of a state of liberation which is not of this world but is a state of peace and tranquility. This state can be achieved even while still alive, as in the schools of Advaita Vedānta and Sāṅkhya-Yoga and in Buddhism and Jainism.

43. This interconnectedness is brought forth by the following Vedic *mantra:*

May there be peace in the sky and space, May there be peace on earth
May there be peace in the waters, May there be peace in the plants
May there be peace in the trees, May there be peace amongst the gods,
May there be peace in Brahman, May there be peace everywhere
May there be peace, peace, peace.

44. This is called a *nāndī* and is the tradition of starting any Sanskrit work by paying respects to the favored deity and the audience and it also invokes blessings for the successful completion of the work begun.

45. The division of space between that of the cities, where the king ruled, and that of the forests, which were sacred spaces for spiritual growth, is well brought out in the play.

46. Not all forest space was reserved for the sages. Some of it was used for hunting, as we see here, Śakuntalā, act 1.

47. All translations from the Sanskrit are my own unless otherwise indicated. A sage draws the attention of Duṣyanta to the sacred spaces thus:

O king, this an *āśrama* deer, it should not be killed
Withdraw therefore your wellstringed bow
To protect the distressed and not to
Hurt the innocent is the use of the bow. Ibid.

48. A good description of a sacred space is found in Bhāsa's play *Svapnavāsavadattam*. Bhāsa lived about the second century B.C.E.:

Deer roam fearlessly due to confidence generated by sacred space
Branches of trees carefully protected are heavy with flower and fruit
Abundance of brown spotted cows and no cultivated fields
Widespread smoke around—undoubtedly this is a penance grove. *Svap.* 1.12.

49. One of the sages accompanying Śakuntalā to the city looks on the city crowded with people as upon a house on fire. Another sage goes further and says:

Cityfolk attached to pleasures I view
Like one who has bathed sees one anointed with oil
Like a clean one sees him who is dirty
Like a wide awake person sees him who is asleep
Like a free one sees him who is bound. Śakuntalā, 5.10–11.

50. Vulgarity of expression and gesture are strictly forbidden in a Sanskrit literary

work. See *Kāvyādarśa,* 1.62. Also see *Bharata's Nāṭyaśāstra* 22.229, where he states that a play is witnessed by father, son, daughter-in-law, and mother-in-law in company, so such acts that are immodest and which bring shame should not be enacted.

51. Śakuntalā, act 1.

52. Has the penance of the great sages been hindered by obstructions
Has anyone behaved improperly with the roaming animals of the *āśrama*
Has the yield of fruits by trees been stopped by my misdeeds
Thus my mind beset with doubts, anxious, is unable to reach a decision.
Ibid., 5.9.

53. Nature is embedded in the life of a Hindu so much so that even the new year and other festivals are important solar and lunar landmarks. See *The Cultural Heritage of India,* vol. 4, ed. Haridas Bhattacharya (Calcutta: Ramakrishna Mission, Institute of Culture, 1983), 479–90.

54. *Mahābhārata, Śāntiparvan,* and *Aśvamedhikaparvan* in particular.

55. *Moral Dilemmas in the Mahābhārata,* ed. B. K. Matilal (Shimla: Indian Institute of Advanced Study, 1989), 58, 60.

56. This Art of Dramaturgy has been created by me
For the purpose of recreation for those suffering from pain
For those tired, undergoing mental anguish
For those engaged in austerities. *Nāṭyaśāstra,* 1.114

57. Love, Humour, Compassion, Terror, Heroism, Fear, Disgust, Wonder are the well-known eight Moods (*rasas*) of a play. Ibid., 1.116

58. This Art of Dramaturgy depends on the actions
Of superior, inferior and middling humans
It entertains, generates courage, gives pleasure
Its purpose is to instruct good behaviour (*dharma*). Ibid., 1.113

59. Śakuntalā, act 7.

60. The name Bhārat (Bhārata) for India is supposed to be taken from this Bharata.

61. For example, *The Merchant of Venice, Julius Ceasar, Hamlet, A Midsummer Night's Dream,* and *As You Like It.*

62. One is reminded of the festival of Pongal in South India where, the day after Pongal, cattle are bathed, decorated with ornaments, fed with choice foods, and paraded in the streets; and of the way homes and the spaces in front of homes are decorated with rice powder (so that birds, ants, and other insects can feed on the decorations).

63. Monier Monier-Williams, *Brahmanism and Hinduism,* 4th ed. (New York: Macmillan, 1891), 350. Cited in *Cultural Heritage of India,* ed. Bhattacharya, 479.

64. Śakuntalā, act 1.

65. Ibid., act 7.28 and 29.

66. O friend if the reed is bent is it due to its own volition or due to the current of the river? Ibid., act 2.

67. Ibid., act 4.3.

68. One forest deity gave a silken garment white as the moon
Another brought forth red pigment soft for the feet
Others with arms extended up to the wrists from trees
In beauty rivalling fresh leaves gave beautiful ornaments. Ibid., 4.5.

69. She who would not drink water till you have drunk
She, though fond of decoration would not pluck tender leaves out of love
That Śakuntalā goes to her husband's home
Please permit her to do so. Ibid., 4.9.

Reply:
Śakuntalā has been allowed to leave
By these trees her forest relatives
By their answer through the sweet
Voice of the cuckoo thus. Ibid., 4.10.

70. Ibid., 4.11.

71. Ibid., 4.12.

72. That little doe whose face hurt by the tip of grass
You treated with wound-healing *iṅgudī* oil
Whom you reared with handfuls of grain
That one, your foster son, obstructs your path. Ibid., 4.14.

73. Ibid., acts 5 and 6.

74. Untutored cleverness of the female
Is seen even amongst non-human species
What to say of the human?
The cuckoo before its young ones take to flight
Gets them reared by other birds. Ibid., 5.22.

75. *Sneham dayām ca saukhyam ca yadi vā jānakīmapi*
Ārādhanāya lokasya muñcato nāsti me vyathā
Love, Compassion, Happiness, even Sītā
Abandon I shall for the sake of pleasing my subjects.
Uttararāmacaritam, 1.12.

76. Some popular works besides Kālidāsa's are: *Kirātārjunīyam* of Bhāravi, *Daśakumāracaritam* of Daṇḍin, *Harṣacaritam* and *Kādaṃbarī* of Bāṇabhaṭṭa, *Nāgānandam* and *Ratnāvalī* of Śrī Harṣa, *Mudrārākṣasa* of Viśākhadatta, *Mṛcchakaṭikam* of Śūdraka, and *Śiśupālavadham* of Māgha.

77. For example, see the five volumes entitled *Prakṛti: The Integral Vision,* published by the Indira Gandhi National Centre for the Arts; writings by Vandana Shiva, such as *Staying Alive;* many works in translation by the Sahitya Akademi; and such works in Malayalam as *Kāvu Tīṇḍile* by Sugata.

78. *Hindustan Times,* 25 February 1999.

79. Ibid.

Reading the *Bhagavadgītā* from an Ecological Perspective

LANCE E. NELSON

> The evocation of the Ecozoic era requires an entrancement within the world of nature in its awesome presence.
>
> —Brian Swimme and Thomas Berry[1]

> They have conquered nature, even here. . . . Therefore, they are established in Brahman.
>
> *Bhagavadgītā* 5.19[2]

Introduction

Just as the feminist revolution has initiated a far-reaching examination of religion for elements that contribute to oppression of women, so the ecological crisis is forcing a new examination of our most cherished religious beliefs and practices. Although the process of reconfiguring culture in the direction of gender equity is not nearly complete, we have at least been trained to avoid sexist language when we write—and to ask questions inspired by feminist analysis whenever we approach religious texts, teachings, and practices. We must now also begin, just as automatically, to ask far-reaching ecological questions of the religious traditions. Just as feminism requires of the religions a rigorous, and sometimes painful, self-examination, so our newfound ecological awareness demands, at least as urgently, a serious criticism of the religious faiths and practices that so profoundly motivate us as human beings. One hopes that such assessments will come primarily from within the traditions themselves. But, especially in view of the urgency of the situation, there should be no objection to respectful questions being asked from without.

When, in reference to the ecological crisis, we suspect that religion may be part of the problem as well as part of the solution, the questioning—as in feminist hermeneutics—should be searching and fearless. When such questioning is undertaken, moreover, the results may be disturbing. But the study of religion and ecology has reached a point, and the external situation we face has become so severe, that a mere superficial searching of the religious traditions for ecofriendly "resources" will not suffice. In the context of ecology, the very term *resources* is of course problematic.[3] More important, the quest for resources obscures the need for the more demanding task, from which it may only distract us. We cannot avoid the need to examine the fundamentals of religious worldviews to determine their impact on ecology. This means that basic categories and deep structures—along with primary scriptures—cannot remain immune from the questioning process.

One must have a standpoint from which to ask questions, so I assume in this paper—without here attempting to justify them, and recognizing that they are primarily Western in provenance—the standards of the emerging ecological paradigm, as identified by Thomas Berry, ecofeminists, the eco-justice movement, proponents of deep ecology, and others.[4] These standards will emerge more explicitly from what follows, but they include notions such as overcoming anthropocentrism as well as androcentrism and recognizing intrinsic value in the natural world and in nonhuman species. They also include the necessity of acting to transform the social, economic, and ideological structures that perpetuate ecological devastation. There seems to be wide consensus among ecothinkers (as the ongoing search for ecofriendly material in the various traditions shows) that these and related ideas must become essential elements of any solution to the contemporary ecological crisis.

This process of questioning the religions in an ecological mode has, of course, already begun, and it has been shown that some among the great traditions that we honor contain elements that seem to subvert environmental goals. The religions of the West, and Christianity in particular, have been singled out in this respect.[5] But what if other religious traditions, by such standards, also fall short? And what if at least some of the elements that so offend are foundational, part of the deep structure that influences—if unconsciously—attitudes to nature in whole cultures? Such negative potentials must be brought to consciousness and examined courageously, and meditated upon with all honesty, in dialogue with the wider community of thinkers, secular and religious, who are concerned about the ecological question. Sweeping them under the rug will not do, if they are indeed antithetic to ecological awareness, for their influence is bound to remain in force, generating consequences, in the cultures that support them. As I have warned elsewhere, "Human nature being as it is, the *negative* outcomes of religious teachings that can be used to rationalize environmental neglect are probably greater than the *positive* influence of those that encourage conservation and protection."[6]

The response of Christianity and Judaism to such ecological critique has thus far been creative. While the process of assessment and reconfiguration is still ongoing, there has been—in addition to the search for ecologically helpful motifs—serious debate regarding the environmental implications of the worldviews involved and a sober holding in consciousness of elements that are ecologically dubious.[7] There is no reason why the Hindu tradition, encompassing nearly one-sixth of humanity, should be exempt from similar examination, nor is there any reason that its response should not be equally sincere, earnest, and fruitful. In this paper I will engage a foundational Hindu text, the *Bhagavadgītā* (*BhG*), asking ecological questions. It will be seen that, while the *Gītā* contains elements that could lend significant support to ecological consciousness, it also carries much that is problematic.

In the present discussion, I do not address the much-debated question of whether the *Gītā*'s message is univalent or multivalent, coherent or contradictory.[8] I undertake the less ambitious task of pointing out significant and influential patterns of thinking in the text that portend important environmental consequences. As I do this, I acknowledge a keen awareness that, in reflecting on a text that is sacred to millions, I am entering on hallowed ground.

The Importance of the Bhagavadgītā

The *Gītā* contains the teachings of Kṛṣṇa, the incarnation, or *avatāra*, of the Supreme Being, delivered in intimate dialogue to his pupil Arjuna, the greatest warrior of his day, on the battlefield just before the beginning of the great *Mahābhārata* war. Although scholars have raised questions about how widely the *Gītā* was known and read in India prior to the nineteenth century,[9] it has long been commented upon and lauded in Hindu literature as a source of wisdom and salvation. Indeed, even in classical times its prestige was such that it became the model for other "copycat" texts, the *Īśvaragītā,* the *Gopīgītā,* and so on. It was considered one of the foundational scriptures of Hindu theology, being counted as one of the *prasthāna-traya,* or "threefold basis," of Vedānta. As such, it was necessary that any teacher claiming to found a school of Vedānta write a commentary on it. Muslim observers of Hindu culture al-Bīrānī (eleventh century) and Dārā Shukōh (seventeenth century) were familiar with the text.[10] In the thirteenth century, the medieval Maharashtrian saint Jñānadev wrote a vernacular rendering and commentary on the text, the *Jñāneśvarī,* that remains immensely popular to this day. With the rise of the Indian nationalist movement of the nineteenth century, the *Gītā*—especially because of its teaching of dispassionate action, or *karma-yoga*—was adopted as a source of religious support for anticolonial activism and as a counter to critiques from missionaries of what they saw as Hindu quietism and fatalism. Mahatma Gandhi hailed it as containing "the essence of *dharma*" and providing "in a nutshell the secret art of living."[11] Convinced of the profound wisdom of its teachings, the patriot and yogic visionary Sri Aurobindo declared, "The *Gītā* will become the universally acknowledged Scripture of the future religion."[12] Having been a central focus of the emergence of the modern Indian consciousness of Hinduism as one of the "great world religions," the *Gītā* today has attained an immense popularity. It is available in hundreds of translations, in Indian vernaculars, English, and other languages, and is read, recited, and not infrequently memorized by millions as part of their daily devotions. It is celebrated through communal readings, and—in the modern context—at specially organized study circles and "Gita camps" both in India and abroad. Nowadays, one may log onto Bhagavad Gita web sites and explore interactive Gita CDs.[13]

Ecological activists in India draw inspiration from the text; indeed, nonviolent campaigns sponsored by the Chipko and other movements have been undertaken to the accompaniment of recitations of the *Gītā.*[14] Widely cited in the literature on Hinduism and ecology, it has been praised as a source of environmentally sensitive ideas, both by

Hindus and by ecothinkers in the West. No less a figure than Arne Naess quotes from it in support of the vision of self-realization that is a foundation of his deep ecology.[15]

Finding Ecological Positives in the Gītā

It goes without saying that the *Gītā,* like other ancient religious texts, does not address contemporary ecological issues directly, for it was composed long before the crisis emerged. In reading the *Gītā* from an ecological perspective, as when with similar concerns we read other classical religious texts, we reframe its teachings in light of problems it was not intended to address.[16]

Much has been made, and rightly so, of the damaging ecological consequences of modern economies and the cultures they spawn, with their consumerist orientation and near religious faith in continuous, open-ended development and growth. There is no doubt that a deeply held belief in the superordinate value of a nonmaterial, spiritual reality has the potential—if rendered sufficiently compelling—to do much to undercut the ecologically damaging ethos of the dominant consumerist, growth-oriented "religion of the market."[17] It is equally clear, as I will be neither the first nor the last to point out, that the *Gītā* provides such a spiritual vision.[18] It teaches of the existence of a transcendent Supreme Being upon whom the universe is dependent for its existence. It teaches in addition of a spiritual self (*ātman*) that dwells within each person. It reminds its devotees that full awareness of the spiritual dimension is indispensable for authentic happiness, and that such awareness cannot be attained without setting strict limitations on the human tendency to live for gratification of desires, "sense pleasures," and ego-indulgence. Indeed, the *Gītā* calls for more than a limitation of wants; it envisions a demanding asceticism in which desires and egoism are to be utterly renounced.[19] There can be no doubt that, as Sadhusangananda dasa has written, the *Gītā*'s ideal is that we be "devoid of any tinge of greed, desire to control, manipulate, or exploit."[20] If an ethic of restraint is wanted, the text certainly has it.

What is interesting here, and important insofar as we are concerned about ecological activism as well as sensitivity, is that the *Gītā* does not at the same time advocate a monastic withdrawal from life, at least for the majority. The asceticism of the *Gītā* is to be combined—as in the model of Arjuna the warrior, whom Kṛṣṇa instructs not to abandon his duty—with a kind of consecrated activity in the world. This is, of course, the famous *karma-yoga* of the *Gītā*. One is to remain in one's work, abandoning egoism, desire, and attachment, acting for the sake of duty or as a form of devotional offering to the Divine. Hence the ideal of "desireless action" (*niṣkāma-karma*), for which the text is well-known. This notion of consecrated action is connected in an ecologically interesting way with the *Gītā*'s reinterpretation of the ancient Vedic sacrifice (*yajña*) and the closely related awareness of the need to maintain good relationships with the Vedic gods, the *deva*s, many of whom are associated with natural phenomena. Although the *Gītā,* as we shall see, is not consistent in its regard for the many gods that it recognizes as informing the world, it is interesting to read, for example, in chapter 3:

> The Lord of Beings, having in the beginning brought forth humankind along with the sacrifice, said, "By this you shall multiply. . . ." Nourish the *deva*s with this, and let the *deva*s nourish you. Thus causing one another to prosper, you will attain the highest good. Nourished by sacrifice, the *deva*s will give you the good things you desire. One who enjoys what is given by the *deva*s without giving in return, is certainly a thief (3.10-12).

The *Gītā* extends this sacrificial metaphor to include all human duties devoutly performed, and eventually shifts the focus from the *deva*s to the one God, Kṛṣṇa, as we shall see. Still, when one considers the strong association of Vedic deities with nature (rain, fire, wind, dawn, etc.), the parallels between this outlook and the ecologically much-prized religious sensibilities of indigenous tribal religions are worth noting.

The *Gītā*'s ideal of social engagement through spiritually disciplined action was the chief basis of the esteem in which the text was held by nineteenth-century Hindu nationalists like Balwantrao Gangadhar Tilak. The latter saw it as a "call to action from God" and sought to recommend *karma-yoga* to his contemporaries under the banner of "Energism."[21] Nowadays, Chipko leader Sunderlal Bahuguna speaks of his struggle to save the forests of the Himalaya as a *dharma-yuddha* (righteous battle), echoing *Gītā* 2.31.[22] If one sees ecological consciousness as entailing activism, as most ecothinkers do, this would appear to be good news.

Even better, it would seem, the motivation for such activity (if desireless action can be said to have any motivation) is declared to be *loka-saṃgraha* (3.20), literally, the "holding together" or "maintenance" of the world, often interpreted for modern audiences as "well-being of the world," "universal welfare," or even "altruism." Just as God acts, out of no need of his own, in order to uphold the sociocosmic order of *dharma,* so should individuals, and for the same reason. Indeed, the *Gītā* provides a powerful model of activism in the image of the Deity's voluntary descent (*avatāra*) to Earth for the very purpose of protecting *dharma* and uprooting *adharma,* the forces of disorder. The Hindu vision of *dharma* involves, as supportive of its vision of a harmoniously ordered cosmos, the idea that human beings must accept certain curbs on their desires so that this order can be preserved. Moreover, Hindu law books, the *dharmaśāstra*s, include numerous prescriptions that we can now consider ecologically friendly (such as protection of trees and rivers). The active concern of the Deity for the preservation of *dharma,* therefore, as well as the admonition to humanity to follow God in that concern—both powerfully evoked in the text—are most significant.

As an ideal, the *Gītā* presents a picture of a devout and frugal lifestyle that has inspired, and may be expected to continue to inspire, ecologically supportive lives of minimal consumption. There is no doubt, furthermore, that Hindus respond powerfully to leaders of social movements like Gandhi and, nowadays, Chipko leader Bahuguna, whose public lives are modeled on and exemplify the asceticism recommended in this scripture.[23] The *Gītā* extolls persons possessed of a "pure" (*sāttvika*) or "divine" (*daivī*) nature. Only those "rise upward" (14.17) who are "established in purity" (*sattva-stha*)—that is, who are oriented toward spirit and are above egocentrism, possessiveness, and selfish desire. It associates such persons with a diet that has been understood, though not explicitly stated, to be vegetarian (17.7–10), another presumed ecological plus. Such portraits of virtue are reinforced by powerful counterimages of "ignorant" and "demonic" persons motivated by the kind of greed and self-indulgence that could nowadays be associated with blind, consumerist acquisitiveness. At least one passage in the *Gītā,* moreover, seems to encourage a profound, Buddhist-like empathy with all other beings in the universe (6.32, on which see the discussion below); another holds up the ideal of sages who "delight in the welfare of all beings" (*sarva-bhūta-hite ratāḥ,* 5.25). Although Gandhi and others have attempted to do so, it is difficult to argue convincingly that nonviolence was a central message of the *Gītā*.[24] The reasons for this will become apparent as we proceed. Still, the Indic ideal of *ahiṃsā* (nonharming), of which much has been made in ecological writings, is mentioned several times in the text (10.5, 13.7, 16.2, 17.14), being included in lists of virtues.

Ecothinkers have lamented the West's loss of the sense of nature's sacredness. The *Gītā,* at least at first reading, provides much material that would support a sacralized view of the world. The cosmos comes forth from God: "Of all manifestations," proclaims Kṛṣṇa, "I am the beginning, the end, and the middle" (10.32). The universe is supported by God: "All this is strung on me, like rows of jewels on a thread" (7.7). Further, there is the suggestion that God may be perceived within the natural world, and vice versa: "One who has been disciplined by *yoga* beholds the Self in all beings and all beings in the Self" (6.29); "I am not lost to one who sees me in all things and sees all things in me" (6.30); "You shall see all beings in the Self and then in me" (4.35); "The *yogin* who, established in oneness, worships me as dwelling in all things, abides in me" (6.31). The divine presence in the world is also evoked in more concrete imagery. Kṛṣṇa declares:

> Entering the earth with my vital force, I support all beings, and becoming the moon, whose essence is sap, I nourish all healing herbs. Becoming the universal fire within all beings, associated with the vital breaths, I digest the fourfold food (15.13–14).

> I am the savor in water . . . the radiance in the moon and sun . . . the pure fragrance in the earth . . . the brilliance in fire . . . the life in all beings (7.8–9).

And several passages are reminiscent of the more impersonal Upaniṣadic panentheism:

> With hands and feet everywhere, with eyes, heads, and mouths everywhere, with ears everywhere—That exists pervading everything in this universe. . . . Without and within all beings, unmoving yet also moving, incomprehensible because of Its subtlety, It is both far and near (13.13, 15).[25]

In chapter 10, "The Yoga of Divine Manifestations (*vibhūti*)," Kṛṣṇa identifies himself, albeit selectively, with elements of the natural universe: sun, moon, the ocean, the Himalaya, the wind, and the Ganges. Although it contains notably few references to nature, chapter 11, "The Yoga of the Universal Form (*viśva-rūpa*)," does provide compelling intimations of the later Vaiṣṇava doctrine of the world as the body of God. "The sun and the moon," says Arjuna in wonder, "are your eyes" (11.19).

Problematics

One could continue to cull ecological "resources" from various passages in the *Gītā* and consider how they might be mobilized to encourage environmental awareness among Hindus in India and elsewhere. My argument in this paper, however, is that this will not take us very far unless we first undertake certain more demanding tasks. Extracting from this or any other text ideas that may possibly be given an ecological slant will not be enough, unless we are interested only in politeness or apologetics. The apparently ecofriendly images and practices cannot be isolated from the contexts in which they are embedded, especially from the underlying worldviews which give rise to them and condition their significance. These too must be examined.

Hierarchical Dualism

Much has been made of the "unitive vision" of Hindu scripture and theology. The Vedānta's doctrine of Brahman as the universal reality present in all things has been presented as the source of the Hindu vision of all reality as sacred and worthy of reverence.[26] Other contributors to this volume affirm that Hinduism teaches the "sanctity of all life," that "all life-forms have equal value," that there is "divinity in nature," and that there is "no dichotomy between matter and spirit." We would do well to ask: Which traditions within the Hindu fold offer such a vision? I have elsewhere examined such claims in relation to classical Advaita Vedānta and shown them to be almost completely unjustified in that context.[27] But what about the *Gītā?*

An image at the beginning of chapter 15 reveals much about the *Gītā*'s valuation of the world of nature. In a well-known and at first glance promisingly organic metaphor, the world is identified as an ancient, many branched *aśvattha* tree. The *aśvattha* (*Ficus indicus*) is the holy fig tree, no doubt. Mary McGee (in this volume) reminds us that it was held in such high regard that the *dharma* texts prescribe the severest of penalties for cutting it down. And indeed, Kṛṣṇa, earlier in the *Gītā* itself (10.26), explicitly identifies himself with this tree. But —alas for those who would grasp at its image in chapter 15 as an ecological "resource"—the great cosmic tree is upside down. Its roots are "above," in the realm of spirit, its branches "below" (15.2). More distressing from an ecological point of view, students of the *Gītā* are advised to do precisely what the *dharmaśāstra*s forbid: to sever the great tree at its roots. We read: "Having cut down this firmly rooted *aśvattha* with the mighty axe of nonattachment, seek that goal most worthy of pursuit, having attained which they do not return" (15.3–4).[28] Evidently, as we shall see more graphically further on, the *Gītā*'s concern for "preserving the world" is ambivalent. Indeed, the commentarial tradition presents us with a revealing, if unscientific, etymology. We are told that the root meaning of *aśvattha,* the tree that here symbolizes the natural world, is "not (*a*-) lasting until tomorrow (*śvas*)."[29]

The "goal from which they do not return" mentioned in the *aśvattha* passage is *mokṣa,* a "liberation" that is defined precisely in terms of escape from the world of nature. Kṛṣṇa promises his devotees their final beatitude when they attain "my being" (*mad-bhāva,* 4.9, 8.5, etc.), "my supreme abode" (*dhāma paramam mama,* 8.21), the "eternal, imperishable place" (*śāśvataṃ padam avyayam,* 18.56), the "supreme Nirvāṇa, the peace resting in me" (*nirvāṇa-paramāṃ mat-saṃsthām śāntim,* 6.15), in which there is a transcendence of the suffering of the world (*duḥkha-saṃyoga-viyoga,* 6.23).[30]

In chapter 13, "The Yoga of the Distinction between the Field and the Knower of the Field," we learn that the vision of the text is anchored in a fundamental dichotomy between matter and spirit, *prakṛti* and *puruṣa*. The realm of *prakṛti,* a term which encompasses and evokes the whole of the natural world, physical and mental, is sharply devalued in relation to spirit.[31] *Prakṛti,* which includes the five elements, is God's lower (*aparā*) nature (7.4–5). In contrast with that which lies beyond, the realm of rebirth constituted by *prakṛti* is the "abode of pain" (*duḥkhālaya,* 8.15), experienced as "transitory and unhappy" (*anityam asukhaṃ,* 9.33). Indeed, the author of the *Gītā* may well see the natural world constituted by *prakṛti* as in some sense an illusion, for he declares, "The unreal never is; the real never ceases to be" (2.16).[32]

In traditional Hindu thought the *guṇa*s are the "strands" or fundamental constituents of *prakṛti* that are understood to be the basic components of the manifest world. Thus, invoked collectively, "the *guṇa*s" becomes a designation for matter or nature as a whole. The *Gītā* repeatedly evokes the *guṇa*s as symbolic, in addition, of life's fundamental problem, identifying them—and *prakṛti*—with *māyā,* the delusive appearance that must be transcended in order to attain liberation. The sage of the *Gītā* is one who is *guṇātīta,* "transcendent to the *guṇa*s," thus:

> The Vedas are concerned with the three *guṇa*s; become free of the three *guṇa*s, free from the pairs of opposites (2.45).

> By these three states of being consisting of the *guṇa*s this whole universe is deluded. . . . This divine *māyā* of mine, consisting of the *guṇa*s, is hard to get beyond. Those who surrender themselves to me alone, cross over this *māyā* (7.13–14).

Association with the world of nature causes the pure spiritual self (*ātman*) to experience itself as the *dehin,* or "embodied one,"[33] which condition is understood as being lamentable, the paradigmatic form of bondage from which escape is an urgent desideratum:

> *Sattva, rajas,* and *tamas*—these *guṇa*s, born of *prakṛti*—bind the imperishable embodied one (*dehin*) in the body. . . . The embodied one, having gone beyond these three *guṇa*s, of which the body is composed, attains immortality, freed from birth, death, decay, and pain (14.5, 20).

Like the pure spiritual self, or *ātman,* the Deity is beyond the "darkness" of *māyā* (8.10), by which the Divine Being is "veiled" (7.25).

As already mentioned, there are some verses in the *Gītā* that suggest the possibility of a world-sacralizing vision. In particular, 4.35 and 6.29–31, which proclaim the possibility of seeing "the Self in all beings and all beings in the Self," have been mentioned above. Arne Naess cites 6.29 as supportive of the "philosophy of oneness" associated with his ecologically inclusive conception of Self-realization.[34] Jacobsen has convincingly shown, however, that none of the classical Vedānta commentators have understood the verses in this way.[35] While modern readers, Hindus or non-Hindus, might justifiably see ecological potential in these texts, it seems important to follow Jacobsen's lead and consider how Hindus have traditionally read the *Gītā*. A complete study of the commentarial traditions—as representative of the various schools of Hindu theology—and the respective potentials of those schools for ecological awareness, must remain beyond the scope of this paper. Nevertheless, it will be instructive to look at how 4.35, 6.29–31, and a number of other possibly "ecofriendly" verses of the *Gītā* were received by representative commentators.

Śaṅkara's reading says much about his valuation of the experience of Brahman but gives no suggestion whatsoever that the particulars of the natural world are to be valued. When 4.35 says, "You will behold all beings in the Self and then in me," Śaṅkara glosses, "The point is that you will see the oneness of the Self and God, as proclaimed in all the Upaniṣads."[36] No nature mysticism there; no mention of the world at all. But this, given Śaṅkara's radical orientation toward transcendence, is not surprising. We might anticipate more from Rāmānuja, whose powerful metaphor of the universe as God's body is typically cited as a source of the world-sacralizing vision of Hinduism.[37] In these passages —where, if anywhere, such a sensibility could find sustenance—he gives no suggestion that nature, or any of its components, is to be perceived as sacred. According to Rāmānuja's interpretation of 6.29, the enlightened sage sees, not the empirical forms of all beings rendered transparent to God, but rather their transcendent, spiritual selves (*ātmans*) as having the same essence when they are separated from matter (*prakṛti-viyukta*), or again, "freed from the defect of association with *prakṛti*" (*prakṛti-saṃsarga-doṣa-nirmukta*), so that their true spiritual constitution can be realized.[38] While this makes sense from the standpoint of traditional Hindu spirituality, it is a far cry from the new ontology of identification of humanity and nature associated with ecological thought. Only the Gauḍīya Vaiṣṇava commentator Viśvanātha Cakravartin comes close to a reading with ecological potential: "You will see all beings—humans, animals, etc.—abiding separately in the Self as adjuncts. And then you will see [them] abiding in me, the supreme cause, as effects."[39]

Another example worthy of discussion is *BhG* 5.18, which reads: "The wise see the same [reality] in a Brahmin endowed with learning and culture, a cow, an elephant, a dog, and an outcaste." Are we here moving beyond anthropocentrism toward a biocentric outlook, in which all beings are regarded as of equal value? The fact that Hindu theologians understand the self or "soul" (*ātman*) within all living beings to be qualitatively identical—and that some, like Śaṅkara, see all selves as metaphysically one—is often cited as evidence of an egalitarian spiritual outlook in Hinduism. The related concept of reincarnation, in which the same soul may appear in different forms, human and nonhuman, has likewise been offered as implying an "organic solidarity between humanity and nature."[40] But are such ideas really to be found in the *Gītā* or the interpretations of its classical commentators?

Here is Śaṅkara's reading of the passage:

> In a Brahmin, in whom *sattva* predominates and who has the best latent mental impressions (*saṃskāra*s), in an intermediate being like a cow, which is dominated by *rajas* and is without [such] impressions, and in [beings] such as elephants, which are wholly dominated by inertia [*tamas*] alone—those wise ones are "equal-visioned" whose habit is to see equally the one immutable Brahman, which is completely untouched by *sattva, rajas,* or *tamas,* or any impressions generated thereby.[41]

Certainly Śaṅkara knows all beings to be the same on the level of the Absolute. He recognizes their oneness in Brahman, but does he abandon consciousness of difference and hierarchy? In his gloss he mentions the holy Brahmin, the middling cow, the inertia-dominated elephant. What happened, one wonders, to the poor dog—and the outcaste (*śvapāka,* literally, "dog-cooker")? Surely, there is a hierarchy to be discerned in Śaṅkara's thinking, and even in the list presented by the *Gītā* itself. What can we make of the fact that the Brahmin is placed first and the outcaste is last, with (or beneath?) the dog, an animal considered extremely polluting by Hindus? Here we run up against the stratification in Hindu thinking about which Anil Agarwal (in this volume) expresses concern.

The idea that nonhuman beings have exactly the same *spiritual* potential as humans—having an *ātman* that is qualitatively the same, with the same potential, in principle, for emancipation—is often cited as a move toward biocentrism. But is it possible that this creates a situation in which some humans are able to construe other human beings as *below* nonhuman animals in social-ritual stature and consequent worth and dignity?[42] (We will be reminded below that the *Gītā* presents Kṛṣṇa as the author and champion of the system of *varṇa*s, or caste-categories.) Abhinavagupta, the Śaiva commentator, acknowledges that the sage sees all beings equally. "But," he says, "they do not behave equally toward them."[43] Viśvanātha Cakravartin, the Gauḍīya Vaiṣṇava commentator, regards Brahmins and cows as being alike in the highest class of beings (*sāttvika-jāti*), while the dog and the outcaste are together at the bottom.[44] This is an interesting kind of solidarity between humans and animals, no doubt, but it is not one in which those concerned with social justice would find much value.

Rāmānuja approaches the text from a slightly different angle:

> Although the selves (*ātmans*) are being perceived in extremely dissimilar forms, the wise know the selves to be of uniform nature. . . . The dissimilarity of the forms is due to *prakṛti,* and not to any dissimilarity in the self.[45]

Here, he does not, like Śaṅkara and the other commentators, recall the hierarchy while seeing the Brahman beyond it. He simply dismisses *all* the modalities of nature as incidental. *Prakṛti,* the material "stuff" of which all natural entities are composed, is again devalued, along with all of its myriad expressions.[46] It is the self (*ātman*) that is important, not nature.

The above should be enough to indicate that we are here encountering, not the "unitive vision" of Hinduism, but rather the same hierarchical, fundamentally dualistic outlook, involving the same elevation of pure spirit above matter, for which feminists and ecothinkers have for some time been faulting Judaism and Christianity. The overarching theism of the *Gītā* carries the potential of resolving this dualism (see, for example, 15.16–18), but it is doubtful whether this resolution is achieved in the text, though some commentators, especially Aurobindo,[47] have seen it there. Too much of the *Gītā*'s discourse is dominated by the language of ancient Sāṃkhya, with its bifurcation of spirit and matter. The self (*ātman*) of the *Gītā,* and of Hindu thought in general, is not the ecological self comfortably constituted by its inalienable connections and interrelationships with the web of life. It has, ontologically speaking, nothing to do with nature. It is a pure spiritual substance whose association with things natural is an epistemological error—and an unnecessary burden. Although a number of writers have tried to find in *karma*-theory an ecofriendly sense of "interconnectedness" with nature, the ties of *karma* are in fact read negatively by the tradition as primary constituents of a situation of bondage (*nibandha*).[48] This dualistic trend of thought is not just a product of the systematic formulations of the commentators. It pervades the text, as we can see from the *Gītā*'s devaluation of the *guṇa*s and *prakṛti,* discussed above.

Any doubt that this is the case can easily be resolved by a reading of chapter 2, where certain disconcerting implications of the text's spirit-matter dualism are graphically evoked. There Kṛṣṇa articulates a well-known string of arguments designed to persuade Arjuna to overcome his qualms against fighting in the *Mahābhārata* war. We learn that the spiritual self is far beyond any merely physical dangers:

> Know that, by which all this is pervaded, to be indestructible (2.17).
>
> It is never born, nor does it ever die. . . . Unborn, eternal, everlasting, ancient, it is not killed when the body is killed (2.20).
>
> It cannot be cut, nor can it be burnt, wetted, or dried (2.24).

We are taught here that only the ignorant imagine that violating the physical existence of other creatures constitutes an assault on their true being, for spirit is unchanging, immutable, untouchable. (This also means, we must suppose, that the real self is impervious to damage from toxic waste, polluted air, and so on.) Again, the implication is that *ātman* is what is important. The physical, on the other hand, is expendable, certainly not worth any emotional distress. "You mourn for those who should not be mourned" (2.11), says Kṛṣṇa. Again: "Knowing this, you should not sorrow" (2.25). And yet twice again:

> Beings are unmanifest in the beginning, manifest in the middle, and unmanifest again in the end. What is there to grieve about in this? (2.28)
>
> The one that dwells in the body of all is eternal, impossible to slay. Therefore you should not mourn for any creature (2.30).

This is an argument to justify killing, at least for warriors engaged in battle. Once one has abandoned the need to mourn—that is, any concern for the merely empirical existence of other beings—one can go into battle and slay the enemy with equanimity (*samatva*). It is true that these verses give solace and comfort to untold numbers of devout believers facing death and tragedy in contexts that have nothing to do with justifications of war. They point powerfully to a transcendent dimension beyond the limitations and pain of our empirical existence.

One must acknowledge with respect their function on that level. But their final implication is that physical harm—whether the destruction of war or, presumably, ecological devastation—however regrettable on the empirical level, does not affect what ultimately matters, namely, spirit.[49]

It has been argued that the Hindu concept of *ātman* (whether singular or plural) reincarnating in different forms, animal and human, supports—in addition to the kind of "biocentrism" already considered—an ethic of *ahiṃsā,* or nonviolence, toward all beings.[50] I would certainly not want to deny that *ahiṃsā* is an important theme in many Hindu texts; it is an important ethical value in the tradition, especially for ascetics. As indicated above, *ahiṃsā* is mentioned several times in the *Gītā* itself, albeit with minimal emphasis, in lists of virtues associated with the yogically advanced sage. The main thrust of the *Gītā*'s discourse, however, runs very much counter to *ahiṃsā.* Reading the argument in chapter 2 in light of our present concern raises a number of disturbing questions about this ethical ideal, much cited in writings on ecology and Indic religion. Important among them must be the suspicion that to extol *ahiṃsā* as an ecological virtue may be to ignore exactly what this text reveals, namely, that *ahiṃsā,* as a value, is articulated for the most part out of concern for the private karmic well-being of the actor. Compassion for others, while a significant factor, is secondary. After all, as we have just seen, damage to the bodies of beings cannot damage their spiritual selves. The disconnection between act and result achieved through *karma-yoga,* on the other hand, is designed precisely to insulate Arjuna from any negative consequences of killing. The one who is free of egoism and attachment, we are told, "even if he kills these people, neither kills [in reality] nor is bound by the act" (18.17).[51] No doubt Gandhi, with some awkward hermeneutical maneuverings, took profound inspiration from the *Gītā,* but so did Nathuram Godse, his assassin.[52] Are we perhaps here touching upon the root of the spiritual "individualism" in India that concerns Agarwal (this volume)? Whether or not the *Gītā*'s analysis helps us understand the final meaning of the *ahiṃsā* doctrine in Hindu ethics,[53] it remains a forceful statement that the physical, including the empirical existence of other beings, does not matter.

At 6.32, however, the *Gītā* does articulate a notion that comes close to a vision of universal empathy: "When one sees the pleasure and pain in all beings as the same in comparison with self . . . one is considered the highest *yogin*." This verse has already been cited as a potential ecological positive. Śaṅkara, to his credit, glosses the verse rather straightforwardly, offering a reflective basis for compassion: "Just as for me pain is both disagreeable and undesired, so is it for all living beings (*sarva-prāṇināṃ*)."[54] Viśvanātha Cakravartin takes a similar tack. But Rāmānuja (remember: the sponsor of the eco-friendly doctrine of the world as the body of God) brings us up short with a stunning reversal of this interpretation:

> One who is cognizant of the similarity between one's own self (*ātman*) and that of others—as being pure, uncontracted knowledge—and who therefore sees everywhere both pleasure, as at the birth of children, and pain, as in their death, to be the same, because of the sameness the disconnection (*asaṃbandha*) [of all selves from both pleasure and pain], . . . that person is considered the highest *yogin*.[55]

Swami Adidevananda, a Ramakrishna monk, explains:

> The idea is to prevent misconstruing the verse as meaning that one shares the joy and misery of all [beings] as his own. It means only that the highest type of yogins understand that the self is unrelated to the pain and pleasures of his own body-mind. He understands also that the same is the case with other selves.[56]

Again, the insignificance of merely empirical calamities is underlined. This may be spiritually elevating for some, but it is ecologically unnerving. If one is to remain unaffected by the death of one's own children, how likely is one to be gripped by a more remote concern, the degradation of the environment?

Related to the hierarchical spiritual dualism of the *Gītā* is its assertive push in the direction of a single-pointed monotheistic devotion. Discussions of religion and ecology in India have tended to look with favor on the world-sacralizing potential of Vedic polytheism, which envisions a multiplicity of deities inhabiting and informing the worlds of nature, with whom humans are enjoined to form mutually beneficial relationships.[57] Indeed, I have already quoted *Gītā* 3.10–12 in this connection. Not only, however, does the *Gītā* say relatively little to suggest the sacral value of nature, but when it does, the source of sacrality is not the indwelling nature deities, but Kṛṣṇa, the one Supreme Being (chapters 9–10). Putting aside the section in chapter 3 already referred to and two verses in chapter 17 (4, 14), the Vedic *deva*s and the religion that cherishes them are, in the majority of passages in which they are mentioned, quite deliberately diminished. In two places, Kṛṣṇa repeats: "The worshiper of the *deva*s goes to the *deva*s; my devotees come to me" (7.23, 9.25). The *deva*s do not know the origin of the one God (10.2, 14); indeed, they are astounded and tremble at the vision of the transcendent Lord of All (11.21–22). Those who worship the *deva*s have "puny intellects" (*alpa-medhas,* 7.23). This is not surprising, since—like every other aspect of nature—the *deva*s are correctly regarded as mere products of *prakṛti.*[58] It is often remarked that Western monotheism has worked the "disenchantment" of nature through its suppression of the deities of nature. A similar process may be detected here, in the heart of "polytheistic" Hinduism.

Antinomian and Deterministic Tendencies

The effort to focus on—and ensure the transcendence of—a single, ultimate reality has lead to an additional tendency that may be worrying from an ecological perspective. In the Upaniṣads and in the *Gītā,* as elsewhere in the tradition (especially in Tantra), there is a marked drift toward an ultimate amoralism (or perhaps transmoralism) in the absolute realm, one that may not bode well for ecological awareness. The knower of Brahman, we discover, is beyond good and evil. Thus, at *Bṛhadāraṇyaka Upaniṣad* 5.14.8 we read: "One who knows this, even if performing much evil, having devoured it all, becomes pure and clean, undecaying and immortal."[59] We have already considered the *Gītā*'s instructions on how to kill with karmic impunity. In what follows, I will not, of course, be arguing that the *Gītā,* when rightly understood, sanctions overtly immoral or destructive behavior. I will be suggesting, however, that the *Gītā* contains elements that could easily be misappropriated to justify diminished moral concern about many things, including the environment.

No doubt, as has been pointed out abundantly by scholars writing on Hinduism and ecology (including Mary McGee and Ann Grodzins Gold in this volume),[60] there is a strong *dharma* tradition in India, one that has much to say about the everyday morality of protection and care of the natural world. But the emphasis on *dharma* is ultimately relativized, and therefore undercut, by the tradition's final *mokṣa* orientation.[61] This relativization of *dharma* is well-illustrated in the

Gītā. Consider 18.66, valorized by the Vaiṣṇava tradition as the *carama-śloka,* the "final stanza" of the text, embodying the essence of its teaching: "Having abandoned all *dharma*s, come to me as your sole refuge. I will liberate you from all sins." The Vedānta commentators are quick to step in with a kind of exegetical damage control, suggesting that this abandonment of *dharma* includes the giving up of *adharma* also. Nevertheless, the idea that *dharma* is secondary to complete abandonment to God is here articulated powerfully. (It is elaborated in the later image of Kṛṣṇa as the adulterous lover of the *gopī*s, a symbol of the God who, as the author of *dharma,* is above it. And, of course, there are related moves in Tantra.)

The *Gītā* teaches, on the one hand, that the enlightened sage is one who has "abandoned both good and evil" (*śubhāśubha-parityāgī,* 12.17) and, on the other, that even the worst malefactors can transcend their *karma* by knowledge and devotion to the Lord (4.36, 9.29). Just as the self cannot be harmed by weapons, so also is it untouched by the effects of any action, positive or negative. It is like the lotus leaf, which floats above, impervious to mud or water (4.14, 5.10, 9.28). "As the all-pervading ether, because of its subtlety, cannot be defiled," we are told, "so the Self, present everywhere in the body, cannot be defiled" (13.32). Likewise, the perfected *yogin* "incurs no evil from doing mere bodily action" (4.21).[62]

In this connection, it is worth remembering Tilak's argument, in his *Gītārahasya,* that the enlightened sage, having attained a state totally free of ego and desire, is beyond the moral law. Tilak can therefore argue that the ethically dubious strategies used by his nationalist hero Shivaji Bhonsle (1627–1680) to overcome his Muslim antagonists cannot be judged by ordinary standards. "Great men," he declares, "are above the common principles of morality. These principles do not touch the ground on which they stand."[63]

If one asks "What is that ground?" the answer, it seems to me, must begin with the *Gītā*'s teaching that the *karma-yogin* should imitate the divine mode of activity revealed by Kṛṣṇa as God (3.22–26). Much of the Supreme Lord's *modus operandi* is shown in chapter 11, the "Universal Vision" (*viśva-rūpa-darśana*) that so terrifies Arjuna. Ecologists concerned with preserving the planet may wonder about these images of a God moving toward world devastation. "Time am I," says Kṛṣṇa, "mighty worker of the world's destruction, turned forth here for the dissolution of the worlds."[64] A frightened Arjuna exclaims:

> As moths rush recklessly into a blazing fire to perish, so indeed do these worlds rush recklessly into your mouths to perish.
>
> With your flaming mouths swallowing all the worlds on every side, you lick your lips (11.29–30).

Here, as Madeleine Biardeau has pointed out, Arjuna's vision is that of one who would observe *pralaya,* the great dissolution of the universe at the end of each cosmic cycle, from some privileged position beyond it. Interestingly, it is just this section of the *Gītā* (11.15–33) that is recited by Hindu monastics as they take their final vows of *saṃnyāsa* (renunciation), which constitute a kind of death to the world.[65] Why not choose such a death, or at least cultivate a thoroughgoing detachment, if the world itself, as directed by God, is inevitably going "down the tubes"? The *Gītā* provides powerful images designed precisely to detach its students from historically oriented hope. God is the source and support of all beings, no doubt, but his concern for *loka-saṃgraha,* the maintenance of the world, only goes so far. Kṛṣṇa reveals himself to Arjuna primarily as the Lord of Dissolution.[66] In the great crush of time, he "devours" the universe (13.16). Like the Tao of the *Lao tzu,* that looks upon all beings as "straw dolls," Kṛṣṇa's ultimate Being is situated beyond limited human concepts of good and evil, and is unrestricted thereby.[67]

The ethical ambiguity of this vision of the final awesome power and transcendent amorality of the Deity (and the universe) is complexified by the elements of determinism that one finds in the *Gītā,* as elsewhere in the *Mahābhārata.*[68] And here we discover the ultimate ground upon which Tilak's "great man" stands. In that awe-inspiring vision of chapter 11, we learn that Arjuna, and through him all human beings, are mere instruments (*nimitta-mātra,* 11.33), subject to the irresistible will of the Deity. Arjuna is told several times that he will be forced to kill his opponents, even if he wills otherwise (18.59–60), or—alternately—that they will die even if he chooses not to act. The outcome has already been ordained by God (11.32–34). Says Kṛṣṇa, "The Lord, O Arjuna, abides in the heart of all beings, causing them to revolve by his mysterious power (*māyā*) as if they were mounted on a machine" (18.61). Viśvanātha Cakravartin correctly sees this as an echo of the Upaniṣadic vision of God as the Inner Controller of all (*sarvāntaryāmin*): "Scripture teaches that the Lord dwells in the heart as the Inner Controller. Doing what? Causing all beings to revolve, to perform their various actions, by *māyā,* [his] innate power."[69] The *Gītā* repeatedly stresses that our sense of individual authorship of actions is an unwarranted conceit (3.27–29, 4.20, 13.29, 14.19, 18.16). The idea of agency (*kartṛtva*) is a notion one ought to abandon. It is the Lord, through his *prakṛti,* who is the universal Doer (*kartṛ*). In this vision, as Will Johnson points out:

> Individual choice is essentially an illusion, since God is the only real chooser, God is the only real actor. . . . Kṛṣṇa subsumes within himself both fate and agency, and as a result all that is left to humans is correctly to attribute the results of their apparent actions to the real agent.[70]

This implies that the moral responsibility of the empirical individual, as constituted by *prakṛti,* is diminished—at least from the final, *mokṣa*-perspective—to that of an instrument, while the accountability of the person's true self (*ātman*) is reduced still further, to zero. Thus, again, the logic effectively insulates the actor from the consequences of action. The Śaiva commentator Abhinavagupta, in his gloss on *BhG* 18.66 ("Having abandoned all *dharmas*, come to me as your sole refuge"), ties the notion of God's sole agency together with other arguments we have considered:

> Having given up the ascription to yourself, "This incidental (*prāsaṅgika*) killing of relatives, etc., in the act of battle, is all my doing," and having mentally abandoned the notion, "Sin will accrue to me, because the killing of teachers, etc., is prohibited," take refuge in me alone, the one Doer of all [action] (*ekam sarva-kartāram*), the Supreme Lord.[71]

This outlook—separated, of course, from its particular use here to justify the killing of war—becomes enshrined as an important foundation of Hindu spirituality. No less than three times in the *Gītā* (3.30, 12.6, 18.57), Kṛṣṇa teaches his followers to "relinquish all action to me" (*sarva-karmāṇi mayi saṃnyasya*). As S. S. Rama Rao Pappu observes, "In Indian religious practice, it is not uncommon for individuals to formally renounce their actions by saying, 'I give away my actions to Sri Kṛṣṇa' as soon as they are performed and before actions start producing consequences."[72] One must also ask whether, along with its negation of individual responsibility, this vision of God as the sole real actor suggests—perhaps even more disturbingly—that the way the world now is (ecological crisis and all), is precisely the way it should be. Are we to regard everything in the world, perhaps even ecological devastation, as part of the cosmic play, or *līlā,* of the Divine?

Apatheia

Given the knower's vision of the ultimate amorality of the universe and the unreality of individual creative initiative, it is not surprising that the *Gītā* encourages its devotees to cultivate toward the world, not the engaged concern that ecological ethics would look for, but an attitude of *samatva*: "sameness" or equanimity. The ethics of the *Gītā* is an ethics of complete detachment (*asakti,* 13.9).[73] Again, Kṛṣṇa himself sets the example; he is detached (*asakta*), "like an indifferent witness" (*udāsīna-vat,* 9.9). *Yogin*s should realize that they, like God, have nothing to gain and nothing to lose (3.18, 22). They should not disturb the world, to be sure, but neither should they be disturbed *by* the world (12.15). Kṛṣṇa says (12.16) that he cherishes those who are "independent" (*anapekṣa*), "indifferent" (*udāsīna*), "untroubled" (*gata-vyatha*). The ideal is "constant even-mindedness (*sama-cittatva*) in the occurrence of the desirable and the undesirable" (13.9). This kind of cultivated indifference leads *yogin*s to regard "a lump of earth, stones, and gold," as well as good and evil persons, as "the same" (14.24). Having abandoned all desire and all hope (*vigata-spṛha,* 2.56; *nirāśī,* 4.21) and given up all intentionality (*saṃkalpa,* 6.2, 6.24), the enlightened are content with whatever experiences life provides (4.22). They have attained the well-known ideal of *karma-yoga:* disengagement from the outcome of one's actions (2.47, 2.48, etc.). The results of activity have all been given over to Kṛṣṇa, who, as we have seen, is the only actual agent.

As an essential feature of the *yogin*'s attempt to be free of any mental agitation, as well as the last vestiges of materiality, the *Gītā* recommends a final disassociation from *sattva,* the most refined of the *guṇa*s, the one that, when dominant in the mind, makes possible both spiritual awareness and moral sensitivity. "Because of its purity, being luminous and free of all ill," we are taught, "*sattva* binds by attachment to happiness and by attachment to wisdom" (14.6). Again, we hear the praises of:

> One who neither despises light, activity, or delusion [i.e., *sattva, rajas,* or *tamas*] when they appear, nor desires them when they are absent. . . .
>
> One who, seated like an unconcerned witness, remains unmoved by the *guṇa*s, who . . . firmly established, is not disturbed (14.22–23).

In order to attain the vantage point of the "knower," one must give up the bondage of even *sattvic* sensibilities, including what might in this context count as the distressing emotionalism of environmental concern.

Static Social Vision

At this point, one could begin to wonder in what sense Hindu reformers like Tilak and Gandhi could have found in the *Gītā* a call to action. This question must be particularly important, of course, to those among us with an environmental agenda, since out of genuine ecological concern must issue forth, it would seem, concrete action to transform social, economic, and even well-established ideological structures.[74] It is well known that the teacher of the *Gītā* does not, despite all the talk about nonattachment and equanimity, recommend passivity. This point has already been made and, I hope, vividly so. Kṛṣṇa urges his pupil, and through him all who would hear, to remain engaged in action and to eschew the temptation to renounce the world in the hopes of attaining a higher degree of spirituality. But what kind of action? This is the key question. The world engagement envisioned by the *Gītā* is of a carefully circumscribed kind: action to support the conventional social order of *dharma,* a hierarchical system the norms of which were defined and enforced by the elite. As Robert Minor points out, the *Gītā* "sought to maintain the structure which would be reinforced by the law books, the *dharmaśāstras,* that would soon appear."[75] Says Kṛṣṇa: "Let the *śāstra* be your authority in ascertaining what ought to be done and what ought not be done. Having known what is said in the ordinance of *śāstra,* you should act here" (16.24). The action that the *Gītā* requires is the action "ordained" (*niyata*) by scripture and social convention: "Perform obligatory action" (*niyataṃ karma,* 3.7); and again, "The renunciation of obligatory action is not proper" (18.7, see also 18.9, 18.47). This saves the teaching from the complete social unworkability of its final antinomian vision: a conventional pattern of behavior is provided and required, even if its rationale is only provisional. But it does not take us very far in the direction of the transformative activism that an adequate response to the ecological crisis seems to call for.

The action given by sacred authority in this case is, above all, the action determined by one's birth in a particular caste category (*varṇa*). Kṛṣṇa is the author of the *varṇa* system (4.13) and sanctions it as a natural expression of the inherent natures (*svabhāva*) of different human beings (18.41), which are determined by their previous existences and their present birth (*paurva-dehika,* 6.43; *sahaja,* 18.48). The whole justification of Arjuna's participation in the bloody battle, as well as the inconceivability of him renouncing this calling, is predicated on fixed conceptions of caste duty. Our hero, as a *kṣatriya,* or member of the warrior caste, is of course duty bound to fight. "Better is one's own [caste] *dharma,*" Kṛṣṇa tells him, "than the *dharma* of another" (3.35, elaborated at 18.45–48). The recurrence of such themes in the text has caused authors such as Surendranath Dasgupta to conclude that "the fundamental idea of the *Gītā* is that a man should always follow his own caste duties."[76]

The world of the *Gītā* was an outgrowth of the ritually constructed cosmos of the Vedic period. As Biardeau has pointed out, the Brahminical creators of that world sought "the realization of a sociocosmic order which promise[d] it eternity or an everlasting renewal."[77] Kṛṣṇa in the *Gītā* is the "undying Guardian of the Eternal Dharma" (*avyayaḥ śāśvata-dharma-guptā,* 11.18). The *yogin*s who, imitating him, act for *loka-saṃgraha* are not acting for "world welfare" as it might be conceived by individual moral agents acting creatively out of conscience, and they are certainly not being called to challenge dominant social and economic conditions for the sake of a better future. They are acting, "without attachment," for the "holding together" (*saṃgraha*) of what the classical texts see as a preordained order. To be sure, the maintenance of general well-being is included in this vision. There is, moreover, the potential of powerful motivations to action if ecological evils can be framed as threats to the dharmic order. Arjuna must, after all, fight *adharma* in the shape of his enemies, the Kauravas. But what if elements of the system of *dharma* itself are implicated, not surely in the origin of the present ecological crisis, but at least in its perpetuation and exacerbation? The *Gītā* provides no mandate for the kind of profound changes in established social, economic, and ideological structures that ecothinkers are saying that we need.

Conclusion

The *Gītā* offers a spirituality that is profound, even awe-inspiring. While it is often claimed that the text makes salvation available to the ordinary person, in reality the main thrust of the text demands much sophisticated understanding and a radical—if this-worldly—asceticism.[78] The picture of the ideal devotee that emerges is one of utter, one-pointed commitment to the spiritual life and all-consuming devotion to God. Everything other than "constant focus on discipline and spiritual knowledge" (*abhyāsa-jñāna-nityatva,* 13.11) is ignorance. As far as one's outer life is concerned, one is called to do no more (or less) than enact, without attachment, as a kind of ritual, one's ordained social role. I have earlier in this chapter outlined a number of aspects of the *Gītā*'s spirituality that could well have positive ecological implications. The significance of these—in particular, the ideal of selfless engagement in activity (*niṣkāma-karma*)—should not be minimized. Nevertheless, the ideal that emerges is hardly one of deep concern about the future of the world. The student of the *Gītā* is called to an awareness that spirit is untouched by mere empirical calamities, which are anyhow ordained by God. The instruction is: Do your sacred duty (*dharma*) well, neither craving rewards nor being anxious about outcomes, and focus your energies on the transcendent. Integration or identification with nature—the experience of humanity as inextricably interrelated with the natural world—does not emerge as a value. Indeed, as we have seen, it is undercut by the text's spirit/matter dualism.

Swimme and Berry tell us that, "The natural world itself is . . . the primary presence of the sacred, the primary moral value."[79] Nature, however, is finally irrelevant to the *Gītā*'s soteriological goals. The Earth Charter speaks of the need to foster a deep sense of belonging to the universe,[80] but the *Gītā* teaches that our true self is outside the world of nature (*nistraiguṇya,* 2.45) and that our true home is elsewhere. Indeed, its ideals are in many ways antithetical to ecological ethics as we know it. As Knut Jacobsen writes: "Environmentalism teaches neither liberation from the world nor the ultimate value of the social order. On the contrary, it has *saṃsāra,* the world of natural processes of birth, flourishing of life, decay and death as its ultimate concern."[81] Not so, most certainly, the *Gītā.*[82] Along with others, Mary Evelyn Tucker has complained of the focus in the Abrahamic religions on God-human and human-human relations, at the expense of human-nature relations. She notes further the tendency in Christianity and Islam toward a "predominantly anthropocentric redemption-orientation, where humans are concerned not so much with nature as with personal salvation from this world."[83] If we are tempted to look to Asia for religious visions that give nature greater significance, however, the *Bhagavadgītā* may not be the best place to start.

I am aware that I am here raising questions about a text that is cherished by millions as the center of their spiritual lives, and that the doubts I am articulating may be disturbing. I am aware, further, that I have touched upon certain issues that were used by nineteenth-century Christian missionaries in their anti-Hindu polemics.[84] So, on several accounts, I find this exercise uncomfortable. But the ecological situation we find ourselves in has the potential of being vastly more uncomfortable; the need of the planet is pressing. There is no point in articulating a *Gītā* "for our times,"[85] I believe, until one has seriously faced the text's profound otherness to the prevailing moods of our day and its possible disjointedness from what we may perceive to be the pressing needs of the future. Any exploration of the *Gītā*'s meaning will certainly be conditioned by present-day concerns. Still, the effort to sound the deep structures of this and other worldviews for their ecological import remains a necessary and, I think, valid endeavor.

As we confront the environmental crisis, all religious traditions, if they are to survive and continue to contribute to solutions, must undergo some degree, perhaps a considerable degree, of reconstruction. Such work cannot be undertaken without a thorough and courageous examination of foundations.[86] To list, marshal, or even carefully orchestrate apparently ecofriendly elements of any religious tradition and then declare the whole tradition ecologically sound, without having critically examined the deeper context, will produce short-lived inspiration at best. The more genuine task will not be easy; the directions in which such reconstructions must move are not immediately apparent. It is likely that we will have to carry in our consciousness for some time searching questions about our most cherished beliefs, as a kind of ecologically mandated meditation (*nididhyāsana*), until there emerge insights that will carry us in authentically new directions, still maintaining continuity with the old.

How is one to regard a sacred text that contains much wisdom, offers a deep spirituality, informs the religious life of millions, and yet teaches (as at 5.18) that we are established in the highest spiritual goal only when we have "conquered nature"?[87] What do this and the other teachings of the *Gītā* that we have considered imply for the future of India—and the earth? And what is the future of those teachings themselves? In the process of investigating these matters, Hindus will move further toward articulating their own core vision for environmental ethics.[88] Meanwhile, they will certainly not be alone in their questioning. In light of the ecological crisis, the validity of spiritualities of transcendence and detachment, in whatever religious tradition they occur, can no longer simply be assumed. Their truth must be evaluated and argued anew.

Notes

The author would like to thank the several persons who commented on drafts of this chapter. In particular, I am grateful to Linda Hess, whose suggestions—although we disagreed on many points—were extremely helpful.

1. Brian Swimme and Thomas Berry, *The Universe Story* (San Francisco: HarperSanFrancisco, 1992), 263.

2. ihaiva tair jitaḥ sargaḥ . . . tasmād brahmaṇi te sthitāḥ, *Bhagavadgītā* (hereafter cited as *BhG*) 5.9. All translations from Sanskrit in this chapter are my own.

3. See Gerald James Larson, "'Conceptual Resources' in South Asia for 'Environmental Ethics,'" in *Nature in Asian Traditions of Thought: Essays in Environmental Philosophy*, ed. J. Baird Callicott and Roger T. Ames (Albany: State University of New York Press, 1989), 269–70.

4. I do not suggest, of course, that there is complete agreement among these trends of thought, neither do I have here the space to articulate the distinctions between them (but see the chapter by Vinay Lal, in this volume). The Earth Charter is a minimal but nevertheless politically significant statement of this new value system (Earth Charter Commission, "The Earth Charter," March 2000. Available [Online]: <http://www.earthcharter.org/>.)

5. Were it not interesting and important in this context to remind ourselves that it was first published more than thirty years ago, one would hesitate to make yet another reference to the seminal article by Lynn White, Jr., "The Historical Roots of Our Ecological Crisis," *Science* 155 (10 March 1967): 1203–7. For examples of more recent work in this vein, see note 7.

6. *Purifying the Earthly Body of God: Religion and Ecology in Hindu India*, ed. Lance E. Nelson (Albany: State University of New York Press, 1998), 5–6.

7. See, for example, Mary Daly, *Gyn/ecology: The Metaethics of Radical Feminism* (Boston: Beacon Press, 1978); *Ecofeminism: Women, Animals, Nature*, ed. Greta A. Guard (Philadelphia: Temple University Press, 1993); Carolyn Merchant, *The Death of Nature: Women, Ecology, and the Scientific Revolution* (San Francisco: Harper and Row, 1980); Rosemary Radford Ruether, *New Woman, New Earth: Sexist Ideology and Human Liberation* (New York: Seabury, 1975), and *Gaia and God: An Ecofeminist Theology of Earth Healing* (San Francisco: HarperSanFrancisco, 1994); H. Paul Santmire, *The Travail of Nature: The Ambiguous Promise of Christian Theology* (Philadelphia: Fortress Press, 1985).

8. See Arvind Sharma, *The Hindu Gītā: Ancient and Classical Interpretations of the Bhagavadgītā* (La Salle, Ill.: Open Court, 1986).

9. *Modern Indian Interpreters of the Bhagavadgita*, ed. Robert N. Minor (Albany: State University of New York Press, 1986), 4.

10. Wilhelm Halbfass, *India and Europe: An Essay in Understanding* (Albany: State University of New York Press, 1988), 26, 34.

11. Quoted by J. T. F. Jordens, "Gandhi and the *Bhagavadgita*," in *Modern Indian Interpreters*, ed. Minor, 93.

12. Sri Aurobindo Birth Centenary Library (Pondicherry: Sri Aurobindo International Center in Education, 1972), 4:57; quoted by Robert Minor, "Sri Aurobindo as a Gita-yogin," in *Modern Indian Interpreters*, ed. Minor, 71.

13. *The Bhagavad Gita Page.* n.d. Available [online]: <http://www.geocities.com/Athens/Acropolis/5294/> [5 April 1999].

14. Ramachandra Guha, *The Unquiet Woods: Ecological Change and Peasant Resistance in the Himalaya* (Berkeley and Los Angeles: University of California Press, 1990), 162, 166, 170.

15. Arne Naess, *Ecology, Community, and Lifestyle,* trans. and revised by David Rothenberg (Cambridge: Cambridge University Press, 1989), 195. See also Knut A. Jacobsen, "*Bhagavadgītā,* Ecosophy T, and Deep Ecology," *Inquiry* 39 (June 1996): 219; and David Kinsley, *Ecology and Religion: Ecological Spirituality in Cross-Cultural Perspective* (Englewood Cliffs, N.J.: Prentice-Hall, 1995), 187.

16. Yet—as Vinay Lal (this volume) points out in the case of Gandhi—nature *was* present to the author of the *Gītā,* unavoidably so, as were problems of disposal of waste, despoliation of natural resources, and so on (as confirmed by the *dharmaśāstra*s' edicts regarding the same), so the text cannot remain entirely blameless if it fails to present an adequate conceptualization of nature.

17. A. Rodney Dobell, "Environmental Degradation and the Religion of the Market," in *Population, Consumption, and the Environment,* ed. Harold Coward (Albany: State University of New York Press, 1995), 229–50; see also David R. Loy, "The Religion of the Market," *Journal of the American Academy of Religion* 65 (summer 1997): 275–90.

18. See Michael Cremo and Mukunda Goswami, *Divine Nature: A Spiritual Perspective on the Environmental Crisis* (Los Angeles: Bhaktivedanta Book Trust, 1995).

19. Nicholas Lash has recently attempted a reading of the *Gītā* from a Christian perspective that argues that the text teaches "purification of desire," not its "abolition." I disagree and am dubious about how far such attempts to tame the otherness of the text will take us in our quest to deal with urgent issues such as the environment. See Nicholas Lash, "The Purification of Desire," in *The Fruits of Our Desiring: An Enquiry into the Ethics of the* Bhagavadgītā *for Our Times,* ed. Julius Lipner (Calgary, Alberta: Bayeux Arts, 1997), 1–9.

20. Sadhusangananda dasa, "The Search for Ecology in the Bhagavad-Gita: Critical Examination of an Academic Study," *On the Way to Krishna,* spring, summer, winter 1998, 6–7, 6–8, 6–8. In this series of three articles, the author, a leader in the Krishna Consciousness movement, undertakes a vigorous and comprehensive critique of a draft version of this chapter. Making reference to the Harvard conference that was the origin of this volume, he writes, "One scholar, a West coast college professor, managed to misinterpret the Gita in one of the most dangerous ways I have heard so far. . . . His interpretation was based on common misunderstandings found in academia" (*On the Way to Krishna,* spring 1998, 6). Needless to say, I am approaching the text from a very different standpoint than Sadhusangananda dasa. While I respect his views, I disagree with most of his critique. Nevertheless, I am grateful to him for his thoughtful response, from which my work has benefited. I should say, moreover, that each issue of *On the Way to Krishna* that I have seen has contained articles on ecological topics, as well as appeals for support for such valuable projects as the ISKCON Rainforest Preservation Partnership. *On the Way to Krishna* is the quarterly newsletter of ISKCON (International Society for Krishna Consciousness) of New England, 72

Commonwealth Avenue, Boston, MA 02116. See <http://www.iskcon.net/ boston/html/projects/newsletter.html>.

21. Robert W. Stevenson, "Tilak and the *Bhagavadgita*'s Doctrine of Karmayoga," in *Modern Indian Interpreters,* ed. Minor, 49–50; Bal Gangadhar Tilak, *Srimad Bhagavadgītā-Rahasya or Karma-Yoga-Sastra,* trans. B. S. Sukthankar, 7th ed. (Poona: Tilak Bros., 1986), 37. "Energism" was Tilak's translation of *pravṛtti,* "engagement."

22. "There is nothing higher for a *kṣatriya* than a righteous battle" (dharmyād dhi yuddhāc chreyo 'nyat kṣatriyasya na vidyate). See Guha, *Unquiet Woods,* 172.

23. Guha, *Unquiet Woods,* 171–72.

24. Gandhi's view was that the *Gītā* is an allegory of the conflict within us between good, represented in the story by the Pāṇḍavas (Arjuna and his brothers), and evil, symbolized by the Kauravas. He believed that the *Mahābhārata,* the epic of which the *Gītā* forms a part, was written in order to teach the futility of war. Thus: "When I first became acquainted with the *Gita,* I felt that it was not a historical work, but that, under the guise of physical warfare, it described the duel that perpetually went on in the hearts of mankind, and that physical warfare was brought in merely to make the description of the internal duel more alluring" (M. K. Gandhi, "Anasaktiyoga," in *The Gospel of Selfless Action or the Gita According to Gandhi,* ed. Mahadev Desai [Ahmedabad: Navajivan Publishing House, 1970], 127). This is not, however, the view of the traditional Vedānta commentators. Much of Gandhi's interpretation, in fact, deviates from tradition. For example, consider his modernist understanding of scriptures as historically conditioned products of human authors and his reconfiguration of the *avatāra* doctrine, which reduced what the *Gītā* and devout Hindus for centuries have regarded as a divine descent into earthly form to no more than a human being who "has been extraordinarily religious in his conduct" (Jordens, "Gandhi," 91; Gandhi, "Anasaktiyoga," 128). For more on Gandhi's views, see note 49. For a contrasting view, see note 67.

25. Cf. *Śvetāśvatara Upaniṣad* 3.16.

26. Christopher Key Chapple, *Nonviolence to Animals, Earth, and Self in Asian Traditions* (Albany: State University of New York Press, 1993), 52; S. Cromwell Crawford, *The Evolution of Hindu Ethical Ideals,* 2d ed. (Honolulu: University Press of Hawaii, 1982), 149–50; Eliot Deutsch, "Vedānta and Ecology," *Indian Philosophical Annual* 7 (Madras: The Center for Advanced Study in Philosophy, 1979), 81–83, and "A Metaphysical Grounding for Natural Reverence: East-West," in *Nature in Asian Traditions of Thought,* ed. J. Baird Callicott and Roger T. Ames (Albany: State University of New York Press, 1989), 265; Lina Gupta, "Ganga: Purity, Pollution, and Hinduism," in *Ecofeminism and the Sacred,* ed. Carol J. Adams (New York: Continuum, 1994), 113; Kinsley, *Ecology and Religion,* 63, 65; Arne Naess, "*Self* Realization: An Ecological Approach to Being in the World," in *Thinking Like a Mountain: Towards a Council of All Beings,* ed. John Seed, Joanna Macy, Pat Fleming, and Arne Naess (Philadelphia: New Society Publishers, 1988), 24–27; Anantanand Rambachan, "The Value of the World as the Mystery of God in Advaita Vedānta," *Journal of Dharma* 14 (July-September 1989), 296.

27. Lance E. Nelson, "The Dualism of Nondualism: Advaita Vedānta and the Irrelevance of Nature," in *Purifying the Earthly Body of God,* 61–88.

28. However, an earlier version of this upside down *aśvattha* metaphor, at *Kaṭha Upaniṣad* 6.3.1, identifies the ancient world tree with Brahman, "the pure, the immortal" (*tad śukraṃ tad brahma tad evāmṛtam ucyate*). See also *Maitrī Upaniṣad* 6.4.

29. Commenting on *BhG* 15.1, Śaṅkara writes: "'It will not last even until tomorrow,' hence [it is called] *aśvattha*. 'It perishes at every moment,' they declare" (na śvo 'pi sthātety aśvatthas taṃ kṣaṇa-pradhvaṃsinam aśvatthaṃ prāhuḥ kathayanti; *Śrīmadbhagavaddgītā,* ed. Wāsudeva Laxman Sāstrī Panśīkar [New Delhi: Munshiram Manoharlal Publishers, 1978], hereafter cited as *BhGŚ*). "The Vedas," explains Rāmānuja in his comments on the same verse, "declare the means of cutting down the tree of *saṃsāra*" (vedo hi saṃsāra-vṛkṣasya chedopāyaṃ vadati; *Sri Bhagavad Gita with Sri Bhagavad Ramanuja's Bhashya,* ed. T. Viraraghavacharya [Madras: Ubhaya Vedanta Granthamala, 1972], hereafter cited as *BhGR*). Again: "Ignorance alone is the basis of this [tree]" (ajñānam eva asya pratiṣṭhā; *BhGR* 15.3).

30. Sadhusangananda dasa writes that the purpose of Kṛṣṇa's *avatāra* is "to convince everyone that our real home is not this temporary material existence but Vaikuntha, the eternal spiritual world" ("The Search for Ecology in the Bhagavadgita: Critical Examination of an Academic Study," *On the Way to Krishna,* summer 1998, 8). See note 20.

31. Vandana Shiva extols the notion of *prakṛti* as the basis of an Indian environmental ethic: "Nature as *prakriti* is inherently active, a powerful, productive force in the dialectic of the creation, renewal, and substance of all life" (*Staying Alive: Women, Ecology, and Development* [London: Zed Books, 1989], 48). This ecological valorization of *prakṛti* as a kind of locus of ultimate concern does not, however, represent any traditional Hindu understanding. It may well be that a new appreciation of *prakṛti* as a value in itself, along with the *guṇa*s and the *mahābhūta*s, will be a key element in Hindu theologies reconstructed to meet the ecological future. But the reconstruction of these notions cannot be done effectively, it seems to me, until the original embeddedness of these concepts as comprising the lower, finally dispensable, term in a hierarchical, dualistic vision of things is openly, directly, and honestly acknowledged. Sadly, overly optimistic and confused readings of the tradition permeate much writing on religion and ecology in the Indian context. O. P. Dwivedi, for example, writes of the five *mahābhūta*s: "They together constitute Brahman" ("Vedic Heritage for Environmental Stewardship," *Worldviews: Environment, Culture, Religion* 1 [April 1997]: 27). Again, while one must acknowledge the vision, inspiration, and effort behind Kapila Vatsyayan's monumental five-volume opus, *Prakṛti: The Integral Vision,* it is an effort seriously undermined by its misreading of foundational conceptions. Vatsyayan, referring to the discussion of the body as "field" (*kṣetra*) in chapter 13 of the *Gītā,* writes:

> Where from do these five elements come? They come from nature, nature here understood by its Sanskrit name *prakṛti*. Is nature dead without attributes? No, there is no absolute dead matter, because nature itself is psycho-physical, psycho-somatic because it is *gunatmaka* (i.e., with attributes and qualities) (Kapila Vatsyayan, gen. ed., *Prakṛti: The Integral Vision,* 5 vols. [New Delhi: Indira Gandhi National Centre for the Arts and D. K. Printworld, 1995], 1:xiv).

This understanding, however, is simply false if we are dealing with a Sāṃkhya-based conception of *prakṛti,* as we are in the *Gītā* (and most everywhere else in Indian thought aside from the realm of Tantra). The effort at rehabilitating this notion as a conceptual support for ecological consciousness will not get far unless we begin with the open recognition that the "vision" of *prakṛti* in classical tradition is anything but "integral," at least in this sense. *Prakṛti*—together with all of its components, the *guṇa*s, even in their expression as mind—is unabashedly proclaimed to be *jaḍa:* insentient, unconscious, an object of consciousness but not conscious itself. Further along in Vatsyayan's *Prakṛti* volumes, one of the contributors, M. P. Rege, sets the record straight: "The totality of *jada* things is called *Prakrti*" ("Samkhya Theory of Matter," in *Prakṛti: The Integral Vision,* vol. 2, ed. Jayant V. Narlikar, gen. ed. Kapila Vatsyayan [New Delhi: Indira Gandhi National Centre for the Arts and D. K. Printworld, 1995], 115). Like Sāṃkhya, the Advaita theologian Śaṅkara regards *prakṛti* and the *guṇa*s as *jaḍa* or *acit* ("unconscious"), as do Rāmānuja, Madhva, Nimbārka, the Gauḍīya Gosvāmins, and other Vaiṣṇava theologians. Thus, the Mādhva teacher Vādirāja proclaims: "*Prakṛti,* called *māyā,* is insentient" (*māyākhyā prakṛtir jaḍā; Yuktimallikā,* quoted by Surendranath Dasgupta, *A History of Indian Philosophy* [Cambridge: Cambridge University Press, 1968], 4:313). And Viśvanātha Cakravartin, in his Gauḍīya Vaiṣṇava commentary on *BhG* 7.5, writes: "This *prakṛti,* which is the power termed 'external,' is lower, i.e., not superior, because it is insentient" (iyaṃ prakṛtir bahir-aṅgākhyā śaktir aparānutkṛṣṭā jaḍatvāt; Śrīmad Bhagavadgītā *with the Commentaries* Sārārthavarṣiṇī *of Viśvanātha Cakravartin and* Gītābhūṣaṇa *of Baladeva Vidyābhūṣaṇa* [Kusumasarovar, Mathurā: Kṛṣṇadāsabābā, ca. 1966–67], hereafter cited as *BhGV*). Even Abhinavagupta, the great exponent of Śaiva tantric nondualism, sees it this way. In his *Gītārthasaṃgraha* on *BhG* 2.19, he writes: "The Ātman is eternal, because it is not an object of knowledge; change belongs to objects of knowledge, which are unconscious" (ātmā tu nityaḥ I yato 'prameyaḥ I prameyasya tu jaḍasya pariṇāmitvaṃ; *Śrīmadbhagavadgītā,* ed. Wāsudev Laxman Sāstrī Panśīkar [New Delhi: Munshiram Manoharlal Publishers, 1978], hereafter cited as *GASA*).

Situating the ecological question properly within the horizon of South Asian consciousness will be a hopeless task if we begin with misreadings. A better start at evoking the ecological potential of the complex of ideas surrounding *prakṛti* might be gained by exploring the relationality inherent in the concept, as has been suggested by Larson: "The notion of *prakṛti* as *triguṇa* (*sattva, rajas,* and *tamas*) is clearly 'systemic and (internally) relational' . . . and environmental ethicists could possibly find powerful conceptual resources for developing 'organic' and/or 'holistic' perspectives on nature within the traditions of Sāṃkhya and Yoga in South Asian thought" ("'Conceptual Resources,'" 269). Another productive avenue is likely to be reflection on the assimilation of *prakṛti* into the idea of *śakti* in Tantra, a move that incorporates the material principle into the ultimate reality. On this, see Rita DasGupta Sherma, "Sacred Immanence: Reflections of Ecofeminism in Hindu Tantra," in *Purifying the Earthly Body of God,* ed. Nelson, 89–131.

32. Abhinavagupta's Śaiva commentary, despite its identification of *prakṛti* as *jaḍa* (insentient), seeks a way beyond this dualism. At *GASA* 13.34, he quotes the verse: "Only the deluded make the distinction between *puruṣa* and *prakṛti;* the perfected ones understand the universe to consist of the untainted Ātman" (pumān

prakṛtir ity eṣa bhedaḥ saṃmāḍha-cetasām । paripūrṇās tu manyante nirmalātma-mayaṃ jagat).

33. Literally, "possessor of a body" (*deha*).

34. Naess, *Ecology, Community, and Lifestyle,* 195.

35. Jacobsen, "*Bhagavadgītā,* Ecosophy T, and Deep Ecology," 222–27.

36. kṣetrajñeśvaraikatvaṃ sarvopaniṣat-prasiddhaṃ drakṣyasīty arthaḥ; *BhGŚ* 4.35.

37. See Klaus K. Klostermaier, "The Body of God," in *The Charles Strong Lectures 1972–84,* ed. Robert B. Crothy (Leiden: E. J. Brill, 1987), 103–20; and Patricia Y. Mumme, "Models and Images for a Vaiṣṇava Environmental Theology: The Potential Contribution of Śrīvaiṣṇavism," in *Purifying the Earthly Body of God,* ed. Nelson, 133–61.

38. *BhGR* 6.29. Cf. *BhGR* 5.19 and also 6.31: "He sees me only, that is, he always sees the *similarity to me* in his own self and in [the selves of] all beings" (mām eva paśyati । svātmani sarva-bhūteṣu ca sarvadā mat-sāmyam eva paśyatīty arthaḥ).

39. aśeṣāṇi bhūtāni manuṣya-tiryyagādīny ātmani jīvātmany upādhitvena sthitāni pṛtak drakṣyasi । atho mayi parama-kāraṇe ca kāryyatvena sthitāni drakṣyasi; *BhGV* 4.35.

40. Rajagopal Ryali, "Eastern-Mystical Perspective on Environment," in *Ethics for Environment: Three Religious Strategies,* ed. Dan Stefferson, Walter Herrscher, and Robert S. Cook (Green Bay: University of Wisconsin Press, 1973), 47–48; as summarized in J. Baird Callicott, *Earth's Insights: A Multicultural Survey of Ecological Ethics from the Mediterranean Basin to the Australian Outback* (Berkeley and Los Angeles: University of California Press, 1994), 49.

41. uttama-saṃskāra-vati brāhmaṇe sāttvike madhyamāyāṃ ca rājasyāṃ gavi saṃskāra-hīnāyām atyantam eva kevala-tāmase hastyādau ca sattvādi-guṇais taj-jaiś ca samskārais tathā rājasais tathā tāmasaiś ca saṃskārair atyantam evāspṛṣṭaṃ samam ekam avikriyaṃ brahma draṣṭuṃ śīlaṃ yeṣāṃ te paṇḍitāḥ sama-darśinaḥ; *BhGŚ* 5.18.

42. For more on Hindu hierarchical notions and their implications for the environment, see Frank J. Korom, "On the Ethics and Aesthetics of Recycling in India," in *Purifying the Earthly Body of God,* ed. Nelson, 197–223; and Sherma, "Sacred Immanence."

43. ata eva samaṃ paśyantīti na tu vyavaharantīti; *GASA* 5.18.

44. *BhGV* 5.18.

45. atyanta-viṣamākāratayā pratīyamāneṣu ātmasu paṇḍitāḥ ātma-yāthātmya-vidaḥ . . . viṣamākāras tu prakṛteḥ, nātmanaḥ; *BhGR* 5.18.

46. See note 31.

47. Aurobindo reads in *BhG* 15.16–18 the key to overcoming the spiritual dualism of the text. The verses describe three *puruṣa*s—the *kṣara* (perishable), equaling *prakṛti;* the *akṣara* (imperishable), equaling *ātman;* and the Puruṣottama (Supreme Person), being God or the "Supermind"—the latter representing the reality that transcends and yet integrates the preceding two. He writes: "Whoever thus knows and sees him [the Supreme Being] as the Purushottama, is no longer bewildered by these two apparent contraries. These at first confront each other here in him as a positive of the cosmic action and as its negative in the Self who has no part in an action that belongs or seems to belong entirely to the ignorance of Nature. . . . [But] he unites the

Kshara and the Akshara in the Purushottama. . . . He still sees the self [*akṣara*/*ātman*] as an eternal and changeless spirit silently supporting all things, but he sees also Nature [*kṣara*/*prakṛti*] no longer as a mere mechanical force . . . but as a power of the Spirit and the force of God in manifestation" (*Essays on the Gita,* 433, 562). Movements toward such a vision may be found in the theologies of Abhinavagupta, Vallabhācārya, and the Gauḍīya Vaiṣṇavas. (See note 32, above.)

48. Whatever karmic connection the self may have with other beings can only be construed as part of its bondage, and indeed as a cause of the perpetuation of that bondage. There is very little basis in the *Gītā* or anywhere else in classical Hindu thought on which to build a positive, ecological reading of karmic interconnectedness. Deutsch writes: "Everything in nature is so interconnected through causal chains and relationships that we ourselves become part of the natural process and are conditioned by it" (quoted by S. Cromwell Crawford, *Dilemmas of Life and Death: Hindu Ethics in North American Context* [Albany: State University of New York Press, 1995], 179). True, no doubt, but there is little if any scope in Hindu traditions for celebration of such connectedness, as might be found in aspects of Buddhist thought. Karmic bonds are just that, fetters to be cut, destroyed, "burnt"—not rejoiced in: "They are released from *karma*" (mucyante te 'pi karmabhiḥ, 3.31); "*karma*s do not sully me . . . One who knows me thus is not bound by *karmas*" (na māṃ karmāṇi limpanti . . . iti māṃ yo 'bhijānāti karmabhir na sa badhyate, 4.14); "the fire of knowledge reduces all *karma* to ashes" (jñānāgniḥ sarva-karmāṇi bhasmasāt kurute, 4.37); "*karma*s do not bind one who has renounced *karma*s by means of *yoga* and is possessed of the self" (yoga-saṃnyasta-karmāṇaṃ . . . ātmavantaṃ na karmāṇi nibadhnanti, 4.41); "bound by your own *karma,* born of your own nature, you will do helplessly that which, out of delusion, you do not wish to do" (kartuṃ necchasi yan mohāt kariṣyasy avaśo 'pi tat, 18.60), and so on.

49. In his introduction to a discourse on chapter 2 of the *Gītā,* Gandhi says: "When I was in London, I had talks with many revolutionaries. Shyamji Krishnavarma, Sarvarkar, and others used to tell me that the *Gita* and the *Ramayana* taught quite the opposite of what I said they did. I felt then how much better it would have been if the sage Vyasa [the traditionally recognized author of the *Mahābhārata*] had not taken this illustration of fighting for inculcating spiritual knowledge. For when even highly learned and thoughtful men read this meaning in the *Gita,* what can we expect of ordinary people? If what we describe as the very quintessence of all Shastras [authoritative religious texts], as one of the Upanishads, can be interpreted to yield such a wrong meaning, it would have been better for the holy Vyasa to have taken another, more effective illustration to teach sacred truths" (M. K. Gandhi, *The Bhagavadgita* [Delhi: Orient Paperbacks, 1980], 23). See note 24.

50. For example: "The Hindu belief in the cycle of birth and rebirth wherein a person may come back as an animal or a bird gives these species not only respect, but also reverence. This provides a solid foundation for *ahimsa*—non-violence (or non-injury) against animals and humans alike" (Dwivedi, "Vedic Heritage," 33). See also Chapple, *Nonviolence to Animals,* 52, 112; Kinsley, *Ecology and Religion,* 65.

51. Consider Dasgupta's summation: "When one can perform an action with a mind free from attachment, greed and selfishness, from a pure sense of duty, the evil effects of such action cannot affect the performer." Again: "If Arjuna fought and

killed hundreds of his kinsmen out of a sense of his caste-duty, then, howsoever harmful his actions might be, they would not affect him" (Dasgupta, *A History of Indian Philosophy,* 2: 503, 508).

52. See Sharma, *The Hindu Gītā,* xv. On Gandhi's interpretation of the *Gītā,* see notes 24 and 49.

53. For a fuller discussion, see Chapple, *Nonviolence to Animals.*

54. yadi vā yac ca duḥkhaṃ mama pratikūlam aniṣṭaṃ yathā tathā sarva-prāṇināṃ duḥkham aniṣṭaṃ pratikūlaṃ; *BhGŚ* 6.32.

55. svātmanaś cānyeṣām cātmanām asaṃkucita-jñānaikākāratayaupamyena svātmani cānyeṣu ca sarvatra vartamānaṃ putra-janmādi-rūpaṃ sukhaṃ tanmaraṇādi-rūpaṃ ca duḥkam asaṃbandha-sāmyāt samaṃ yaḥ paśyati . . . sa yogī paramaṃ mataḥ . . . ; *BhGR* 6.32.

56. *Śrī Rāmānuja Gītā Bhāṣya,* ed. and trans. Swami Adidevananda (Madras: Sri Ramakrishna Math, n.d.), 230.

57. See O. P. Dwivedi and B. N. Tiwari, *Environmental Crisis and Hindu Religion* (New Delhi: Gitanjali Publishing House, 1987); and Marta Vannucci, *Ecological Readings in the Veda* (New Delhi: D. K. Printworld, 1994).

58. *BhG* 18.40 (see also 2.42–46, 4.12, 7.20–22, 11.52). At *BhGR* 15.1, Rāmānuja speaks of the "*prakṛti* that has transformed into the forms of the gods, etc." (*devādy-ākāra-pariṇata-prakṛti*). Once one realizes that the *Gītā* identifies deities of that level as themselves products of *prakṛti,* there is perhaps less force in arguing, as Dwivedi does, that the "*mahābhūtas* have been deified in the Vedic and Puranic literature" ("Vedic Heritage," 27).

59. Cf. *Taittirīya Upaniṣad* 2.9.1: "One who knows the bliss of Brahman does not fear anything anywhere. Verily, such a one is not tormented [by thoughts such as], Why did I not do good? Why have I done evil?"(ānandam brahmaṇo vidvān na bibheti kutaścana. etaṃ ha vā va na tapati, kim ahaṃ sādhu nākaravam iti, kim aham pāpam akaravam iti). See also *Bṛhadāraṇyaka Upaniṣad* 4.3.22, 4.4.22–23.

60. See also Vasudha Narayanan, "'One Tree Is Equal to Ten Sons': Hindu Responses to Problems of Ecology, Population, and Consumption," *Journal of the American Academy of Religion* 65 (summer 1997): 291–332.

61. Purusottama Billimoria comments regarding Upaniṣadic ethics: "It appears almost as if *dharma* could be dispensed with" ("Indian Ethics," in *A Companion to Ethics,* ed. Peter Singer [Oxford: Blackwell Reference, 1993], 48). Despite its emphasis on detached participation, this possibility always hovers around the edges of the *Gītā*'s exposition. The *BhG* never reclaims the world of *dharma* as having any ultimate value or soteriological relevance in itself. Noting that the masses of India have historically had little direct access to the "*tattva* texts" of Hinduism (those dealing with the nature of ultimate reality, or *tattva,* and the quest for *mokṣa*), Vasudha Narayanan has recently argued eloquently for a "paradigm shift" in the discussion of religion and ecology in India from "from reality/*mokṣa* texts to works and practices of *dharma* if our enterprise is to have any success." There is some wisdom in this, indeed a good deal, but I still think Narayanan overstates her case, making an "either/or" out of what more reasonably should be a "both/and." To say that the *tattva* texts "do not necessarily trickle down," so that they have "limited power over ethical behavior," is—it seems to me—like saying that the writings of Descartes and Locke

have little impact on the moral ethos of modern Americans. No doubt these texts may be little read among the masses, but their influence is surely there, shaping cultures, worldviews, and unconsciously held attitudes (Narayanan, "'One Tree Is Equal to Ten Sons,'" 295–99).

62. See note 51. Commenting on *BhG* 12.13–17, Bilimoria ("Indian Ethics," 50–51) wonders, "If good and evil are transcended and the distinction obliterated can there any longer be an ethic to speak of? (Can we each be like Nietzsche's superman?)."

63. Quoted by Stevenson, "Tilak," 58.

64. kālo 'smi loka-kṣaya-kṛt pravṛddho lokān samāhartum iha pravṛttaḥ; *BhG* 11.32.

65. Madeleine Biardeau, "Études de mythologie hindoue," pt. 3, in *Bulletin de l'École Français d'Éxtrême-Orient* 58 (1971): 50–54, summarized and partially translated by Alf Hiltebeitel, *The Ritual of Battle: Krishna in the* Mahābhārata (Ithaca, N.Y.: Cornell University Press, 1976), 114–18.

66. Hiltebeitel, *The Ritual of Battle*, 118.

67. "We must look existence in the face if our aim is to arrive at a right solution, whatever that solution may be. And to look existence in the face is to look God in the face. . . . Nature creates indeed and preserves but in the same step and by the same inextricable action slays and destroys. It is only a few religions which have had the courage to say without any reserve, like the Indian, that this enigmatic World-Power is one Deity. . . . We must acknowledge Kurukshetra [the battle-field of the *Gītā*]; we must submit to the law of Life by Death before we can find our way to the life immortal; we must open our eyes, with a less appalled gaze than Arjuna's, to the vision of our Lord of Time and Death and cease to deny, hate, or recoil from the universal Destroyer" (Sri Aurobindo, *Essays on the Gita* [Pondicherry: Sri Aurobindo Ashram, 1980], 41–42). Cf. Gandhi's interpretation, notes 24 and 49.

68. See Sukumari Bhattacharji, *Fatalism in Ancient India* (Calcutta: Baulmon Prakashan, 1995), but also, for a balancing view, Christopher Key Chapple, *Karma and Creativity* (Albany: State University of New York Press, 1986), 55–59.

69. śruti-pratipādita īśvaro 'ntaryyāmī hṛdi tiṣṭhati, kiṃ kurvvan? sarvāṇi bhūtāni māyayā nija-śaktyā bhrāmayan tat-tat karmāṇi pravarttayan; *BhGV* 18.61.

70. Will Johnson, "Transcending the World? Freedom (*mokṣa*) and the *Bhagavadgītā*," in *The Fruits of Our Desiring: An Inquiry into the Ethics of the* Bhagavadgītā *for Our Times*, ed. Julius Lipner (Calgary, Alberta: Bayeux Arts, 1997), 95, 99. Sri Aurobindo seems to agree: "Not only the desire of the fruit, but the claim to be the doer of works has to be renounced in the realization . . . of all works as simply the operation of the universal Force, of the Nature-Soul, Prakriti, the unequal, active, mutable power. Lastly, the supreme Self has to be seen as the supreme Purusha [Person] governing this Prakriti, . . . by whom all works are directed, in a perfect transcendence, through Nature" (*Essays on the Gita,* 34).

71. yad idaṃ yuddha-karaṇe prāsaṇgika-bandhu-vadhādi tasya sarvasyāhaṃ kartety ātma-dharmatāṃ parityajya tathācāryādi-hanana-kriyā-niṣedhe mamādharmo bhaviṣyatīti manasā vihāya mām evaikaṃ sarva-kartāraṃ parameśvaraṃ svatantraṃ śaraṇaṃ sarvasva-bhāvādhiṣṭhāyakatayā vraja.

72. S. S. Rama Rao Pappu, "Detachment and Moral Agency in the *Bhagavad*

Gītā," in *Perspectives on Vedānta: Essays in Honor of Professor P. T. Raju,* ed. S. S. Rama Rao Pappu (Leiden: E. J. Brill, 1988), 156. In the same article (156–57), Rama Rao Pappu elaborates: "Kṛṣṇa urges Arjuna to act as His agent. In other words, though the material accountability for discharging the responsibilities of his role as a warrior fell on Arjuna, the formal, and the ultimate, responsibility of Arjuna's actions belong[s] to Lord Kṛṣṇa. . . . [T]he detached individual . . . is not responsible for the consequences of his actions in two senses: (a) His responsibilities, if any, are not 'actor-responsibilities' but 'agent-responsibilities.' (b) He is not an 'actor,' too, since his Self is not invoked in the production of (desirable) consequences."

73. "The detached life is the most admired, desired, and sought after way of life to the Indian mind" (Rama Rao Pappu, "Detachment and Moral Agency," 157). See also *BhG* 3.7, 19, 25; 5.21; 13.14; 18.49.

74. As Jacobsen indicates, "The sixth and eighth point of the deep ecology platform says: '(6) Policies must therefore be changed. These policies affect basic economic, technological, and ideological structures. The resulting state of affairs will be deeply different from the present. (8). Those who subscribe to the foregoing points have an obligation directly or indirectly to try to implement the necessary changes' " ("*Bhagavadgītā,* Ecosophy T, and Deep Ecology," 238, n. 83, quoting Arne Naess, "The Deep Ecological Movement: Some Philosophical Aspects," *Philosophical Inquiry* 8 [1986]: 14).

75. *Modern Indian Interpreters,* ed. Minor, 4.

76. Dasgupta, *History of Indian Philosophy,* 2:502; see also 2:513 and Stevenson, "Tilak," 57. Dasgupta comments further: "Should a Śūdra think of performing sacrifices, *tapas* or gifts or the study of the Vedas, this would most certainly be opposed by the *Gītā,* as it would be against the prescribed caste-duties" (2:514).

77. Madeleine Biardeau, *Hinduism: The Anthropology of a Civilization* (New York: Oxford University Press, 1989), 3.

78. At *BhG* 7.2, Kṛṣṇa declares: "Among thousands of persons, perhaps someone strives for perfection, and even among those perfected ones who strive, perhaps someone knows me as I truly am" (manuṣyāṇām sahasreṣu kaścid yatati siddhaye | yatatām api siddhānāṃ kaścin māṃ vetti tattvataḥ).

79. Swimme and Berry, *The Universe Story,* 255.

80. Earth Charter Commission, "The Earth Charter."

81. Jacobsen, "*Bhagavadgītā,* Ecosophy T, and Deep Ecology," 220.

82. See *BhGR* 5.19, "Abidance in Brahman is, verily, conquest of *saṃsāra*" (brahmani sthitiḥ eva hi saṃsāra-jayaḥ).

83. Mary Evelyn Tucker, "The Emerging Alliance of Ecology and Religion," *Worldviews: Environment, Culture, Religion* 1 (1997): 19.

84. See Eric J. Sharpe, *The Universal Gita: Western Images of the Bhagavad Gita* (LaSalle, Ill.: Open Court Pub. Co., 1985), 32–46, and Sarvepalli Gopal, *Radhakrishnan: A Biography* (London: Unwin Hyman, 1989), 16–19. For a contemporary example of a serious Christian critique of (but not a polemic against) Hinduism, see Hendrik Vroom, *No Other Gods: Christianity in Dialogue with Buddhism, Hinduism, and Islam* (Grand Rapids, Mich.: Eerdmans Publishing Co., 1996), 43–82.

85. See *The Fruits of Our Desiring,* ed. Lipner.

86. This kind of work will also have to involve, it seems to me, a difficult but

thoroughgoing taking of responsibility on the part of the traditions for the shadow elements they contain, ecological and otherwise. Rather than simply dismissing damaging applications of one's tradition as arising from unfortunate "misinterpretations"—or worse, ignoring them—one must ask whether the original doctrines, images, or practices contain elements that promote (or lend themselves to) such misappropriation, and to what extent these "misinterpretations" have even been enshrined at some levels within the tradition itself. In the Jewish and Christian case, I think of, as one immediate example, the notorious verses of Genesis offering humanity dominion over, and a calling to subdue, nature. In the Hindu case, one must reflect on the extent to which a civilization has been based on an ontology and ethics designed for and by world-renouncers. The *Gītā* itself recognizes this problem and this danger, if only from the viewpoint of the renouncers themselves: "The wise (*vidvān*) should not unsettle the minds of the ignorant, who are attached to action" (3.26).

87. See note 2.

88. If Hindu thinkers seek an adequate response to defend the *Gītā* against some of the questions that have been raised here on the basis of an implicit postmodern environmentalist ethic, they must at the same time cease arguing so enthusiastically that Hinduism embodies that currently fashionable ethic even more thoroughly than other traditions. If the tradition is to respond creatively to critiques based on such standards, an alternate environmental ethic must be brought forward, one that takes into account the "otherness" of Hindu worldviews in many respects to the ecological paradigm being fashioned in the West. This would be a welcome development, for it would force Western environmental ethicists to reexamine and defend their assumptions. A Hindu critique, for example, of Swimme and Berry's *The Universe Story*—with its uncritically Western assumptions of "fulfillment of the human personality" through "intimate communion with the natural world" and the triumph of the "dominant time-developmental mode of consciousness" (3–4)—would be very instructive.

Can Hindu Beliefs and Values Help India Meet Its Ecological Crisis?

ANIL AGARWAL

My contribution to this volume comes from the perspective of an activist and scientist, not as a specialist in the religious traditions of India. I do not fully understand the importance that the Hindu religion holds out for the environment, but I definitely know something of how Hindus behave toward their environment.

Like most modern Indians, I, too, am a child more of Macaulay and Nehru, deeply steeped in Western thinking. Like many of my peers, I was encouraged from a young age to study science and gained my education as an engineer at the prestigious Indian Institute of Technology in Kanpur, the equivalent of the Massachusetts Institute of Technology in the United States.

My young student days at IIT-Kanpur had a major influence on me. The institute was located in the midst of a large number of extremely poor villages of the Indo-Gangetic Plains. Even though our Western-trained faculty would repeatedly tell us that we were going to be the generation to lead India into modernity, it was clear to anyone who cared that none of us knew anything about dealing with the problems of the villages sitting cheek by jowl with our high-tech institution.

Most of my colleagues emigrated, taking up offers of professional studies in North America and Europe. I decided to stay in India and make a difference by using my skills and training to help meet the many challenges of India. In order to understand India, I became a journalist. And, in 1974 I came across an extremely evocative movement called the Chipko movement in the high Himalayan ranges of

Uttar Pradesh. The village women there were threatening to hug the trees if the government allowed them to be cut. For me, it became a story that determined my life. It showed how important the environment was for the poor of India: it was their source of daily survival, the foundation of their economy and culture.

As my understanding of the extraordinary diversity of India's environment grew, I began to see the people living in those diverse habitats—the fisherfolk dependent on the rivers and wetlands, the nomads dependent on the vast grasslands of the Thar Desert and the upper Himalayan pastures, and the forest-dwelling people dependent on the diverse forest types of India. And the more I understood the way people lived with their environment, the better I understood their culture. Slowly, India's extraordinary cultural diversity and ecological diversity led me to feel much more an Indian, and my regard for Mahatma Gandhi as a political leader grew strong. It was an intellectual journey that took me deeper and deeper into India and its cultural recesses, making me despise more and more, as Jean-Paul Sartre would put it, the "whitewashing" of modern Indians.

The State of India's Environment: *The First Citizens' Report*

Over twenty years ago, I helped establish the Centre for Science and Environment, a nongovernmental organization dedicated to educating the public on the most pressing ecological issues facing the Indian subcontinent.[1] In 1982, we published *The First Citizens' Report,* a comprehensive overview of the state of India's environment.[2] The issues that we identified then continue to be important within India today: land, water, forests, dams, atmosphere, habitat, people, health, energy, wildlife, and governance. These concerns will be summarized below.

Perhaps the most basic environmental challenge facing India can be found in the first three issues discussed in the *Citizens' Report:* land, water, and forests. The future of India will depend on its ability to feed its people. Neglect of the soil could lead to disaster; good soil management can guarantee food sufficiency. Clean water is only obtained by less than one-third of the population of India. The Institute of Hygiene and Public Health in Calcutta has noted that "In poor communities, diarrhoea and dysentery have become a part of life."[3]

Community waste and industrial waste have fouled most of India's rivers, which require intensive cleanup. India's forests have been viciously attacked and destroyed during the past century. Between 1951 and 1972, India lost 3.4 million hectares of forest to dams, crop lands, roads, and industries. Today, no more than 12 percent of the country's land surface has a good forest cover. With loss of forest cover, erosion sets in and the hydrological system is disturbed, leading to increased droughts and loss of prime agricultural lands.

Dams, encouraged by Jawaharlal Nehru in the years following India's independence, create great benefits for the Indian people, in terms of irrigation and hydroelectric power generation. However, dams also pose major threats to biological species, particularly in Kerala, and to tribal peoples. Just as harnessing the power of water has created a complex of issues for India, so also the burning of fossil fuel and the growth in vehicular traffic have darkened the atmosphere over India's burgeoning cities. The Centre for Science and Environment has lobbied rigorously for the adoption and enforcement of pollution controls for industries and private vehicles. It has also worked toward raising awareness regarding energy consumption. Nearly half of the energy consumed in India is still spent on cooking food. Energy planners often skew their planning priorities in favor of the larger energy consumers. Fast-growing species planted along farm boundaries can meet all the cooking fuel requirements and reduce the hours in a woman's day taken up in the collection of twigs, branches, and leaves used for cooking fires.

This brings us to the broad issues of habitat, people, and health. In the thirty years between 1951 and 1981, the populations of Delhi and Bangalore quadrupled; their growth since 1981 has escalated at an even greater rate. With urbanization, housing, water, and adequate transportation are in short supply. Additionally, 98 percent of India's rural population do not have toilets and 70 percent do not have access to safe drinking water. Pastoral nomads, who comprise over 5 percent of India's population, have been crowded out as pasture lands have been converted to farmland. For India's 44 million tribals, destruction of forests has meant a cultural and social death. Health services have failed to meet the needs of the majority of the population. Most old diseases continue to be rampant, while new ones are making rapid strides. Every third person who dies in India is a child younger than five, the victim of a vicious combination of poverty, malnutrition, unsanitary environment, and unclean drinking water. The increased use of cigarettes and the greater use of chemicals due to industrialization result in a half million or more cancer deaths each year.

Before moving on to a discussion of possible remedies, I want to include mention of wildlife. With only 2 percent of the world's landmass, India possesses around 5 percent of the known living organisms on earth. Significant portions of the 15,000 plant species and 75,000 animal species found in India are threatened by the pressure of human activity on land and forests. India lies at the confluence of the African, European, and Southeast Asian biological systems; it possesses interesting components from each of these realms. The fauna particular to India includes the sloth bear, blackbuck, four-horned antelope, and snakes belonging to the family *Uropeltidae*. With over 1,200 species and about 900 subspecies of birds, India's avian diversity is unmatched except by Latin America. The cheetah, the pink-headed duck, and the Jerdon's courser have already disappeared from India. The Wildlife Protection Act of 1972 listed 133 species as either rare or highly endangered, a figure that undoubtedly has grown. The lion-tailed macaque, which inhabits the evergreen rain forests of South India, is among the world's most endangered primates.

At the time this work was completed, we thought that we had adequately summarized India's environmental problems. However, in this early work, one central issue was neglected: the overwhelming effect of pollution in India's burgeoning cities. This was first dramatically felt in the mid-1990s, when the cumulative result of increased automotive traffic took hold. In the United States, there are two cars for every three people. If India continues to follow the development model set forth by the United States, it will have 600 million cars. India does not have them today because it is poor, but this is indeed India's dream. As it grows more wealthy, India will acquire more and more motorized vehicles, including the highly polluting scooters. The government is simply not prepared to deal with the control of the pollutants from this army of vehicles, or with the other forms of pollution that accompany industrial development.

Many years ago, I signed a Statement of Shared Concern, published in *The First Citizens' Report,* which has served as the foundation for much of the information given in the above paragraphs. The statement concludes with the following observation:

> The environmental crisis, born out of a global maldevelopment, affects all countries of the world, north or south, east or west. But no country can wait for the rest of the world to find a solution. In India, particularly, the problems are too pressing. We strongly believe that a sustainable egalitarian development is possible if we keep in mind what Mahatma Gandhi once said: "There is enough in the world for everyone's need but not enough for everyone's greed."[4]

The Centre for Science and Environment has worked diligently for the past several years, employing the best information from the scientific community and appealing to India's secular sensibilities. We have worked tirelessly, through journalistic endeavors, lobbying, and the support of direct action. We have cooperated with various government agencies, including the Ministry of Environment.

However, over the years I have seen various examples of the powerful religious imagery of India providing metaphors and motivations that transcend the logic and rationality of the secular, scientific, governmental modalities. I would like to share some of these stories, in the hope that they can contribute to the emerging theories that link religion and ecology.

Kṛṣṇa the Cowherd and the Pastoral Lands of Rajasthan

Nearly a decade ago I was invited to address a public meeting in Bhinasar, a rich village near the desert town of Bikaner, Rajasthan. Local activist and journalist Shubhu Patwa had fought a battle with an influential local businessman who wanted to encroach on the village grassland (locally known as *gauchar,* a place where cows graze) and bring it under a private eucalyptus plantation.

All over the Thar Desert, the most densely populated desert in the world, grasslands have shrunk as plantations and irrigated agriculture have spread across the region. Yet, the region's traditional economy was built around grasslands and animal care. Nomadism was common and even those who practiced agriculture saw it as an occupation secondary to animal care. Agriculture gave fruit only in years when there was adequate rainfall and, in those years when there was insufficient rain, the population would fall back on returns from animal care and other occupations, such as handicrafts, to survive. Most people lived off a multiple occupational base. Trees, shrubs, and grasses like *khejri* (*Prosopis cineraria*), *jharberi* (*Zizphus mummularia*), and *sewan* (*Lasirius sindicus*) were never uprooted, even on farmlands, because they provided valuable fodder for the animals. The farmer would guide his plough around the *khejri* sapling and clumps of *sewan* and *jharberi* growing on his farms and let them be. They ensured his survival. The cereals and the minor millets did not.

But the modernization programs pushed by the state did not respect the local economy. As a result, grasslands decreased and with more and more animals trying to survive on them, their productivity began to plummet. While some farmers gained from the new agricultural developments, many others, particularly the pastoral nomads, lost out.

For many Gandhians who had been involved with *goseva* (service of the cow) programs and for environmentalists, the destruction of grasslands was a sore point. They wanted programs that enhanced the productivity of grasslands and strengthened the traditional local economy instead of the promotion of high-yielding but unsustainable programs that converted grasslands to water-intensive croplands and exotic plantations. They wanted innovative schemes that used the available irrigation waters to increase and stabilize the productivity of the rich grasslands of the Thar and enriched the traditional economy.

Aware of this debate in the region, I went to Shubu's meeting prepared to talk about the importance of grasslands in desert ecology and other such high-sounding ecological principles. Just as I finished my talk I saw that behind me, on the dais, was a picture of the god Kṛṣṇa. I failed to see its significance at that time, but it soon dawned on me how irrelevant my talk had been. The importance of protecting the *gauchar* was all there in that picture of Kṛṣṇa, a *gwala* (cowherd) who tended the *gau* (cow). If the *gwala* and the *gau* were sacred, then why not the *gauchar,* the grasslands set aside for the cow? *Gauchar*s indeed have a certain sacredness attached to them in the Thar region, as do trees used for fodder. By recognizing the continuity between the god Kṛṣṇa and modern day *gauchar*s, a symbol system emerges that values the age-old practice of living in harmony with the local ecosystem.

The Bishnoi village of Khejadali is famous for an incident several hundred years ago when people died trying to save a few hundred *khejdi* trees from being felled by the king's men. Ramachandra Guha and many others have written about this famous incident. The *khejdi* trees provide the respite of shade under the hot desert sun. Local women lop branches from these hardy desert trees to feed their goats. Even today, Bishnoi farms in Rajasthan are full of *khejdi* trees.

With this local history and the image of the pastoral Kṛṣṇa, I really did not need any alien scientific or ecological metaphors to explain to the local people the importance of protecting grasslands in desert ecosystems. The popular metaphors were all there in the local faith. But, as an engineer who had grown up in secular, modern India, that is not how I had been trained.

Religion and the Land

Secularism under the Nehruvian vision had come to mean disregard for all religions, contrary to the Gandhian vision of secularism which meant respect for all religions. My own work and consciousness in the field of environmentalism originally arose from the Nehruvian vision. Unfortunately, however, it left little room for understanding the people-ecology interface. I gradually realized that in a densely populated, poor country like India, people were dependent for their daily survival on the gross nature product, the benefits that nature gives them for free, rather than the gross national product. In India, no understanding of ecology can be complete without understanding the relationship between the people and their landscape. Nomads cannot be removed from the context of their grasslands. Tribals cannot be separated from the relationship with the forest. Fisherfolk cannot be seen without taking into account rivers or the oceans.[5]

This led me to see the interface between culture, people, and ecology. I could not understand ecology without understanding people, and I could not understand people without understanding their culture, including their religious faith. My respect for all cultures, including their faiths and their secular practices, began to grow. I began to find in them an innate wisdom that arose from these people living within their environment. Over the years, people had used trial and error in their innovations and in trying to exploit the environment. The land—whether grasslands, forest, or water—soon taught them what was sustainable and what was not. People kept the sustainable practices, while the unsustainable practices were weeded out. In many communities, these practices were codified into religious beliefs and stories, especially in Hinduism, where it is difficult to separate religious practice from secular practice.

Hinduism gave rise to numerous ecologically sound tenets. Rivers and their origins were declared sacred. The cow was (and is) sacred. The Indian basil plant, *tulsi,* which has many medicinal applications, took on a central religious function, as Madhu Khanna explains in her chapter in this volume. Recycling was given importance, as Frank Korom describes in Lance Nelson's book *Purifying the Earthly Body of God.*[6] Cow dung became an important part of daily life, as cooking fuel and as an astringent. One can go on and on citing examples from daily life in India, including the many references included in this book. Hindus clearly believe in and act upon a form of "utilitarian conservationism," as opposed to any sort of protectionist conservation: they value and protect those features of nature that have gained significance within the ritual cycle of human flourishing.

The sad fact is that these practices and beliefs are rapidly breaking down under the onslaught of Western-style technological modernization and social concepts of secularism. Rather than sustainability, the modern ethos driven by science and technology calls for high productivity. Rather than seeing people within the context of a life-style that has been developed over hundreds or thousands of years, the modern world seeks to rush them into a money economy, driven, in Gandhi's words, not by need but by greed.

Is Hinduism Culpable?

Since I am writing about Hinduism, I must say that certain practices of Hinduism itself contribute to this breakdown. First, Hinduism is a highly individualistic religion. It looks into the self, emphasizing the *ātman* as the key to spiritual ascent. *Dharma,* or social responsibility, focuses first on oneself, emphasizing one's own behavior. The consequences of one's behavior on others plays a secondary role; the primary concern is to do one's own *dharma* for the sake of one's own well-being. Mahatma Gandhi's own battle techniques of *svarāj,* or self-rule, were highly individualistic, emphasizing one's own contribution. They greatly endeared him to the nation. This culture of self-centeredness, from which arose India's modern civil society, failed to create a culture in which institutions could grow. The dismal state of

Gandhian institutions today is a fine example. It was probably not Gandhi's failure alone, but, to the extent that he tapped into the concept of isolation of self (*puruṣa-kaivalya*) as a cornerstone of Hindu philosophy, he laid the ground for relegating the greater social good to a secondary status.

In India today, we have numerous environmental heroes and heroines, and everyone knows their names. Whenever I come to the West, I hear only of institutions, like Greenpeace and the Sierra Club. No one seems to know who started them or who keeps them going. Yet, they form the backbone of the Western environmental movement.

In my observations, it is a sad fact that, among Hindus, the linkages of one's individual behavior with the rest of society are less emphasized and less important than the state of one's own sense of well-being. Under the onslaught of modern-day secularism, this has brought out the worst type of individualism in Hinduism. Hinduism believes in the existence of pollution, but only to the extent that it affects one's self directly. It must be avoided at all costs. But the concern focuses on keeping the pollution away from one's own being, not on making certain that the pollution does not hurt someone else. Respect for the person outside has disappeared. It is commonly said today that, while Indians will wash and bathe with great diligence every day, and keep their houses clean as well, they do not blink an eye at throwing garbage out into the street. On the inside, the house is clean. Step outside a Hindu home and you will see garbage. The streets around Hindu upper-caste settlements are often dirtier than poor tribal villages. As a result, the public spaces in Indian cities are becoming synonymous with garbage heaps.

The highly stratified and hierarchical structure of Hindu society has further promoted this individualism and callousness. Hindus may have no problems in dealing with cow dung, but they have great problems in dealing with human excreta. Human excreta in the hierarchical structure becomes a problem that outcaste or lower-caste people have to deal with. In the absence of sewage systems, this becomes very apparent. Eventually, some people just shut their eyes to the problem and concentrate on keeping themselves clean in a dirty environment. Gandhi often complained about how sad he felt upon entering any village. On the outskirts would be the effluent from the morning purgations and ablutions, casually strewn about, including excrement. Gandhi used to say that all that it would take to keep things clean would be for each person to dig a small hole in the ground, and then to cover it with leaves. But nobody would listen to him, not even the purest of Brahmins.

Hinduism's primary focus lies on the self, one's immediate family, and one's caste niche, to the neglect of the larger society and community. I know of no other human group that has more structured forms of private interpersonal relations than the Hindus. If you examine any anthropological study of Hinduism, such as the works of Lawrence A. Babb, C. J. Fuller, or Madeleine Biardeau,[7] you will find a complex description of how to conduct oneself as a Hindu: what to eat, when to eat, with whom to eat, how to behave with one's mother-in-law, spouse, father, mother, brother, sister. All the rules and roles are clearly laid down. For instance, Rāma and Lakṣmaṇa knew what they had to do, as a king and as a brother, and modeled behaviors toward women that are followed in India even today.

Whereas the private sphere is carefully scripted in Hindu tradition, public life in India borders on and often descends into chaos. Public interpersonal relations in India present some of the most unstructured behavior patterns in the world. A Hindu may go down to the Ganges River to purify himself or herself. The next moment, the same person will flush the toilet and discharge effluent into the very same sacred river. Kelly Alley's chapter in this book discusses this profound paradox, articulating the elusive difference between ritually pure and scientifically unclean. The externalization is so deep that one cares about one's own pollution but does not even consider how the sloughing off of one's own defilement can pollute others and, eventually, even oneself.

This leads to other problems in modern society, in an environment where even the mighty Ganges (Gaṅgā) cannot purify itself. No rule of law is respected. The state is an agency that commands no respect, either from those who work for it or from those who have to obey its rules. No father tells his son about the evils of corruption and bribery.

The modern state in India has failed largely because it has been blind to this behavior pattern and has failed to structure itself accordingly. When we look at the India of just three to four hundred years ago, it was one of the world's richest countries and one of the most urbanized and literate. There were hundreds and thousands of sacred groves, which were protected by community fiat. The king had nothing to do with it.

Communities throughout India built hundreds and thousands of water harvesting structures, building a reliable body of water in a monsoonal climate, where rains fall for just one hundred hours in a year. This was considered a more important and sacred act than having one thousand sons. The water catchment systems often had religious sites associated with them. In Rajasthan, the catchment area of the tanks was sacred. People could not ease themselves into the catchment area (*agor*), nor could they go into it wearing shoes.

In premodern India, nomads respected other nomads. Each group would follow its own route. Fisherfolk and hunting groups had their own rules. There were seasons in which certain fish were not to be caught and certain animals were not to be hunted. An enormously structured society operated at the community level and respected community rules.

The modern state was promoted first by a colonizing nation, England, and then by an independent elite with a socialist mind-set, starting with Nehru. Both systems disregarded governance from the bottom and tried to create governance at the top. Probably no other country has more rules and laws than India. But all these dictums remain at the stratospheric level and almost nothing reaches the bottom. Even if a rule does percolate downward, it faces entrenched behavior patterns that try to bypass or find a shortcut through the rules, as seen in the recent response to antipollution legislation. Thus, the traditional governance system that was structured like a pyramid has taken today the shape of a teetering, inverted pyramid.

As a result, India's ecology is in deep trouble. Its rivers are extremely polluted. Yamunā, the daughter of Yama, the god of death, is today a river that lives up to its name. The Gaṅgā is no better. Indeed, no river, once it hits the plains, remains clean, as explained in the chapters on river systems in this book. Groundwater is being overexploited across the country, and in urban industrial areas it is becoming irreversibly polluted.

Nearly one-third of the country's land lies bare, due to abuse and mismanagement. Biomass shortages are acute, and women can spend eight to ten hours just collecting basic necessities like firewood, fodder, and water. In such situations, the young girl immediately is recruited as an assistant to the mother in her daily chores. This means that female literacy programs can never succeed unless the work burden on the mother is removed.

The air in most Indian cities is becoming literally unbreathable. Automobiles, most without proper emissions controls, clog the streets of all major cities, but particularly New Delhi, Calcutta, and Mumbai. The air quality of New Delhi rivals that of Mexico City as the most polluted in the world. Tens of thousands of people die prematurely in New Delhi each year due to environmentally induced causes.

Environmental Victimization: A Personal Story

As an environmental activist and writer, I have tried for years to promote nationwide concern about the deteriorating state of the Indian ecosystem. In 1994, I began to experience an obstruction in my vision —black lines inside my left eye—so that I could hardly see with it. I was referred to the National Eye Institute in the United States, whose scientists diagnosed ocular and central nervous system non-Hodgkin's lymphoma. At the National Cancer Institute (NCI), I learned that the black lines in front of my eyes were cancer cells that had formed a sheet in front of the retina.

Fortunately, doctors at NCI had an experimental chemotherapy for the disease. After a highly invasive treatment, I had a year's blissful remission. After a year, the cancer returned. I was once again faced with the prospect of blindness, neurological disorders, and death. This time, the physicians recommended bone marrow transplant, which I went through in mid-1996. I hope that I have finally gotten rid of the disease. That, however, is something that only time will tell. Meanwhile, I will keep praying that some medical scientist somewhere will find a simple, less horrifying, and more definitive cure for this rare disease.

Why should this story of an individual cancer patient be of interest to anybody in a large and growing nation like India? Individuals are, after all, mere statistics. My case, however, is instructive because it represents the scale of life-threatening and destructive processes that we are inflicting on ourselves today. My cancer, like most other cancers, is deeply related to environmental pollution—an issue on which, ironically, I have written numerous articles and books, given lectures, and made films to increase public awareness of the threats we face. Therefore, I feel a sense of moral responsibility for continuing my efforts to educate.[8]

The poor, naturally, suffer more than the rich from environmental degradation. However, at least the powerful urban middle- and upper-classes were intelligent and self-indulgent enough to try and protect themselves and moderate the impact of environmental destruction on their own lives—or so we had thought. That theory has proved to be a total chimera. The elite of our nation have failed to internalize the ecological principle that every poison we put into the environment comes right back to us, in our air, water, and food. These poisons slowly seep into our bodies and take years to show up as cancer, as immune system disorders, or as hormonal or reproductive system disorders, affecting even the fetus.

Is it not imperative, therefore, for a society to find a way to balance its urge for economic growth and material comforts with the requirements of its natural and human health? Is this not something that we owe to ourselves and to our children?

Conclusion

The self-centered, materialist modern person of India desperately wants economic benefits. However, these benefits come at a tremendous cost. The average person does not understand or appreciate the dangers to life and to future generations that accompany an unquestioning embrace of the industrial model and its attendant pollutants. Furthermore, most people seem unwilling to pay the additional price to ensure that their own vehicles meet minimum air-quality standards.

Clearly, it would be wrong to say that all these problems arise only because of the nature and structure of Hinduism or only because of the nature and structure of the modern Indian state. The interaction between religious values and political structures must be reformed.

First, India needs a major overhaul of its structures of state. A sincere effort must be made to create systems that engender respect for rules and regulations at the community level. Participatory, local democratic systems must be created to replace the current exclusive dependence on a highly centralized electoral democracy. Gandhi had said that a free India should be a country of 560,000 village republics. In other words, each village should be a republic. Unfortunately, that has not happened.

Second, Hindus need to reexamine, ruthlessly, their religion. The good must be strengthened and the bad must be weeded out. Gandhi, an extremely devoted Hindu, fought extremely hard to reform the ills of the Hindu religion. Religious tenets that promote self-centeredness and hierarchy within the family and in the public sphere must change. No romantic view of Hinduism will work. Hinduism needs real reform.

Given the eclectic nature of Hinduism, there is a vast reservoir of tenets, practices, and beliefs that can help Hindus to reform Indian society. This will entail creating respect for different communities within the vast web of hierarchies. It will require striving to achieve a balance between the accumulation of material wealth and the protection of the bounteous beauties of nature. At the personal level, a balance is needed between material aspirations and the greater good of simple living. A government can never effectively legislate these balances. Only the people's own faith and beliefs and public education in the form of examples set by their leaders can help the society achieve a balance.

Democratic systems can foster more earth-friendly modes of industrialization, but only if people are prepared to pay the full cost for their consumption and only if they are prepared to respect rules and regulations. Clean technology, to the extent of its availability, is expensive. A reformist Hindu movement can meet the "challenge of balance" by promoting simple living, respect for each other, and respect for nature. It could also encourage people to make personal sacrifices for the health and well-being of the greater community. In turn, the society must move to transform the systems within which the religion works, through de-emphasizing materialism, hierarchy, and self-centeredness. Singing the praises of religion, as many Hindu scholars do, will not work by itself.

Notes

1. Centre for Science and Environment website: www.cseindia.org

2. Anil Agarwal and Sunita Narain, *The First Citizens' Report 1982: State of India's Environment* (New Delhi: Centre for Science and Environment, 1982).

3. Ibid., 17.

4. Ibid., 191.

5. Anil Agarwal, "Human-Nature Interactions in a Third World Country," in *Ethical Perspectives on Environmental Issues in India*, ed. George A. James (New Delhi: A.P.H. Publishing Corporation, 1999), 31–72.

6. Frank Korom, "On the Ethics and Aesthetics of Recycling in India," in *Purifying the Earthly Body of God: Religion and Ecology in Hindu India,* ed. Lance Nelson (Albany: State University of New York Press, 1998).

7. See, for example, Lawrence A. Babb, *The Divine Hierarchy: Popular Hinduism in Central India* (New York: Columbia University Press, 1975); C. J. Fuller, *The Camphor Flame: Popular Hinduism and Society in India* (Princeton: Princeton University Press, 1992); Louis Dumont, *Homo Hierarchicus: The Caste System and Its Implications,* trans. Mark Sainsbury (Chicago: Chicago University Press, 1970); and Madeleine Biardeau, *Hinduism: The Anthropology of a Civilization* (Delhi: Oxford University Press, 1989).

8. Anil Agarwal. "My Story Today, Your Story Tomorrow: An Environmentalist Searches for the Genesis of His Own Cancer," *Down To Earth: Science and Environment Fortnightly* 5, no. 13 (30 November 1996): 27–37.

Gandhian Philosophy and the Development of an Indigenous Indian Environmental Ethic

Too Deep for Deep Ecology: Gandhi and the Ecological Vision of Life

VINAY LAL

Gandhi and the Indian Environmental Movement

In a lecture given in 1993, the Indian historian Ramachandra Guha proposed to inquire whether Gandhi could be considered an "early environmentalist."[1] Though Gandhi was a vociferous critic of modern industrial civilization, he had, when one considers the sheer range and volume of his writings, relatively little to say about nature or its perception and representation, whether in art, literature, or science, by human beings. No doubt his writings are littered with remarks on man's exploitation of nature, and his views about these matters can reasonably be inferred from his famous pronouncement that the earth has enough to satisfy everyone's needs but not everyone's greed. He had strong views about "human nature," as does everyone else, and some circles seek to remember him for nothing more than his radical advocacy of "nature cure."

Still, when "nature" is viewed in the conventional sense, Gandhi was rather remarkably reticent on the relationship of humans to their external environment. His name is associated with innumerable political and social reform movements, extending from his famous *satyāgraha* campaigns in defiance of British rule to his efforts to abolish untouchability and open Hindu temples to what were then known as the "depressed classes," but it is striking that he never explicitly initiated an environmental movement, nor does the word "ecology" appear in his writings. Again, though commercial forestry had commenced well before Gandhi's time, and the depletion of Indian forests would persistently provoke peasant resistance, Gandhi himself was never associated with forest *satyāgraha*s, however much his name was invoked by peasants and rebels.

Guha observes also that "the wilderness had no attraction for Gandhi."[2] His writings are singularly devoid of any celebration of untamed nature or rejoicing at the chance sighting of a wondrous waterfall or an imposing Himalayan peak; and, indeed, his autobiography remains utterly silent on his experience of the ocean, over which he took unusually numerous journeys for an Indian of his time. A Melville, Conrad, or Jack London would have been, one might justifiably surmise, unintelligible to him. In Gandhi's innumerable trips to Indian villages and the countryside—and seldom had any Indian acquired so intimate a familiarity with the smell of the earth and the feel of the soil across a vast land—he almost never had occasion to take note of the trees, vegetation, landscape, or animals. As we shall see, he was by no means indifferent to animals, but he could only comprehend them in a domestic capacity. Students of Gandhi certainly are aware not only of the goat that he kept by his side and of his passionate commitment to cow protection, but of his profound attachment to what he often described as "dumb creation," indeed to all living forms.

The modern environmental movement was, of course, still several decades distant from being inaugurated in Gandhi's time, and it may be no more than wishful thinking to expect that Gandhi could have been an environmentalist a generation before that became a political possibility. But it is indubitably certain that Gandhi at least cannot be constrained or exculpated by that conventional and tedious yardstick with which so much scholarship sadly contents itself: namely, that he was a man of his times, and that an environmental sensibility was not yet positioned to intervene significantly in the shaping of society. Gandhi was an ardent exponent of vegetarianism, nature cure, and what are today termed "alternative" systems of medicine well before these acquired the semblance of acceptability in the West; he was a dedicated practitioner of recycling before the idea had crept into the lexicon of the liberal consciousness; he was a trenchant critic of modernity before the Frankfurt school, not to mention the postmodernism of Jean-François Lyotard, had provided some of the contours of modern thought; and he was, needless to say, an advocate of nonviolent resistance long before uses for such forms of resistance were found in the

United States, South Africa, and elsewhere. No one suspects that Gandhi was merely a man of his times: so it is not unlikely that Gandhi could have been an environmentalist and more, anticipating in this respect, as in many others, modern social and political movements.

Indeed, the general consensus of Indian environmentalists appears to be that Gandhi inspired and even, perhaps, in a manner of speaking, fathered the Indian environmental movement. He cannot, however, be likened to John Muir or Aldo Leopold, and much less to Thoreau: Gandhi was no naturalist, and it is doubtful that he would have contemplated with equanimity the setting aside of tracts of land, forests, and woods as "wilderness areas," though scarcely for the same reasons for which developers, industrialists, loggers, and financiers object to such altruism. The problems posed by the man-eating tigers of Kumaon, made famous by Jim Corbett, would have left less of a moral impression upon him than those problems which are the handiwork of men who let the brute within them triumph. It is reported that when the English historian Edward Thompson once remarked to Gandhi that wildlife was rapidly disappearing in India, Gandhi replied: "wildlife is decreasing in the jungles, but it is increasing in the towns."[3]

Though Ramachandra Guha has noted some limitations in viewing Gandhi as an "early environmentalist," such as his purportedly poor recognition of the "distinctive social and environmental problems" of urban areas, Guha readily acknowledges, as do most others, that the impress of Gandhian thinking is to be felt in the life and works of many of India's most well-known environmental activists.[4] It was Gandhi's own disciples, Mirabehn and Saralabehn, who came to exercise an incalculable influence on Chandi Prasad Bhatt, Vimla and Sunderlal Bahuguna, and others who have been at the helm of the Chipko agitation, a movement to ensure, in the words of women activists, that Himalayan forests continue to bear "soil, water, and pure air" for present and future generations.[5] Similarly, Baba Amte and Medha Patkar, the most well-known figures associated with the more recent Narmada Bachao Andolan, a movement aimed at preventing the construction of one of the world's largest dam projects and the consequent dislocation and uprooting of the lives of upwards of one hundred thousand rural and tribal people,[6] have been equally generous in acknowledging that their inspiration has come in great part from Gandhi. It may be mistaken to speak of these movements as "Gandhian," since any such reading perforce ignores the traditions of peasant resistance, the force of customary practices, and the appeal of localized systems of knowledge, but the spirit of Gandhi has undoubtedly moved Indian environmentalists.

Thus far, then, it appears that Gandhi presents something of a difficulty to those who would propose to describe him as the author or father of Indian environmentalism. It is undoubtedly possible to see the environmentalist in him, particularly if one is willing to engage in certain hermeneutic exercises, but one hesitates in describing him as an environmentalist. Similarly, if I may multiply the layers of this anomaly, Gandhi was a lover of animals without being a pet-lover, a warrior who absolutely forsook arms, an autocrat deeply wedded to democratic sentiments, an admirer of the *Rāmāyaṇa* who rejected the dogmatism of many of its verses, a follower of the *sanātan dharma,* or eternal faith, who in his later years would only bless intercaste weddings, and a traditionalist whose apparent allegiance to hideous traditions led him to counsel the rejection of all authorities except one's own conscience. His close friend and associate, the renowned poet and rebel Sarojini Naidu, pointed to this seemingly enigmatic aspect of Gandhi's personality when she once quipped that "it takes a great deal of money to keep Bapu living in poverty."[7] Though his pronouncements spoke of the conventional division of labor between men and women as "natural," in his own ashrams he insisted that all the members were to partake equally of all the tasks, and no differentiation was permitted, in matters of either labor or morality, between men and women; moreover, the kitchen, the toilet, the viceroy's palatial residence, and the prison were all equally fertile arenas for testing the truth of one's convictions. These circumstances constitute the grounds, as I shall endeavor to argue, for viewing Gandhi as a man with a profoundly ecological view of life, a view much too deep even for deep ecology.

What Is Deep Ecology?

Gandhi's own views would perhaps be deemed to have the closest resemblance, among the various strands of radical ecology encountered today, to the philosophical presuppositions of deep ecology. It is no coincidence that the Norwegian philosopher Arne Naess, with whose name "deep ecology" is preeminently associated, was an ardent student of Gandhi's thought and work before he turned his attention to the problems of the environment, and that in Gandhi he found a political philosopher who most clearly shows the way to the resolution of group conflicts.[8] In an extended interview that he gave a few years ago, Naess described himself as having fallen under the "influence" of Gandhi in 1931; and as war came to Europe some years later, Naess was to counsel nonviolent resistance to his pacifist friends.[9] From Gandhi Naess divined the importance of all work as a form of self-realization, and it is Gandhi who, as he was to write in a study of Gandhi's mode of conflict resolution, provided him with the assurance that

> the rock-bottom foundation of the technique for achieving the power of non-violence is belief in the essential oneness of all life. . . . More than a few people, from their earliest youth, feel a basic unity with and of all the human beings they encounter, a unity that overrides all the differences and makes these appear superficial. Gandhi was one of these fortunate people.[10]

In a short paper published in 1973, Naess was to distinguish between the "shallow" and "deep" approaches to environmentalism, and so pave the way for the "deep ecology" movement.[11] The exponents of the shallow view of environmentalism, Naess maintained, are bound to an anthropocentric view of the universe. It is not merely that they would privilege humans over other living beings, since clearly humans are endowed with faculties that are lacking in animals, such as the ability to reason and distinguish between good and evil; more fundamentally, they have no intrinsic commitment to the preservation of nature, but are only interested in nature insofar as it affects the interests of humans. Their worldview has room enough for viewing nature as something other than merely the repository of wealth to be extracted and exploited for use and profit; but if nature is not merely an instrument to some better end, it is emphatically not an end in itself. Though shallow environmentalists might take delight in clean air and water and heavily wooded forests (who wouldn't), they are desirous of having a safe and clean environment insofar as it provides a minimum assurance of healthy and comfortable living for humans. While not necessarily beholden to an economistic framework, they are by no means averse to cost-benefit analyses: thus, they would deplore pollution not only on the grounds that it fouls the air, contaminates the soil and the food that is put on the table, and renders unsafe our supply of drinking water, but because it leads to numerous other costs that outweigh any benefits that might be generated by industries that release pollutants in the air. Thus, the shallow environmentalists would insist on factoring in the costs of treatment for respiratory and skin diseases, the expenditure on research aimed at providing solutions to problems created by pollution, and so on. They would be sensitive to the fact that smoke from industries in the vicinity of the Taj Mahal has eroded the pristine quality of the marble and rendered somewhat obscure the marvelous hues of the inlaid gems and stones, and they would undoubtedly have agreed with the judge who ordered the relocation of these industries. But they may too readily ignore the fact that such relocation jeopardizes the livelihood of many people and introduces a new set of class relations. As Naess notes, "if prices of life necessities increase because of the installation of anti-pollution devices, class differences increase too."[12] Moreover, if I may hazard the proposition in this provocative form, shallow environmentalists prize museums more than they do living cultures.

Shallow environmentalists, as can now be surmised, have no intrinsic objection to industrialism, but only to its excesses: they are advocates, in the clichéd phrase, of "development with a human face," or "sustainable development" as it is known in the scholarly literature, though this perhaps slightly overstates their compliance with bourgeois models of human engineering.

Naess has also objected to shallow environmentalists on the grounds that they are largely concerned with the fate of the affluent or post-industrial nations, this concern having arisen as a consequence of the rapid depletion of nonrenewable natural resources. Though shallow environmentalists are not without democratic sentiments, they have always envisioned an upward leveling: the rest of the world was to be raised to a higher standard of living, but no decrease in their own standards of living was to be contemplated. Thus, when faced with an oil crisis and increased pollution, the shallow environmentalists would not necessarily have countenanced the elimination of automobiles, but only their more efficient use. The shallow environmentalists are among those who would support research on battery-powered automobiles, but their enthusiasm for mass-transport systems, such as battery-powered trains, is less frequently expressed. They are agreed that the problems created by technology are best resolved by improved technology. The ingrained presupposition is that technology can invariably resolve, if necessary with the aid of ethics, sociology, and the applied sciences, its own shortcomings.[13]

In contrast to "shallow" environmentalism, Naess and his supporters posit an ecological view of the world that is less wedded to technocratic and managerial solutions, short-term panaceas, and an instrumentalist (though not necessarily exploitative) view of nature. What is distinctive about "deep ecology," quite simply, is that it asks "deeper questions." Where shallow environmentalism, or what may be termed (after Thomas Kuhn) normal ecology, is reticent about asking "what kind of society would be the best for maintaining a particular ecosystem"[14]—such a query being seen as falling within the provenance of politics, ethics, value theory, and sociology—deep ecology is intrinsically committed to the proposition that it is not possible to alter humankind's relationship to nature without altering humans' relations to other humans and even the relationship to self.[15] Deep ecology entails, in Naess's words, the "rejection of the human-in-environment image in favor of the relational, total-field image":[16] humans are viewed as being not merely "in" the environment, but "of" it; and where the environment takes precedence, humans and all other species receive their just due.

The elaboration of the deep ecology movement is to be found in what are called the "platform principles." These principles command us to recognize that the "well-being and flourishing of human and nonhuman Life on Earth have value in themselves," and that these "values are independent of the usefulness of the nonhuman world for human purposes." Human beings are enjoined to respect the "richness and diversity of life forms," which are not to be compromised "except to satisfy vital needs"; the "quality of life," rather than a "higher standard of living," is to be accorded primacy"; and this "quality of life," for human and nonhuman species alike, is described as not being achievable except through a "substantial decrease of the human population." The platform principles decry the increasing interference of humans with the nonhuman world, and call for policy changes that would affect the "basic economic, technological, and ideological structures" that are today widely accepted.[17]

Deep ecology, unlike shallow environmentalism, recognizes the intrinsic worth of the nonhuman world, just as it recognizes the importance of conserving resources for the use of all species, not only human beings. Domestic animals are valued not merely because they make for good pets and serve as companions in increasingly fragmented societies, or because they satisfy our aesthetic impulses or desire to nurture those who are weaker than us, but because they have an invaluable place in the moral order. Deep ecology celebrates wilderness and an untethered nature. If the principles of diversity, symbiosis, and "biological egalitarianism" undergird deep ecology,[18] no less important is its insistence on spiritualism and religious values: as Gandhi might have put it, we are only God's trustees on earth. Deep ecology rejects the claim that growth is an intrinsically good economic end, but unlike the "growth specialists" who claim to take the long-term view but are interested in no more than resource management and profit maximization, deep ecologists feel responsible for future generations. Thus, deep ecology recognizes that overpopulation in so-called advanced countries is no more acceptable than overpopulation in developing countries;[19] and it would even go so far as to acknowledge that the vastly higher per capita consumption, whether of goods or resources, in industrialized countries places greater pressures on the environment than does overpopulation in the developing world.[20] To this extent, deep ecology can be said to have sensitivity to issues of class, and it certainly does not appear to countenance a world order where the health and well-being of the affluent nations become the predominant criteria by which policies are framed.

The proponents of deep ecology would, then, go far beyond the shallow environmentalists in the manner in which they address problems posed by the degradation of nature. If the most radical proponents of shallow environmentalism would be prepared to go no further than to advocate exclusive spending on mass-transit systems, deep ecologists must be prepared to offer a critique of automobile pollution of an altogether different ontological order. Such a critique must begin with the complex social history of the automobile, its relationship to the design and planning of American megalopolises such as Los Angeles and Houston, and the culture of fast food, drive-in theaters, and shopping malls that emerged from automobiles. This social history would also encompass the relation of the automobile to the creation of the American suburb and the rise of advertisements: "pollution" itself must be seen as taking on new meanings. All this and a great deal more, however, falls as much within the expertise of any accomplished sociologist, cultural geographer, or student of urban planning, and the deep ecologist must go much further in expounding a different world-view. If, analytically, one might ask how the automobile alters conceptions of time and space, and how it gives rise to new ideas of leisure and changing conceptions of "home," the deep ecologist must also inquire what inverse relation the automobile has to the ethos of walking. What does the decline of walking as a once widely recognized activity for the mind and body portend for our culture, and what different conception of self does the peripatetic mode suggest? Must we go only as far as our hands and feet take us, as Gandhi was to argue in *Hind Swaraj,* and what sort of transgression of limits is entailed by the automobile?[21] Have our bodies, as a consequence of automobiles, become unfit for experiencing other modes of reality?

There is something self-evidently ecological about walking, no doubt, but here commonsense understanding, or even the interpretive framework of the "expert," will not suffice to suggest why it is that the peripatetic mode signifies a different symbolic and cultural order of being. It is a telling fact that, in the English language at least, politicians run (or even stand)—but do not walk—for elections; and it is equally significant that no Indian had walked across the breadth and length of India as much as did Gandhi, just as he never ran for office. With the attainment of independence and the creation of the nation-state, the space for those who would rather walk than run had appeared to narrow. Gandhi's life was marked by an extreme regularity, and prominent in his daily regime of subversive discipline—if I may so entertain an oxymoron which has never been explored—was the daily walk of ten kilometers. It is on these walks that Gandhi encountered the poverty of a nation, and so came face to face with the village India that had all but disappeared from nationalist discourse;[22] it is on these walks that Gandhi was flanked on both sides by his secretaries, who took down his dictation and so enabled him to reply to each and every one of the tens of thousands of letters that he received; and it is on these walks that Gandhi kept pace with the time of India and the rhythms of his own body.[23]

From the perspective of deep ecology, the whole can never be encompassed by the sum of the parts. It requires no great imagination to critique technology on the grounds that it displaces human labor and so leads to anomie, just as it is transgressive of limits, but it does require an ecological vision to be able to hint at the principle of compensation that underlies the moral universe we inhabit. When the gain is easily perceived, the heart must be moved to apprehend the loss; and when the loss is patently before our eyes, we must train ourselves to perceive the gain. How many of us have even momentarily thought, as the typewriter collapsed before the onslaught of the computer, that the typewriter required a spirited defense on the grounds that the computer surrenders possession of the primal sound? What relation does the aesthetics of sound bear to the flow of ink and the stream of thought? And what of the typewriter's own predecessors, the writing brush or the humble lead pencil which no one other than Henry David Thoreau did more to develop before he moved to the next phase of his adventure in the woods?[24] The novelist Junichiro Tanizaki wonders what the history of Japan might have been if the fountain pen had been invented by the Chinese or Japanese. "The ink would not have been this bluish color but rather black, something like India ink," he writes, "and it would have been made to seep down from the handle into the brush." Japanese paper would still have been in vogue; and Japanese literature and thought might not have been so imitative. "An insignificant little piece of writing equipment, when one thinks of it," Tanizaki concludes, "has had a vast, almost boundless, influence on our culture."[25]

It is no less than the "vast, almost boundless, influence on our culture" that a pencil exercises on us that Gandhi had in mind on the occasion when he misplaced a two-inch stub of a pencil. One of Gandhi's associates, Kaka Kalelkar, has noted that at an annual session in Bombay of the Indian National Congress, the preeminent body of nationalist opinion, he found Gandhi frantically searching for something one evening. When his inquiry revealed that it was no more than a pencil, he offered Gandhi his own pencil and pleaded with him not to waste his time. But Gandhi insisted that he could not have any other pencil, and added: "You don't understand. I simply must not lose that little pencil! Do you know it was given to me in Madras by Natesan's little boy? He brought it for me with such love! I cannot bear to lose it."[26] If, today, we effortlessly substitute one pencil for another, what prevents us from substituting something else in its place tomorrow? What are the limits of substitutibility? If we recognized that we hold even a pencil in trust, would we not treat the earth more gently? And when this trust is betrayed, how do we calibrate the nature and extent of that betrayal? Another one of Gandhi's associates, Jehangir Patel, tells us, to evoke a yet more complex pencil story, that one morning he found Gandhi examining the tiny stub of a pencil "which had been put ready for his use." Gandhi commented that whoever had sharpened the pencil was "very angry. See how roughly and irregularly the wood has been scored and cut." Jehangir replied that he didn't find much wrong with it, but if Gandhi was so particular about this matter, he could perhaps make an inquiry. At breakfast, Gandhi looked around the table, and as soon as his eyes fell on Manu, he asked her: "Manu, you sharpened my pencil this morning, didn't you, and you were feeling angry when you did it?" "Yes, I was," she replied. "Well, " said Gandhi, "please don't sharpen my pencil while you are angry, it distresses me."[27] To deep ecology's concern for spiritualism and idolization of "value," Gandhi would no doubt have added, in the fullest sense of these terms, the insistence on truth and nonviolence.[28]

The Critique of Deep Ecology

So far "deep ecology" has appeared as a movement that might receive our sympathetic if not unequivocal assent, though I have already hinted at the beginning of a critique. As is now well known, it has been subjected to more systematic criticisms by exponents of social ecology and, more recently, ecofeminists. In keeping with my endeavor to pave the way for Gandhi from deep ecology, with which its proponents believe he would have had considerable affinity, I will suggest only the outlines, and that too briefly, of the principal critiques of deep ecology, since one can imagine that Gandhi would have shared in these critiques to some extent. A more exhaustive study of the critiques of deep ecology, or of the numerous variants of the ecological movement, such as bioregionalism, is well beyond the scope of this paper. From there I will move, finally, to a discussion of how Gandhi might be seen as having an ecological view of life, though perforce that does not make him an environmentalist.

The critics of deep ecology have described it as an ideology and movement that, in its resolute ecocentrism, expects human beings to reorganize their societies around the laws of nature. They have pointed to deep ecology's misanthropic tendencies, and it has not helped that such prominent ecologists as Dave Foreman, who has been prominent in the Earth First! movement, have described humans as a "cancer on nature."[29] The principal organ of the Earth First! movement, *Earth*

First! Journal, has frequently been known to espouse neo-Malthusian positions, and in the early years of the AIDS crisis, its pages aired the view that this epidemic was a blessing in disguise, since it promised to diminish human population and so relieve the pressure on the earth's resources. Deep ecology's agenda to contain the human population has been even less favorably received, as one can imagine, by some feminists. They perceive this objective, to put the matter simply, as yet another patriarchal attempt to take control of women's reproductive powers, and in particular to render third world women subservient to the interests of both indigenous and first world elites. Though the celebration of nature's fertility receives fulsome expression in the writings of deep ecologists, they seem considerably less enthused by human fertility. This tendency points to more than what feminists axiomatically assume to be the male fear of female sexuality, and to the declining emphasis on female fertility: it suggests the continuing inability of Western culture to treat children with dignity, as something other than incomplete or miniaturized versions of adults, and to exult in the joy of children.

It is the same ecocentrism of deep ecology that, as some scholars and critics have suggested, renders it oblivious to the fact that its agenda cannot be transplanted to third world countries without aggravating the social inequities that exist in those societies. The establishment of wilderness areas—a widely agreed upon objective of the American ecological and conservation movement—in India, it has been argued, often involves the displacement of local populations and the loss of traditional homelands; elsewhere, as among the Ik people of Uganda, who were expelled to make way for the Kidepo National Park, the consequences have been more catastrophic, including famine, begging, the rise of prostitution, and the total collapse of traditional societies.[30] The American model of national parks, many of them set up in areas which are very sparsely populated, and where in any case there was little conflict between people and resources, was transplanted wholesale to countries such as India where the relationship between people and their environment has been much closer, and where animals and people continue to have a symbiotic, though scarcely conflict-free, relationship with each other.[31] All of this was opaque to the members of the Indian "conservation establishment," who are inclined to see "'ordinary people' and 'conservation' [as] irreconcilably opposed," and who hold to the view that, unless one wishes to surrender to the experience of peasants and tribals, only so-called experts are qualified to speak on this matter.[32]

Even the much applauded Project Tiger, which is among the more successful conservation projects in India, has come in for this criticism, and as Ramachandra Guha writes, "the emphasis on wilderness is positively harmful when applied to the third world."[33] Guha suggests that such an emphasis amounts, in effect, to a transfer of wealth from the poor to the rich, just as it ignores the more pressing problems of environmental degradation that affect the poor, from scarcity of key natural resources to air and water pollution. If there is an intellectual poverty in deploying the conservation ethos of industrialized nations in countries that have not similarly been able to fatten themselves on the exploited wealth of others, there is perhaps also a failure in the deep ecology movement as a whole to recognize adequately the "structural" nature of environmental problems. The brunt of the Marxist critique is none other than the assertion that it is the logic of capitalism which leads to environmental degradation, and that the establishment of a "wilderness cult" not only creates rifts between environmentalists and all those who are involved in innumerable other social and political struggles, but that it signals a descent into reactionary politics. The vitriolic Murray Bookchin writes:

> The moral cant that marks the recent reworking of the ecology movement into a wilderness cult, a network of wiccan covens, fervent acolytes of Earth-Goddess religions, and assorted psychotherapeutic encounter groups beggars description. For all their talk about "self-empowerment," theistic "immanence," "care," and "interconnectedness," such mystics actually manage to navigate themselves away from the serious social issues that underlie the present ecological crisis and retreat to strategies of personal "self-transformation" and "enrichment" that are predicated on myths, metaphors, rituals, and "green" consumerism.[34]

Though Bookchin's own denigration of what may be termed poetic modalities, his caricature of certain strands of feminism, and similarly his equation—which he has rendered explicit elsewhere—of deep ecology with gnosticism and witchcraft, sheer woolliness about preliterate human's supposed oneness with nature, and oriental forms of "mysticism"—which are better recognized in the West than they are in the East—all equally point to the acute limitations in his own thinking, the force of his criticism cannot but be acknowledged by those who are conversant with the history of how radical movements and philosophies are almost invariably denuded of their political force in the United States.

Indeed, Bookchin and many others have gone much further in denouncing deep ecology for its fascistic tendencies. It is alleged that deep ecology's valorization of "rootedness in the soil," its excessive biocentrism, and its undifferentiated love of animals make it the companionable mate of the "nature mysticism" of National Socialism. It was Nazi Germany that, in November 1933, passed the first law in the Western world calling for the explicit protection of animals as beings-in-themselves—in other words, a law which "would recognize the right which animals inherently possess to be protected in and of themselves."[35] One philosopher finds in Nazi legislation and deep ecology "a shared revalorization of the *primitive* state against that of (alleged) civilization," the "same *romantic and/or sentimental* representation of the relationship between nature and culture."[36] Of course, this criticism is little more than the tiresome rejoinder that we all are, or ought to be (barring the reticence of some obdurate primitives), the children of the Enlightenment, and that no critique of modernity is to be permitted except in categories and terms rendered permissible by the Enlightenment's own structures of thought. Those critiques which seek to lay bare the purported fascism of deep ecology seem woefully unaware of their own oppressive parochialism.

By far the most sustained critique of deep ecology, however, has emanated from within ecofeminism. Where social ecology finds deep ecology inadequately grounded in an awareness of the nature of modern social relations, ecofeminism finds deep ecology deeply embedded in the same patriarchal assumptions which men generally hold, and which as a consequence render them sharply deficient in political awareness. The outlines of the feminist critique were first noted in a short albeit powerful paper by the Australian feminist Ariel Kay Salleh. She observed that the formulations of deep ecology use "the generic term *Man* in a case where use of a generic term is not applicable." This is no minor matter, for women's experiences of menstruation, pregnancy, childbirth, lactation, and menopause already ground their consciousness "in the knowledge of being coterminous with Nature." What the deep ecologist seeks to introduce as an "abstract ethical construct," namely, the desirability of a communion with nature on the principle of shared living, already constitutes part of women's experiences.[37]

Though deep ecology purports to celebrate life-affirming values, its advocacy of strict population control constitutes an intervention in natural life processes, and to this extent it partakes of the same rationalist and technicist worldview that it otherwise critiques. While deep ecology recognizes the fact of oppression, and deplores man's exploitation of man and of nature, it does not recognize man's oppression of woman. It is attuned to the suffering and oppression of animals, but the everyday oppression of women escapes its watchful attentiveness.[38] Salleh insists, quite rightly, that to assimilate man's oppression of woman into a general class of exploitative acts perpetrated by man upon man is to overlook the political meaning of violence against women and the iconic meanings attached to "woman" in various discursive formations. The ecofeminist is reserved about deep ecology's otherwise commendable anticlass postures because "in bypassing the parallel between the original exploitation of nature as object-and-commodity resource and of nurturant women as object-and-commodity resource, the ecologist's anti-class stance remains only superficially descriptive, politically and historically static. It loses its genuinely deep structural edge."[39]

Though appreciative of deep ecologists' endeavors to be more humane and caring, Salleh characterizes their objective, the "spiritual development of 'personhood,'" as the "self-estranged male reaching for the original androgynous natural unity within himself." Deep ecology is a largely "self-congratulatory reformist move; the transvaluation of values it claims for itself is quite peripheral"; and it represents "a spiritual search for people in a barren secular age."[40] It is surely no accident that it is the most secular versions of Eastern spiritual traditions, such as Zen Buddhism and the thought of Krishnamurti, that have attracted the largest followings in the West. And if deep ecology's most sustained contribution is to reject the "instrumentalist pragmatism of the resource-management approach to the environmental crisis," even here Salleh finds deep ecology's shortcomings ominous. She makes the very pointed remark that the constant references to "implementation of policies," "exponential growth of technical skill and intervention," and the like betray the fact that "the masculine sense of self-worth in our culture has become so entrenched in scientistic habits of thought, that it is very hard for men to argue persuasively without recourse to terms like these for validation." Naess's own writings are pervaded by terms such as "rules," "postulates," "hypotheses," and "policy formulations," and his "overview of ecosophy is a highly academic and positivized one, dressed up in the jargon of current science-dominated standards of acceptability."[41]

It was another feminist writer who suggested that ethics has been discussed primarily in the language of the father. This is the language of fairness, justice, and rights. Perhaps, if ethics deigned to speak in the language of the mother, the language of human caring, and of the memory equally of caring and of being cared for, deep ecology might become truly deep.[42] It is this language that, as we have seen in the expression of Bookchin's outrage, hard-nosed realists will seek to mock, and not always without cause. Caring, too, in the manner of everything else, has become an industry with its management specialists, professionals, and various other staffers; it has also become, in a world saturated by the media, pop psychology, and political correctness, a substitute for thought, reflection, spiritual discipline, and equanimity. Nonetheless, the contamination of the ethic of caring by marketing and crude psychological reductionism does not entirely vitiate the possibility that we can yet render ourselves ecological in more ways than deep ecology can possibly imagine. The "deep ecology movement will not truly happen," writes Ariel Salleh, "until men are brave enough to rediscover and to love the woman inside themselves."[43] No ecology, howsoever deep, can give us pregnant fathers. How far beyond deep ecology, then, does Gandhi take us?

Gandhi and the Ecological View of Life

Though Gandhi was no philosopher of ecology, and can only be called an environmentalist with considerable difficulty, he strikes a remarkable chord with all those who have cared for the environment, practiced vegetarianism, cherished the principles of nonviolence, resisted the depredations of developers, or accorded animals the dignity of humans. It is useful to recall that the word 'ecology' is derived from 'economy' (from Greek, *oikonomia*), which itself has little to do, in its primal sense, with inquiries made by those who are now styled economists; rather, economy was understood to pertain to the most efficient and least costly management of household affairs. It is largely an application of this meaning that Thoreau had in mind when, in titling the opening chapter of *Walden* "Economy," he described the manner in which he reduced his needs as much as his wants to the bare minimum, and so lived life to the fullest. It is the same economy of life-style—and indeed of conduct, speech, and thought—that Gandhi ruthlessly put into practice in his various ashrams. From there, then, one can follow the trajectory from 'economy' to 'ecology'. The *Oxford English Dictionary* defines ecology as the "science of the economy of animals and plants," and this implies the imperative to look after animals, plants, and the environment to which they bear a relation. Ecology consequently means, in the first instance, that we are commanded to economize, or render less wasteful, our use of the earth's resources. To do so, we have to use our own resources, howsoever narrowly conceived, with wisdom and with the utmost respect for economy. On no other grounds can we explain the many apparently enigmatic, and some would say bizarre or idiosyncratic, practices of thought and conduct in which Gandhi engaged.

A recent study of Gandhi, which describes him as "a practicing ecological yogi," makes the point that Gandhi bound himself to the observance of a certain set of rules of conduct.[44] Some of these rules, or *niyama*s, prescribe what a human being should do, while others (*yama*s) are injunctions to abstain from forms of conduct harmful to humans and, in some cases, other living beings. T. N. Khoshoo suggests that it is from these environmental and ethical principles, which variously counsel us to practice austerity, introspect on the self, cultivate contentment, learn self-reliance, renounce possessions beyond our needs, and always keep in mind the interests of the weakest and the poor, that Gandhi derived his political movement; and it is from these same principles, argues Khoshoo, that Gandhi worked to develop his ideas of "sustainable development."[45]

It is doubtful, however, that Gandhi spoke at all in the language of "development," much less in the language of "sustainable development," since the very idea of development owes a great deal to the politics of knowledge in the post-World War II period.[46] Besides, ethics, ecology, and politics were all closely and even indistinguishably interwoven into the fabric of his thought and social practices. If, for instance, his practice of observing twenty-four hours of silence on a regular basis was a mode of conserving his energy, entering into an introspective state, and listening to what he called the still voice within, it was also a way of signifying his dissent from ordinary models of communication with the British and establishing the discourse on his own terms. Similarly, Gandhi deployed fasting not only to open negotiations with the British or (more frequently) various Indian communities, but to cleanse his own body, free his mind of impure thoughts, feminize the public realm, and even to partake in the experience of deprivation from which countless millions of Indians suffered. Gandhi deplored the idea of waste, and fasting was a sure means of ascertaining the true needs of the body and preserving its ecological equanimity.

In considering Gandhi in relation to ecology, then, his entire life opens up before us, a life documented, moreover, in almost excruciatingly minute detail. One is confronted at once with the anomaly that, whatever Gandhi's propensity to be ecological in thought and conduct, he was an extraordinarily prolific writer. Admittedly, by far the greater majority of the pieces collected together in the gargantuan one-hundred-volume set of his collected writings are models of brevity, and Gandhi did not waste a word. To write poorly was to do violence to the language and to the recipient of one's missives, and Gandhi chose his words with great care. At the same time, he was quite adamant that nothing was to remain of his writings upon his death. "My writings should be cremated with my body," he wrote, adding: "What I have done will endure, not what I have said or written."[47]

Indeed, what is remarkable about Gandhi's life is that, unlike most public figures with a political career, whose social practices are tradition-bound even when their pronouncements are radical, Gandhi was extremely conservative in his pronouncements while being radical in his conduct. As I have noted previously, though he thought that men would continue to be the principal bread-winners for the family, his ashrams were run firmly on the principle that women and men would share equally in all the work. For someone who ate very modestly, Gandhi himself spent an inordinate amount of time in the kitchen. Though spinning might, in village India, be a task undertaken by women, Gandhi himself spun every day, and made of it, and the respect that it implied for bodily labor as the source of one's daily bread, a religion to be followed by all those who aspired for independence. If he was insistent that women were to remain chaste, he was even more adamant that such chastity was incumbent on men, who had rendered women into sexual objects.[48] The profoundly ecological impetus of his style here demands recognition: promising little, he was generous in the fulfillment of his word. Nature may appear to be niggardly, but its rewards are rich and deeper than we habitually imagine.

Moving from Gandhi's writings to his social practices and conduct, to which anecdotes from his life are perhaps the best guide, suggests innumerable ways, of which I shall mention only a few, in which Gandhi can be seen as having an ecological vision of life.

First, as nature provides for the largest animals as much as it provides for its smallest creations, so Gandhi allowed this principle to guide him in his political and social relations with all manner of women and men. Gandhi's close disciple and attendant, Mirabehn, wrote that while he worked alongside everyone else in the ashram, he would carry on his voluminous correspondence and grant interviews. "Big people of all parties, and of many different nations would come to see Bapu, but he would give equal attention to the poorest peasant who might come with a genuine problem."[49] In the midst of important political negotiations with senior British officials, he would take the time to tend to his goat. It is this aspect of Gandhi's personality that his contemporary, the short-story writer Acharya Chatursen Shastri, captured when, in a story about Gandhi, he showed him peeling potatoes while in a conversation with a little boy.[50] He remained supremely indifferent to considerations of power, prestige, and status in choosing his companions; similarly, he was as attentive to the minutest details as he was to matters of national importance. One of his associates has reported—and such stories are legion—that when news reached Gandhi of the illness of the daughter of a friend, he wrote to her a long letter in the midst of an intense political struggle in Rajkot, detailing the medicines that she was to take, the food that she was to avoid, and the precautions she was to exercise. Though he was notoriously thrifty, writing even some of his letters on the back of envelopes addressed to him, he did not begrudge spending a large sum of money to send her a telegram.[51] His own grandniece, pointing to the meticulous care with which Gandhi tended to her personal needs, all the while that he was engaged in negotiations for Indian independence, perhaps showered him with the most unusual honor when, in writing a short book about him, she called it *Bapu—My Mother*.[52] The interpreters of Gandhi have chosen to describe his equanimity as an illustration of the teachings of the *Bhagavadgītā,* which counsels us to perform our duty and remain equally indifferent to joy and sorrow, praise and censure; but his life here can just as well be seen as illustrative of the working of laws of nature. The banyan tree gives its shade alike to the prince and the commoner, the vendor and the businessman, the saint and the crook.[53]

Second, without being an advocate of wilderness as that is commonly understood today, Gandhi was resolutely of the view that nature should be allowed to take its own course. Arne Naess has written that he "even prohibited people from having a stock of medicines against poisonous bites. He believed in the possibility of satisfactory co-existence and he proved right. There were no accidents. . . ."[54] His experiments in nature care are well-known, as is his advocacy of enemas and mud baths, but there is more to these narratives than his rejection of modern medicine. Mirabehn has reported that one day, as Gandhi worked in a tent in the afternoon heat of 110 degrees, she and some other workers became exasperated at their inability to keep away the hordes of flies. "I'm told they have come down from the tree tops for shade, Bapu," said Mirabehn, whereupon he replied: "Yes. It is not for me to blame them. If God had made me one such, I should have done exactly the same."[55] Gandhi scarcely required the verdict of the biologist, wildlife trainer, or zoologist to hold to the view that nature's creatures mind their own business, and that if humans were to do the same, we would not be required to legislate the health of all species. On occasion a cobra would come into Gandhi's room: there were clear instructions that it was not to be killed even if it bit Gandhi, though Gandhi did not prevent others from killing snakes. "I do not want to live," wrote Gandhi, "at the cost of the life even of a snake."[56] In some rendering of these stories, the cobra would often be described as rearing itself up before Gandhi and placing its hood above his head, as if in homage to the emperor.[57] The hagiographic tone of these accounts does not, however, detract from the fact, for which there is ample evidence, that Gandhi was quite willing to share his universe with animals and reptiles, without rendering them into objects of pity, curiosity, or amusement. He described himself as wanting "to realise identity with even the crawling things upon earth, because we claim descent from the same God, and that being so, all life in whatever form it appears must essentially be so,"[58] but it is altogether improbable that he would have followed some deep ecologists in treating animals, insects, and plants as persons.

Though it may be reasonable to infer that it was Gandhi's adherence to nonviolence that would have prevented him from taking the life of a snake, such an interpretation ignores the critical primacy accorded to *satya* (truth) over *ahiṃsā* (nonviolence) in Gandhian thinking, much as it overlooks the fact that Gandhi was an advocate of the mercy killing of animals. The incident when he had a young calf in his ashram put to death with an injection when she could not be saved from an extreme illness is well-known; less known is the incident of the stray dogs.[59] In 1926 Ambalal Sarabhai, a textile magnate in Ahmedabad and friend of Gandhi, rounded up sixty rabid dogs on his properties and had them shot; subsequently, feeling repentant, he approached Gandhi to share his anguish with him. Gandhi comforted him with the remark, "What else could be done?"[60] When the Ahmedabad Humanitarian Society came to know of this, it sought an urgent explanation from Gandhi; and thereafter, for the next three months, as Gandhi himself took this issue to the public, he was plummeted with letters accusing him of cruelty to animals and of forsaking his commitment to *ahiṃsā*. Throughout, while admitting that he might have erred, Gandhi explained his position with consistency and clarity. There was no course open to Sarabhai "but the destruction of rabid dogs. At times we may be faced with the unavoidable duty," wrote Gandhi, "of killing a man who is found in the act of killing people."[61] Roving dogs, particularly a swarm of them, were a "menace" to society; the multiplication of them was quite unnecessary; and those who now counseled their protection on the grounds of religion, even at cost to the life and safety of humans, were to be reminded that to practice "the religion of humanity" required also the recognition that "we offend against dogs as a class by suffering them to stray and live on crumbs or savings from our plates that we throw at them and we injure our neighbours also by doing so."[62]

Gandhi unequivocally rejected the argument that protection must always entail "mere refraining from killing"; quite to the contrary, "Torture or participation, direct or indirect, in the unnecessary multiplication of those that must die is *hiṃsā* [violence]."[63] As he was to reiterate in another rejoinder, "Merely taking life is not always *hiṃsā,* one may even say that there is sometimes more *hiṃsā* in not taking life."[64] He advised those who had become agitated about this matter to learn from the West, where the people had "formulated and perfected" a "regular science of dog-keeping." It was mistaken to believe that people in the West were lacking in humanity: as Gandhi put it trenchantly, "The ideal of humanity in the West is perhaps lower, but their practice of it is very much more thorough than others. We rest content with a lofty ideal and are slow or lazy in its practice. We are wrapped in dark darkness, as is evident from our paupers, cattle and other animals. They are eloquent of our irreligion rather than of religion."[65]

Third, Gandhi transformed the idea of waste and rendered it pregnant with meanings that were the inverse of those meanings invested in it by European representational regimes. As the complex scholarship around the practices of colonialism has now demonstrated, almost nothing was as much anathema to European colonizers as the idea that the vast lands lying before their gaze, whether in largely barren areas of Australia and Canada or in the densely inhabited parts of India, were entirely unproductive or certainly not as productive as they thought desirable. To render them fertile, they had first to render them productive of meaning, as something other than realms of emptiness (and hence of nothingness), which was only possible by construing them as wastelands which required the brain, will, and energy of white men to effect their transformation.

Gandhi, on the other hand, was inclined to the view that humans were prone to transform whatever they touched, howsoever fertile, fecund, or productive, into waste. His close disciple and associate, Kaka Kalelkar, narrates that he was in the habit of breaking off an entire twig merely for four or five *nīm* leaves he needed to rub on the fibers of the carding-bow to make its strings pliant and supple. When Gandhi saw that, he remarked: "This is violence. We should pluck the required number of leaves after offering an apology to the tree for doing so. But you broke off the whole twig, which is wasteful and wrong."[66] He also described himself as pained that people would "pluck masses of delicate blossoms" and fling them in his face or string them around his neck."[67] Yet this alone was not wasteful: there was also human waste, around the disposal of which an entire and none too savory history of India can be written. While it was a matter of shame that Indian society had set apart a special class of people to deal with the disposal of human excrement, whose occupation made them the most despised members of society,[68] Gandhi found it imperative to bring this matter to the fore and make it as much a subject of national importance as the attainment of political independence and the reform of degraded institutions. Unlike the vast majority of caste Hindus, Gandhi did not allow anyone else to dispose of his waste. His ashrams were repositories for endeavors to change human waste into organic fertilizer. Moreover, during the course of the last twenty years of his life, he was engaged in ceaseless experiments to invent toilets that would be less of a drain on scarce water resources.

Fourth, and this is a point that cannot be belabored enough, Gandhi did not make of his ecological sensitivities a cult or religion to which unquestioning fealty was demanded. His attitude toward meat is illustrative in this respect. Though he was himself a very strict vegetarian, he was not insistent that everyone else, even in his presence, should be forbidden from eating meat. Khoshoo credits him with the saying, "I am a puritan myself but I am catholic towards others," and rightly rejects the notion that Gandhi might have been a "puritanical vegetarian."[69] But this is a testimony only to Gandhi's liberality, not to that ecumenical feature of his thinking which is based on a different notion of largesse. Once, when he had a European visitor at his ashram who was habituated to meat at every meal, Gandhi had meat served to him. He himself partook of milk and milk products, unlike those who style themselves "vegans" in the United States, and his reverence for life and respect for animals did not border on that fanaticism which is only another name for violence. One of his disciples, Jehangir Patel, has written that one day Mirabehn came running to him in an agitated state of mind. "Bapu won't be able to eat his breakfast," she said. "Some one has put meat into the fridge where his food is. How could you allow such a thing?" The cook, Ali, explained that he had gotten the meat for the dogs, and offered to remove it at once. Jehangir asked him to let the meat remain there, and Gandhi himself was fetched. Jehangir then apologized to Gandhi: "I did not think of speaking to Ali. I did not realise that this might happen." Gandhi replied, "Don't apologise. You and Ali have done nothing wrong, so far as I can see." Gandhi took some grapes lying next to the meat, and popped them into his mouth; turning then to Mirabehn, he said: "We are guests in our friend's house, and it would not be right for us to impose our idea upon him or upon anyone. People whose custom it is to eat meat should not stop doing so simply because I am present."

Similarly, though Gandhi championed prohibition, he would not prevent anyone from drinking alcohol, and he condemned altogether the principle of drinking on the sly; as he told Jehangir, "I would much rather you were a drinker, even a heavy drinker, than that there should be any deceit in the matter."[70]

Gandhi's entire life, I would submit, constitutes an ecological treatise, and it is no exaggeration to suggest that he left us, in his life, with the last of the Upaniṣads, or "forest books." He dispersed wisdom, but not from a mountaintop; he even waded through human waste as he walked around riot-torn villages, but he retained his equanimity. The grounding for his own ecological vision was clearly furnished by what he understood, perhaps with some naïveté, as the ecological wisdom of India's epic and religious literature,[71] just as it is amply clear that in his practice of simple living and nonviolence and advocacy of *satya* and *brahmacarya,* Gandhi sought to put the principles of an ecologically aware life into motion.

But these are truisms that shall have to be inflected in more than the ordinary fashion, and yield more than the clichéd observations that Gandhi was the "prophet of nonviolence" or an astute political campaigner unusually interested in moral questions, if we are to be fully cognizant of the profound manner in which Gandhi's entire life functioned much like an ecosystem. This is one life in which every minute act, emotion, or thought was not without its place: the brevity of Gandhi's enormous writings, his small meals of nuts and fruits, his morning ablutions and everyday bodily practices, his periodic observances of silence, his morning walks, his cultivation of the small as much as of the big, his abhorrence of waste, his resort to fasting—all these point to the manner in which the symphony was orchestrated. Though the moralists, nonviolent activists, feminists, journalists, social reformers, trade union leaders, peasants, prohibitionists, nature-cure lovers, nudists, critics of Western medicine, renouncers, and scores of others will all find in Gandhi something to sustain them in their aspirations and objectives, Gandhi will remain elusive unless the deeply ecological foundations of his life are recognized.

Notes

1. Ramachandra Guha, *Mahatma Gandhi and the Environmental Movement,* The Parisar Annual Lecture 1993 (Pune: Parisar, 1993), 2. I am grateful to my research assistant, Ben Marschke, for his help with library work.

2. Ibid., 20.

3. Cited by T. N. Khoshoo, *Mahatma Gandhi: An Apostle of Applied Human Ecology* (New Delhi: Tata Energy Research Institute, 1995), 18.

4. Guha, *Mahatma Gandhi and the Environmental Movement,* 20.

5. For a short account of the Chipko movement, which highlights in particular its Gandhian impetus, see J. Bandyopadhyay and Vandana Shiva, "Chipko," *Seminar* 330 (February 1987): 33–39; a more systematic account is furnished by Anupam Mishra and Satyendra Tripathi, *Chipko Movement: Uttarakhand Women's Bid to Save Forest Wealth* (New Delhi: People's Action for Development with Justice, 1978). For a short account by Sunderlal Bahuguna, which acknowledges the inspiration he received from Gandhi's life, see *The Chipko Message* (Silyara Tehri, Garhwal: Chipko Information Centre, [1987]), esp. 22–26; also useful is Sunderlal Bahuguna, "The Crisis of Civilization and the Message of Culture in the Context of Environment," *Gandhi Marg* 9, no. 8 (November 1987): 451–68, though Bahuguna reads much better in Hindi. These might well be considered "partisan" accounts of the Chipko movement; for a more scholarly and detached treatment, the reader would do well to turn to Ramachandra Guha, *The Unquiet Woods: Ecological Change and Peasant Resistance in the Himalaya* (Delhi: Oxford University Press, 1991), 153–84. Mirabehn's work in the Himalayan region is ably and touchingly evoked in Krishna Murti Gupta, *Mira Behn: Gandhiji's Daughter Disciple,* Birth Centenary Volume (New Delhi: Himalaya Seva Sangh, 1992). On the women of Chipko, see Vandana Shiva, *Staying Alive: Women, Ecology, and Survival in India* (New Delhi: Kali for Women, 1988), 67–77.

6. The figure of one hundred thousand displaced people is based, as Madhav Gadgil and Ramachandra Guha have noted, on the "outdated 1981 census," and others have furnished much higher numbers. See their *Ecology and Equity: The Use and Abuse of Nature in Contemporary India* (Delhi: Penguin Books, 1995), 73.

7. Cited by Geoffrey Ashe, *Gandhi* (New York: Stein and Day, 1968), 267. Some attribute the saying to Gandhi himself: see David Rothenberg, *Is It Painful to Think? Conversations with Arne Naess* (Minneapolis: University of Minnesota Press, 1993), 170.

8. Arne Naess, *Gandhi and the Nuclear Age* (Totowa, N.J.: Bedminster Press, 1965).

9. Rothenberg, *Is It Painful to Think?* 103–5. Elsewhere, Naess wrote of himself as a "student and admirer since 1930 of Gandhi's non-violent direct actions in bloody conflict," "inevitably influenced by his metaphysics." See Arne Naess, "Self-Realization: An Ecological Approach to Being in the World," *Trumpeter* 4, no. 3 (1987): 35–42; reprinted in *The Deep Ecology Movement: An Introductory Anthology,* ed. Alan Drengson and Yuichi Inoue (Berkeley, Calif.: North Atlantic Books, 1995), 22.

10. Arne Naess, *Gandhi and Group Conflict: An Exploration of Satyagraha. Theoretical Background* (Oslo: Universitietsforlaget, 1974), cited by Rothenberg, introduction to *Is It Painful to Think?* xix.

11. Arne Naess, "The Shallow and the Deep, Long-Range Ecology Movement: A Summary," in *The Deep Ecology Movement,* ed. Drengson and Inoue, 3–10.

12. Ibid., 5.

13. For an engaging critique of "technicism," see Ashis Nandy, "From Outside the Imperium: Gandhi's Cultural Critique of the West," in his *Traditions, Tyranny, and Utopias: Essays in the Politics of Awareness* (Delhi: Oxford University Press, 1987; reprint, Oxford India Paperbacks, 1992), 127–62.

14. Stephen Bodian, "Simple in Means, Rich in Ends: An Interview with Arne Naess," in *Deep Ecology for the Twenty-first Century: Readings on the Philosophy and Practice of the New Environmentalism,* ed. George Sessions (Boston: Shambhala, 1995), 27.

15. I have used the expression "man's relationship to nature" deliberately, since the brunt of ecofeminist criticism, which I shall consider at greater length subsequently, is precisely that deep ecology is just as patriarchal as environmentalism or the other philosophies that it critiques.

16. Naess, "The Shallow and the Deep, Long-Range Ecology Movement," 3.

17. Arne Naess and George Sessions, "Platform Principles of the Deep Ecology Movement," in *The Deep Ecology Movement,* ed. Drengson and Inoue, 49–50.

18. See Ariel Kay Salleh, "Deeper than Deep Ecology: The Eco-Feminist Connection," *Environmental Ethics* 6, no. 4 (winter 1984): 340.

19. Bodian, "Simple in Means, Rich in Ends," 29.

20. Arne Naess, "The Deep Ecological Movement: Some Philosophical Aspects," in *Deep Ecology for the Twenty-first Century,* ed. Sessions, 72–73.

21. M. K. Gandhi, *Hind Swaraj or Indian Home Rule* (1909; new ed., 1938; reprint ed., Ahmedabad: Navajivan Publishing House, 1982), 44–46.

22. I merely wish to call attention to the criticism, without considering it at length, that Gandhi idealized "village India," and consequently endorsed the caste system which has historically been the basis of village life. The further extrapolation, namely, that the eradication of untouchability, which cannot be understood other than in relation to the caste system and village life, demands the embrace of all that stands in opposition to "village India"—urbanization, industrialization, scientific planning, technological growth—suggests how fundamentally at odds Gandhi's worldview may be with the position advocated by radical Dalits. Doubtless, the Dalits (formerly known as the Untouchables, described in official reports as the "scheduled" or "depressed" castes, and "christened" as "Harijans," or "children of God," by Gandhi) appear as ardent advocates of modernization and industrialization. A proper consideration of the issues at stake here would take us far astray into more detailed readings of Gandhi's views on science, modernity, and development, as much as into the heated debates surrounding Indian developmental politics. It is also erroneous to suppose that, because the Dalits are hostile to the idea of "village India" and to Gandhi's thinking, they lack an ecological sensibility. Oppressed people who have survived on as little as have the Dalits are extraordinarily ecological in their habits and thinking. For an interesting illustration of some of these debates, which have made their way into Indian newspapers, see Gail Omvedt, "Why Dalits Dislike Environmentalists," *Hindu,* 24 June 1997, 17, and a rejoinder of sorts by A. Ranga Rajan, "Why Dalits Should Become 'Environmentalists,'" *Hindu,* 22 July 1997, 17.

23. The cultural politics of walking has not received any attention, and the huge literature on Gandhi has singularly failed to comprehend the place of this phenomenon in Gandhi's life; as might be expected, however, Gandhi's famous march to the sea at Dandi has been the focus of many scholarly studies, and also of major artistic representations, such as the portraits of Gandhi by the Bengali painter Nandlal Bose. Rabindranath Tagore had a glimpse of what it meant for Gandhi to walk from one village to another when he wrote of him: "He stopped at the threshold of the huts of the thousands of dispossessed, dressed like one of their own. He spoke to them in their own language. Here was living truth at last, and not only quotations from books. For this reason the Mahatma, the name given to him by the people of India, is his real name." See his "The Call of Truth," *Modern Review* (October 1921); there are slight variants in this widely reproduced article. A unique study by Anne D. Wallace, *Walking, Literature, and English Culture: The Origins and Uses of the Peripatetic in the Nineteenth Century* (Oxford: Clarendon Paperbacks, 1994), points to some of the ways in which a study of Gandhi's association with walking might be attempted.

24. For an arresting account of the history of the pencil and the part played by Thoreau in creating what is essentially its modern shape, see Henry Petroski, *The Pencil: A History of Design and Circumstance* (New York: Alfred Knopf, 1990).

25. Junichiro Tanizaki, *In Praise of Shadows,* trans. Thomas J. Harper and Edward G. Seidensticker (New Haven: Leete's Island Books, 1977), 7–8.

26. Kakasaheb Kalelkar, *Stray Glimpses of Bapu* (Ahmedabad: Navajivan Publishing House, 1950; 2d rev. ed., 1960), 26–27. G. A. Natesan, founder and editor of the *Indian Review,* became a firm supporter of Gandhi, and it is his publishing house that released the first major anthologies of Gandhi's writings.

27. Jehangir P. Patel and Marjorie Sykes, *Gandhi: His Gift of the Fight* (Rasulia, Madhya Pradesh: Friends Rural Centre, 1987), 107–8.

28. Few people in the deep ecology movement, apart from Naess himself, appear to have studied Gandhi's works and life; and among those who have, they seem to have gone no further than Naess's own books on Gandhi. One of the principal figures in the movement, Bill Devall, admits that the "term *nonviolence* creates problems for some supporters of deep ecology because they equate nonviolence with passivity or non-resistance. Nothing could be further from the meaning of nonviolence as used by Gandhi, Martin Luther King, Jr., and by Greenpeace." See Bill Devall, *Simple in Means, Rich in Ends: Practicing Deep Ecology* (Salt Lake City: Gibbs Smith Publisher, 1988), 141. His own discussion of Gandhi, under the heading "Direct Action, Monkeywrenching and Ecotage," betrays considerable ignorance of the complexity of Gandhi's thought and the common susceptibility to sloganeering and impoverished classificatory schemes (pp. 138–50). "Ecotage," presumably from ecology + sabotage, would not likely have met with Gandhi's approval; and fancy neologisms do not go very far toward making a nonviolent revolution.

29. Dave Foreman, "Beyond the Wilderness," *Harper's Magazine,* April 1990, 48. Murray Bookchin, in the revised edition of *The Ecology of Freedom: The Emergence and Dissolution of Hierarchy* (Montreal and New York: Black Rose Books, 1991), says of deep ecology and other "mystical ecologies" that they "often view the human species as an evolutionary aberration—or worse, as an absolute disaster, a 'cancer' on the biosphere" (xxi).

30. See Ashish Kothari, Saloni Suri, and Neena Singh, "People and Protected Areas: Rethinking Conservation in India," *Ecologist* 25, no. 5 (September-October 1995): 188–94, at p. 190.

31. On the Sundarbans Tiger Preserve, which has the largest tiger population in the world, see the charming book by Sy Montgomery, *Spell of the Tiger: The Man-Eaters of Sundarbans* (Boston: Houghton Mifflin, 1995). I by no means wish to suggest that conflicts over resources between animals and humans are unknown in India; indeed, they are all too common, and widely reported in Indian newspapers and specialized magazines, such as *Sanctuary* and *Down to Earth*. Nor is my observation of the "symbiotic" relationship of animals to humans based on the Hindu reverence for the cow, or on the fact that cows roam Indian streets with abandon. The question partly revolves around understanding how far animals are or are not invested with distinct identities, their place in the cultural imaginary, and the dependence of animals and humans on each other and the closeness of their relations.

32. Kothari, Suri, and Singh, "People and Protected Areas," 190.

33. Ramachandra Guha, "Radical American Environmentalism and Wilderness Preservation: A Third World Critique," *Environmental Ethics* 11, no. 1 (spring 1989): 75.

34. Bookchin, *Ecology of Freedom,* il.

35. Cited by Luc Ferry, *The New Ecological Order*, trans. Carol Volk (Chicago: University of Chicago Press, 1995), 91–100.

36. Ibid., 93. See also the discussion in Michael Zimmerman, *Contesting Earth's Future: Radical Ecology and Postmodernity* (Berkeley and Los Angeles: University of California Press, 1994), 170–83.

37. This paragraph draws on Salleh, "Deeper than Deep Ecology," 340–41.

38. The argument is reinforced in Ariel Salleh, "The Ecofeminism/Deep Ecology Debate: A Reply to Patriarchal Reason," *Environmental Ethics* 14, no. 3 (fall 1992): 204.

39. Salleh, "Deeper than Deep Ecology," 341.

40. Ibid., 344–45.

41. Ibid., 342–43.

42. Patsy Hallen, "Making Peace with Nature: Why Ecology Needs Feminism," in *The Deep Ecology Movement,* ed. Drengson and Inoue, 208, citing Nel Noddings, *Caring: A Feminine Approach to Ethics and Moral Education* (Berkeley and Los Angeles: University of California Press, 1984), 1.

43. Salleh, "Deeper than Deep Ecology," 345.

44. Khoshoo, *Mahatma Gandhi: An Apostle of Applied Human Ecology,* 1–2.

45. Ibid., 2, 8.

46. Cf. Arturo Escobar, *Encountering Development: The Making and Unmaking of the Third World* (Princeton, N.J.: Princeton University Press, 1995).

47. Cited by Sunil Khilnani, "A Bodily Drama," *Times Literary Supplement,* 8 August 1997.

48. The recent anthology by Puspha Joshi, *Gandhi on Women* (Ahmedabad: Navajivan Publishing House, 1990), provides extraordinary insights into Gandhi's views on chastity, the division of labor between the sexes, and the public role of women; for his experiments in the kitchen and thoughts on food, see M. K. Gandhi,

Key to Health, trans. Sushila Nayar (Ahmedabad: Navajivan Publishing House, 1948), 10–24.

49. Gupta, *Mira Behn: Gandhiji's Daughter Disciple,* 286–87.

50. Acharya Chatursen Shastri, "Lauhapurusha" (in Hindi), translated into English by Vinay Lal as "The Iron Man," unpublished ms.

51. Kalelkar, *Stray Glimpses of Bapu,* 165–66.

52. Manubehn Gandhi, *Bapu—My Mother* (Ahmedabad: Navajivan Publishing House, 1948).

53. It has been suggested to me that the image of the "banyan tree" is "unfortunate," "as nothing grows under it." As the reviewer noted, however, the image is not entirely inapposite, since Gandhi did not give birth to a "new generation of leaders" who truly were cast of the same clay. Some of the political leaders who style themselves "Gandhian" are an embarrassment, and even crooks have profited by deploying Gandhi's name toward the achievement of ends that stand distinctly in opposition to his teachings and ideas. Shahid Amin's masterful analysis, "Gandhi as Mahatma," *Subaltern Studies* 3, *Writings on South Asian History and Society,* ed. Ranajit Guha (Delhi: Oxford University Press, 1984), 1–61, suggests how the "Mahatma" became a floating signifier, in whose name even violence could be committed by the votaries of nationalism. After much deliberation, and with full knowledge of the fact that historians appear to have a predilection for metaphors of "seed" and "tree" in their work, I have decided to retain the image of the banyan tree: it resonates in Indian culture in myriad ways, and points the way to more semiological, hermeneutic, and nuanced readings of the trope of the banyan tree in the study of Indian culture. In this case, considering that this paper is on ecology, the historian in me might perhaps be indulged.

54. Naess, "Self-Realization: An Ecological Approach," 28.

55. Gupta, *Mira Behn: Gandhiji's Daughter Disciple,* 120.

56. M. K. Gandhi, *Truth Is God* (Ahmedabad: Navajivan Publishing House, 1959), 102.

57. Kalelkar, *Stray Glimpses of Bapu,* 54–55. See also Mukulbhai Kalarthi, *Anecdotes from Bapu's Life,* trans. from Gujarati by H. M. Vyas (Ahmedabad: Navajivan Publishing House, 1960), 22–23.

58. Gandhi, *Truth Is God,* 50.

59. Louis Fischer, *The Life of Mahatma Gandhi* (New York: Harper and Brothers, 1950), 239.

60. M. K. Gandhi, "Is This Humanity?–I," in *The Collected Works of Mahatma Gandhi,* 100 vols. (New Delhi: Publications Division, Ministry of Information and Broadcasting, Government of India, 1969), 31:486 (hereafter cited as *CWMG*). Gandhi penned eight articles on this subject under the title "Is This Humanity?"; all citations are from these articles. A short account of the exchange between Gandhi and his correspondents is furnished in Fischer, *Life of Mahatma Gandhi,* 236–39.

61. *CWMG,* 31:487.

62. Ibid., 505–6.

63. Ibid., 505.

64. Ibid., 525.

65. Ibid., 32:16.

66. Kalarthi, *Anecdotes from Bapu's Life,* 31. Mirabehn has recounted having had a similar experience and being reprimanded by Gandhi for plucking too many leaves, that too at night when trees are resting (see Gupta, *Mira Behn: Gandhiji's Daughter Disciple,* 130).

67. Gupta, *Mira Behn: Gandhiji's Daughter Disciple,* 130.

68. This subject is given a moving and poignant treatment in the 1935 novel by Mulk Raj Anand, *Untouchable* (reprint, Harmondsworth: Penguin Books, 1995).

69. Khoshoo, *Mahatma Gandhi: An Apostle of Applied Human Ecology,* 19.

70. Patel and Sykes, *Gandhi: His Gift of the Fight,* 103–4.

71. The subject of what India's epic and religious literature has to say about the environment, and how far early texts can be viewed as documents encompassing an "ecological wisdom," is far too complex to entertain in this paper. It is even possible to adopt a contrary position and argue that the epics are not merely indifferent to ecological thinking, but hostile to the ecological vision of life. The *Rāmāyaṇa* begins with the wanton killing of a bird engaged in love-play with its mate; in the other epic, the *Mahābhārata,* the Khandava Forest is burned down so that the Pāṇḍavas can make good their escape. My own view is that these are not very nuanced readings of the epics, which I am inclined to consider as repositories of ecological insight; and I would refer the reader to Ramchandra Gandhi, *Sita's Kitchen: A Testimony of Faith and Inquiry* (New Delhi, 1992), for an ecologically radical and inspired reading of the *Rāmāyaṇa.*

The Inner Logic of Gandhian Ecology

LARRY D. SHINN

Hug the Trees

The border war between China and India in 1962 brought national attention from the Indian government to the mountainous region of Uttar Pradesh called Uttarakhand. As the next decade unfolded, new policies of the Indian government opened this lush mountain region to heavy logging. Chandi Prasad Bhatt, a resident of a village in the region, organized a local economic and political cooperative association called the Dashauli Gram Swarajya Sangh (DGSS) to assist local workers in participating in the new forest-based economy. Chandi Prasad's movement had been inspired by a speech given a decade earlier by the Sarvodaya leader Jayaprakash Narayan. Villagers of this region had always used the forest as a source of building materials and firewood, but they now sought to build roads, engage in logging for profit, and develop secondary wood processing businesses that would retain some of the economic benefits from the forest economy for their local villages. At every point of engagement, however, India's central government thwarted the local co-ops' economic initiatives.

As large corporate conglomerates worked with the government to strip the nearby forests of all their trees, Chandi Prasad and the DGSS warned that ecological disaster was a possibility, given this deforestation. In 1970 torrential rains came, and so did the predicted ecological and human devastation. The Alakanandā River rose sixty feet in the narrow mountain valleys, and the flooding carried top soil more than one hundred miles downriver, clogging many tributaries. Over two hundred persons were killed in flash floods. The DGSS wrote a well-documented report of this avoidable tragedy and presented it to the central government. The report received no response.

In October 1971, members of the DGSS appeared at a government forestry meeting and engaged in a peaceful civil protest of the logging and other policies that continued the deforestation of their region. In early 1972, the government refused the DGSS their annual allotment of ash trees, from which a local DGSS co-op made handles for tools, and instead gave the ash allotment to the Simon Company, a manufacturer in the plains of India that made cricket, tennis, and other sports rackets. When Simon employees came to mark the ash trees for removal in March 1973, Chandi Prasad declared, "Let them know we will not allow the felling of a single tree. When their men raise their axes, we will embrace the trees to protect them."[1] (This incident gave the name *Chipko,* usually translated as "hugging," to Chandi Prasad's environmental movement in the Uttarakhand region.) Confronted by such organized and persistent opposition, the government rescinded its permit to the Simon Company and gave the local DGSS co-op its ash tree allotment. Further protests were required periodically to keep the Simon Company at bay.

The government next decided to grant permits to nonresident lumber companies for the logging of twenty-five hundred trees per year from the Reṇi Forest in Uttarakhand. Chandi Prasad and members of the Chipko movement engaged in protest meetings, pamphlet writing, and other nonviolent methods in an attempt to stop the logging of Reṇi Forest. Rather than confront the mountain villagers directly, the government enticed all of the men of the region to the distant town of Chamoli, where they were to receive long-disputed war-reparation payments from the 1962 China-India skirmishes that displaced many villagers in the region.

After the village men had left for Chamoli, the lumber company men arrived to begin logging the Reṇi Forest. When the village women saw the lumberjacks arrive in buses and disappear into the Reṇi Forest, thirty women marched into the woods to confront the loggers. The loggers argued that it was their right to log the allotted timber. Finally, a women named Gaura Devi stepped forward, bared her breast, and said, "This forest is like our mother. You will have to shoot me before you can cut it down."[2] Though some drunken loggers with guns threatened the women with physical harm, cooler heads prevailed and the loggers turned back without cutting a single tree. Then, the women dislodged a large boulder and blocked the forest road, effectively preventing further intrusion from unwanted loggers. Thirty women from the Chipko collaborative had turned back the threat to the Reṇi Forest. When their husbands returned, the women urged the Chipko leaders not to report the harassing and drunken behavior of the loggers lest their company fire them for such behavior.[3]

In referring to the Chipco's crusade to protect the region of Uttarakhand from ecological abuse, Chandi Prasad said:

> Our movement goes beyond the erosion of land, to the erosion of human values. . . . The center of all of this is humankind. If we are not in a good relationship with the environment, the environment will be destroyed, and we will lose our ground. But if you halt the erosion of humankind, humankind will halt the erosion of the soil.[4]

Such is the philosophy of Chandi Prasad's Chipko movement, which draws its inspiration from Gandhian *sarvodaya* ("for the good or welfare of all") teachings and practices. This essay will explore the Gandhian logic that has directly and indirectly inspired the Chipko and similar movements (such as *sarvodaya*). In so doing, we will seek the "inner logic" of Gandhian ecology.

The Inner Logic of Nonviolent Relationships

It is with a certain amount of trepidation that I attempt to construct a systematic statement of Gandhi's thought regarding ecology. First, Gandhi never attempted to be a systematic philosopher or theologian. As K. S. Bharathi has said, "Gandhiji was not a philosopher in the academic sense of the term. . . . He was a practical man."[5] Joan Bondurant goes even further when she claims that "Gandhi was a political activist and a practical philosopher, he was not a theorist. His writings abound with inconsistencies."[6] Because Gandhi often wrote short, self-contained essays on complex topics (such as *svarāj,* or "self-rule") in response to specific personal experiences or political events, it has usually been left to devoted followers or scholars to pull together all his ideas on a particular topic.

Second, Gandhi was a public activist and spokesperson for more than half a century. Not surprisingly, his ideas about truth (*satya*), nonviolence (*ahiṃsā*), self-rule (*svarāj*), and religion changed considerably over his lifetime. One example is Gandhi's increasing reliance in his *satyāgraha* (nonviolent) campaigns upon prayer, fasting, and ecumenical religious practices. In this paper, I will attempt to present mostly the "mature" thought of Gandhi, while recognizing that no simple chronological periodization of his thought is possible.

Third, the subject of ecology is not one that Gandhi himself addressed often in his writings. While some of his writings speak directly, or indirectly, to the topic, many of the formulations of this paper are constructive and synthetic attempts to create a systemic voice for Gandhi where none exists. I have tried to be faithful to the consistent and core teachings of Gandhi while recognizing that other people might express these same Gandhian insights in different terms.

Fourth, Gandhi was a modern Hindu. That is, he was raised in India at the height of British rule and studied English from an early age. He studied law at London University and was as familiar with Western thinkers like Ruskin and Tolstoy as he was with Indian philosophers. Gandhi was courted by Christian missionaries in England and South Africa and was deeply impressed by the teachings of Jesus as presented in the New Testament. Only when he returned to India and South Africa as a young barrister did Gandhi turn his attention intently back to his own Hindu faith and Indian customs. As a modern Hindu, Gandhi presented the perspectives of a world citizen even as he embraced the particularity of his own cultural and religious traditions. Consequently, Gandhi's thought is more broadly applicable than might be expected. Though Gandhi died over fifty years ago, his ideas often reach across time and cultures in their contemporary relevance.

With these four contextual frames in mind, let us now turn to one clear, interconnected set of ideas and principles that directed Gandhi's life and work throughout his lifetime. These interrelated principles constitute what I understand to be the "inner logic" of Gandhi's nonviolent activism (*satyāgraha,* or "soul force") and provide a foundation for a Gandhian ecology.

Truth (Satya)

Satya, or truth, is a foundational principle for Gandhi's thought and living. Gandhi said:

> But for me, truth is the sovereign principle which includes numerous other principles. This truth is not only truthfulness in word, but truthfulness in thought also, and not only the relative truth of our conception,

> but the Absolute Truth, the Eternal Principle, that is God. There are innumerable definitions of God, because his manifestations are innumerable. . . . But I worship God as Truth only. I have not found him, but I am seeking after him.[7]

The mature Gandhi drew upon manifold Indian notions of truth. The major religious traditions in India often equate truth with spiritual wisdom (variously, *vidyā, jñāna,* etc.). In popular folklore and scriptural traditions alike, truth has traditionally been understood to be a cosmic and ontological power to be utilized by religious leaders and laypersons alike. For example, the Buddha reportedly saved a shipload of merchants from a destructive storm by standing in the bow of the boat and proclaiming: "I am not conscious of ever having deliberately injured a single living creature. If these words be true may the ship return in safety." And the ship made the forty-day trip in one! Such "acts of truth" are reported in Hindu scriptures as well. The prostitute Bindermati astounded King Aśoka with her ability to make the Ganges River flow upstream by her "act of truth" of treating all her clients, rich or poor, with equal regard.[8] Truth in Indian philosophical and religious texts and communities is a real and cosmic power governed by spiritual practice and ethical behavior.

The word *satya* derives from the Sanskrit root *sat,* which means "to be" or "to exist." *Satya,* or truth, therefore, is not simply a quality of speech (as in much Western usage), but a qualitative and integral dimension of the natural order itself. As an ontological concept, *satya* is linked to *ṛta,* the fundamental order and structure of the natural world itself. In classical India, the Vedic understanding of proper livelihood and action is that "by truth the earth is made firm" (*satyenottabhita bhūmiḥ*).[9] When Gandhi named his nonviolent theory and method of action *satyāgraha,* he understood it to be "the force that is born to Truth and Love or non-violence."[10] Throughout the years of his public life, Gandhi sought not only to speak truthfully, but to act truthfully. For Gandhi, truth was an essential power to be called upon in all situations, whether public and political or personal and spiritual.

Gandhi asserted that we human beings can never know Absolute Truth, or God, fully. Gandhi said, "There are innumerable definitions of God, because his manifestations are innumerable."[11] Because we can never know the Absolute Truth, we must seek earthly approximations in our daily lives. Gandhi entitled his autobiography *Experiments with Truth* because he saw all of his life and work as an attempt to know better the Absolute Truth by experimenting with its manifestations in private and public behavior.[12] "Practical truth" is a phrase Gandhi sometimes used to describe actions or events in this world that sought to discern divine or Absolute Truth (for example, the objectives of the many *satyāgraha* campaigns through which he and his followers sought to remove an unjust law or condition). Truth was always for Gandhi a "vital force" that was forged in spiritual practice and moral actions—that is, a truth that was lived. For Gandhi, it was the case that "by truth the earth is made firm."

Finally, for Gandhi, "Truth is like a vast tree, which yields more fruit, the more you nurture it."[13] Service to others, both public and private, is the orchard in which the trees of truth grow best. "That is why," says Gandhi, "my devotion to truth has drawn me into the field of politics. . . . Those who say that religion has nothing to do with politics do not know what religion means."[14] For Gandhi, then, truth is not relegated to personal or spiritual quests alone, but is intended for collective political and economic strivings as well. Truth is a force for good in a world where less noble ideas and actions often hold sway (as in the Chipko movement, which had both practical and moral dimensions).

Because Absolute Truth is understood and realized in practical truth experiments, even the most localized nonviolent experiments—such as the Ahmedabad mill strike *satyāgraha* campaign in 1917—offer opportunities for moral growth and increasing one's understanding of the boundaries and power of truthfulness. In that campaign Gandhi used fasting as a mode of self-sacrifice and moral persuasion, but he was accused of coercion by those who claimed that his friendly relationship with the mill owners put unfair pressure on them. Gandhi's insistence on the inextricable relationship between truth (*satya*) and nonviolence (*ahiṃsā*) left him open to that charge—that is, if fasting is based on truthful self-sacrifice, it will exert a force on those who have yet to see that practical truth. Consequently, Gandhi most often linked human seeking after truth to nonviolent action.

Nonviolence (Ahiṃsā)

Indian religious traditions agree less often on the efficacy and morality of violence (*hiṃsā*) and nonviolence (*ahiṃsā*) than they do on the nature and power of truth (*satya*). On the one hand, some Indian religious traditions make *ahiṃsā* a central feature of their religious life. *Ahiṃsā* as nonviolence in all aspects of living is the primary means of liberation for followers of the Jaina tradition. For Buddhists, *ahiṃsā,* or noninjury, is an ethical goal for monks and laypersons alike. In Hindu traditions, however, *ahiṃsā* is a universal ethical ideal only for ascetics and those who renounce the world; whether or not it is an ethical ideal for a particular layperson depends on his or her caste *dharma* or duties. Hindu priests are permitted to kill animals for the Vedic sacrifice. Soldiers and kings are charged to kill their enemies when necessary to protect their kingdoms or promote the welfare of the people.

As Gandhi searched his own Hindu traditions, he found the most obvious contradictions to the ethic of universal nonviolence in the foundational epic the *Mahābhārata* and one of its central episodes, the *Bhagavadgītā*. In one battle scene, the god Kṛṣṇa exhorts the warrior Arjuna to fight and kill his cousins who opposed him (*Gītā* 2:3–7). After his attempts to allegorize the *Gītā* as a nonviolent treatise failed, Gandhi concluded:

> I have admitted in my introduction to the *Gita* . . . that it is not a treatise on non-violence, nor was it written to condemn war. Hinduism . . . has certainly not condemned war as I do. . . . I hold that the logical outcome of the teaching of the *Gita* is decidedly for peace at the price of life itself. . . . The immortal author of the *Mahabharata* . . . has shown to the world the futility of war by giving the victors an empty glory, leaving but seven victors alive out of the millions said to have been engaged in the fight.[15]

Gandhi focused upon the *Gītā*'s call to sacrifice oneself for others and offer devotion to God as he came to believe that *ahiṃsā* was not just a way of living, but an eternal quality of truth itself. He said: "*Himsa* [violence] does not need to be taught. Man as animal is violent but as spirit is non-violent. The moment a man wakens to the spirit within he cannot remain violent."[16] It is not that violence is never an option, as Gandhi says in *Non-Violence in Peace and War:*

> I do believe that, where there is only a choice between cowardice and violence, I would advise violence. . . . But I believe that non-violence is infinitely superior to violence. . . . The religion of non-violence is not meant merely for the *rishis* and saints. It is meant for the common

> people as well. Non-violence is the law of our species as violence is the law of the brute. The spirit lies dormant in the brute, and he knows no law but that of physical might. The dignity of man requires obedience to a higher law—to the strength of the spirit."[17]

For Gandhi, *ahiṃsā* as a higher, spiritual law is "the basis of the search for Truth."[18] Nonviolence is the only means that can produce truthful ends. Furthermore, *ahiṃsā* is clearly an eternal attribute of Absolute Truth. As such, according to Shashi Sharma, for Gandhi, *ahiṃsā* is not merely the absence of violence, "it is an active engagement in every field of life due to the dynamic force of love."[19] For Gandhi, compassion is a synonym for *ahiṃsā. Ahiṃsā* as compassion is an active force for good, for service, for concern for the other. *Ahiṃsā* is both the outcome ("the ends") and the means to *satya,* the absence of injury and the practice of compassion, a divine attribute and a practical guide to living. Raghavan Iyer observes that Gandhi

> regarded both *satya* and *ahimsa* as inherent in nature and in man, underlying the constant working of a cosmic law and constituting the only common basis of human aspiration and action in the midst of society. No society can survive without a measure of *satya* and *ahimsa*.[21]

If nonviolence and/or compassion is the essential attribute of Ultimate Truth and the practical expression of the spiritual dimension of all human beings, how is the way of brutish violence to be diminished and the spirit of *ahiṃsā* to be nourished? For Gandhi, the answer is found in self-purification (*tapas*) and self-rule (*svarāj*).

Self-Purification (Tapas) *and Self-Rule* (Svarāj)

According to Gandhi, the capacity to live a life of nonviolence does not come from timidity or cowardice but from strength of character and an indomitable will.[21] As a nonviolent political method, *ahiṃsā* is both a personal and a collective strategy of a group of *satyāgrahi*s, or "truth-seekers." To acquire such collective and practical results, individual self-purification (*tapas*) is required. *Tapas* is an ancient concept that pervades the major religious traditions of India. On one level, *tapas* works as a cosmic and spiritual "economic system." When you pay an extreme price through penance, prayer, fasting, or other spiritual practice, you receive corresponding material or spiritual benefits. Pārvatī's fasting led to her marriage to Śiva. Arjuna's penance resulted in the acquisition of a divine weapon. Jajali's ascetic compassion for a family of birds led ultimately to his freedom from rebirth.

For Gandhi, *tapas,* or self-purification, is the key to uncovering the spiritual depth of a person that enables compassion (*ahiṃsā*) and truthfulness (*satya*) to emerge in daily living. Gandhi says of fasting and prayer (two forms that *tapas* takes):

> Fasting is a hoary institution. A genuine fast cleanses body, mind, and soul. It crucifies the flesh and to that extent sets the soul free. A sincere prayer can work wonders. It is an intense longing of the soul for its even greater purity. Purity thus gained when it is utilized for a noble purpose becomes a prayer. Fasting and prayer therefore are a most powerful process of purification, and that which purifies necessarily enables us the better to do our duty and to attain our goal.[22]

If violence is the normal expression of the brutish and often physical dimension of a person, then a spiritual dimension must be uncovered. Prayer and fasting, therefore, is the personal avenue to self-purification that enables corporate nonviolent action. However, one engages in fasting not for oneself alone, but for others' purification as well. According to Gandhi:

> The fact is that *all* spiritual fasts *always* influence those who come within the zone of their influence. That is why spiritual fasting is described as *tapas*. And all *tapas* invariably exert purifying influence on those in whose behalf it is undertaken.[23]

Fasting and prayer represent for Gandhi "the yearnings of a soul striving to lay his weary head in the lap of his Maker."[24] In looking back over his life, Gandhi remarked that prayer did not come early or easily to him. It was only later in life that he concluded that "as food is indispensable for the body, so was prayer indispensable for the soul."[25]

Gandhi's religious faith is not easy to understand or to explain since at times he speaks in universalistic and impersonal ways about the Divine and at other times speaks in quite personal and intimate terms of God. Regardless of his particular expression of God/Truth, Gandhi consistently links prayer and fasting (*tapas*) to the cultivation of compassion/nonviolence (*ahiṃsā*). He makes this connection directly in an essay in the *Harijan:* "Non-violence is an active force of the highest order. It is the soul-force of godhead within us. We become Godlike to the extent we realize non-violence." Self-purification (*tapas*) through fasting and prayer unleashes the spiritual power of *satya* in the form of nonviolent action (*ahiṃsā*). This is the inner logic of Gandhian nonviolent living that connects us to other humans and all beings on Earth.

Gandhi connects personal purification with nonviolence toward all of God's creation in this way:

> Identification with everything that lives is impossible without self-purification; without self-purification the observance of the law of *Ahimsa* must remain an empty dream; God can never be realized by one who is not pure of heart. Self-purification therefore must mean purification in all the walks of life. A purification being highly infectious, purification of oneself necessarily leads to the purification of one's surroundings.[26]

Self-purification (*tapas*) thus understood is a necessary condition of political, economic, and personal "self-rule" (*svarāj*). When *svarāj* is understood in the context of *satya, ahiṃsā,* and *tapas,* the inner logic of Gandhian ecology begins to emerge.

The Village as an Ecological Model of *Svarāj*

An Interdependent Cosmos

To deduce a Gandhian ecology, one must set the inner logic of nonviolent truth-seeking against an Indian cosmological backdrop that asserts a homologous and interdependent relationship between humans and all other animate and inanimate aspects of creation. In his book *Classifying the Universe: The Ancient Indian "Varna" System and the Origins of Caste,* Brian K. Smith reveals how subtle and pervasive the notion of *varṇa* (literally, "color"or "caste") is as an organizing principle in Vedic thought.[27] Using more than a dozen cosmological accounts from a wide variety of Vedic scriptures, Smith shows how *varṇa,* as a hierarchical pattern of four discrete categories of creation (and "modes of being" in the world), organizes all aspects of creation: cosmos, time, space, flora, fauna, human beings, and even the gods themselves. This fourfold, ranked pattern of creation is observed in the four ages of time/creation (the *yuga*s), which began with the "Golden Age" and have devolved to our present "Dark Age" (Kaliyuga).

The fourfold *varṇa* system is revealed explicitly in the fourfold social system of caste, which includes, in descending order of rank, priests, rulers/warriors, merchants/farmers, and workers/servants. This pervasive pattern of ranked relationships establishes cosmological interdependencies that connect human beings to all other aspects of creation. Bruce Lincoln takes Smith's observation further: "Similarly, proper human action consists of the attempt, in ritual and elsewhere, to set things in this pattern, thereby preserving or even renewing the original and essential order of the cosmos."[28] Christopher Chapple, in an essay entitled "Hindu Environmentalism," concurs that this understanding of human intimacy, interdependence, and continuity with nature is both a traditional and a modern Hindu conception.[29]

The homologous webs of relationships among human beings and all other dimensions of creation explain not only the dietary, social, economic, and other conventions that most Hindus must observe, but also how each person's duty is determined by his or her "birth position" (*jāti*) in relation to that of other persons and animals. While Gandhi rejected the formal social and economic distinctions of caste, he clearly accepted the traditional Indian worldview that assumes the continuity and interdependence between humans and the natural world. (In fact, an interesting avenue of exploration that goes beyond the scope of this paper is the extent to which Gandhi's understanding of "truth-force" (*satyāgraha*) is consistent with the traditional Vedic worldview of homologous relationships that are found in classical notions of "acts of truth.") In any case, it is clear that Gandhi accepted a fairly traditional Hindu understanding of the integral relationship and interdependence between humans and nature, and that this understanding forms a backdrop for our understanding of Gandhian ecology.

The Village and Svarāj

Gandhi says simply: "Just as the whole universe is contained in the self, so India is contained in the villages."[30] According to Hindu cosmologies, human beings are interdependent with all other aspects of creation, and Gandhi believed that one place where these basic "natural" homologies are revealed is in the village. Indeed, the inner logic of nonviolent living (*satya, ahiṃsā,* and *tapas*) is best realized in the village, according to Gandhi:

> I hold that without Truth and Non-violence, there can be nothing but the destruction of humanity. We can realize Truth and Non-violence only in the simplicity of village life. . . . The essence of what I have said is that man should rest content with what are his real needs and become self-sufficient.[31]

This brief statement contains some basic elements of a Gandhian ecology. First, the quest for *svarāj* (self-rule) is located preferentially in the village. Second, *svarāj* as an economic and moral goal is expressed in terms of self-sufficiency (*svadeśi*). Third, *svarāj* is given both economic and spiritual meanings in Gandhi's notion of the "simple life" rooted in one's basic needs.

According to Gandhi, political notions of *svarāj* are best understood as an extension of a nonviolent approach to living that is best realized in self-reliant villages. While he acknowledged the nobility of India's desire to shed British rule and assume political "self-rule," Gandhi made clear that the real significance of the term was misunderstood when expressed only as a national political goal. In his early series of essays on *svarāj,* first published in the *Indian Opinion* in South Africa and collected into book form in 1908, Gandhi concluded that: "1. Real home-rule is self-rule or self-control. 2. The way to it is passive resistance: that is soul-force or love-force. 3. In order to exert this force, *Swadeshi* [self-reliance] in every sense is necessary."[32] In this reflection on the Congress Party's campaign for India's political independence from Britain, Gandhi moved quickly from an understanding of *svarāj* as a political movement to an understanding that *svarāj* must first be cultivated and manifested personally and economically.

Gandhi argues that the nonviolent, self-sufficient village life is an antidote to the "brute force" of capitalism and is a precondition of any form of national political or economic self-rule. If individual *satyāgrahi*s could not control their own material passions and greed, how could they improve upon British rule of India, if given a chance? Likewise, if individuals and villages were reliant upon a centralized economic system (capitalism) for their basic needs (food, shelter, clothing), how could they truly be said to be free of external rule?

Such questions, and the answers they imply, led Gandhi to identify India's villages as the locus for political, economic, and personal *svarāj*.

Gandhi said: "Our cities are not India. India lives in her seven and a half lacs of villages, and the cities live upon the villages."[33] To rephrase an earlier Gandhi comment, the whole universe is contained in the village. He thought it possible to create an ideal village that would be a model of *svarāj* for the whole country:

> My ideal village will contain intelligent human beings. They will not live in dirt and darkness as animals. Men and women will be free and able to hold their own against anyone in the world. There will be neither plague, nor cholera, nor small-pox; no one will be idle, no one will wallow in luxury. Everyone will have to contribute his quota of manual labour.[34]

Gandhi also said that "you cannot build non-violence on a factory civilization, but it can be built on self-contained villages."[35] The political and economic expression of Gandhi's ideal village culture would be a circle of villages working collaboratively with each other and bound to the cities and other village collectives in republican forms of government.[36] The center of the village circles would be nonviolent individuals ready to give selflessly for the good of the village "republic."

Given his notion of a self-reliant republican village life focused on the welfare and equality of all of its inhabitants, it is no wonder that Gandhi criticized the capitalist economic system in India that centralizes the means of production and tends to accumulate wealth in the hands of a few who have access to sufficient capital. Gandhi's desired village-based economy would decentralize the production of the basic necessities of life (food, clothing, and shelter) and would, thereby, promote self-sufficiency (*svadeśi*).

Therefore, a second basic condition for *svarāj* is economic self-reliance (*svadeśi*). Gandhi based this conclusion upon his assessment that industrialization had made human beings slaves to machines and to the centralized system of manufacturing that often resulted. It is not the purpose of this essay to debate Gandhi's view of technology and machines. While it is true that Gandhi sometimes spoke as though he were opposed to all modern technologies and machines, in general what Gandhi protested most vigorously was modern capitalism's over-reliance on machines and technologies that put thousands of

Indians out of work and thrust them into even greater poverty. Subrata Mukherjee summarizes Gandhi's view of technology and industrialization as one in which "machinery for man not man for machinery has to be the cardinal principle of mechanized production. . . . Industrialization involving mass production, centralization of initiative, power, authority, and policy formation is undesirable."[37]

Gandhi was concerned that industrialization, with its consequent centralization of capital and production and its reliance on expensive machines and technologies, had concentrated wealth in the hands of a few, displaced millions of Indian workers, and left the unemployed bereft of an ability to provide for even their basic needs of food, clothing, and shelter. Gandhi summarizes his complex views of machines and technologies this way:

> What I object to is the craze for machinery, not machinery as such. The craze is for labour-saving machinery. . . . I want to save time and labour not for a fraction of mankind but for all. I want the concentration of wealth, not in the hands of the few, but in the hands of all. Today, machinery merely helps the few to ride on the backs of millions.[38]

In a later response to the charge that he was against the "machine age," Gandhi replied, "I am not against machinery as such, but I am totally opposed to it when it masters us."[39] Gandhi advocated village industries as an antidote to the ills of industrialization. He saw village-based industries as opportunities for Indians to use appropriate "intermediate technologies" that allow individuals and villages to control the production of basic goods and services that benefit all who live in a village. In Nagpur, for example, such intermediate technologies as electrified oil presses enabled local villages to maintain the labor-intensive seed collection process while removing the backbreaking work of manual oil presses. Gandhi wanted villages to develop a system of nonviolent occupations[40] which would result in "production by the masses *versus* mass-production."[41]

Self-reliance (*svadeśi*) in such a system of village industries would provide considerable self-sufficiency with regard to the staples of food, clothing, and shelter, but would still leave villages open to exploitation in the production of technology itself.[42] It would be left to collaborative federations between village republics and the cities to create economic justice and independence in the technological areas, according to Gandhi. The ideal Gandhian village economy is one "where each one has an opportunity to work, where exploitation is totally eschewed, where almost all wants are met through local production, where exchange is resorted to in respect of commodities not locally produced."[43]

The Simple Life and the Spinning Wheel

In the mature thought of Gandhi, village economic life is the best arena for expressing one's moral life, one's search for truth. He says in a *Young India* essay in 1927: "whereas religion, to be worth anything, must be capable of being reduced to terms of economics, economics, to be worth anything must also be capable of being reduced to terms of religion or spirituality."[44] Be it at the personal, the village, or the national level, self-rule (*svarāj*) requires a tempering of material desires that can only result from placing one's tastes and actions in a larger, truth-seeking context. Such living emerges only when one subdues the material and sensual greed of the brute in each of us.[45] Gandhi says:

> I know that it is argued that the soul has nothing to do with what one eats or drinks, as the soul neither eats nor drinks; . . . my firm conviction [is] that, for the seeker who would live in fear of God and who would see Him face to face, restraint in diet both as to quantity and quality is as essential as restraint in thought and speech.[46]

Here Gandhi applies the notion of *svarāj* to personal dietary habits that, if exercised, would result in modest and egalitarian personal lifestyles.

Gandhi extends his notion of personal *svarāj* to a collective ideal. In a *Harijan* essay published in 1946, Gandhi likens *svarāj* as a relationship between the individual and the village to "an oceanic circle whose centre will be the individual, always ready to perish for the circle of villages till at last the whole becomes one life."[47] Jagdish Prasad describes Gandhi's human-centered and village-based economy this way:

> Gandhiji's ideas such as creativity in work, simplicity in economic activity, vitality in rural life and an organic relationship between man and nature calculated to satisfy his need rather than his greed will assume a

> new significance and relevance once the Gandhian way for the revival of village industries is adopted.[48]

In the concept of "the simple village life," where persons live in harmony with each other and the natural world, we find the meeting place of Gandhi's inner logic of truth-seeking and our human responsibility to each other and to our environment. The basic ecological challenge for Gandhi is essentially a spiritual one—how we can live in simplicity and at peace with ourselves and others and, by extension, with the natural world around us.

In summarizing Gandhi's religious-based economic views, D. B. Kerur says, "Wants by nature are limitless, and resources are limited. . . . Failure to discern between need and greed and reckless use of resources will soon lead society to the brink of ruin."[49] Gandhi's advice to the wealthy and elite capitalists of his day is to live and do business with restraint so that the consumption of the earth's resources is sustainable. Gandhi concludes, "Earth provides enough to satisfy every man's need, but not every man's greed."[50]

Only by devoting one's life to self-purification (*tapas*) and spiritual growth can one gain true liberation from the passions that inhibit self-rule (*svarāj*). Such self-rule is the inner logic of a Gandhian ecology that pertains equally to the political, economic, and religious life of an individual and, by extension, of a nation. For Gandhi, the spinning wheel and *khadi,* the hand-spun cloth it made, are multifaceted symbols of the simple life.

Gandhi states: "The spinning wheel is a force in national regeneration. If we wish for real *swaraj,* we must achieve economic independence."[51] Gandhi did not evoke the spinning wheel and *khadi* cloth as mere symbols of political defiance, as the British sometimes thought. Rather, Gandhi meant *khadi* production to be a practical solution to the twin problems of unemployment and idleness that resulted from industrialization and its labor-saving machines.[52] The spinning wheel was a form of intermediate technology that put economic power back in the hands of individuals and the villages in which they lived.

For Gandhi, the process of making hand-spun cloth was a form of truth-seeking that led to true self-reliance (*svadeśi*). As Gandhi said, "To me the economic and religious aspects of *Swadeshi* are far more attractive than the political."[53] On the economic level, *khadi* production could engage many villagers in the production of a vital necessity for living. It was a basic expression of economic independence. It was also a way to involve women in the central village economy. Gandhi said: "In hand spinning is the protection of women's virtue, the insurance against famine, and the cheapening of prices. In it is hidden the secret of *swaraj*."[54]

On the spiritual level, spinning drew villagers back to a more simple and reflective life and offered opportunities for personal growth as well. During morning prayers, the spinning wheel was often used as a meditational vehicle. Participants in *satyāgraha* campaigns used the spinning wheel as a way to develop concentration on the campaign's objectives. Gandhi wanted to develop what Subrata Mukherjee calls a "*khadi* mentality" by which "spinning would purify the body and the soul of the spinners and would lead to spiritual progress."[55]

Gandhi saw the spinning wheel as the inclusive symbol for the political, economic, and spiritual self-rule of India. It was also a nonviolent enterprise that developed personal and economic self-reliance while nurturing personal spiritual reflection and growth, if done properly. In this multifaceted sense, spinning and *khadi* production are symbolic expressions of egalitarian and life-giving relationships among truth-seekers, among self-reliant villages, and between human beings and nature. It is a manifestation of the inner logic of *svarāj*. Gandhi concludes: "Real *swaraj* will come not by acquisition of authority by a few but by the acquisition of the capacity by all to resist authority when it is abused."[56] Such an understanding of both personal and collective self-rule forms the inner logic of Gandhian ecology.

Outlines of a Gandhian Ecology

Gandhi seldom wrote systematic treatises on any topic—and certainly not on the topic of ecology. However, the foregoing presentation of the inner logic of *svarāj*, or self-rule, as a personal and collective strategy for economic, political, and religious life gives us a strong sense of what Gandhi might say to those who seek to understand the relationship of Indian spiritual traditions to ecology. Here, by "ecology" I mean the relations among people as well as those among people and their natural environment. Because ecology requires us to respect linkages between religious, social, economic, and natural processes,

Gandhi's multifaceted understanding of *svarāj* is especially relevant to our investigation.

A Simple Life

To live in an ecologically sustainable world structured by true *svarāj,* each person, village, city, and nation would have to adopt Gandhi's dictum that "Earth provides enough to satisfy every man's need, but not every man's greed." Self-rule means that we must, as individuals and as political and economic communities, master our passions, including avarice. Gandhi offered India a village-centered life and economy as an ideal model:

> I believe that independent India can only discharge her duty towards a groaning world by adopting a simple but ennobled life by developing her thousands of cottages and living at peace with the world. High thinking is inconsistent with complicated material life based on high speed imposed on us by mammon worship.[57]

He believed that each person should engage each day in what he called "bread labor," that is, physical labor that provides one's food. Intellectual labor is not to be a substitute, since it is only by physical labor that one maintains a connection to the body and to the earth. Here Gandhi's egalitarian sentiments are expressed in the extreme since the goal of "bread labor" is to remove distinctions between rich and poor, touchable and untouchable.[58] The simple life can be seen especially in Gandhi's call for a *khadi* industry in every home and village. Spinning is simultaneously an economic, political, and spiritual act and, therefore, a symbol of self-rule (*svarāj*) and self-reliance (*svadeśi*)—the simple life.

Tolstoy Farm in South Africa gives a hint of how Gandhi's ideal village looked early on in his truth-seeking.[59] This farm was donated in 1910 by a friend, Hermann Kallenbach, for use by *satyāgrahi*s who were displaced from their homes and jobs by the various nonviolent political campaigns Gandhi led in South Africa. The Indians who lived on Tolstoy Farm came from many different areas of India and from many faiths (such as Hindu, Muslim, Christian). At the community's beginning, the residents had to decide whose interests to accommodate when it came to diet (vegetarian or not?), the assignment of tasks (should all residents be required to engage in manual labor?), and other such fundamental aspects of communal living. For economic and religious reasons, a strictly vegetarian diet was decided on. Likewise, it was determined that each person had to engage in some physical labor in order to help provide food, clothing, and shelter. A key feature of Tolstoy Farm was that it was virtually self-reliant in terms of constructing residences, raising nearly all of the food eaten, making the clothes that were worn, and collaboratively educating the children. All policy decisions were made by the group.

Gandhi engaged in many experiments that were novel for that day. He wanted persons of all faiths to feel at home and thus conducted religious services that contained aspects of each faith represented. Education was deemed complete only if it blended academic subjects, manual labor, and spiritual exercise. Looking back on this experiment, Gandhi thought the most extraordinary aspect of Tolstoy Farm was the practice of having boys and girls sleep next to each other, dormitory-style, with only Gandhi as a guardian. All of these experiments were deemed to have worked in this case, but Gandhi questioned his own sanity in even trying the co-ed sleeping arrangement!

Tolstoy Farm served as a haven for *satyāgrahi* families for nearly ten years, and it provides us one glimpse of "the simple life" that Gandhi and his followers not only preached, but lived. The Gandhi ashrams (retreats) in India provide other such glimpses of what ideal Gandhian villages might be. On a visit in 1972 to the central Indian ashram called Sevagram, I experienced a life very much like the one described on Tolstoy Farm. The residents, including the students, raised all of the food consumed at the ashram. Meals consisted of simple grains, fruits, and vegetables in season. Milk came from the ashram dairy, which sold its excess milk. The farm stored its cow manure in large vats in the ground to create bio-gases used for heating and cooking.

It is difficult to imagine large numbers of Indians, let alone Americans, abandoning their lives of comfort—usually in cities and suburbs —for such self-disciplined, simple living. However, can any of us imagine our world sustaining its current urban cultures and levels of resource depletion even fifty years from now? Gandhi's dictum, "Earth provides enough to satisfy every man's need, but not every man's greed," argues for simple living that promotes a sustainable future in human/natural world terms—whether in a third world village or in a large city. One real challenge in developing a contemporary Gandhian ecology would be to translate his village-based model to one of "neighborhoods" in our suburban and urban twenty-first-century world. We would need to reconnect individuals and families to each other and to the natural world that sustains us, for without some sense of our connectedness to the natural world, how are we to understand the size and impact (positive or negative) of the ecological footprint we leave?

To understand the Gandhian simple life in terms of the sustainability of our economic and political actions and lives is a good way to frame the enormity of the challenge before us. If citizens in first world, high-tech cultures accepted the challenge to live a sustainable life of connectedness to each other and the earth (to heed Gandhi's call to live a simple life), the economic impact would be to reduce the widening gap between the highly educated "haves" and the working-class "have-nots," and the ecological impact would be to reduce the drain on irreplaceable natural resources. Simply put, Gandhi's call to a life connected to physical labor and the earth is a call to be in touch with the animals and the earth that sustain us. However we conceive or adapt this message, it does raise questions about how sustainable our large, impersonal cities are and how we might choose (or be required) to restructure them when we can no longer raise enough food to feed ourselves. Gandhi would call us to the simple life before we or our children are forced to adopt diminished life-styles by necessity.

Decentralized, Local Economies

A second element of a contemporary Gandhian ecology would be an emphasis on local and self-reliant economic communities that are decentralized in terms of the locus of control. Just the opposite type of economic system is sweeping the world today in our "global economy." Because of the rapid speed of various modes of transportation and communication, virtually every market and economy is a global one. The General Agreement on Trade and Tariffs (GATT), North American Free Trade Agreement (NAFTA), and other recent trade agreements have encouraged the spread of multinational corporations. One example of our global economic age is Japan's Tokico Corporation, which manufactures automobile parts. Tokico imports iron ore from South America, smelts the ore and forges steel castings in its factories in Japan, machines and assembles brakes, parts, and shock absorbers from the steel castings in a Tokico factory in Berea, Kentucky, and then sends the finished brakes and shock absorbers to Mexico to be assembled on "American-made" Ford and Chrysler cars.

More than half a century ago, Gandhi spoke forcefully against the poverty and exploitation caused by centralized, capital-intensive forms of industry that exploit third world countries like India. In a recent book called *The Case against the Global Economy and a Turn toward the Local,* more than forty authors discuss the negative economic impact and the environmental devastation caused by exploitation of the third world's labor force and natural resources by multinational, first world corporations.[60] In his chapter, "Global Economy and the Third World," Martin Khor describes how the initial economic benefits of the green revolution—modern fishing techniques, logging of tropical forests, and other such global enterprises in third world countries—resulted ultimately in poisoned well water, barren seas, and soil erosion—the ultimate legacy of these once-profitable economic activities.[61]

In a television series called *Race to Save the Planet,* the actor Roy Scheider narrates a segment called "In the Name of Progress,"[62] in which vivid visual examples are given of how India's attempt to compete in a global economy has resulted in the government displacing thousands of villagers in the Singrali district to make way for ever-expanding, coal-burning power plants. Not only have the villagers lost their farmlands, but the air they now breathe is chemically polluted, the water is undrinkable, and their health is deteriorating. One villager asks: if the British needed control of nearly 40 percent of our globe to create her modern economy, how many earths will it take for India to do the same? Gandhi's need versus greed dictum quickly comes to mind.

Gandhian ecology urges small community self-reliance (the village in India; possibly neighborhoods, towns, or small cities in the United States) in economic and political terms so that there can be harmony between residents/workers, the natural environment, and economic modes of production. Centralization of the modes of production usually leads to large-scale, capital-intensive industries that are divorced from local social and human needs and interests.

The Union Carbide disaster in Bhopal in 1984 is a classic example of what can happen when local control (*svarāj*) of economic production is lost.[63] Union Carbide (India) Limited (UCIL) originally produced batteries in India. With the emergence of the green revolution in the 1960s in India, UCIL laid plans for entering the agricultural chemical market. In 1969 it opened a new plant in Bhopal; by 1974 the plant manufactured extremely poisonous pesticides. Although the municipal government protested the citing of the UCIL plant in an area zoned for light and nonhazardous industries, UCIL had considerable political clout on the state and national levels and prevailed in putting its plant in a residential area of the city. It is clear that Bhopal's authorities did not have "self-rule"—that is, political control over their own city's economic development.

The market competition for agricultural products became intense in India, and by the early 1980s the Bhopal Union Carbide plant was losing money each year. Like its parent company in the United States, UCIL faced the need to undergo restructuring of financial and human resources in 1982 and 1983. This left the Bhopal plant with fewer personnel to run the plant, diminished resources to develop competitive products, and decreasing economic viability—all of which demoralized the company's workforce. Some of the best managers left UCIL, which left new employees in control of the sophisticated technological production process of hazardous agriculture chemicals. Add to these company-specific conditions an influx into Bhopal of villagers from rural regions of Madhya Pradesh, which created large slum encampments all around the UCIL facility, and the ingredients for disaster were present.

On the night of 3 December 1984, Suman Dey, a control-room operator in the Bhopal pesticide plant, saw gauges indicating that pressure was rising in a refrigerated storage tank where the dangerous pesticide ingredient methyl isocyanate gas (MIC) was stored. Dey ran outside to investigate and heard a loud rumbling sound at the same time that he saw a cloud of gas rising out of the storage-tank stack. Dey and his supervisor tried to activate all of the emergency safety systems; when these actions failed, he and others in the plant fled in panic. Although the silent and toxic gas killed many people in their sleep, thousands of people in the vicinity of the UCIL facility fled when their neighbors ran through the streets shouting in distress. At the end of one week, over three thousand people and two thousand animals had died and more than three hundred thousand persons were affected by the poisonous gas.

The Bhopal crisis was, on one level, a technological one. It was also clearly caused by the economic, political, and social consequences of divorcing local needs and interests from UCIL's national and international political influence and control of the mode of production. Gandhian ecology applies not only to the simple life lived in a remote village; its nonviolent logic of living recommends that self-rule (*svarāj*) be understood in an economic and political sense on the small community, town, or city level as well. Even large cities like Bhopal create a chasm between those who rule the city and those affected by industrialization or other local environmental conditions. The Union Carbide disaster in Bhopal bolsters Martin Khor's claim that third world countries often pay the price for developed countries' unrestrained consumption—or for third world countries' attempts to achieve first world economies—which leads to loss of local control of economic modes of production. Khor's vision is essentially a Gandhian one of 1) reduction of consumption in developed countries, 2) redistribution of wealth so that local economies can be self-sufficient, 3) sustainable uses of renewable resources, and 4) local control of the production of basic goods and services.[64] The fundamental difference between Gandhi's and Khor's visions is that Khor believes that some very negative economic, political, or natural calamity will be needed to effect such changes. Gandhi believed that one could appeal to noncoercive or nonviolent means to develop self-rule and thereby ensure truthful or good ends.

A Nonviolent, Compassionate Ecology

The Chipko movement of Chandi Prasad and Vandana Shiva provides perhaps the best insight into the third structural dimension of Gandhian ecology: its steadfast adherence to a nonviolent and self-reliant ecological philosophy. First of all, the Chipko movement stresses harmonious and sustainable relations between humans and nature. Though its early foes were nonresident, large corporations, Chandi Prasad also chastised local villagers for their abuses of the forests around their own villages. Humans must respect nature as an extension of themselves and, therefore, interact with the natural "communities" with respect and compassion. Second, the Chipko movement encourages local control of basic modes of economic production and recommends village-based industries so that villagers can control their own economic destiny. Third, the Chipko leaders, male and female alike, urge a nonviolent respect and compassionate concern for those persons and institutions that threaten local self-rule and control of natural resources.

The Reṇi Forest campaign reveals a multilevel commitment to nonviolence. As in typical Gandhian *satyāgraha* campaigns, the first level of opposition to the governmental and corporate intrusions into the local forests was to communicate the Chipkos' concern for their natural resources and their opposition to outsiders' control. When that failed, nonviolent sit-ins were held. Finally, after a list of demands was made, physical confrontation of the loggers by village women escalated the opposition. Throughout the whole process, however, the government and company representatives were never considered "enemies." The women who confronted the drunk and angry loggers urged the village men not to tell the company that the loggers had been drinking lest the loggers lose their jobs. Such is the "truth-force" that saved Reṇi Forest and such is the compassion that marked the nonviolence of Chipko.

Chandi Prasad reveals the inner logic of Gandhian nonviolent ecology when he says:

> Our movement goes beyond the erosion of land, to the erosion of human values. . . . The center of all of this is humankind. If we are not in a good relationship with the environment, the environment will be destroyed, and we will lose our ground. But if you halt the erosion of humankind, humankind will halt the erosion of the soil.[65]

Whatever else a Gandhian ecology may be, it will begin with individual and collective self-rule (*svarāj*) premised upon truth (*satya*), nonviolence (*ahiṃsā*), and self-sacrificial actions (*tapas*). It will insist upon respect and compassion for all creatures and for nature itself. It will encourage economic self-reliance and self-sufficiency at the local (village, town, or neighborhood) level. Finally, a Gandhian ecology will interweave religious, economic, and political dimensions of life on both the personal and corporate levels. While many called Gandhi an unrealistic dreamer, he and his followers overcame a foreign oppressor by wedding individual spiritual discipline with collective political and economic action. As the Chipko movement reveals, a Gandhian ecology is both an economic and a political mode of living that is expressed in practical environmental actions that are grounded in ultimate values and truth-seeking. Such will be the inner logic of a Gandhian ecology that addresses the practical environmental and economic issues of our day.

Some Final Thoughts

Throughout his lengthy political and public career, Gandhi had many detractors and critics. Some said his nonviolent views were too idealistic; others criticized their practical application. The inner logic of Gandhian ecology that I have outlined above does lend itself to similar questions and criticisms. Can we hope for or rely upon individuals' self-purification (for example, overcoming greed) as a practical means for reducing material consumption of dwindling resources? Must we await such personal transformations for collective opposition to environmental threats to be effective? Can such personal or collective will be engendered without a leader like Gandhi, Vinobe Bhava, or Chandi Prasad? Gandhi himself was forced to address similar questions in his day. What is clear is that nonviolent truth-seeking is a journey Gandhi and his followers were—and are—willing to make even before all our philosophical or theoretical questions can be answered. Therefore, a Gandhian ecology will be one that engages environmental issues on a practical level through nonviolent means as a way to find ultimately peaceful and sustainable solutions.

Gandhian ecology will not simply oppose environmental abuse through nonviolent means, as has the Chipko movement; rather, it will seek new forms of the simple life to create sustainable futures. A Gandhian ecology so conceived meshes well with other ecological worldviews centered on sustainable living and spiritual awakening. One such expression, derived from a Christian theological framework, is that of John B. Cobb, Jr., in his book *Sustainability: Economics, Ecology, and Justice.*[66] Cobb, like Gandhi, locates our ecological crisis in a milieu of materialism where greed, not need, creates consumer cultures seeking "universal affluence."[67] Cobb understands "ecojustice" as an interweaving of economic, ecological, and moral concerns. And his solution is predicated both on individual transformation of attitudes and on actions based upon Christian compassion and self-sacrifice.[68] An economically and ecologically sustainable future is the desired end.

Nonetheless, the question still remains, for both Gandhi and Cobb, whether we humans can sensitize ourselves through personal piety and collective action to motivate ourselves and others to live a more simple life before we are forced to do so. Will it take an ecological catastrophe to motivate us to move toward the simple life that awaits affluent humans sometime in the next fifty years and less advantaged earth sojourners before then? A Gandhian ecology offers an alternative journey of truth-seeking through experiments with such a sustainable simple life.

Notes

Researching many of the extended sources for this paper (e.g., in Marxist views of Gandhian economics) would not have been possible without the thoughtful and reliable efforts of Olivia Bolyard, an English major at Berea College. Olivia worked for half a summer assisting me in bibliographic and library work and prereading certain texts. I owe Olivia a deep debt of gratitude for her work as an undergraduate researcher.

1. See "Hug the Trees," in Mark Shephard, *Gandhi Today: The Story of Mahatma Gandhi's Successors* (John Cabin, Md.: Seven Locks Press, 1987), 69.
2. Shephard, *Gandhi Today,* 75.
3. Ibid., 79. Christopher Key Chapple also provides a brief summary of the perspective of a well-known Chipko woman spokesperson, Vandana Shiva, in his essay "Hindu Environmentalism: Traditional and Contemporary Resources," in *Worldviews and Ecology,* ed. Mary Evelyn Tucker and John A. Grim (Lewisburg, Pa.: Bucknell University Press, 1993), 119–21. From its beginning, the Chipko movement has incorporated women and men primarily from the villages that seek economic and environmental justice. In many respects the movement is a faithful Gandhian *sarvodaya* —"for the good of all"—movement.
4. Shephard, *Gandhi Today,* 80.
5. K. S. Bharathi, *Thoughts of Gandhi and Vinoba* (New Delhi: Concept Publishing Company, 1995), 11.
6. Joan V. Bondurant, *Conquest of Violence: The Gandhian Philosophy of Conflict* (Berkeley and Los Angeles: University of California Press, 1967), 7.
7. Mahatma Gandhi, *An Autobiography, or, The Story of My Experiments with Truth* (1925), *The Selected Works of Mahatma Gandhi,* vols. 1–2, ed. Shriman Narayan (Ahmedabad: Navajivan Publishing House, 1968), xxi.
8. W. Norman Brown, *Hindu Act of Truth* (n.p., n.d.), 37.
9. Ibid., 40.
10. Mahatma Gandhi, *Satyagraha in South Africa* (1928), *The Selected Works of Mahatma Gandhi,* vol. 3, ed. Shriman Narayan (Ahmedabad: Navajivan Publishing House, 1968), 151.
11. Gandhi, *My Experiments with Truth,* xxi.
12. Ibid., xix–xxiii.
13. Ibid., 326.
14. Ibid., 753.
15. *Gandhi: Selected Writing,* ed. Ronald Duncan (New York: Harper Colophon Books, 1971), 40–41, 60–61.
16. Raghavan Iyer, *The Moral and Political Thought of Mahatma Gandhi* (New York: Oxford University Press, 1973), 211.
17. *The Essential Gandhi,* ed. Louis Fischer (New York: Vintage Books, 1962), 156–57.
18. Gandhi, *My Experiments with Truth,* 411.
19. Shashi Sharma, "Technology and Man," in *Gandhian Model of Development and World Peace,* ed. K. P. Misra (New Delhi: Concept Publising Company, 1988), 77.
20. Iyer, *Moral and Political Thought of Mahatma Gandhi,* 224.

21. *The Essential Gandhi,* ed. Fischer, 157.

22. M. K. Gandhi, *My Religion,* comp. and ed. Bharatan Kumarappa (Ahmedabad: Navajivan Press, 1955), 83.

23. Ibid., 67.

24. *Harijan,* 18 April 1936, quoted in M. K. Gandhi, *The Village Reconstruction,* ed. and published by Anand T. Hingorani (Bombay: Bharatiya Vidya Bhavan, 1966), 77.

25. Gandhi, *My Religion,* 90.

26. Gandhi, *My Experiments with Truth,* 753.

27. Brian K. Smith, *Classifying the Universe: The Ancient "Varna" System and the Origins of Caste* (New York: Oxford University Press, 1994).

28. Bruce Lincoln, review of *Classifying the Universe,* by Brian K. Smith, *History of Religions* 36, no. 4 (May 1997): 394.

29. Chapple, "Hindu Environmentalism," 114–17. See also Ranchor Prime, *Hinduism and Ecology: Seeds of Truth* (London: Cassell Publishers Limited, 1992), for a good description of human/nature interdependence and traditional Hindu values that are still relevant today.

30. Gandhi, *Village Reconstruction,* ed. Hingorani, v.

31. Press Report, 30 June 1944, in ibid., 41.

32. Mahatma Gandhi, *Hind Swaraj,* in *The Selected Works of Mahatma Gandhi,* vol. 4, ed. Shriman Narayan (Ahmedabad: Navajivan Publishing House, 1968), 201.

33. Gandhi, *Village Reconstruction,* ed. Hingorani, 8.

34. Ibid., vi. For a good summary of Gandhi's economic thought as connected to the Indian village, see Prime, *Hinduism and Ecology,* 58–69.

35. *Harijan,* 3 November 1939, quoted in Gandhi, *Village Reconstruction,* ed. Hingorani, 125.

36. *Harijan,* 28 July 1948, quoted in ibid., 117.

37. Subrata Mukherjee, *Gandhian Thought: Marxist Interpretation* (New Delhi: Deep and Deep Publications, 1991), 31.

38. *Young India,* 13 November 1934, quoted in ibid., 44.

39. *Harijan,* 1937, quoted in N. S. Siddharthan, "Gandhi and Technology," in *Gandhi and Economic Development,* ed. B. P. Pandy (New Delhi: Radiant Publishers, 1991), 185.

40. Bharathi, *Thoughts of Gandhi and Vinoba,* 89.

41. Jagdish Prasad, "Gandhi's Views on Village Industries: Relevance to India's Unemployment and Poverty Situation," in *Gandhi and Economic Development,* ed. Pandy, 203.

42. *Gandhian Model of Development and World Peace,* ed. R. P. Misra (New Delhi: Concept Publishing Company, 1989), 58.

43. D. B. Kerur, "Social Dynamics of Marx and Gandhi," in ibid., 45.

44. *Young India,* 15 September 1927, quoted in Gandhi, *Village Reconstruction,* ed. Hingorani, 29.

45. *The Essential Gandhi,* ed. Fischer, 89.

46. Gandhi, *My Experiments with Truth,* 406.

47. *Harijan,* 28 July 1946, quoted in Prasad, "Gandhi's Views on Village Industries," 202.

48. Prasad, "Gandhi's Views on Village Industries," 206.

49. Kerur, "Social Dynamics of Marx and Gandhi," 50.

50. Shephard, *Gandhi Today,* 63.

51. *Young India,* 6 July 1921, quoted in Gandhi, *Village Reconstruction,* ed. Hingorani, 4.

52. Gandhi, *Village Reconstruction,* ed. Hingorani, vi.

53. *Young India,* 1919, quoted in ibid., 3.

54. *Young India,* 1921, quoted in ibid., 6.

55. Mukherjee, *Gandhian Thought,* 39.

56. Quoted in ibid., 33.

57. *Harijan,* 9 January1946, quoted in Bharathi, *Thoughts of Gandhi and Vinoba,* 86–87.

58. Bharathi, *Thoughts of Gandhi and Vinoba,* 87.

59. For a full description of Tolstoy Farm, see Gandhi, *Satyagraha in South Africa,* 316–52.

60. *The Case against the Global Economy and a Turn toward the Local,* ed. Jerry Mander and Edward Goldsmith (San Francisco: Sierra Club Books, 1996).

61. Martin Khor, "Global Economy and the Third World," in *The Case against the Global Economy,* ed. Mander and Goldsmith, 47–59.

62. *Race to Save the Planet* (Boston: WGBH, 1990).

63. For a full accounting of the Bhopal disaster and its multiple causes, see Paul Shrivastava, *Bhopal: Anatomy of a Crisis,* 2d ed. (London: Paul Chapman Publishing, Ltd.), especially 1–51.

64. See Khor, "Global Economy and the Third World," 57–59.

65. Shephard, *Gandhi Today,* 80.

66. John B. Cobb, Jr., *Sustainability: Economics, Ecology, and Justice* (Maryknoll, N.Y.: Orbis Books, 1992), 20–33.

67. Ibid., 28.

68. Ibid., 30.

Forests in Classic Texts and Traditions

The Natural History of the *Rāmāyaṇa*

DAVID LEE

As Rāma set out in the early morning with Saumitri, he began to speak to lotus-eyed Sītā. "Look, Vaidehī, the *kiṃśuka* trees are in full blossom now that winter is past. Garlanded with their red flowers they almost seem to be on fire. Look at the marking-nut trees in bloom, untended by man, how they are bent over with fruit and leaves. I know I shall be able to live. Look at the honeycombs, Lakṣmaṇa, amassed by honeybees on one tree after another. They hang down large as buckets. Here a moorhen is crying, and in answer to it a peacock calls through delightful stretches of forest richly carpeted with flowers. And look, there is Citrakūṭa, the mountain over there with the towering peak, teeming with herds of elephants and echoing with flocks of birds." So the brothers and Sītā proceeded on foot and reached the delightful mountain, charming Citrakūṭa.

Ayodhyākāṇḍa 50:5–11[1]

Thus, Rāma, Sītā, and Lakṣmaṇa, the central figures of the *Rāmāyaṇa,* began their fourteen-year exile in the southern forest of Daṇḍaka (figure 1). The story of Lord Rāma was put into verse by the sage Vālmīki in approximately 600 B.C.E. The story—and Vālmīki's poetry —has inspired countless other renditions,[2] and the *Rāmāyaṇa* continues to inspire today, not only in India but in countries throughout tropical Asia. I personally became acquainted with the epic through the shadow puppet plays of southern Thailand and Malaysia, the classical music of Java, and the dance drama of Bali. The *Rāmāyaṇa* conveys the values and the civic and cultural life of that time, and much has been written about the central role of *dharma,* upheld by Lord Rāma.[4] The story, with its vivid portrayal of forests and gardens, also tells us

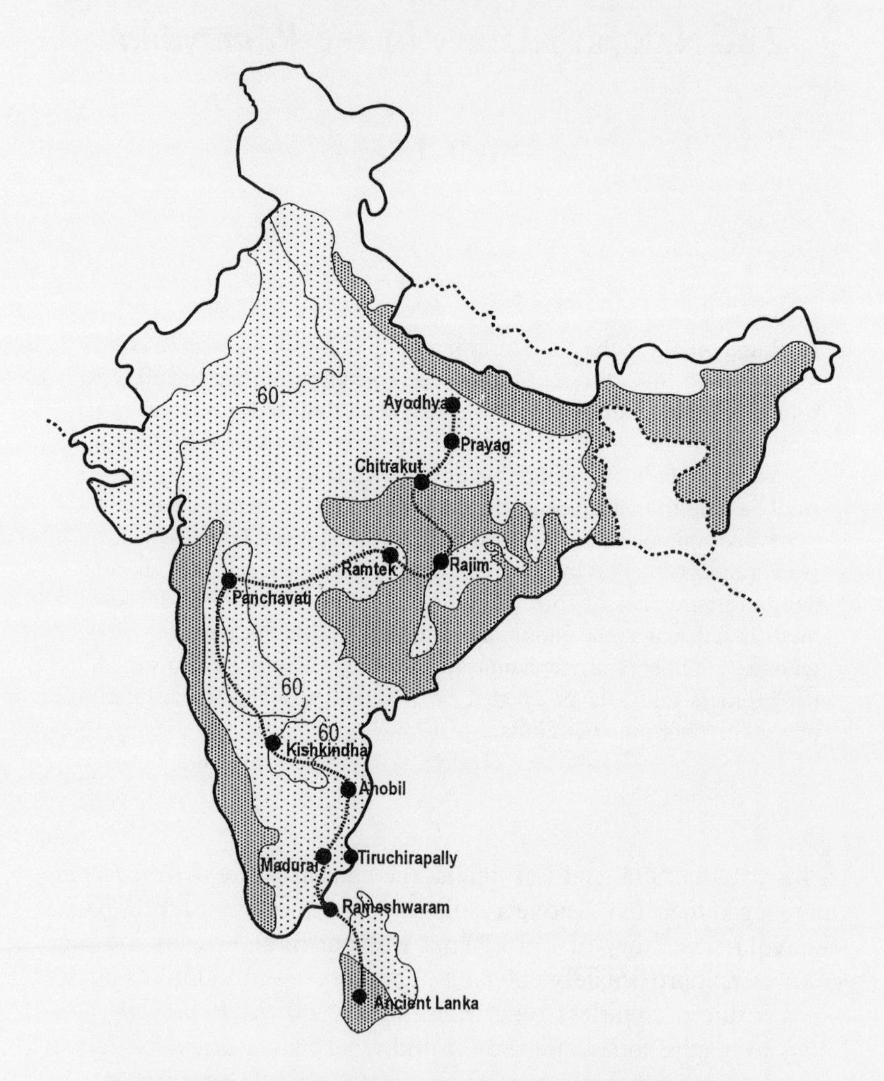

Figure 1. Map of India with annual rainfall isopleths of 600 and 1,000 mm and distributions of major vegetation types.[3] The travels of Rāma are also indicated, with significant halting places.

much about classical knowledge of, and attitudes toward, nature—the subject of this essay. I write this as an ecologist, with considerable firsthand experience in India's forests, so this essay is certainly not one of literary criticism.[5] There are lessons to learn from the great epic that can help us in at least two ways to preserve and restore India's wonderful natural heritage. First, the epic tells us about classical knowledge of forest natural history and attitudes toward nature. Second, the contemporary popularity of the epic makes it a vehicle for popularizing messages about nature and the preservation of natural history.

Before we explore the the descriptions of nature in the *Rāmāyaṇa,* it is useful to define the two ways in which our discussions are set. In one sense nature includes the organisms which inhabit the world, little influenced by human civilization. The relationships that explain how these organisms associate with each other and the physical world are the subject of ecology. These organisms live together in communities, influenced by the physical environment, what ecologists call ecosystems. Ecosystems are defined by these interactions, particularly the flow of energy and cycling of nutrients among the physical and biotic components. Much of the intellectual interest in ecology concerns how the patterns of diversity influence the functioning of the various ecosystems and how ecosystems have influenced the patterns of evolution of this diversity. In this view ecology is not destroyed, but is changed by human impacts to the environment. In fact, such "perturbations" allow us to learn how ecosystems function. Thus, the pollution of such a river as the Yamunā (as saddening and vile as it appears)[6] does not mean the permanent destruction of an ecosystem, because the organisms could reappear from unpolluted tributaries if the river were restored. In Ohio the inflammable Cuyahoga River was restored to support fish. Ecologists are particularly concerned about threats to the biota (the plants and animals). Their loss, through extinction, irrevocably changes the ways ecosystems function. Wholesale destruction of ecosystems, such as forests, may also mean the extinction of their individual components, the plants and animals. Many ecological problems—the pollution of air and water—are reversable, but the extinction of organisms is not.

Such a definition of ecology sounds far removed from the intellectual understanding of people, tribals (or *ādivāsi*) and those of rural Hindu villages, who use the remaining forests. What we are really talking about concerning energy flow is "what eats what." Embedded in the knowledge of some tribal and village elders is a detailed knowledge of the animals and plants in the surrounding forest, their behaviors (for animals, their courtships and times of birth, and for plants, their flowering and fruiting times), their dietary preferences (where animals are likely to feed and where plants grow). This "embedded ecology," or ethnoecology, is an important part of any traditional culture, whose survival depended (and still depends in a few cases) on such knowledge. A colleague, Steve Young, once told me that the Eskimo elders he worked with on Saint Lawrence Island, in the Bering Strait, held knowledge of natural history in their surroundings equivalent to a Ph.D. in ecology. Ecologist Kent Redford wrote: "To live and die with the land is to know its rules. When there is no hospital at the other end of the telephone and no grocery store at the end of the street, when there is no biweekly paycheck nor microwave oven, when there is nothing to fall back on but nature itself, then a society must discover the secrets of the plants and animals."[7]

This ecological knowledge has become embedded in the religious beliefs of indigenous cultures, enforcing behavior that may make extraction of resources more sustainable. Ecological interactions are celebrated ritually in many traditional cultures through sacrifice and *pūjā,* an act of gratitude for the abundance of these trophic networks.[8] This knowledge and respect would have been held by Rāma and Lakṣmaṇa, members of a hunting society attuned to the cycles of nature. We may also feel gratitude knowing that virtually every carbon atom in our bodies is assimilated from the air by a plant, and most nitrogen atoms in our DNA are fixed by bacteria living in the roots of plants.

In another sense nature is something different from us, particularly distinct from the artifacts of our culture. Thus ecology, or the "environment," is something that we have degraded, something that we need to care for better. We need to understand ecology in the first—and scientific—sense in order to take better care of organisms and ecosystems. This article attempts to interpret the understanding of organisms and ecology conveyed in the *Rāmāyaṇa* and to discuss the meaning of environmental understanding and perceptions of the *Rāmāyaṇa* for contemporary problems.

Diversity

> "There are lovely trees on the mountains, *sāla, tālas, tamālas, kharjūras, panasas, āmrakas, nīvāras, timiśas,* and *puṃnāgas;* thick groves of various other trees—mangoes, *aśokas, tilakas, campakas,* and *ketakas,* and flowers, vines, and shrubs; sandalwood trees, too, and *syandanas, nīpas, panasas, lakucas, dhavas, aśvakarṇas, khadiras, śamīs, kiṃśukas,* and *pāṭalas.* This is a pure and holy spot teeming with birds and beasts, and here is where we will live. . . ."
>
> *Araṇyakāṇḍa* 14:16–18[9]

The Vālmīki *Rāmāyaṇa* describes a great variety of plants and animals, particularly during Rāma's exile and wanderings in the Daṇḍaka Forest. In the quote above, Rāma names twenty-five plants, virtually all trees, in his vivid description of the forest surrounding Pañcavaṭī, their ashram-in-exile (see table 1). Since the text was copied and modified in different parts of India over the centuries, names could have been changed according to the ecology of the region. However, the production of an authoritative version by the Oriental Institute of Baroda, and a new English translation almost finished by the Princeton Library of Asian Translations, has removed many of these additions. The exact number is also blurred by synonymy in some of the Sanskrit names. For instance, there are twenty-three Sanskrit names for the important Ayurvedic plant *aśvagandha* (*Withania somnifera*),[10] and many of the Sanskrit names lack scientific equivalents. Some 200 plant species are described in the text.[11] The entire flora of India is estimated at approximately 15,000 species, and some 4,900 of these taxa are listed in *The Wealth of India.*[12] Of the 4,100 native species described in this massive study, some 680 have Sanskrit names. S. C. Banerjee published 956 Sanskrit plant names from a scriptural survey, but only 262 biological species could be associated with those names.[13] Thus, the *Rāmāyaṇa* lists a great diversity of plants, using many names described in earlier scriptures. In a similar way, it describes a variety of animals, including most of the large game animals and an impressive variety of birds. In contrast, The Old and New Testaments of the Bible name 110 plants. Eleven of those are not indigenous to the region, and many are crop plants.[14] Few of the plants named during Rāma's sojourn in the forest are cultivated species; almost all are wild forest trees. They were undoubtedly collected for use in medicine and supplementary sources of nutrition. We can conclude that the author of the *Rāmāyaṇa,* and his readers, knew much of the natural diversity of classical India (table 1).

Table 1. Names, distributions, and uses of plants in the forest of Pañcavaṭī mentioned in this article.[15]

Names		Uses		Sacred	
Sanskrit	Scientific	Medicinal	Economic	value	Distribution
amra	*Mangifera indica*	•	•	•	Ne, cult
āmraka	*Spondias pinnata*	•	•		N, Pen
aśoka	*Saraca indica*	•	•	• Indra	Ne, Pen
aśvakarṇa	*Dipterocarpus alatus?*	•	•		Ne, Sw
badari	*Zizyphus mauritiana*	•	•		N, Pen
bilva	*Aegle marmelos*	•	•	•	N, Pen
campaka	*Michelia champaca*	•	•		N, Pen
candana	*Santalum album*	•	•	•	Sw
dhava	*Anogeissus latifolia*		•		N, Pen
kakubha	?				
ketaka	*Pandanus odoratissimus*	•	•		Cult
khadira	*Acacia catechu*	•	•	• Gaṇeśa	N, Pen
kharjūra	*Phoenix dactylifera*	•	•		Pen, cult
kiṃśuka	*Butea monosperma*	•	•	• Brahmā	N, Pen
lakuca	*Artocarpus lakoocha*	•	•		N, Sw
nīpa	*Anthocephalus chinensis*	•	•	• Kṛṣṇa	N, Pen
nīvāra	?				
panasa	*Artocarpus heterophyllus*		•		Sw, cult
pāṭala	*Stereospermum suaveolens*	•	•		N, Pen
puṃnāga	*Calophyllum inophyllum*	•	•		S, coasts
sāla	*Shorea robusta*	•	•		N
śamī	*Prosopis cineraria*	•	•	•	Nw, Pen
syandana	?				
tāla	*Borassus flabellifer*		•	•	Pen, cult
tamāla	*Garcinia xanthochymus*	•	•		N, Sw
tilaka	*Symplocos racemosa*	•	•		N
timiśa	*Ougeinia oojeinensis*	•	•		N, Pen
tinduka	*Diospyros peregrina*	•	•		Pen

Distributions are summarized as Ne = northeast India; Pen = peninsula; N = northern India and foothills; Sw = Western Ghats. The symbol • indicates positive.

Ecology

Events in the *Rāmāyaṇa* occur in most of the terrestrial ecosystems of the region, with the exception of the Indian desert (figure 1). Hanūmān collected medicinal plants in alpine regions of the Himalayas, and Sītā was held captive in Laṅkā, surrounded by tropical evergreen forests. However, the majority of the narrative, including the early days in Ayodhyā and the exile in Daṇḍaka, occurred in the single ecosystem that spans much of the Indian subcontinent: tropical deciduous forest.[16] The extent of this forest is defined by the southwest monsoon, which arrives in Kerala and moves up the peninsula. The heaviness of the rains depends on location and topography, but the total rainfall varies dramatically at most locations from year to year. The amount of rainfall determines the richness of the forest, along with soil type, elevation, and slope. However, many species, such as *kiṃśuka, khadira,* and *pāṭala,* grow throughout the distribution of this biome. It is clear from this description that Vālmīki viewed the Daṇḍaka Forest in the context of his own residence on the northern edge of this ecosystem, in the shadow of the Himalayas. When Lakṣmaṇa described to Rāma the beauty of the forest in winter, he included the dusting of snow and the morning frost—never to occur in Daṇḍaka. Vālmīki listed *sāla* (*Shorea robusta*), which is a dominant in the north but is replaced by teak (*śāka, Tectona grandis*) further south. Yet, he also knew of trees of the south from their products, such as the aromatic sandalwood (*candana*) and the medicinal bark of *puṃnāga* (*Calophyllum inophyllum*). Although he placed cultivated plants, such as the pomegranate, in a few forest descriptions, Vālmīki was consistent in his use of a diversity of wild trees in describing the forest. He also vividly described many animals, with their unique behaviors—such as the elephants rubbing against the trunks of trees—native to this ecosystem.

Vālmīki included long lists of trees, like that quoted above, in several forest descriptions in the *Rāmāyaṇa*. It might be argued that such lists are less descriptive of ecology and more representative of the practice of using lists for poetic effect or as mnemonic devices in classical epics.[17] It is true that the *Mahābhārata* included a similar list in its vivid description of the forests of Mount Gandhamādana (although in that work lists are less frequent and varieties of trees fewer than in the *Rāmāyaṇa*).[18] However, the consistency of species placed in these lists suggests an extensive knowledge of forest natural history.

Francis Zimmermann has argued that the archaic vegetation of most of this region was thorn scrub, and not forest.[19] Such vegetation is consistent with the classical meaning of *jāṅgala*. Almost all of the subcontinent, excluding the far west, has fifty-year annual precipitation averages of more than 600 mm[20]—rainfalls that would support dry deciduous forest.[21] Vālmīki described a forest, probably with mature trees of the varieties he listed attaining heights of twenty to twenty-five meters. The animals described in the forest journeys of Rāma, particularly the spotted deer and elephant, are consistent with a forest ecology.

The most vividly described scenery in the *Rāmāyaṇa* usually includes a water element, a lake or stream. In the attractive forest setting, lakes are full of such aquatic plants as lotus and water lilies of varied colors. Most of the birds described in the *Rāmāyaṇa,* such as the *krauñca* (or saraus crane), naturally frequent these bodies of water. The water in the lakes and ponds is clear and sweet to drink. Water is a central element in the ecology of the tropical deciduous forest. The timely arrival of rains, beautifully described in the *Kiṣkindhyākāṇḍa,*[22] ensures that seeds produced during the dry season will germinate, that animals will reproduce, that life will continue. Vālmīki was also aware of the dramatic changes in the forest after the monsoon. He beautifully described the flowering of the *kiṃśuka* tree (*palāśa, Butea monosperma*) in the winter months following the ending of the monsoon rains, a scene we can witness today throughout the distribution of the tropical deciduous forest ecosystem. The thick forest is also a conserving element in the hydrological cycle, ensuring that the streams continue to flow during the dry season, that they run more clearly during the heavy rains. The strong connection between water and the forest mirrors an ecological reality.

Vālmīki was a natural historian as well as a poet. He knew much about the ecology of his home, although he had to guess about areas he had not visited.

Landscape

> But soon they came to a trackless, dreadful-looking forest, and Rāma Aikṣvāka, son of the best of kings, asked the bull among sages: "What a forbidding forest this is! Echoing with swarms of crickets, it swarms

> with fearsome beasts of prey and harsh-voiced vultures. It is filled with all sorts of birds, screeching fearsome cries, as well as lions, tigers, boars and elephants. It is full of *dhava, aśvakarna, kakubha, bilva, tinduka, pāṭala,* and *badari* trees. What dreadful forest is this?"
>
> *Bālakāṇḍa* 23:11–14[23]

Not all of Vālmīki's descriptions of the Daṇḍaka Forest were enraptured exclamations of the beauty of the trees and animals. In some passages, as in the one above, depicting Rāma's early reaction to the Daṇḍaka wilderness, the forest is seen as a dangerous and forbidding place. Such forests are impenetrable, full of thorny shrubs, noisy with the raucous cries of birds and other animals, and "echoing with swarms of crickets." Three of the trees of this gloomy forest are crowded with thorns, and one (*tinduka*) has a somber dark green crown. In such a forest Rāma invariably encountered a fierce *rākṣasa,* or demon; it was a dangerous place. Contrast such a frightening place with Vālmīki's exuberant description of a forest vista:

> As they traveled on with Sītā, they saw varied mountain landscapes, forests, lovely rivers with cranes and sheldrakes upon the sandbanks, ponds covered with lotuses and thronged with water birds, dappled antelopes massed in herds, rutting horned buffaloes and boars, and elephants butting at trees.
>
> *Araṇyakāṇḍa* 10:2–4[24]

How do we reconcile these totally different descriptions of the Daṇḍaka Forest? Certainly the forest is the setting for an epic, and the dramatic episodes yield much from literary study.[25] Analysis of the landscapes of the forest also reveals something about Vālmīki's and his readers' attitudes toward nature. The threatening forest was closed in and impenetrable. Such a forest could be seen as dangerous because a traveler could not receive advance warning of a predator or enemy; neither could he escape from it. Attractive forest scenes are always accompanied by long vistas, across a lake or of a distant mountain. Furthermore, in the pleasant forest, the path toward the destination is clearly seen.

Considerable research in environmental psychology, in a variety of cultural and geographic contexts, has shown that human preferences are for landscapes similar to those Vālmīki described.[26] We prefer landscapes that are open, with a clear path and distant vista, over those that are closed. A survey of villagers in Bali reported similar biases.[27] The Balinese expressed their preferences by selecting photographs and taking their own snapshots; their choices were not significantly different from those of a control group of Australian tourists. Balinese are Hindu in religious practice; the *Rāmāyaṇa* was imported early, via Java, and plays a part in Balinese popular arts to this day.[28] Gordon Orians has hypothesized that such "univeral" preferences are a legacy of human origins in the savannah-like open forest vegetation of east Africa, and that these preferences could be "hard-wired" in an adaptive way. They could also be deep-seated cultural norms.[29] Thus, Vālmīki's descriptions may mirror an attitude present in the society of his time and still prevalent today.

Nature Modified

> Soon after entering the vast wilderness, the wilderness of Daṇḍaka, Rāma, the self-disciplined and invincible prince, saw a circle of ashrams where ascetics dwelt. *Kuśa* grass and bark garments were strewn about it, and, flooded with brahmanical splendor, it was as luminous and blinding to the eye as the sun's circle in heaven. It was a place of refuge for all creatures; its grounds were always kept immaculate, and troupes of *apsarases* ever paid homage there and danced. Spacious fire-sanctuaries made it beautiful, so too the sacrificial implements, the ladles and all, hide garments and *kuśa* grass, bundles of kindling, pitchers of water, and roots and fruit. Tall forest trees encircled it, holy trees that bore sweet fruit.
>
> *Araṇyakāṇḍa* 1:1–5[30]

The attractive forest landscapes of the Vālmīki *Rāmāyaṇa* included people. Rāma, Sītā, and Lakṣmaṇa encountered well-populated ashrams throughout their wanderings in the wilderness. In addition, they met with magical animals, *kiṃnara*s (beings with human bodies and horses' heads), *apsaras*es (semidivine female beings), and *rākṣasa*s. Jataus, the courageous vulture, eloquently tried to dissuade Rāvaṇa from kidnapping Sītā. In Rāvaṇa's flight south to Laṅkā, numerous towns were revealed below. Much has been written about the identity and historical basis for the creatures in the wilderness.[31] Whether they represent aboriginal tribal people—Adivasi—or not, such dwellers would have populated the forest vastness, just as they live in the remaining remnants of forest today. The history between Adivasi and orthodox Hindu has long been a thorny one, and perhaps their general absence (or disguise) in the *Rāmāyaṇa* is emblematic of this ambiguous relationship. Since the classical Hindu society was agrarian, the forest was the background for their cities and farms. For the Adivasi, the forest *was* the home, and their religion—sometimes influenced by Hinduism—was an integral part of their relationship to the forest.[32] Even when practicing agriculture, as shifting cultivation, the Adivasi also hunted and gathered plants of food and medicinal value from the forest. Rural Hindus also used these plants.

Hints of these relationships to the forest landscape are revealed in the Vālmīki *Rāmāyaṇa*. The vistas, the tree diversity, and the lofty fruit trees surrounding spacious ashrams all suggest modifications of the natural environment. The vistas and open avenues in the forest suggest a history of modification through fire. Zimmermann has suggested that the early historic environment of much of the subcontinent was of open scrub, but Michael Dove has convincingly argued that these areas are more likely the result of human impacts due to dry-season fires, our most effective tool in modifying habitats.[33] The dry *jāṅgala* landscape was the result of human habitation and not its antecedent. Our previous view of the "virgin" tropical forests of the planet has been shattered by discoveries of charcoal layers and pottery shards in many forests, such as those in Amazonia, indicating that they were well-populated prior to Western contact, and were modified by fire.[34] In the Vālmīki *Rāmāyaṇa* we see descriptions of some of the forest beginning to be modified by humans, probably principally along river valleys and ridge tops.

An additional means of "humanizing" these forests was by changing the composition of trees to improve their usefulness. Again, there are good examples of indigenous people transporting and introducing desirable plant species in tropical forests.[35] Increasing the variety of edible and medicinal plants would have made them more valuable to Adivasi and rural Hindus. Indeed, most of the plants listed in the Vālmīki *Rāmāyaṇa* are of medicinal importance. Ethnobotanical studies have revealed their value to the Adivasi, who may have shared this knowledge with early Hindus, who in turn may have incorporated such knowledge into the classical system of Ayurveda.[36] The mango (*amra*), mentioned frequently throughout the *Rāmāyaṇa,* originated further east, in Indochina, where the distribution of the genus,

Mangifera, is centered; it was domesticated from thin-fleshed resinous ancestors on the subcontinent.[37]

The attractive open vistas and populated fruitful landscapes of the Vālmīki *Rāmāyaṇa* are an early description of relationships to nature and wilderness characteristic of Asian cultures. In this relationship the wilderness is a world in which humanity is involved. People lived in the forest, collecting fruits, root vegetables, fibers for clothing, and medicines. Again, most of the plants listed by Vālmīki have other economic uses besides the medicinal ones (table 1).

Tending Nature

As the dominant religion, Hinduism should have much to say about the conservation of nature in India. The roots of Hinduism reveal a potential unity between humanity and nature.[38] Natural diversity is frequently valued in Hinduism's scriptures and epics, as we have seen for the Vālmīki *Rāmāyaṇa*. The particular value of the various *Rāmāyaṇa*s in this arsenal of teaching and traditions is the vividness of the story—and its fascination for the people of tropical Asia. This includes those of other religious faiths, such as the Muslims of Java, Malaysia, and southern Thailand. In the *Rāmāyaṇa* the forest is a special place, where the seeker finds enlightenment. It is a place full of plants of medicinal and economic value. It is inhabited by magical beings. Modified by human habitation, and at its most scenic, it is a place of great beauty. The *Rāmāyaṇa* conveys the value of forests, and thus supports the preservation of those remaining.

People in the Forest

Excluding people from natural areas in order to preserve these lands is a recent idea imported from the West; it has not worked well in India and the rest of tropical Asia.[39] The *Rāmāyaṇa* is a story occurring within the forest. Given India's population density, its occupancy of the best agricultural lands, and its small remaining forest estate (see appendix 2, table 6, of this volume), people must live in these areas unless they are too climatically hostile. The Narmada River controversy illustrates the impossibility of finding suitable land for those removed from the valley[40]—as much a gross injustice to poor rural people as an environmental problem. The challenge in the preservation of India's natural resources is not to exclude people from the resources, but to encourage inhabitants living in forests to live in a fashion that sustains the use of the natural resources and the protection of the biota. It is not enough simply to let things be; forest and rural dwellers may need help in finding strategies for living in balance with those resources. There is evidence that some traditional people have lived in such a balance, yet others have had dramatic and long-term effects—tarnishing the long-held notion of the "noble savage."[41] Ann Grodzins Gold has observed the rapid degradation of a forest by rural people following the collapse of local monarchy in Rajasthan.[42]

Throughout India the agencies regulating the use of forest resources have been the state forest departments. The forest inhabitants generally despise the officers, who are uniformed, have powers of arrest, come from other regions, and frequently give in to graft. These departments have tended to blame the rural inhabitants for most of the deforestation (firewood collecting obviously plays an important role; appendix 2, table 7). In Himachal Pradesh, the Gaddi sheepherders were accused by state forest officers of damaging the mountain pastures, even though objective studies revealed no such effects.[43] The ultimate goal of these officers was to remove the sheepherders from their traditional grazing lands, destroying their way of life.

Nongovernmental organizations in India have been working with forest dwellers living in game preserves to help them earn income through the selling of minor forest products, maintaining the ecosystem integrity for wildlife. Communities of the Soligas, traditional hunter-gatherers and shifting agriculturists, live in the Biligiri Rangaswamy Temple Wildlife Sanctuary in Mysore District of Karnataka. Members of the tribe are being assisted by a Large Scale Adivasi Multipurpose Society (LAMPS) to form their own society to regulate the collection of minor forest products to provide a sustainable income for its members.[44] Many of the plants they collect, such as amla, haritaki, beleric myrobalan, soap nut, and rock bee honey, were mentioned in the Vālmīki *Rāmāyaṇa*. The challenge is to encourage the Adivasi to be independent, and yet to help ensure that the rate of collection does not degrade the forest environment. LAMPS are now assisting other rural groups in India, a hopeful development that may help maintain India's remaining forest. It does seem that attitudes about coexistence in natural areas are improving in India, and this may be a fundamental change in a policy that dates back to the British in the nineteenth century.

Sacred Forests

The *Rāmāyaṇa* stresses the sacredness of forests in two ways. First, the forest is the magical place for the pursuit of spiritual perfection. Second, the story of the *Rāmāyaṇa* itself has made many places—and the forests of these places—sacred. Today, sacred forests preserve some of the most endangered ecosystems in India, in tracts of varying size. Economic forces that have degraded most forest lands have faltered against the religious traditions. Some of these forests have been sacred since before the arrival of Vedic religion (frequently dedicated to a mother goddess); others have gained sacredness as sites for events mentioned in the classical epics and stories, such as the *Rāmāyaṇa*. The forested mountain above a site I studied in Thana District, Maharashtra, was held sacred as the place where Sītā dropped her jewelry after being kidnapped by Rāvaṇa.[45] Frédérique Apffel-Marglin and

Figure 2. Moist deciduous forest during the dry season, Thana District, Maharashtra. (Photograph by author)

Purna Chandra Mishra describe a sacred forest in Orissa, dedicated to the goddess Bali Haracandi, and how its preservation sustains local residents.[46] The prohibitions on uses of this, and other, sacred forests preserve it and allow the provision of numerous ecological services throughout the year. Such sacred forests have been surveyed in the Western Ghats as well as in the Kathmandu Valley of Nepal.[47] Protection of such forests from all cutting has preserved their more mesic character, which has been lost from surrounding tracts degraded by tree felling and lopping for firewood. Sacred forests frequently contain rare species not present in the degraded forests. Villagers continue to resist economic pressures to fell these forests, although some are being lost—and some even damaged by hordes of largely urban pilgrims.[48] Certainly, the continuing existence of such sacred forests throughout India is evidence that Hindu religious traditions can help to preserve intact ecosystems and protect endangered species.

We may view the Silent Valley National Park, established in 1984, as the modern equivalent of such a sacred forest. The upper valley of the Kunthipuzha River was deemed sacred because of its association with the *Mahābhārata*. The Pāṇḍava brothers allegedly visited the valley during their exile, and the river was named after their mother,

Figure 3. Silent Valley, Kerala, looking up the valley, above the proposed dam site. (Photograph by author)

Kuntī. This protected valley in the Western Ghats of Kerala contains magnificent forests and extraordinary biotic richness. Because of the high rainfall and elevation above the plains, Silent Valley is the ideal candidate for a hydroelectric project. The Kerala State Electricity Board proposed such a project in the 1960s, preparation of the site began in 1976, and the fight intensified. The Kerala Sastra Sahitya Parishat (KSSP), a public science organization, fought the project because of the damage to the forest and loss of biodiversity, and many sectors of society and government became involved. Even the question of whether the forest was truly "virgin" became an issue. The matter was finally settled by Prime Minister Indira Gandhi, and the area was declared a national park in 1980. Her reasoning must have been complicated by the arguments of both sides, but undoubtedly the decision was swayed by her love of nature, encouraged by her father and particularly by her schooling at Santiniketan, the university established by Rabindranath Tagore.[49] Tagore established this remarkable school, believing that education should assimilate some of the traditional values of the forest ashrams, described in the Daṇḍaka Forest.

Sacred Gardens

The Vālmīki *Rāmāyaṇa* also described nature as present in resplendent gardens. The most important of these was the Aśoka Vana, the garden in Laṅkā where Rāvaṇa kept Sītā captive. Beyond the innumerable sweet-scented flowering vines and herbs, seventeen trees are described in this garden.[50] Such sacred gardens as this were also the inspiration for the worship of nature in temple gardens. Auspicious plants, inspired by scriptures and connected to certain deities (table 1), are commonly planted around the temples.

In the state of Karnataka, during the 1980s the Forest Department had implemented a policy of replacing native forests with exotic trees (particularly the infamous *Eucalyptus deglupta*). This conversion to exotic plantations reduced the forest value for local villagers, and a grassroots protest arose in the North Kanara region—inspired by the Chipko movement in the Himalayas. The Appiko movement, coupled with academic and press criticism, embarrassed the Forest Department, which began to experiment with native tree species and then established plantations with those natives. It also began to use the scriptural tradition to justify the preservation of local forests. Near Sirsi, the district forest officer worked with local Brahmin priests to establish the Vana Degula, a garden based on the sacred uses of plants. This garden is being duplicated in other areas of Karnataka. It is actually a series of smaller gardens based on various sacred themes.[51] Descriptions of the Aśoka garden of Sītā's exile inspired the establishment of an Aśoka Vana with special plants grouped together. I saw the garden soon after its establishment in 1988, and in 1997 it was lush in comparison with the grazed areas outside. Thus, the Karnataka Forest Department is incorporating scriptural references in support of its programs in reforestation, but this is just the begining of what could be done.[52]

Valuing Biodiversity

The hour is late for preserving India's natural resources. India's population will soon pass one billion, at some 350 individuals per square kilometer. One-third of its population will live in cities. It will use some two-thirds of its total land area (including deserts and high mountains) for housing, manufacturing, and food production (see appendix 2, table 5). Nineteen percent of its original forest will remain, less than 8 percent with extensive tree cover, and only 3.5 percent will have not been heavily exploited.[53] Humans now use or alienate more than 80 percent of the terrestrial plant production in India.[54] What is left over consists of an amazingly rich flora that supports a diversity of wildlife. Although only three animals are known to have become extinct (the forest owlet, pink-headed duck, and Himalayan mountain quail), 1,256 plants, 71 birds, and most of the large mammals are threatened with extinction (see appendix 2, table 15). Much of this diversity is concentrated in the evergreen forests of the southwest and northeast, the latter under the most severe pressure. Virtually all of the animals described in the *Rāmāyaṇa* are threatened. India's biological diversity teeters on the edge of a tragic crash.

How can Hinduism, and specifically our reading of the *Rāmāyaṇa,* reinforce values to protect the remaining diversity? First, the Vālmīki *Rāmāyaṇa* helps us understand traditions that may have contributed to contemporary attitudes toward nature. We see an ancient love for forests from a society that flourished on their margins. The forests provided much of value to this society, including edible, medicinal, and sacred plants. Second, contemporary readings of the *Rāmāyaṇa* are a powerful cultural force. They may help us improve attitudes to sustain nature. If the *Rāmāyaṇa* can continue to influence attitudes toward the environment, toward whom should this message be directed? India is still primarily a rural society, dominated by agriculture. This is the population that lives most closely with India's biological diversity, that can most directly affect it. Because of the reduced natural areas available for resource extraction and greater rural populations, the pressures increase. However, the populations that have the greatest impact on the welfare of the remaining natural areas are urban ones, because of their greater political and economic influence. It is primarily the urban markets that justify the industrial exploitation of natural resources.

I believe that the *Rāmāyaṇa* can have the greatest influence on urban populations. It is still an enormously popular story among Hindu and other segments of India's population. Furthermore, it is an epic that is continually being modified to remain relevant to the needs of different segments of the population. Its release on television in the late 1980s was seen by more than 80 million Indian viewers, in a society and time when few owned television sets. The *Rāmāyaṇa* "of Doordarshan" included ecological messages about the value of planting trees; the story had already been modified for environmental purposes.[55] It is these urban populations that are most estranged from India's natural environments and which are ignorant of its plants and animals. How can ecology be valued in the midst of such ignorance? A society can only value what it knows, and the *Rāmāyaṇa* and other epics and scriptures can still be the means of teaching about natural history.

In Miami we face a similar problem about the value of indigenous biodiversity in an increasingly urbanized and pluralistic society. How could its citizens value the Everglades if they knew nothing of it? Educators developed a strategy to teach about native plants by targeting school children. We taught teachers how to plant gardens on school grounds with native plants. These gardens attracted butterflies, birds, and other wildife. This urban wildlife has become an inspiration for students and a powerful means for teaching about local natural history. In a similar way plants important in the *Rāmāyaṇa* could be planted in the corners of school properties (miniature Aśoka Vanas!)

for teaching about tradition and natural history. Perhaps sectarian concerns can be reduced by the great popularity of the story of Rāma, and by the shared cultural value of the same plants in different religious traditions (such as similar purposes assigned to plants in the Siddha and Unani medical systems).

The vast forest of Daṇḍaka, of Lord Rāma's exile, has shrunk into small isolated pockets. What is left is still a cradle of most of India's natural diversity. Seeing it less as infinite wilderness and more as a very finite and fragile garden may help us become more responsible in its care.

Notes

1. Sheldon I. Pollock, *The Rāmāyaṇa of Vālmīki: An Epic of Ancient India,* vol. 2, *Ayodhyākaṇḍa* (Princeton: Princeton University Press, 1986), 190–91.

2. *Many Rāmāyaṇas: The Diversity of Narrative Tradition in South Asia,* ed. Paula Richman (Berkeley and Los Angeles: University of California Press, 1991). The authors in this collection emphasize the power of the epic in continually reinventing itself for the needs of its contemporary settings.

3. Rainfall isopleths and vegetation distribution are modified from H. G. Champion and S. K. Seth, *A Revised Survey of the Forest Types of India* (Delhi: Manager of Publications, 1968), 8; and G. S. Puri, V. M. Meher-Homji, R. K. Gupta, and S. Puri, *Forest Ecology,* vol. 1, *Phytogeography and Forest Conservation* (New Delhi: Oxford and IBH Publishing Company, 1984), 264. The voyage route was adapted from K. M. Balapure, J. K. Maheshwari, and R. K. Tandon, "Plants of Ramayana," *Ancient Science of Life* 7 (1987): 77.

4. The huge literature on interpretation of the *Rāmāyaṇa* is summarized in A. D. Pulsakar, "The Ramayana: Its History and Character," in *The Cultural Heritage of India,* vol. 2, ed. S. Radhakrishnan (Calcutta: Ramakrishna Mission, 1962), 14–31; and in Swami Nihsreyasananda, "The Culture of the Ramayana," ibid., 32–70; and more recently by Robert Goldman, *The Rāmāyaṇa of Vālmīki: An Epic of Ancient India,* vol. 1, *Bālakāṇḍa* (Princeton: Princeton University Press, 1984), 1–59.

5. I have benefited much from careful reading of Philip Lutgendorf's article in this volume, "City, Forest, and Cosmos: Ecological Perspectives from the Sanskrit Epics," and from his "The Secret Life of Rāmcandra of Ayodhya," in *Many Rāmāyaṇas,* ed. Richman, 217–34. There is also a summary of literary interpretations of the forest in the *Rāmāyaṇa* and the *Mahābhārata* in Thomas Parkhill, *The Forest Setting in Hindu Epics* (Lewiston, N.Y.: Mellen University Press, 1995).

6. Kelly D. Alley, "Separate Domains: Hinduism, Politics, and Environmental Pollution," and David L. Haberman, "River of Love in an Age of Pollution," both in this volume.

7. Kent Redford, "The Ecologically Noble Savage," *Orion* 9, no. 3 (1990): 24–29; and Frédérique Apffel-Marglin and Pramod Parajuli, "'Sacred Grove' and Ecology: Ritual and Science," in this volume.

8. Examples of the nexus of such beliefs among many Indian tribals are provided in *Nature-Man-Spirit Complex in Tribal India,* ed. R. S. Mann (New Delhi: Concept Publishing Company, 1981). The importance of sacrifice in an ecological context has been analyzed in Paul Shepard, *Traces of an Omnivore* (Washington, D.C.: Island Press, 1996), 11–26, which explains the seeming violence described by Laurie Patton, "Nature Romanticism and Sacrifice in Ṛgvedic Interpretation," in this volume. For the ecological significance of *pūjā* see Madhu Khanna, "The Ritual Capsule of Durgā Pūjā: An Ecological Perspective," in this volume.

9. Sheldon I. Pollock, *The Rāmāyaṇa of Vālmīki,* vol. 3, *Araṇyakāṇḍa* (Princeton: Princeton University Press, 1986), 118–19.

10. R. N. Chopra, I. C. Chopra, K. L. Handa, and L. D. Kapur, *Indigenous Drugs of India,* 2d ed. (Calcutta and New Delhi: Academic Publishers, 1958), 436.

11. Balapure, Maheshwari, and Tandon, "Plants of Ramayana," 76–84.

12. *The Useful Plants of India,* ed. S. P. Ambasta (New Delhi: *Publications and Information Directorate,* CSIR, 1986). This book summarizes information on the 5,000 plant species described in the first edition of the eleven volumes of *The Wealth of India: A Dictionary of Indian Raw Materials and Industrial Products* (Delhi: Council of Scientific and Industrial Research, 1948–).

13. S. C. Banerjee, *Flora and Fauna in Sanskrit Literature* (Calcutta: Naya Prokash, 1980).

14. Michael Zohary, *Plants of the Bible* (Cambridge: Cambridge University Press, 1982).

15. See *The Useful Plants of India,* ed. Ambasta, for scientific names and uses. See Kareem M. Abdul, *Plants in Ayurveda* (Bangalore: Foundation for Revitalization of Local Health Traditions, 1997), for scientific equivalents of Sanskrit names. And see Dietrich Brandis, *Indian Trees* (Dehra Dun: Bishen Singh Mahendra Pal Singh, 1906; reprint, Dehra Dun: International Book Distributors, 1984), for distributions.

16. Champion and Seth, *A Revised Survey of the Forest Types of India,* 174. Their classification includes a number of subtypes based on species associations and latitude. It has been more recently revised to reduce the importance of latitude by Puri et al., *Forest Ecology,* 1:387. Both works agree that forests occur at annual precipitations in excess of 600 mm.

17. E. Washburn Hopkins, "Mythological Aspects of Trees and Mountains in the Great Epic," *Journal of the American Oriental Society* 30 (1910): 347–74, argued that these lists are for poetic effects only. Frank Korom, in response to the original version and oral presentation of this paper at the conference, 3 October 1997, argued for their mnemonic importance.

18. J. A. B. van Buitenen, trans., *The Mahābhārata,* book 3, *The Book of the Forest* (Chicago: University of Chicago Press, 1975), 517.

19. Francis Zimmermann, *The Jungle and the Aroma of Meats* (Berkeley and Los Angeles: University of California Press, 1987), 74. Zimmermann argued for the widespread distribution of thorn scrub throughout the subcontinent, but he errs on the side of xericism.

20. Champion and Seth, *A Revised Survey of the Forest Types of India,* 8. The fifty-year averages indicate that most areas of the subcontinent have 700 mm or more precipitation, with many over 1,000 mm.

21. Puri et al., *Forest Ecology,* 1:264. There is little evidence for significant climatic changes during the historical period, with the possibility of trends toward greater dryness because of human influences.

22. *Kiṣkindhyākāṇḍa* 28:1–66, *Srimad Vālmīki Rāmāyaṇa,* pt. 2 (Gorakhpur: Gita Press, 1969), 1168–76.

23. Goldman, *The Rāmāyaṇa of Vālmīki,* vol. 1, *Bālakāṇḍa,* 170.

24. Pollock, *The Rāmāyaṇa of Vālmīki,* vol. 2, *Āraṇyakāṇḍa,* 104.

25. See note 5 above.

26. Rachel and Stephen Kaplan, *The Experience of Nature: A Psychological Perspective* (New York: Cambridge University Press, 1989). Much of the large body of research on landscape preferences was pioneered by the Kaplans and their students and is summarized in their volume. An insightful review has also been written: Roger Ulrich, "Biophilia, Biophobia, and Natural Landscapes," in *The Biophilia Hypothesis,*

ed. Stephen R. Kellert (Washington, D.C.: Island Press, 1993), 73–137.

27. R. B. Hull and G. R. G. Revell, "Cross-cultural Comparison of Landscape Scenic Beauty Evaluations: A Case Study of Bali," *Journal of Environmental Psychology* 9 (1989): 177–91.

28. Miguel Covarrubias, *Island of Bali* (Kuala Lumpur: Oxford University Press, 1974), 217.

29. Gordon Orian's savannah hypothesis was reviewed in J. H. Heerwagen and G. H. Orians, "Humans, Habitats, and Aesthetics," in *The Biophilia Hypothesis,* ed. Stephen R. Kellert (Washington, D.C.: Island Press, 1993), 138–72. Different sides of this contentious issue were discussed in other articles in the volume, particularly those by Paul Shephard ("On Animal Friends," 275–300), Roger Ulrich ("Biophilia, Biophobia, and Natural Landscapes," 73–137), and Jared Diamond ("New Guineans and Their Natural World," 251–71).

30. Pollock, *The Rāmāyaṇa of Vālmīki,* vol. 2, *Āraṇyakāṇḍa,* 87.

31. Some of this literature is reviewed succinctly in Pollock, *The Rāmāyaṇa of Vālmīki,* vol. 2, *Āraṇyakāṇḍa,* 68.

32. Mann, *Nature-Man-Spirit Complex in Tribal India;* Madhav Gadgil and Ramachandra Guha, *This Fissured Land: An Ecological History of India* (Berkeley and Los Angeles: University of California Press, 1992), 15–27.

33. Michael Dove, "The Dialectical History of 'Jungle' in Pakistan: An Examination of the Relationship between Nature and Culture," *Journal of Anthropological Research* 48 (1992): 231–53. Thorn scrub is only described as a human-produced vegetation type in Puri et al., *Forest Ecology,* 1:325. Bridget Allchin, "Early Man and Environment in South Asia 10,000 B.C.–A.D. 500," in *Nature and the Orient,* ed. Richard Grove, Vinita Damodaran, and Satpal Sangwan (Delhi: Oxford University Press, 1998), 29–50; and George Erdosy, "Deforestation in Pre- And Proto-Historic South Asia; in *Nature and the Orient,* ed. Grove, Damodaran, and Sagwan, 51–69. The latter two articles emphasize the ambivalence of paleoecological data in determining vegetation history.

34. Betty J. Meggers, "The Prehistory of Amazonia," and Karl L. Hutterer, "The Prehistory of the Asian Rain Forests," in *People of the Tropical Rain Forest,* ed. Julie Denslow and Christine Padoch (Berkeley and Los Angeles: University of California Press, 1988) 53–62, 63–72.

35. Darrell Posey, "Alternatives to Forest Destruction: Lessons from the Mêbêngôkre Indians," *Ecologist* 19 (1989): 241–44, and "Science of the Mêbêngôkre," *Orion* 9, no. 3 (1990): 16–24. Posey has documented the widespread modification of deciduous forests in the Amazon by the Mêbêngôkre (Kayapó), who use both fire and seed planting to modify forests.

36. C. R. Karnick, "Ethnobotanical Records of Drug Plants Described in Valmiki Ramayana and Their Uses in the Ayurvedic System of Medicine," *Quarterly Journal of Crude Drug Research* 13 (1975): 143–54. J. K. Maheshwari, K. K. Singh, and S. Saha, *The Ethnobotany of the Tharus of Kheri District, Uttar Pradesh* (Lucknow: National Botanical Research Institute, 1981); J. K. Maheshwari, K. K. Singh, and S. Saha, *The Ethnobotany of Tribals of Mirzapur District, Uttar Pradesh* (Lucknow: National Botanical Research Institute, 1981). Approximately half of the plants described in these treatises have medicinal uses.

37. J. M. Bompard and R. J. Schnell, "Taxonomy and Systematics," in *The Mango,* ed. R. E. Litz (New York: CAB International, 1997), 37. It was not until the sixteenth century that the vegetatively propagated superior varieties began to appear.

38. O. P. Dwivedi and B. N. Tiwari, *Environmental Crisis and Hindu Religion* (New Delhi: Gitinjali Publishing House, 1987); O. P. Dwivedi, "Dharmic Ecology," and K. L. Seshagiri Rao, "The Five Great Elements (*Pañcamahābhūta*): An Ecological Perspective," both in this volume.

39. Gadgil and Guha, *This Fissured Land,* 113–80. Ramachandra Guha, "Radical American Environmentalism and Wilderness Preservation: A Third World Critique," *Environmental Ethics* 11 (1989): 71–83. Walter Fernandes, Geeta Menon, and Philip Viegas, *Forests, Environment, and Tribal Economy* (New Delhi: Indian Social Institute, 1988), 175–248. *Nature and the Orient,* ed. Grove, Damodaran, and Sangwan, various chapters.

40. William F. Fisher, "Sacred Rivers, Sacred Dams: Competing Visions of Social Justice and Sustainable Development along the Narmada," in this volume.

41. Redford, "The Ecologically Noble Savage," 24–29. Fernandes, Menon, and Viegas, *Forests, Environment, and Tribal Economy,* 148–74. Theoretical issues regarding sustainable resource use by traditional peoples are discussed by Joel Heinen and Bobbi Low, "Human Behavioural Ecology and Environmental Conservation," *Environmental Conservation* 19 (1992): 105–16; and Joel Heinen, "Applications of Human Behavioural Ecology to Sustainable Wildlife Conservation and Use Programmes in Developing Nations," *Oryx* 29 (1995): 178–86.

42. Ann Grodzins Gold and Bhoju Ram Gujar, "Wild Pigs and Kings: Remembered Landscapes in Rajasthan," *American Anthropologist* 99 (1997): 70–84; and Ann Grodzins Gold, "'If You Cut a Branch You Cut My Finger': Court, Forest, and Environmental Ethics in Rajasthan," in this volume.

43. Vasant Saberwal, "Pastoral Politics: Gaddi Grazing, Degradation, and Biodiversity Conservation in Himachal Pradesh, India," *Conservation Biology* 10 (1996): 741–49

44. K. S. Murali, Uma Shankar, R. Uma Shaanker, K. N. Ganeshaiah, and Kamaljit S. Bawa, "Extraction of Non-Timber Forest Products in the Forests of Biligiri Rangan Hills, India. 1. Impact of NTFP Extraction on Regeneration, Population Structure, and Species Composition," *Economic Botany* 50 (1996): 252–69. Sharachchandra Lele, K. S. Murali, and Kamaljit S. Bawa, "Biodiversity Conservation through Community Enterprise: An Experiment in the BRT Wildlife Sanctuary of Karnataka, India," *Proceedings of the UNESCO Regional Workshop on Community-Based Conservation: Policy and Practice* (New Delhi: Indian Institute of Public Administration, 1997), 449–66.

45. David W. Lee, "Canopy Dynamics and Light Climates in a Tropical Moist Deciduous Forest in India," *Journal of Tropical Ecology* 5 (1989): 65–79. The Warli villagers who minded my research plot revered the mountain above it; on its other side was the site of a former sacred forest described by Madhav Gadgil and V. D. Vartak, "The Sacred Groves of Western Ghats in India," *Economic Botany* 30 (1975): 152–60.

46. Frédérique Apffel-Marglin and Purna Chandra Mishra, "Sacred Groves: Regenerating the Body, the Land, the Community," in *Global Ecology: A New Arena of*

Political Conflicts, ed. Wolfgang Sachs (London: Zed Books, 1993), 197–207. See also Frédérique Apffel-Marglin and Pramod Parajuli, "'Sacred Grove' and Ecology: Ritual and Science," in this volume.

47. Gadgil and Vartak, "The Sacred Groves of Western Ghats in India," 152–60. Joe Robert Mansberger, "Ban Yatra: A Bio-cultural Survey of Sacred Forests in Kathmandu Valley" (Ph.D. diss., University of Hawaii, 1991). M. D. Subash Chandan, "Shifting Cultivation, Sacred Groves, and Conflicts in Colonial Forest Policy in the Western Ghats," in *Nature and the Orient,* ed. Grove, Damodaran, and Sangwan.

48. Some of the sacred forests listed by Gadgil and Vartak in Thana District, Maharashtra, no longer existed when I conducted my survey in 1989. For the degradation of the Bhimashankar Sanctuary near Pune, see Renee Borges, "Bhimashankar: Forests of the Gods," *Hornbill,* 1994, no. 2:21–23.

49. For a good overview of the controversy, see Darryl D'Monte, *Temples or Tombs* (New Delhi: Centre for Science and Environment, 1985), 29–89; T. M. Thomas Isaac and B. Ekbal, *Science for Social Revolution* (Trichur: Kerala Sastra Sahitya Parishat, 1988); Dom Moraes, *Indira Gandhi* (Boston: Little, Brown and Company, 1980), 66–68; and Krishna Dutta and Andrew Robinson, *Rabindranath Tagore: The Myriad-Minded Man* (New York: St. Martin's Press, 1996), 323–34.

50. *Srimad Valmiki Ramayana,* pt. 2, 962–74.

51. *Sacred Plants* (Dharwad: Karnataka Forest Department, 1988). This booklet describes the layout of all gardens and lists all plants in them.

52. M. G. Chandrakanth and J. Romm, "Sacred Forests, Secular Forest Policies, and People's Actions," *Natural Resources Journal* 31 (1991): 741–56.

53. *The State of the Forest Report* (Dehra Dun: Forest Survey of India, 1994), 7. Analysis of satellite images in the 1980s revealed a large discrepancy between forest department estimates of forest cover, largely based on areas gazetted as forest (many of which were treeless), and vegetation in the images. This led to the establishment of the Forest Survey of India and its more careful assessment of forest cover.

54. Peter Vitousek, Paul Ehrlich, Ann Ehrlich, and Pamela Matson, "Human Appropriation of the Products of Photosynthesis," *Bioscience* 36 (1986): 368–73. Their methods gave an estimate of human appropriation of 40 percent of photosynthetic production for the entire planet.

55. Philip Lutgendorf, "Ramayan: The Video," *Drama Review* 34 (1990): 127–76; and Lutgendorf, "City, Forest, and Cosmos."

City, Forest, and Cosmos: Ecological Perspectives from the Sanskrit Epics

PHILIP LUTGENDORF

"Only when Rāma went to the forest did he become Rāma."
Hindi saying[1]

Books of the Forest

We live in an era of accelerating deforestation and proliferating scholarship. The two are not altogether unrelated, since scholarly productivity (such as the critical editing and exegesis of Sanskrit epics) consumes, like Agni, copious amounts of wood. Hence I hesitated, when first asked to write about the classical epics for this volume, to fell any more trees.[2] The contrast between the paradigmatic landscapes of "city" and "forest" is a familiar narrative convention of the *Rāmāyaṇa* and *Mahābhārata* that has been briefly noted by scholarly translators like J. A. B. van Buitenen and Sheldon Pollock.[3] It is clear enough that Indian epic heroes, destined for the eventual rule of city-states, only grow to their full stature through a period of exile and wandering in a "wild" landscape that is filled both with dangers and with magical and spiritual forces—an archetypal narrative pattern that is hardly limited to South Asia but may be found in the *Märchen* and sagas of many cultures. My own understanding of the city-forest dichotomy and its centrality to the classical Indian epics has developed through many years of reading, thinking about, and teaching these stories, in their early Sanskrit versions (read in English translation) as well as in their later retellings through regional vernacular texts, oral performances, and iconographic representations. In the present paper I would like to sketch briefly some characteristic features of what I will term the "narrative ecosystem" of the *Mahābhārata* and *Rāmāyaṇa,* with the eventual aim of identifying therein enduring cultural attitudes toward the natural environment as well as cultural resources that might be applied to understanding and responding to the environmental crisis of our time.

A striking feature shared by these two narratives, the *Mahābhārata* and *Rāmāyaṇa,* which are in many respects so different in character that they may even be seen as opposites, is that each contains a "Book of the Forest" (the former's *Araṇyakāṇḍa;* the latter's *Vanaparva or Āraṇyakaparva*), which sends its heroes into the wilderness for a period which, in chronological if not narrative terms, accounts for a major portion of the central story (fourteen and twelve years, respectively). Moreover, the invocation of the forest as a contrasting landscape to that of the royal court and city-state is not confined to these "forest books," but is pervasive of the stories, and indeed of much of the copious narrative tradition of the Sanskrit and vernacular Purāṇas. I suggest some ramifications of these contrasting literary landscapes in the chart below, wherein it should be understood that "city" refers not only to urban settlements, but to the royal city-state envisioned in the epics, consisting of the towns, villages, fields, and hinterlands directly controlled by the king—in short, to "civilization" in a broad sense. The landscapes connoted by the signifier "forest" will be further clarified below. Many of the elements of this chart will be readily familiar to students of the epics; others may appear more speculative. Certainly, the majority (within the sprawling universe of Hindu narrative) will admit to exceptions. I hope to elucidate the relevance of most of them in the pages that follow.

Although much of the chart consists of balanced binary pairs, it is immediately apparent that the "forest" ecosystem sustains a denser population of supernatural beings, as well as more (albeit more "liminal") lifestages. This reminds us that, since the majority of tellers and hearers of epic tales are householders who do not, in fact, live in the forest, it becomes for them a landscape of "otherness" and a fertile ground for ideas and images perceived as relatively more remote from

	"City"	"Forest"
human inhabitants	the four *varṇa*s	"outsiders" (tribals, ascetics)
superhuman inhabitants	*deva*s (when ritually summoned)	*deva*s *siddha*s *yakṣa*s *gandharva*s *kiṃnara*s *apsaras*es *rākṣasa*s *nāga*s, etc.
animals	domesticated (horses, cows, etc.)	wild (deer, monkeys, etc.)
foods	cultivated, cooked (grains, legumes)	uncultivated, (often) raw (tubers, wild fruits)
paradigmatic lifeways	king, householder ("*kṣatriya*") violence procreation self-expression action consumption	sage, ascetic ("*brāhmaṇa*") nonviolence celibacy self-control contemplation conservation
primary pursuits (*puruṣārtha*s)	*artha, kāma*	*dharma, mokṣa*
lifestages	householder	student retiree renunciant
other categories	"culture" self Agni, Indra	"nature" other Soma, Śiva
attempted "compromise" or synthesis	royal-sage warrior-ascetic Viṣṇu	

the mundane. I must also note that the chart's static dualities fail to suggest the aesthetic and emotional variety that characterizes the representation of forest in the epic traditions, a variety that is often expressed through abrupt shifts in mood. Again, it would appear that, as the less-marked "other" to the settled landscape of the city-state, the forest was an ideal screen for the projection of heightened aesthetic emotion. To elucidate the latter idea below, I will use some of the terminology of *rasa* theory.[4] Here I must add that, whereas the literary trope of the urban settlement displays an often-noted consistency throughout much of Sanskrit literature (wherein every royal city tends to sound much like every other), "the forest" as described by epic poets—despite my generic label—is not a single environment, but rather encompasses many, marked by different natural and supernatural features and characterized by different moods. I will examine several of these paradigmatic "forest" environments below, relating each to the dialectic (and its attempted resolution) suggested by my chart.

But first, a note on terminology. The two most common Sanskrit words for "forest"—*vana* and *araṇya*—appear to be nearly synonymous in usage. This is suggested by their parallel occurrence in numerous compounds in which they function identically to qualify the second member of the compound as "wild" or "of the forest" (e.g., *vana-kadalī, araṇya-kadalī:* "wild plantain"; *vana-cara, araṇya-cara:* "living in the forest," i.e., woodsman, wild animal). *Vana,* by far the commonest word, has the additional sense of "plenty, abundance," and can refer to a "quantity of lotuses or other plants growing in a thick cluster."[5] That the ancient forest was understood to be dense or thick is also suggested by the less common adjectival epithet *gahana* ("deep, thick, impervious, impenetrable," hence, as a noun "a thickness, a density"), also sometimes applied to forest settings.[6] *Araṇya* can also mean more generically "wilderness" or even "desert," though its use in compounds, as already indicated, generally suggests a high density of wild vegetation; interestingly, its most archaic sense is said to be of "a foreign or distant land" (cf. English "forest," derived from Latin *foris,* denoting land that is "outside" human habitation and cultivation)—a derivation that may support the ecohistorical reading that land-clearing for pastoralism and agriculture gradually drove the forest and many of its denizens to the peripheries of the increasingly settled Indo-Gangetic plain. Other, less common epic labels for forested settings include *vipina* ("stirring or waving in the wind"), *kānana* ("a forest, grove"; sometimes synonymously compounded with *vana* as in *kānanaukas = vanaukas:* "forest dweller, monkey"), and *kāntāra* ("forest, wilderness, waste").[7]

It should be noted that all these words are quite different in meaning from *jāṅgala,* which likewise carried a connotation of "wild" (as modern Hindi *jaṅgal* and its English derivative "jungle" still do), but which primarily referred to a landscape that was "arid, sparingly grown with trees and plants (though not unfertile . . .)."[8] As Francis Zimmermann and Michael Dove have persuasively argued in recent studies, the ancient *jāṅgala* was not the "dense" and relatively "distant" forest to which exiles and ascetics repaired, but rather was the preferred, encompassing ecosystem of early Indo-Aryan culture—a tree-studded, grass-covered savanna that was considered ideally suited to pastoralism, small settlements, occasional hunting, and limited cultivation.[9] Although this ecosystem may already have been yielding, on the Indo-Gangetic plain, to more intensive agricultural and urban patterns by the period of composition of the epics, the archaic sense of *jāṅgala* survives in them; thus, in the *Mahābhārata*, the name *Kuru-jāṅgala* refers simply to the "land" or "kingdom" of the ruling Kuru dynasty, not to the denser forest tracks beyond its borders, into which the heroes are driven in exile. I will return to this linguistic distinction, and to Zimmermann's and Dove's insights, in my conclusion.

Motifs and Moods

Sporting Kings

Enforced exile (or voluntary retirement in old age) is not the sole motive that drives kings to the forest in the epics; more commonly, indeed, almost constantly, they repair there temporarily for recreational purposes, primarily to hunt. Very often they have adventures, get separated from their companions, wander or are led from the outskirts of the wood (where they seem more comfortable and indeed more kingly) into its depths, where they lose their way and sometimes their identities, and where they meet powerful figures (demons, sages, beautiful and mysterious women) who change their lives. That kings often get into trouble through the excitement and hubris of hunting—as when Pāṇḍu slays a mating buck who proves to be a sage, or when Daśaratha aims at the sound of an "elephant" and fatally wounds an ascetic boy—may suggest Brāhmaṇa disapproval of Kṣatriya violence. Yet the epic bards, who also recognize that the king is socially sanctioned to commit necessary and world-maintaining acts of violence (through warfare, punishment, and sacrificial ritual, as well as through hunting), do not mind, at times, reveling in it. That this has serious consequences for the forest and its denizens seems, at such moments, quite beside the point. Like the directors of a modern action film, the poets cheerfully unleash a spectacle of carnage that seems primarily intended to evoke the aesthetic mood of *vīrya* ("heroism"). Thus, in the *Mahābhārata*'s subsidiary narrative of Śakuntalā, King Duṣyanta sallies forth for a hunt, accompanied by a small army, and showered with flowers and amorous glances by the women of his capital. Far from the city, he enters "a wood like Indra's paradise" and proceeds to wreak havoc, slaying "many families of tigers," "thousands of deer," and "killing wild game and fowl with javelin, sword, mace, bludgeon, harberd." The tone seems cheerfully celebratory, even when describing frantic herds of deer dropping unconscious as they vainly try to escape, and maddened elephants, "dropping dung and urine and streaming with blood," running amok and trampling the king's own men (*Mahābhārata* 1.7.63).[10] It just as abruptly changes to a mood of ascetic peace, and then to one of erotic abandon (for both, see below), yet in my reading at least, this does not negate the essentially positive vision of the imperial hunt: as husband of the earth (a theme that the later courtly poet Kālidāsa would greatly amplify in his famed retelling of the story), King Duṣyanta is entitled and even expected to cull the big game of the forest.

Nor is any "environmental impact statement" offered for Prince Bharata's expedition to Citrakūṭa to meet with his exiled brother Rāma, accompanied by nearly the entire population of Ayodhyā. This brief excursion—the narrative purpose of which is to affirm the brothers' mutual devotion and to confirm Rāma's decision to live out his term of exile—gives the *Rāmāyaṇa* poet an opportunity to describe the construction of a royal highway (*Rāmāyaṇa* 2.74), as "surveyors and men trained in measurement, powerful excavators who were zealous in their work; engineers, laborers, craftsmen, and men skilled in machinery; carpenters, road-levelers, woodcutters, well-drillers, pavers, cane-weavers, and capable guides all set out from the city" (2.74.1–3). Like the builders of modern expressways, the workmen fell trees, flatten hillocks, divert streams, and, in lieu of roadside "rest areas," construct multistoried mansions. Here the intended mood appears to be one of admiring wonder (*adbhūta*) at the royal capacity to decisively transform a natural landscape "in no time at all" (2.74.11).[11]

I could offer other examples of spectacular and egregious destruction of the environment for apparent aesthetic effect—the *Mahābhārata*'s superman Bhīma ascends Mount Gandhamādana to obtain a magical lotus for Draupadī, and proceeds Rambo-like, slaying thousands of animals and uprooting countless trees, for no apparent reason other than to show his immense strength and impatience (*Mahābhārata* 3.146.20–60); and the *Rāmāyaṇa*'s counterpart and half-brother to Bhīma, Hanūmān, destroys a forested mountain and all its denizens while leaping toward Laṅkā (*Rāmāyaṇa* 5.1.10–42)—but none is perhaps more troubling to modern sensibilities than Kṛṣṇa and Arjuna's burning of the Khāṇḍava Forest at the close of Book One of the *Mahābhārata*. Ordered by the voracious Agni himself, the conflagration seems to be indirectly linked to the expansion, through forest-clearing, of the Pāṇḍavas' new kingdom of Indraprastha (though its capital city of the same name has already been built elsewhere), and it is difficult to dismiss altogether the mythohistoric explanation that the story may reflect ancient practices of deliberate burning that (according to Michael Dove) regularly transformed thick thorn-tree forest into Ārya-friendly savanna.[12] But Agni doesn't simply desire wood and land; he wants "the fat of the living," and so Kṛṣṇa and his friend cheerfully use their warrior prowess to seal off the perimeters of the forest, conduct a "vast massacre" of its fleeing creatures, and hurl their bodies back into the flames. They allow virtually no living beings to escape (although, like every holocaust in the *Mahābhārata,* this one has an eventual handful of survivors—a not inconsequential point in the larger ecology of the epic, and one to which I will return). Meanwhile, the poet indulges in a grisly and detailed description of the creatures' panic, their attempts at flight, and their cries of agony as their eyes, wings, fat, and marrow catch fire, transforming them into "living torches," while "the huge flames of the happy Fire jumped up to the sky and caused the greatest consternation among the Gods" (1.19.217).[13] Although modern readers may find it difficult to feel "happy" while reading such descriptions (and those who, like Irawati Karve, favor an ethnohistorical interpretation may recoil in horror from the supposed genocide of "nāga" tribals),[14] I suspect that early epic audiences felt, and were meant to feel, no more reprehensible delight than that experienced by modern cinema-goers watching, say, Los Angeles highrises explode in the 1997 disaster movie *Volcano*. Together with the "heroism" of Kṛṣṇa and Arjuna (which the poet shamelessly celebrates), the aesthetic sentiments intended here are seemingly those of (vicarious) fury, terror, and disgust (*raudra, bhayānaka, bībhatsa*). The forest is utilized as a setting for the spectacle of disaster—befalling expendable others at an aesthetically safe distance—that seems to be enduringly human to enjoy.

Voluptuous Women

In the *Mahābhārata,* King Duṣyanta's recreational carnage ceases abruptly when he enters "another wood," albeit still "in search of deer." Here it is immediately apparent that a different aesthetic mood holds sway: that of *śṛṇgāra,* or eros. The scene opens with descriptions of natural beauty and abundance (trees laden with blossoms and fruits, swarms of bees and birds), but the landscape is rapidly eroticized by visions of intertwining branches, breezes that waft pollen to the trees "as though to make love to them," and of a woodland "grown in the embrace of a river" (*Mahābhārata* 1.7.64).[15] All such metaphors (and the last carries special import for the *Mahābhārata,* wherein dynasts regularly mate with river goddesses or their symbolic substitutes) suggest an erotically feminized forest, to which the epic king responds as the hypermale and world-husband he invariably is. The heady passage thus prefigures the approaching vision of the virginal beauty Śakuntalā, and the king's declaration of love for her.

An equally romantic, albeit less erotically charged, vision of nature occurs in the (generally more austere and propriety-minded) *Rāmāyaṇa*. Although Rāma has earlier warned Sītā of the hardships of forest life, she has nevertheless anticipated the "pleasures I shall share with you, my mighty husband, in the honey-scented forests" (*Rāmāyaṇa* 2.24.10).[16] For a time, these expectations are indeed fulfilled, and though the couple's happy life at Citrakūṭa and Pañcavaṭī is only a briefly hinted at interlude in Vālmīki's version (2.50; 3.14–15; later Sītā recalls it to Hanūmān at 5.36.12–15), the descriptions of nature are lush and suggestive, and many later retellings have expanded this portion of the story into a long, erotic idyll, wherein Rāma and Sītā, tirelessly guarded by Lakṣmaṇa, sport for twelve years in an earthly paradise.[17]

Rampaging Rākṣasa*s*

Rāma's admonitory refrain to Sītā in *Ayodhyākāṇḍa*—"the forest is a place of utter pain" (*Rāmāyaṇa* 2.25)—will of course prove prophetic for both of them, but his warnings of the dangers of the forest (e.g., lions, snakes, scorpions, and thorn trees), reflect a dread of the wilderness that was already present in the anomalous *Ṛgveda* hymn dedicated to the "lady of the forest" (*araṇyāni*), upon whom a frightened wayfarer, lost in the forest at twilight and imagining the terrors lurking in its shadows, calls for protection (*Ṛgveda* 10.146). The terrors of the deep woods are best represented in the epics by the horrific and cannibalistic *rākṣasa*s who lurk in them, although the portrayal of these creatures is itself marked by a certain ambivalence—e.g., Vālmīki's distinction between the "wild" *rākṣasa*s of the forest and their more refined, though still deadly and anthropophagous, cousins in Laṅkā; and the legions of *rākṣasa*s in the *Mahābhārata* who seem, true to the apparent derivation of their name, ferocious "guardians" of remote natural beauty spots favored by the gods (e.g., Kubera's lake on Mount Gandhamādana, in *Mahābhārata* 3.33.151–52). The entry of epic heroes into deep forest is frequently signaled by terrifying encounters with these guardian beings; a fine example is Virādha, whom Rāma and his companions encounter in the second chapter of the *Araṇyakāṇḍa*, immediately after their entry into "the wilderness of Daṇḍaka."

> Sunken-eyed, huge-mouthed, his belly deformed, he was massive, loathsome, deformed, gigantic, monstrous, a terror to behold; clad in a tiger skin dripping with grease and spattered with blood, as terrifying to all creatures as Death with jaws agape. On an iron pike he held impaled three lions, four tigers, two wolves, ten dappled antelopes, and the massive head of an elephant, complete with tusks, and smeared with gore. And he was roaring deafeningly. (*Rāmāyaṇa* 3.2.5–8)[18]

That such figures become, particularly in the *Rāmāyaṇa,* a locus for the poet's projection of the reprehensible and sociopathic traits of irreducible "otherness" has been well analyzed by Sheldon Pollock.[19] I will only add that such forest scenes also provide for the evocation of moods of terror and disgust (the *rasa*s of *bhayānaka* and *bībhatsa*); the heroes' response to such apparitions is, of course, to slay them with a characteristic display of *vīrya*.

Serene (and Scary) Sages

Although lions do not lie down with lambs in Sanskrit epics, tigers do rest beside deer in one forest setting beloved to epic bards: the *āśrama* ("place of exertion," i.e., through asceticism and Vedic study), or hermitage, of a sage. Such forests-within-the-forest appear as veritable ecopreserves, abounding in trees of many species and herds of tame animals, and also as protocampuses, full of students diligently engaged in Vedic recitation and study. These settings, evocative of the mood of tranquility (*śānta*), are frequently juxtaposed with scenes of destructive Kṣatriya exhuberance; thus, in the *Mahābhārata,* the gory excess of King Duṣyanta's hunt yields, within fifteen verses, to the tranquil groves and spotless sacrificial halls of the sage Kaṇva's hermitage—a paradise on earth "that mirrored the world of Brahmā"—and which evokes in the king "the purest joy." It evokes humility as well, for the king who just majestically supervised the slaughter of the forest's big game, now dismounts from his chariot, abandons his armor and escort, and walks alone into the protected enclosure to pay homage to its ruler (*Mahābhārata* 1.7.64).[20]

Vālmīki's *Bālakāṇḍa* offers a comparable vision of the forest hermitage of the immortal Vasiṣṭha, family preceptor to the Ikṣvāku dynasty (*Rāmāyaṇa* 1.50.22–27). Its crown jewel is the magical cow Śabalā, and when visiting King Viśvāmitra attempts to seize her (since, as he puts it, "all gems belong to the king"; 1.52.9), the mood of *śānta* quickly yields to *vīrya,* as sage and king commence a deadly and protracted battle to determine who is ultimately more powerful. In the process, the peaceful hermitage is trashed in an orgy of royal violence that easily equals anything wrought by *rākṣasa*s (1.54.21–28).[21] Such scenes, recapitulated, for example, in the *Mahābhārata*'s Bhārgava narrative of King Kārtavīrya's destruction of the sage Jamadagni's hermitage (*Mahābhārata* 3.33.116), bring the contrasting paradigmatic life-styles and values of king and sage into their starkest juxtaposition.

It should be added that the sages' serenity and wisdom are not the sole traits they derive from their association with the forest. Often there is also a quality of wildness that verges on the nonhuman and suggests the feral, extrasocial, and even unsettling dimension of the less-settled lands. This quality is particularly evident in some of the forest-dwelling sages of the *Mahābhārata.* Thus, when Vyāsa, the epic's author and the master of all sacred lore, is mentally summoned to the city of Hāstinapura by his mother Satyavatī in order to father sons on her widowed daughters-in-law, he brings not only his supernatural knowledge and fecundity, but also an all-too-natural stench and a form and visage so fearful that the princesses close their eyes or faint in his presence (*Mahābhārata* 1.7.99–100). More feral still is Kiṃdama, "a hermit of unequaled austerities" but somewhat unusual sexual tastes—for the ill-fated Pāṇḍu accidentally slays him while Kiṃdama is in the guise of a deer, engaged in mating with a doe, "because I shy away from humans, and as a deer I live with deer in the depths of the forest" (*Mahābhārata* 1.7.109). We may also recall the strange "deer-horn" sage Ṛśyaśṛṅga, whose mother is indeed a doe and who grows up without human companions, save his father, Kāśyapa, who is himself described as having "the tawny eyes of a lion . . . with hair as far down as his nails" (*Mahābhārata* 3.33.111). In the case of both Kiṃdama and Ṛśyaśṛṅga, we may note the sages' association with the quintessential wild game animal, whose range serves as a boundary-marker for the "forest," and who repeatedly initiates epic adventure by luring royal hunters out of their settled realms and into its mysterious and protean depths.

It may be useful to summarize briefly some attitudes toward the forest landscape that are suggested by the above examples.

The forest is vast and exploitable. Although kings are occasionally admonished to plant trees (especially along royal roadsides), their more characteristic concern with the forest is in clearing it or securing its abundant game. The forest is for them primarily a zone for exploitation and consumption, and there is no sense in the epics of the modern notion of the "fragility" or endangerment of the forest ecosystem. On the contrary, the destruction of huge tracts of forest or of the wildlife it contains is grist for entertaining narrative and is no more alarming to the bards and their presumed audiences than was the felling of North American forests to nineteenth-century European settlers.[22]

The forest is cherished by sages. If there is a kind of protoconservationist ethic in the epics, it is revealed in the attitudes of ashram-dwellers to their forest environment. Unlike kings who enter the forest in exile or for an occasional recreational hunt, sages choose to reside there, and they cultivate an attitude of nonviolence that transforms stretches of forest into an earthly paradise. Their behavior is set in stark contrast to that of kings, although the stories clearly celebrate both.

The forest is evocative of high emotion. As the landscape of "otherness" to the (in theory) rigidly coded realm of the śāstraically governed kingdom, the relatively more chaotic forest is the ideal locale for the evocation of intense emotional moods (terror, disgust, eroticism) that may be disruptive or incongruous in a more mundane setting.

The forest is a place of transcendence. It is generally in the forest, rather than in the city-state, that transcendence of human limitations becomes possible and that communication occurs between heaven and earth; it is here that gods most often descend to communicate with human beings, and that the latter ascend to heaven, and it is here that superhuman beings and demigods are most often encountered.

The forest teems with life. Here I offer an observation which relates to all the forest locales noted above, and to which I will return below: the epic forest swarms with a superdensity of plant and animal species, is seldom devoid of human inhabitants (generally tribals), and not uncommonly also contains the superhuman life-forms noted earlier. It is a "crowded cosmos" that indeed suggests the complexity and abundance of the South Asian ecosystem, even as it permits epic poets to indulge in their characteristic penchant of packing their *śloka*s with impressive lists.[23]

The forest is a state of mind. Although the poets display moments of keen observation and accurate description of natural phenomena, their forest landscape cannot, on the whole, be said to be naturalistic. Its particulars are always subservient to the demands of aesthetic mood, resulting at times in ecological incongruity, as when (a commonplace) the poets place peacocks, elephants, and huge fig trees on high Himalayan peaks.[24] Their forest is a stage-set, with props that can be moved around as narrative atmospherics require.

The King in the Forest

It is my view that one of the implicit and overarching projects of the classical epics, especially apparent in the subsidiary and seemingly "digressive" stories incorporated into their introductory books (which form, in each case, a sort of overture or prelude to the main narrative), is the attempt to resolve the dialectical tension between the paradigmatic realms of "city/kingdom" and "forest" and their respective human life-styles through integrative characters who combine elements of both. Such attempted integration cannot escape the tension inherent in the dialectic, however, and is itself often provisional or inconclusive. It occurs relatively more harmoniously in Vālmīki's *Bālakāṇḍa,* whose principal subnarratives (largely and appropriately concerned with ancestors of Rāma) feature what I like to term "upwardly mobile" Kṣatriyas, who aspire to become sages or even gods (e.g., Bhagīratha, Viśvāmitra, Triśaṅku), and sages who become royal householders or warriors (e.g., Ṛśyaśṛṅga, Paraśurāma). The comparable series of introductory narratives in the *Ādiparva* of the *Mahābhārata* are, as one might expect, more ominous, violent, and savvy of the *realpolitik* dimensions of the Brāhmaṇa-Kṣatriya relationship, and story after story depicts the unhappy outcome of the martial elite's apparently never-ending challenge to the authority of *āśrama*-based priest-scholar lineages. Here it is often the case that a tenuous balance is achieved only through the near-annihilation of the offending party—e.g., the repeated theme of the sacrificial holocaust, one of the reigning motifs of the *Mahābhārata* as a whole.

Yet, together with the fundamentally Brāhmaṇa bias of both Sanskrit epics in their surviving written forms, both also reveal a theological preoccupation with Viṣṇu, displayed particularly in his literal embodiment in key personalities. As a celestial god-king who seems to have historically absorbed many of Indra's more positive attributes, Viṣṇu emerges in the epics as a Kṣatriya god who is adored by, and who adores, Brāhmaṇas. His favorite human incarnations, both Kṣatriya god-kings, assume a broadly mediatory role between the contending exemplars of city and forest: the priest-sage and warrior-householder. As Rāma, Viṣṇu himself becomes a sage-king who protects ascetics in the forest, yet refuses (in the city) to protect his own interests through the violent *kṣatriya dharma* advocated by his younger brother Lakṣmaṇa. As Kṛṣṇa, Viṣṇu functions in a different (and characteristically more devious and "dark") manner, achieving his goal of world renewal indirectly through the Pāṇḍava brothers headed by Yudhiṣṭhira, and by means of a sacrificial bloodbath that lifts the earth's burden of violent, overweening Kṣatriyas. Like Rāma, Yudhiṣṭhira ultimately takes the throne as a Brāhmaṇa-friendly sage-king, whose inclinations to non-violence (except when force is unavoidable to uphold *dharma*) have been nurtured by long years of residence in forest ashrams. In each case, achieving some kind of balance requires an extended period of residence in the forest, as if to become imbued with its characteristic values. Significantly, it is the "balanced" Kṣatriya heroes who will at times manifest a conservationist ethic: Yudhiṣṭhira is visited in a dream by weeping deer, who ask him to move the Pāṇḍavas' encampment, lest their dwindled herds—"the seed of the future"—be hunted to extinction; he willingly complies (*Mahābhārata* 3.40).[25]

Epic Landscape and the Real World

In a recent article entitled "The Dialectical History of 'Jungle' in Pakistan," ecologist Michael Dove expands on and critiques the analysis of the Sanskrit term *jāṅgala* that appeared in Francis Zimmermann's groundbreaking 1987 study of Ayurveda as an intellectual system, *The Jungle and the Aroma of Meats*. Zimmermann showed that ancient *jāṅgala* was neither city nor forest, but a semiarid savanna of grasses and intermittent, low trees that was considered the ideal terrain for pastoralism and light agriculture, human settlement, ritual activity, and indeed somatic health.[26] Although Zimmermann supposed the *jāṅgala* ecosystem to have been native to much of northern South Asia, Dove's more recent fieldwork indicates that *jāṅgala* was in fact an "anthropogenic landscape" that arose from and was maintained by periodic human intervention, especially through purposely set brushfires, to prevent it from rapidly reverting to the thornwood climax forest that he argues is naturally indigenous to the region. It was thus "an active zone of tension between society and nature . . . a mediator between . . . village and forest"—ecologically speaking, a compromise between more concentrated human activity and its absence or cessation. In late Vedic times, this "compromise" savanna ecosystem, the human-engineered habitat of the black antelope, was considered the landscape most suited to the Aryan sacrificial lifeway—itself an effort at negotiating a precarious equilibrium between contending cosmic forces: Agni and Soma, order and disorder, life and death. As Zimmermann shows, its characteristics, set in contrast to those of the less-favored *ānūpa* ("marshy" and heavily forested) lands, developed into an elaborate paradigm for ancient legalists and physicians. Whereas Zimmermann is primarily interested in the rather cerebral proliferations of a Brāhmaṇical ideology which originally grew out of the observed features of a landscape, Dove insists that ideology and other forms of human intervention *actively shape* the natural landscape, although ideology characteristically seeks to conceal its hand by implying that things were "always already" in the form which it has, in fact, itself gradually molded. He argues that, as a tangible ecosystem, the *jāṅgala* was gradually eliminated through population pressure resulting in part from the preference of rulers and centralized states for intensive cultivation and settlement over pastoralism and "extensive" patterns of land use (involving light or intermittent farming). As a result, the term itself gradually changed its meaning, taking on connotations of "wilderness" and "wildness," indeed sometimes of dense vegetative cover (all alien to its archaic meaning) that it retains in its modern Romanized incarnation as "jungle."[27]

I would argue that, taken as a whole (if one may venture such a stupendous conceit), the two Sanskrit epics present what may be termed a "narrative ecosystem" that, like the *jāṅgala* of Dove's analysis, is similarly a "zone of tension." Their complex stories—whose cyclical and fugal narrative structure nineteenth-century European scholars sometimes disapprovingly likened to a "jungle" in the latter-day sense of impenetrable tropical growth—in fact present us with a savanna-like sociocultural landscape in which the balance of power between kings and sages is maintained through periodic confrontations, and indeed, conflagrations. At the same time they offer, through some of their most cherished characters and at the deepest levels of their theology, an attempt at a more satisfying balance attained through mediation and synthesis. The heroes of these stories must fulfill their destinies as the rulers of city-states, but they are paradigmatic culture-heroes precisely because they have been to the forest, faced its terrors, and lived among (and, to a degree, as) its sages—an observation nicely encoded in the modern Hindi epigram with which I commenced this essay. Unlike prescriptive texts, the epics achieve a less stable resolution to their fundamental tension between civilization and wildness, exploitation and conservation, king and sage. Although sage-kings like Rāma and Yudhiṣṭhira can be regarded as integrative ideals, the narrative continues periodically to challenge the stability of each (e.g., with the disturbing yet inevitable tests and reversals of the concluding books of each epic—the *Rāmāyaṇa*'s *Uttarakāṇḍa* and the *Mahābhārata*'s *Svargārohaṇika parva*). Yet, this precariously balanced "narrative ecosystem" has itself been maintained with considerable stability through millennia of epic and Purāṇic retellings.

What resources may a modern reader or listener draw from this "narrative ecosystem" that might be applied to healing the imperiled environment of the late twentieth century? Most obviously, perhaps, the notion of balance and compromise between the competing life-styles of kings and sages—between "consumers" and "conservers" of the environment. Like the *jāṅgala* as described by Dove, the epic "forest" is less a pristine wilderness (such as modern conservationists often strive to maintain) than a zone of interaction between often competing forces. It is (to borrow the slogan of the U.S. National Forest Service) a "land of many uses," and these do not exclude—especially in the archaic context of boundless wild lands beyond the periphery of human settlement—extensive exploitation. Yet, I would insist that the presiding ideal of both epics is the sage-king who understands and governs according to *dharma,* and in each case his ultimate character is seen to blend the qualities of the Kṣatriya and the Brāhmaṇa and to have been molded through extended experience of extraurban and indeed extrasocial realms. The resultant personality significantly tempers self-indulgence with self-control. Could one argue, in other words, that a king like Rāma would favor recycling and renewable resources over excessive waste and environmental exploitation? I am inclined to think so—although I note with concern that opposing attitudes may more often claim epic models: modern political parties that espouse *Rāma-rājya* and that combine a reactionary social stance with economic policies favorable to big industries and polluters.

It has often been observed that the universe of traditional Hindu narrative—to which the epics subscribe and which they, indeed, help establish—is inherently and literally "conservative": a "closed system" in which nothing is ever truly lost, but in which all things—energy, tangible substances, souls, and even stories—are endlessly recycled and transformed. That such a worldview might have tangible resonances may seem evident to anyone who has ever marveled at the inventive reuse of objects and resources—from sandals soled with tire treads to cows contentedly grazing on yesterday's headlines—that, even today, marks much daily life in India, especially for the poor and lower middle classes. Yet, as an ever-rising flood of factory-made goods and disposable packaging feeds middle-class and elite consumerism, we would do well to invoke another prevailing epic vision which I cited earlier: that of a "crowded cosmos." What is most significant here is not the mere fact of the abundance of life-forms which swarm through every tier of the epics' multistoried universe, but the underriding notion of balance and of the necessity for mutual survival. Epic holocausts—of deer, of Kṣatriyas, or even of venomous snakes and menacing *rākṣasa*s—invariably stop short of total annihilation and reveal a kind of extrahuman perspective according to which all life-forms are seen to have a place and a right to exist. If such a world-view has a pragmatic, economic parallel, it seems to me that it is assuredly in the old-fashioned Indian bazaar, in which scores of tiny shops—many of which stock much the same merchandise—coexist and, yes, "compete," though without the notion of maximizing profit through destroying competitors, such as we find in the "slash-and-burn" economics of modern chain store retailing.

Earlier I proposed that the epic "forest" was often an imagined landscape—more a state of mind than a tangible environment. The research of Zimmermann and Dove suggests that *jāṅgala* remained a powerful ideological paradigm even after the landscape on which it was based had ceased to exist, or in places where it had never existed. Likewise, the deep forest (*vana, araṇya*), with its terrors and transcendent possibilities, would remain an appealing narrative topos even as the balance in the physical environment of South Asia shifted decisively toward urbanization and intensive cultivation, and as real forests and their wildlife were exploited almost to extinction. Mythology, too—and the classical Indian epics in particular—offers a rich terrain for imaginative journeying and introspective discovery that can proceed quite independent of circumstances in the physical environment. At the same time, the Indian epics, in their broadest tradition of interpretation, make a claim to be not simply marvellous allegorical entertainments, but also charters for an ancient and enduring ideal of the good life: for the remaking of the world according to a higher order of harmony and beauty. Understanding that such a life necessarily requires the precarious balancing of the competing claims of consumption and conservation, and that it is situated within a complex web in which all organisms have a place and are entitled to survive, can help us to identify the relevance of their message for our times.[28]

Today—as developers bulldoze hillsides in Rishikesh to erect apartment blocks that will offer retirees from Delhi a little piece of Himalayan "*vanaprastha*" (the third Hindu life-phase of "withdrawal to the forest")—people in ever greater numbers are claiming the right to enjoy the best of two increasingly colliding worlds: to consume and exploit the landscape as kings and to revere and protect it as sages. But in an age in which rampaging commercial interests pose a greater menace than *rākṣasa*s, the challenge to modern South Asians—and the rest of us—is to preserve and restore enough verdant and biodiverse terrain not only to meet our pragmatic and material needs, but also to insure that future heroes (and heroines) have a place to go to "become themselves."

Notes

I am grateful to Christopher Key Chapple, Arvind Sharma, Mary Evelyn Tucker, and the other organizers of the conference "Hinduism and Ecology," held at Harvard University's Center for the Study of World Religions, for the invitation that led to the writing of this paper. My Iowa colleagues Jael Silliman and Paul Robbins provided encouragment and valuable references. The present version has also benefited from the thoughtful comments of panel discussant Francis X. Clooney.

1. *Jab rām ban gaye, tab hī rām ban gaye.* The saying's pun results from the fact that, since in Eastern Hindi "b" often replaces "v," the compound verb *ban jānā* ("to become, to fulfill a role or acquire a status") can also be read as *ban* (*van*) *jānā* ("to go to the forest").

2. I use the term "classical epics" to distinguish the two pan-Indian narratives from more region-bound traditions, such as the Hindi Ālhā cycle or the Rajasthani Pābhujī epic, although these have also achieved the status of local "classics."

3. See Sheldon I. Pollock, *The Rāmāyaṇa of Vālmīki,* vol. 2, *Ayodhyākāṇḍa* (Princeton: Princeton University Press, 1986), 3–4, and *The Rāmāyaṇa of Vālmīki,* vol. 3, *Araṇyakāṇḍa* (1991), 12–13; and J. A. B. van Buitenen, *The Mahābhārata,* book 2, *The Book of the Assembly Hall;* book 3, *The Book of the Forest* (Chicago: University of Chicago Press, 1975), 173–76.

4. For the sections that follow, I am indebted to an unpublished paper by Robert Menzies on the characterization of the forest in both Sanskrit epics, written for my class on the *Mahābhārata* in spring, 1997. Noting the apparent "inconsistencies" in the epic's portrayal of the forest, Menzies identified a series of six conventional landscapes (e.g., love-bower, royal hunting ground, hermitage, demon-lair, etc.), giving examples of each. The paper drew on but considerably extended some ideas I had presented in class; I in turn gratefully borrow and expand on some of Menzies's identifications and examples below.

5. Monier Monier-Williams, *A Sanskrit-English Dictionary* (Oxford: Oxford University Press, 1899; reprint, Delhi: Oriental Publishers, n.d.), 917.

6. Ibid., 352.

7. Ibid., 972, 270, 271.

8. Ibid., 417, 408–9.

9. See Francis Zimmermann, *The Jungle and the Aroma of Meats* (Chicago: University of Chicago Press, 1987), esp. 1–19; and Michael R. Dove, "The Dialectical History of 'Jungle' in Pakistan," *Journal of Anthropological Research* 48 (1992): 231–47. I thank geographer Paul Robbins for alerting me to the latter essay.

10. J. A. B. van Buitenen, *The Mahābhārata,* book 1, *The Book of the Beginning* (Chicago: University of Chicago Press, 1973), 157–58.

11. Pollock, *The Rāmāyaṇa of Vālmīki,* 2:239–40.

12. Dove, "The Dialectical History of 'Jungle'," 236–37.

13. van Buitenen, *The Mahābhārata,* 1:417–18.

14. See Karve's essay on this passage in *Yugānta* (Delhi: Sangam Books, 1974; reprint, Hyderabad: Disha Books, 1991), esp. 114–17.

15. van Buitenen, *The Mahābhārata,* 1:158–59.

16. Pollock, *The Rāmāyaṇa of Vālmīki,* 2:135.

17. See Philip Lutgendorf, "The Secret Life of Rāmcandra of Ayodhya," in *Many Rāmāyaṇas: The Diversity of a Narrative Tradition in South Asia,* ed. Paula Richman (Berkeley and Los Angeles: University of California Press, 1991), 217–34. The forest as a preferred setting for erotic play may have its *locus classicus* in the ancient epics, but its extraordinary staying power in popular Indian art is suggested by the abrupt but obligatory (and sometimes unexplained) cut to a forest setting for the love songs of modern Hindi film musicals. The featured couple may be driving on a city street or dancing in a discotheque, but when they begin singing of their love, one can be sure that they will soon be cavorting on mountain slopes or on the wooded shores of a river, coyly peeking at each other from behind treetrunks or splashing in waterfalls; the verdant forest (which Bombay crews now often go to Switzerland to film) remains the preeminent landscape of *śṛṅgāra.*

18. Pollock, *The Rāmāyaṇa of Vālmīki,* 3:89. Although this vision may seem hyperbolic enough, it cannot compare with the baroque exuberance of the Tamil poet Kamban's extended retelling; see George L. Hart and Hank Heifetz, *The Forest Book of the Rāmāyaṇa of Kampaṉ* (Berkeley and Los Angeles: University of California Press, 1988), 38–45.

19. Pollock, *The Rāmāyaṇa of Vālmīki,* 3:68–84.

20. van Buitenen, *The Mahābhārata,* 1:159.

21. Robert P. Goldman, *The Rāmāyaṇa of Vālmīki,* vol. 1, *Bālakāṇḍa* (Princeton: Princeton University Press, 1984), 224, 229.

22. The research of both Zimmermann and Dove suggests the essentially negative valuation of thick forest in ancient India, and helps us to understand the ominous dimensions of forest-exile; as the Pāṇḍavas manifestly know, being sent to the forest is *not* a good thing—it takes a paragon of positivity like Rāma to see it as such, and even he occasionally wavers.

23. E.g., even the "trackless, dreadful-looking" wilderness laid waste by the demoness Tāṭakā in Book One of the *Rāmāyaṇa* is characteristically described as abounding in life-forms (albeit ominous ones): "Echoing with swarms of crickets, it swarms with fearsome beasts of prey and harsh-voiced vultures. It is filled with all sorts of birds, screeching fearsome cries, as well as lions, tigers, boars, and elephants. It is full of *dhava, aśvakarṇa, kakubha, bilva, tinduka, pātala,* and *badari* trees." (*Rāmayaṇa* 1.23.12–14; Goldman, *The Rāmāyaṇa of Vālmīki,* 1:170.)

24. See, e.g., *Mahābhārata* 3.155.26–70 for an extended description of the lush, subtropical forests the Pāṇḍavas find on the highest ridges of "Mount Gandhamādana," after their ascent of the Himalayas.

25. van Buitenen, *The Mahābhārata,* 2:699.

26. Zimmermann, *The Jungle and the Aroma of Meats,* 1–19.

27. I have some difficulty with Dove's and Zimmermann's linguistic arguments, since I feel that both give too much weight to the modern English sense of "jungle" as translated back into (especially urban) Hindi-Urdu. In my own experience, although *jaṅgal* is now sometimes used as a synonym for *van* ("forest"), it is also often used by rural people to refer to grazing land on the village outskirts, or to (even temporarily) uncultivated and overgrown tracts, a usage more consistent with the archaic sense of *jāṅgala.* Moreover, urban people use the adjective *janglī* to characterize boorish rustics, as well as "wild" animals and plants. Since I question the alleged "reversal" in

meaning of the word, the two authors' conflicting explanations for it (Zimmermann's bemused observation that it is an "extraordinary misunderstanding," and Dove's insistence that it reflects "a meaningful historical process") seem of less consequence than their compelling and largely complementary arguments for the term's archaic meaning. The point on which Dove may significantly correct Zimmermann's work, however, is in the recognition that the archaic *jāṅgala* ecosystem was itself not wholly "natural," but reflected substantial human intervention in the landscape. The understanding that this was in fact land forcibly wrested from the indigenous forest actually enhances Zimmermann's analysis of the legalists' enduring dislike for the heavily forested *ānūpa* regions. It also suggests, perhaps, why the internal challenge to Vedic hyperscholasticism and its attendant ideology came initially from "forest teachings" (*āraṇyaka*s) promulgated by ascetics living beyond the periphery of, or in surviving forested pockets within, the more ritually pure savannas.

28. Interestingly, the recent reincarnations of the *Rāmāyaṇa* and *Mahābhārata* as lengthy television serializations incorporated many contemporary concerns; notably the Indian government's desire to promote "national integration" and other civic and pragmatic objectives. There was also, in each, a brief bow to "reforestation." In Ramanand Sagar's *Ramayan* serial, Sage Vasiṣṭha, who is educating Rāma and his brothers in an idyllic ashram, instructs the boys to plant a mango seed, and to protect and nurture its seedling; this leads to a brief discourse on the merits of trees and the importance of protecting them (*Ramayan,* 1987, Episode 2). B. R. Chopra's *Mahabharat* contains an ecofriendly reversal of the Khāṇḍava Forest story: here the "wilderness" the Pāṇḍavas receive as a kingdom from their Kaurava cousins is not a forest but a desert. Hence no burning occurs; instead, Kṛṣṇa assists them in transforming it, through tree-planting, cultivation, and construction, into a park-like suburban landscape, bristling with rows of young saplings that look like the handiwork of the government Forestry Division (*Mahabharat,* 1990, Episode 39).

“Sacred Grove” and Ecology: Ritual and Science

FRÉDÉRIQUE APFFEL-MARGLIN
and PRAMOD PARAJULI

> The environmental crisis is making the modern mind obsolete.
>
> Donald Worster[1]

Introduction

The network of groves that covered the subcontinent so impressed Sir Dietrich Brandis, the first inspector general of forests in colonial India, that he urged a system of forest reserves and preserves modeled upon it.[2] In the 1880s Brandis was already lamenting the destruction, under the British system of forest management—of what he, along with others, called “sacred groves”—a destruction that has continued unabated and perhaps even accelerated since independence.[3] “Sacred groves” are supposed to be pre-Vedic in origin (about 3000 B.C.E.). They certainly predated the Gautama Buddha, who was born in a sal grove dedicated to the goddess Lumbini in the Himalayan foothills.[4] According to Madhav Gadgil and Ramachandra Guha, “one of the best known ancient state-sponsored forest conservation efforts was carried out by the emperor Aśoka after his conversion to Buddhism.”[5]

Adivasis in India have named their religion the religion of the Sacred Grove (*sarna*).[6] The practice of preserving uncut patches of original forest in which dwells a deity is by no means exclusive to the Adivasis. Gadgil and V. D. Vartak counted over four hundred groves in Maharashtra and Madhya Pradesh. In the Sarguja district of Madhya Pradesh, every village has a grove of about twenty hectares where both plants and animals receive absolute protection.[7] In the hills of Nepal, where one of us grew as a child, one cannot fail to notice at least a dozen patches of protected forests in each and every settlement. While some groves are meant to protect the water sources, others host a variety of gods and goddesses. Many belong to individual peasants who use that particular patch for household use. Ecologically speaking, the idea that the gods and goddesses live in or beneath huge trees on the top of hills contributes greatly to protecting the whole hilly landscape from erosion. As in the hilly range of middle India, in Nepal, too, each mountain pass is imbued with sacredness. The sources of rivers and lakes therein are periodically visited and protected. These practices transcend the Hindu/Buddhist, Aryan/non-Aryan divides.

In this paper we will look closely at the practices of coastal smallholder peasants and craft fisherfolk in Orissa (Puri district) during their most important festival of the year, which takes place (in part) in a grove dedicated to Goddess Bali Haracandi (Haracandi of the Sand). These coastal dwellers are caste Hindus (although Brahmins do not observe this festival). What we will show is that their practices share a great deal with those of Adivasis, as well as those of Muslims and Buddhists. We will highlight the role of Muslims in the festival and draw parallels from Nepal and the Jharkhand region. Rather than emphasize the historical links between these communities, we will emphasize what they share, namely, a place and a way of acting in that place and of relating to the nonhuman collectivity of that place.

We will pay close attention to who participates as well as to what is done and said by whom; this will enable us to raise issues concerning the categories of "Hinduism," of "ecology," of "conservation" and "science," and of "ritual." In so doing we will seek to formulate an alternative to both a religion-based ecological movement and one based on the modern Western model of parks, preserves, and the scientific ontology.

Transcending Religious Labels

The wooded hill of Bali Haracandi near the market town of Brahmagiri (between Puri town and Chilika Lake) is located some fifty kilometers from the Dhauli hill where the famous Aśokan edict still stands, in which the following words can be read:

> The king with charming appearance, the beloved of the gods, in his conquered territories and in the neighboring counties, thus enjoins that: medical attendance should be made available to both man and animal; the medicinal herbs, the fruits trees, the roots, the tubers, are to be transplanted in those places where they are not presently available, after being collected from those places where they usually grow; wells should be dug and shadowy trees should be planted by the roadside for enjoyment both by man and animal.[8]

Buddhism had a long and influential presence in Orissa, being the state religion until the tenth century and maintaining an important presence there as late as the sixteenth century.[9] The later Tantric Buddhist teachers of the sixteenth century resided in the cut rock caves of Dhauli and nearby Khandagiri and Udaygiri that had been Buddhist centers ever since the second century B.C.E. This later tantric Buddhism was absorbed into the local form of Vaiṣṇavism.[10] Even today, the sovereign of the land, namely, Lord Jagannātha's ninth incarnation, is considered to be the Buddha.[11] In any case, the early Buddhist and Jain archaeological sites near Bhubaneswar signal that wet rice cultivation has been practiced in coastal Orissa at least since that time. The peasants of the area of Brahmagiri are thus heir to a continuous tradition in which rice cultivation and the uncut groves are key features, evincing what by any standard must be recognized as a sustainable way of life.

Popular Practice and a Plurality of Religions

Buddhism is not the only ingredient in the multireligious and multiethnic traditions of coastal Orissa. The tribal element is very strong in the cult of Jagannātha.[12] This can nowhere be better appreciated than in the annual peasant and fisherfolk festival of Raja Parba, celebrating the menses of women, of the earth, and of the sea. This festival is celebrated all over rural Orissa. Bali Haracandi is one of these sites and it draws peasants and fisherfolk from some sixty villages around it.[13] It is a non-Brahmin festival involving all the other castes, including the so-called untouchables as well as the Muslims of the area.

Raja Parba is situated in time at the onset of the monsoon, at the articulation between the dry/hot season and the rainy season. It lasts four days, just like women's menstrual observances. The goddess/earth/sea is menstruating during the first three days and on the fourth she takes her purificatory bath. Men congregate in the grove where all the men from one village sleep and eat in the same tent, regardless of caste or untouchability. The women celebrate in the villages where they take over all public spaces and where young women play on swings hung in the branches of trees. All agricultural and fishing (in general, all productive) activity, is suspended, and there is an intensification of gift exchanges from husbands and brothers to their wives and sisters.

Menses are spoken of as the fallow time of women—who must do no work during their menses and during the festival—and of the earth. During the hot season, the earth lies fallow, hot and dry (in the sense of not giving fruit), like women at their menses. The yearly rhythm of the seasons, of the alternation between dry and wet, synchronizes with the rhythms of women's menses, their alternation between fallow times (menses) and productive times. Men as well synchronize their activities with those same rhythms. They abstain from any agricultural/productive activity during the four days of the festival. These seasonal alternations are synchronized in time and space by actions in which the human and the nonhuman collectivities participate. This participation involves exchanges and reciprocities between the two collectivities. It is this collective action engaged in by both human and nonhuman collectivities that brings about—if carried out successfully—the continuity and regeneration of both the natural and the cultural world.

This corresponds closely to practices of the Hill Marias, shifting Adivasi cultivators of Bastar in Madhya Pradesh. For the Hill Marias, the meaning of

> fallowness and of the year of "no cultivation" is established by the correspondence between the menstrual cycle of a woman, the fallow land and forest regeneration. . . . The rhythms of the household routines are set with reference to menstruation. This is similar to the way rhythms of cultivation work cycles are set with reference to the fallow periods. Both menstruation and "fallowness" signify recuperation of fertility. . . . In a forest the dry seasons are barren and the wet seasons are fertile. However, the dry season is necessary for clearing the forest and for the

> ripening of the crop, just as menstruation is necessary for the regeneration of fertility.[14]

Although with permanent rice cultivation there is no cycle of forest regeneration, in both cases we have an articulation between cycles of natural regeneration and cycles of human regeneration. This is also true of fisherfolk who do not fish during certain times of the year to allow for the regeneration of the fish.

The Hill Marias of whom Savyasaachi speaks are part of the Adivasi belt of middle India stretching from Calcutta in the east to Bombay in the west. The Adivasis have in the recent past named their religion Sarna Dharam (the religion of the sacred grove).[15] By this, the Adivasis do not intend to convey that they have rules, institutions, and texts but that they share a certain way of relating to nature and of using nature. They have become aware that this way is increasingly coming into conflict with developmentalist practice and the spread of the global market. What we find in the practices of coastal caste Hindus and their Muslim neighbors is that there is a great deal of overlap between Adivasi, caste Hindus, Muslims, and Buddhists in their way of being in nature and making a living from nature.

The most surprising realization was that Muslims not only participate in the festival at Bali Haracandi but that Muslim girls undergo the exact same menarche ceremony and menstrual observances as non-Muslim women.[16] In 1992 one of us spoke with the owner of a *pān* stall at the *melā* at Bali Haracandi. Mr. Fariduddin lived in a Hindu-Muslim village two kilometers from the grove of Bali Haracandi and had created a youth club to help keep law and order at the fair in the grove, along with other non-Muslim youth clubs. He told one of us the following:

> We try to see that everything goes well at the *melā*. We join with our Hindu brothers. They also join us at our festivals. We keep good relations with the Hindus, we exchange [gifts] with them. . . . The Hindus say that their Goddess is at her periods but we do not understand this. We don't ask about this, it may offend them. We observe *raja* by stopping our ploughs. Some of our girls and women play on swings and wear new clothes.

Walking with Mr. Fariduddin back to his village we met a Muslim farmer, Sheikh Imabul who told us that during *raja*

> we give rest to the land, the bullocks, and to men. It is the way people do, it is a custom. We also make *pithas* [special cakes made by all women at *raja*]. The people stop work. If we cultivate at that time, people will surely feel bad, so why should we do it?

Asked whether he thought, like the Hindus, that the earth menstruates during that time, he replied: "No, there is no such thing in our book. . . . But this earth is Mother; only because of that we give rest to her." In these statements is expressed not only solidarity with Hindu neighbors but the view that the earth is Mother and that she should be given rest. Thus, the practices of not ploughing and of stopping work (among others) are not only engaged in to keep harmonious relationships with their Hindu neighbors but also because of their way of viewing the earth as Mother.[17]

Ecological Ethnicities

We propose to use the concept of *ecological ethnicity* to refer to such small peasants and fisherfolk, irrespective of caste, class, or religion.[18] Ecological ethnicity is a social category that designates those people who have developed a certain respectful use of the bounty of nature and, consequently, a commitment to create and preserve a technology that interacts with the place and its nonhuman collectivity in a sustainable manner. It is a land- (or sea-) based ethnicity that in most cases is not isomorphous with a homogeneous blood-, caste-, religious-, or linguistic-based grouping. Even in a place as central to Hinduism as Puri Jagannath (Bali Haracandi guards the southern boundary of Shamkha Khyetra, the holy pilgrimage place on the eastern seaboard), the festival in the grove of Haracandi shares essential elements with both Adivasi and Buddhist practices. It is also celebrated by the local Muslims.

The concept of ecological ethnicities includes peasants and other ecosystem peoples,[19] such as fisherfolk, tribals, forest dwellers, nomadic shepherds, and a host of people marginalized by development projects and the programs of environmental modernization. We argue that these ecological ethnicities should be theorized within the ecological field, although not ecology as practiced by scientists, but what we call a *moral ecology*. The concept of moral ecology transcends both the domains of religion and of science. We find that the boundary setting that usually comes with the various religions of the subcontinent conspires to render invisible what the ecological ethnicities share among themselves. On the other hand, what they share sets them apart from secular modernists since it does not partake in either the amorality of economics or the amorality of science.

The concept of ecological ethnicity captures the fact that the practices of the rural peoples of the subcontinent who are heavily dependent for their livelihood and survival on their immediate environment (and on its biomass) share a great deal in common, despite differences at the level of articulated "beliefs" or "worldviews." The latter tend to be inflected by the literate religious traditions which, more often than not, are inscribed by intellectuals not directly engaged in living off the land.[20] The gap between articulated "beliefs" and popular practice can be marked. The literate classes seem to have more of a vested interest in articulating their differences and less in a direct involvement with the soil or the sea. This is particularly so with the modernizing literati who either belong to or aspire to belong to what Gadgil and Guha have called the "omnivore" class. When the "war over resources" intensifies in the increasingly globalized economy of the subcontinent, the tension between the omnivore class and the ecological ethnicities will also increase.

When ecological ethnicities' access to crucial biomass is threatened by developmental activities such as mining, damming of rivers, industrial agriculture, and others, they have tended to band together. This has happened especially in the Jharkhand area and the rest of middle India that has been so heavily industrialized at the expense of local inhabitants who are Adivasi, Hindu, Muslim, and Christian peasants and artisans.[21] In the Narmada valley, too, what we witness is a new configuration of alliances between the landholding Patidar peasants and other ecosystem peoples—Adivasis, fisherfolk, and other artisanal groups who, in many respects, are dependent on the landholding class. We are not trying to convey that such alliances are "heavens of harmony and equality," but they are definitely the source of ethno-ecological formations with new opportunities for social critique as well as regeneration.

Let us take a closer look at coastal Orissa and its long ethno-ecological history. The different peoples that have come and shared this place have recognized the indwelling beings of the land and sea. The founding myth of the cult of Jagannātha recounts how he was originally a deity in the forest worshiped by the Adivasis.[22] Buddhist iconography gives form to the multiple sentient beings of the trees, the air, the waters, and the earth. The Muslims who now till the soil speak of the earth as a Mother who needs periodic rest. Local history is the outcome of this interaction between the varied human collectivities and the nonhuman collectivities. It is only rather recently—that is, since the birth of modernity around the seventeenth century in Western Europe—that history has been recorded as an exclusively human achievement. Coastal Orissa never seems to have been so anthropocentric.

The label "Hindu" for the practices we are writing about renders invisible all the other people who share them; it is too exclusive a term. This holds equally so whether we speak of Nepal or the Jharkhand region. Another issue raised by the term Hindu is that it refers to a religion. For many, this association means a separation between a religious/spiritual sphere and a secular/utilitarian sphere. We seek rather to highlight in the practices of these ecological ethnicities a way of being-in-the-world, of pursuing a livelihood in that world, of acting with that world, that thoroughly blurs the boundary between what is considered "spiritual" and what is considered secular or profane. It is clear that the activities, daily as well as during special occasions such as festivals and other rituals, of the ecological ethnicities that participate in Raja Parba have everything to do with their making a living from the land or the sea. It has everything to do with their ability to survive, to continue to live in that place. This is made clear by the local fisherfolk and peasants around Chilika Lake who are fighting Tata's proposed shrimp factory. Like their counterparts in Jharkhand, or the ecological ethnicities fighting the Narmada damming project, or the Adivasis of Ghandhamardan fighting the copper mines, or the Ho Adivasis trying not to be displaced by the damming of the Subarnareka River, or the Mahato and Ghatwal Hindu peasants trying to stop the mines of Hazaribagh, they are fighting to preserve a way of life that cannot be captured by any of the religious labels.

Scientific Ecology versus Moral Ecology

The way of life of ecological ethnicities is a way of acting and living in their place that is radically distinct from the scientific way. It is a way of life undergirded by an ontology no less arbitrary than the one that undergirds science and modern technology, namely, that of an inert and mechanical nature radically separated from humans. In earlier works we have used the terms "styles of cognition" to capture these distinct ways.[23] We now feel that the term "cognition" gives undue emphasis to intellectual activities; the collective practices we speak of are actions that are not necessarily consciously held in the mind and articulated.[24] This is in great part due to the fact that the knowledge embedded in these actions is not individual knowledge. Collective action—as philosopher Kathryn Pyne Addelson has shown[25]—involves a non-individualistic form of knowing, one embodied in collectivities as well as in institutions, whereas science is an individualistic form of knowing. In the case at hand, these collectivities include nonhuman ones. One of us has argued that the post-Descartian form of rationality not only requires an ontological divide between the individual's *ratio* and the world but necessitates the existence of an *individual* knower.[26] It is not only that knowledge, or even more generally cognition, is disembedded from the human collectivity, it is also disembedded from the place, that is, from the nonhuman collectivities that make up a place.[27]

Science and the Separation of Fact and Value

Anthony Giddens has generalized Karl Polanyi's notion of the disembeddedness with which the latter characterized the modern market economy as *the* characteristic of modernity.[28] To speak of re-embedding the individual in the place and the community, or re-embedding knowledge in place and community, does lead us to a reconfiguration of boundaries between science on the one hand and values/ethics/politics on the other. If we want to argue for a moral ecology, it might be necessary to revisit the birth of the separation between fact and value.

The separation of fact and value was a founding principle of science.[29] Bruno Latour, along with others, has argued that this separation was invented as a keystone feature of the political program in both England and France in the seventeenth century to restore an order from above, one that placed the state and the elites firmly in control.[30] Focusing on the debate between Robert Boyle, the inventor of the experimental method, and Thomas Hobbes, Latour writes:

> [Boyle and Hobbes] are inventing our modern world, a world in which the representation of things through the intermediary of the laboratory is forever dissociated from the representation of citizens through the intermediary of the social contract. . . . They are like a pair of Founding Fathers, acting in concert to promote one and the same innovation in political theory: the representation of nonhumans belongs to science, but science is not allowed to appeal to politics; the representation of citizens belongs to politics, but politics is not allowed to have any relation to the nonhumans produced and mobilized by science and technology.[31]

The separation between "facts of nature" and politics is a political innovation. In other words, the worldview that holds that facts are "discovered" in nature is part of a political project. This debate took place "while a dozen civil wars were raging";[32] the stakes were high indeed. The new "constitution" (Latour's term) is a thoroughly political enterprise to reign in the ragtag masses of levelers, diggers, ranters, and others who were seeking a radically new world. Nevertheless, the separation between representation of (natural) facts and representation of citizens is arbitrary; it is not given in the world, rather, it is made. What the body of Latour's ethnographic work on laboratory life shows is that "the facts are fabricated."[33] What one is dealing with is not the direct knowledge of things-in-themselves but the creation of hybrids, mixtures of nature and culture. Latour argues that modernity is the denial of the proliferation of hybrids that science generates, through a constant process of "purification" that keeps nature and culture separate.

The Boyle-Hobbes debate led to political and social action in England. Similar action took place in France where institutions were created to keep the two realms separated and to keep both science and politics in the firm grasp of elite groups that would not trespass on each other's domain. The French Academy, reorganized in 1666, and the Royal Society in England had close ties with each other and were key to legitimizing the separation between matter and spirit or mind and body.[34] As Stephen Toulmin has argued, the new knowledge was the road to the construction of a new certainty that would create (a certain type of) order out of the chaos of religious and civil wars.[35]

This all too brief excursion into the history and anthropology of science is necessary, it seems to us, to clarify that advocating something like a moral ecology in which human and nonhuman collectivities act in concert and communicate is not a fuzzy-minded romantic return to superstitious and archaic modes of being. As Latour has shown, the perception of the birth of modernity as a rupture with the *ancien régime,* and simultaneously with all other societies that instantaneously became premodern (or "traditional"), can only be maintained by keeping the domains of knowledge separate and dividing them in turn from religion and politics. It is based on denial, hence his title "We have never been modern."

Our notion of ecological ethnicities highlights what the more conventional categories based on either religious, caste, or class identification tend to obscure. The notion of moral ecology transcends the divide between science and religion/ethics/politics. The former notion seeks to circumvent or preempt the dangers of fundamentalist politics. The latter notion seeks a new discourse transcending the divides between science and religion, modernity and premodernity (or "tradition").

It seems to us that Addelson's notion of collective action and of the outcome of collective action has the enormous advantage of not only circumventing the individual and the separation between mind and body, thought and action, but also of bridging the rift between ethics, politics, and knowledge. It also opens the possibility of collective action by the nonhuman collectivity. Bruno Latour has argued that in Louis Pasteur's work the microbes were as much actors that shaped the outcome of Pasteur's work as Pasteur himself and the laboratories and technology he and his associates used.[36]

Is Sacred to Profane as Nonutilitarian Is to Utilitarian?

The festival of the menses (Raja Parba) is a moment, an articulating moment, within a larger cycle of time and takes place in spaces that are also located in larger spaces. This articulating moment, which we call a festival or a ritual, is at once separate—in the sense of being emphatically different from daily time and place—as well as part and parcel of the rest of the yearly cycle. The timing of the festival marks the end of the hot and fallow period of the earth and the beginning of the fertile and wet period of planting. The Oriya (and Sanskrit) word *ṛtu* means both the articulation of the seasons and a woman's menses. We argue that the collective actions undertaken by the human collectivity in interaction with the nonhuman collectivity bring about the regeneration of the human community by synchronizing it with the regenerative cycles of the seasons, the earth, the sea, the animals, and the plants. In other words, such collective action is efficacious in its impact not only on the human collectivity but on the environment, or more properly on the nonhuman collectivities (the word environment is too anthropocentric in our context). Its efficacy pertains to the fact that it ensures the continuity of life, in other words, the survival of the people and the place, and cannot therefore be separated from a utilitarian outcome.

The regenerative cycles of nature are the collective actions of the beings that make up the nonhuman world, and ecological ethnicities do not place themselves in a dimension radically separate from these. Both collectivities communicate, exchange, and reciprocate. The ontology of a nonhuman world made up of many different live beings is no less arbitrary than the mechanistic Baconian/Cartesian ontology of a nonhuman world made up of inert particles that are moved solely by external force. The latter ontology has produced many technological wonders but has a much less stellar record in terms of sustainability. The former ontology has given rise to remarkably stable, sustainable ways of life with a low level of technology.

The construction of the nonhuman world as an inert mechanism has enabled an extractive attitude. The speed of production and extraction attained in the latest phase of industrial capitalism, namely, globalization, violates the biogeochemical cycles of nature necessary to allow regeneration and renewal. Or, in Ilmar Altvater's words, economic time outdoes biological time and, as we have shown, has become the impetus for the formation of politically active ecological ethnicities.[37]

The regenerative cycles of nature have been seen as the guarantors of continuity at least since Vedic times if not well before that. *Ṛtu* in the Vedic literature refers to the articulating activity which creates *ṛta,* the ordered cosmos.[38] We prefer the term "orchestration," since it is both collective and dynamic. This orchestration is a dynamic rhythm in which the sun articulates by its movements a well-orchestrated continuity. It is a dynamic activity that humans do not observe from a position outside of it; rather, it is one in which humans play an essential role. Without the performance of *ritu*-als (a word appropriately derived etymologically from the Sanskrit *ṛtu,*[39] the orderly movement of the cosmos is threatened. This has tended to be viewed as a ritualistic attitude whose lack of material efficacy is contrasted with the efficacy of rational scientific/technological action. Such a view is predicated on the assumption that the ontology of science is a given rather than a construction. If the ontology shifts to one where both human and non-human collectivities of beings make up the world, the divide between efficacious rational/technological action and ritual action vanishes. Without necessarily positing a continuous historical link between Vedic times and contemporary peasants and fisherfolk of coastal Orissa, the performance of rituals is seen by them in exactly the same terms as those posited for Vedic times.

We moderns see both the cycles of the sun and of the seasons, as well as the monthly cycles of women, as part of nature, activities that are devoid of mindfulness. This is not the case with the peasants and fisherfolk who participate in Raja Parba. The earth and the sea is a woman and she actively *bleeds* rather than being the passive vehicle of some wholly foreign biology.[40] The word *ṛtu* refers not only to the bleeding but to the activities that men and women, humans and non-humans engage in at that time, when all refrain from work and productive activities (sexual intercourse is one such, being considered a productive activity). Men are as involved in these actions as women; although they do not bleed they must refrain from doing many things, such as touching women or plowing the earth at that time or making their bullocks work. The body and the place are not separated from what one might call mindfulness or consciousness. To call the personification of various natural phenomena, such as mountains, rivers, the sea, rocks, trees, the earth, the anthropomorphizing of nature seems to us to be rooted in the assumption that nature is wholly unlike and wholly outside humans, devoid of agency or sentience, inert, and mechanical. In other words, it is based on the scientific ontology.

In the words of one of the participants in the festival of the menses: "the Mother, the earth, and women are the same thing in different forms." This phrase could not be clearer. The Mother (*mā,* the goddess here) does not symbolize the earth in the form of a woman. All three are one in different aspects, different forms. For the fishermen, both those who fish in Chilika Lake and those who fish in the sea (whether they are Telugu speakers or Oriya speakers), Goddess Haracandi is their main deity and she is the sea. Ramacandra Nayak, a fisherman, told us the following during the festival: "We fish in the sea. Goddess Haracandi protects us from all sufferings; she is the Lady of the sea and protects us from floods and famine."

What appears, then, is that there is no essential identity for Goddess Haracandi. She can be at once the earth, the sea, the Mother, women in different circumstances. She can also be the trees of the grove, as one of us learned one of the times she visited the festival and found that one of the trees next to Haracandi's temple was garlanded, encircled with cloth and marked with *sindūra*. The priest on duty explained that before the temple was built, the goddess resided in the trees and she still does. Her anthropomorphic image in the temple is only one of her manifestations.[41]

This does not exhaust the list of Haracandi's multiple and fluid forms. During the four days of Raja Parba, Goddess Haracandi is considered to be Draupadī, the heroine of the *Mahābhārata,* and specifically, Draupadī separated from her five husbands, whose temples are in five different villages in the area. At other times of the year, some of these husbands visit Haracandi, but they do not do so during Raja Parba and this is because at that time Draupadī is menstruating.[42] Fishermen, farmers, and artisans all come to the grove during Raja Parba; all observe this time of rest and do no work.

For farmers and related artisans one important aspect of the goddess is the earth from which comes their livelihood. As we know, artisans gain their livelihood from a portion of the harvest in the farmers' fields to whom they supply the needed services or implements (this is known as the *jajmani* system). For fishermen, a major aspect of Goddess Haracandi is the sea upon which they depend for their livelihood. For most, during the period of the festival she is Draupadī at her menses and in general she is also women and the grove itself, that is, the trees that are never cut. For all, including the Muslims, she is the Mother who must now rest.

What emerges from this list of some of the most important aspects of Haracandi is that it would be wrong to identify Haracandi solely with the grove or forest preserves. The grove is indeed a sort of forest preserve, a traditional form of forest conservation, as Dietrich Brandis understood it. Here one should point out that our use of the phrase "sacred grove" does not derive from a translation of local usage but rather from hallowed use in the literature.[43] It would be wrong to say that this grove is sacred because the trees are the goddess, and cannot be cut for use. The trees are indeed the goddess but so is the earth and the sea and these are definitely used by farmers and fishermen. The earth and the sea can also be qualified by the word "sacred" since they are worshiped not only in the form of the goddess Haracandi in her temple but also directly, with offerings in the fields at various moments of the agricultural cycles or offerings to the sea. Thus, sacredness cannot be equated with a nonutilitarian attitude. Here lies the profound difference between forest preserves, parks, wildlife sanctuaries, and other biodiversity preserves, on the one hand, and the so-called sacred grove, on the other. Although the "sacred grove" shares with all manner of modern preserves the fact of being set aside and not used for human consumption, what it emphatically does not share with these is its relation with those parts of the environment that are used for human ends.

Thus, rather than a sacred/nonutilitarian versus profane/utilitarian categorical dichotomy, we see a temporal/spatial alternation between rest/fallow and active/productive, both being suffused with rituals and worship, in other words, by what has usually been characterized by the word "sacred."

"Sacred Groves" and Wildlife Preserves

Although it is well known that both the earth and the sea—among other aspects of nature—are the recipients of offerings and worship in the subcontinent, the term "sacred earth" (and, even less so, "sacred sea") is not one that is in wide use in the literature. The term is somewhat redolent of more marginal positions, such as those of the women's spirituality movement. One might even say that the widely used term "sacred grove" and its ready acceptance in the literature may have something to do with the origin of nature parks and forest preserves in the second half of the nineteenth century in the United States. As William Cronon has pointed out, by the end of the nineteenth century the traditional Euro-American view of wilderness as a wasteland, signifying the biblical cursedness of the earth after the Fall, had experienced a complete turnabout. When John Muir arrived in the Sierra Nevada in 1869, he declared, "No description of Heaven that I have ever heard or read of seems so fine." And Cronon pithily remarks that "[f]or Muir and the growing number of Americans who shared his views, Satan's home had become God's temple."[44] Such a radical turnabout had everything to do with several circumstances of that time and place. One of the most important was the closing of the frontier and with it the beginnings of the realization that natural resources were finite and could no longer be exploited to exhaustion since no more of it lay further west.

For the concept of wilderness to acquire such profound cultural influence in America, it had to become sacred. But the sacredness of wilderness depended entirely on its being "wild," namely, devoid of human utilitarian activities. This, as Cronon so incisively argues, could not happen as long as the native inhabitants of the land were a real and threatening presence, furthermore, one strongly associated with the wild and the savage. Thus, one of the key ingredients in this sacred brew was the end of the Indian wars in the second half of the nineteenth century.[45]

The enclosure of the Indians on reservations and the disappearance of the frontier were necessary and key ingredients in the making of the sacredness of wilderness preserves. These two events highlight the fact that wilderness preserves and other related parks and biodiversity preserves are predicated on the implicit assumption that human use of these habitats is incompatible with their status as (sacred) preserves. A nature park or wilderness preserve is almost by definition a nonutilitarian space and hence the dogged insistence by the machinery of the modern bureaucratic state of keeping the people who live in and use this space out of it. As Gadgil and Guha have shown, this has often been at the expense of the very lives not only of the people who lived in these spaces but even of some of the wildlife supposedly to be preserved in them.[46]

Thus, it seems that the long accepted usage of the term "sacred grove" owes a great deal to the history of nature conservation in the United States.[47] Although the grove at Bali Haracandi houses the goddess's temple, is the site of the festival of Raja Parba in the month of Jyeṣṭha (mid-June to mid-July), and is not put to any utilitarian use, the ordinary fields, the bullocks, the carts, the tools and fishing gear, the land, the sea, the rivers, the sun, the moon, and a great variety of village trees, plants, animals (wild and domestic), rocks, and many other beings and things can also be labeled sacred since they are the recipients of offerings, worship, and reverence. The divide between being used and not being used for human purposes does not correlate with something that one might call sacred.

Human and Nonhuman Alternations

The salient distinction between the grove and other members of the nonhuman collectivity is not one of sacred versus nonsacred but rather one of active versus resting, of fallow versus productive, of menstruating versus nonmenstruating. These do not oppose categories having essential attributes, such as sacred or profane/utilitarian. These refer to a dynamic movement in which the movement between different times or places is what counts. Alternations are one such dynamic movement and there must surely be others. The rhythm of the seasons, from the daily rising and setting of the sun and moon cycles to the larger cosmic cycles, to animal and plant and human lifecycles, to water cycles, all involve alternations between hot and wet, between fallow and productive, between light and dark, between touching or joining and not touching or separating, between rest and activity.

In other words, orchestrated alternations between states that more often than not are dramatically different from, even opposed to, each other (such as dry and wet, light and dark, active and inactive) are the actions by the human and nonhuman collectivities whose outcome—if successfully carried out—is the continuity of life. These alternations are rhythmic, and the point in space and moment in time at which one phase or mode articulates with another, opposed phase or mode are marked by special activities which have been called rituals.

The point that seems to emerge in clear outlines from the collective activities (both human and nonhuman) during Raja Parba is that they synchronize or harmonize or articulate human alternations with nonhuman alternations. The technologies used by the people are such that they do not disturb these rhythms. This is precisely what the linear, accelerated, extractive activities of global industry are totally oblivious to, thus exhausting the capacity of the nonhuman collectivities (as well as many human collectivities, such as ecological ethnicities) to regenerate themselves, be they trees, soils, fish stocks, water cycles, atmospheric cycles, or ecological ethnicities. These high technology activities depend on science as a form of knowledge in which the knower is disembedded or decontextualized not only from the human collectivity but from the nonhuman collectivity, the place.

Conclusion: Collective Responsibility and Moral Ecology

As Guha and Gadgil, B. D. Sharma, and one of us have argued, the rising environmentalism in India on the part of the elites who call for more biodiversity preserves and more nature and wildlife parks cannot be an answer either in India or anywhere else.[48] To separate humans' daily activities of making a livelihood from "the preservation of nature" and ethical behavior toward the nonhuman collectivity simply perpetuates a context in which the pursuit of livelihood can justify all sorts of excesses, since care and preservation happen somewhere else. It is also based, as Cronon has shown us, on the erasure of those who live(d) in these preserves.

This attitude on the part of urban elites is no different from the situation in a country such as the United States. This segment of the population does not, however, exhaust the urban landscape. Although urbanites do not depend directly on nature for survival—as the ecological ethnicities do—many engage in practices, such as the drawing of *kōlam*s described by Vijaya Nagarajan, or certain rituals that encode an ecological wisdom, such as the ritual to the goddess discussed by Madhu Khanna.[49] It remains to be seen whether engaging in such practices excludes the sort of attitude that characterizes the urban omnivores whose solution to environmental degradation is to advocate for more wildlife parks. Such practices are the sedimentation of an ancient agricultural civilization whose preindustrial urban civilization was entirely dependent on agriculture and trade. It remains to be seen whether in today's urban centers in India such practices correlate with what could be called a "moral ecological" awareness as distinct from a "scientific ecological" awareness.

Scientific ecology carries along with it the separation between fact and value that constructed science in seventeenth-century England and France. What we need today is clarity about "the modern mind" and a recognition that the times require a new configuration. After the work of Latour and all the other social, anthropological, philosophical, and historical studies of science, we can no longer call such reconfigurations "going backwards" or "romantic" since we now know that the self-proclaimed rupture between premodern and modern is itself a political program.

Although the ways the Hindu texts treat and write about the Himalayas and rivers, plants and animals, sacred places and pilgrimages is admirable, Hinduism (or any other religion) cannot offer the much needed alternative to the ways the combined crises of nature and of social justice have engulfed the contemporary world. We come to that conclusion for several reasons:

1. As we argued above, religion is generally perceived as separating the sacred or spiritual from the profane or utilitarian. The practices we have been speaking about have everything to do with the daily, mundane, utilitarian activities of making a living, of surviving. The reverence, worship, and rituals engaged in by the human collectivity are communications and acts of reciprocity with the members of the nonhuman collectivity. Reverence is shown to them as an expression of gratitude for the recognition that the continuity of the life of the human collectivity and of the place where it lives not only totally depends on the continued bounty of the earth, the plants, the animals, the sea, and so on, but on the recognition of the mutuality of this dependence. Social scientists have often highlighted the awareness of interdependency in what James Scott has called the moral economy to refer to the reciprocities and exchanges that human collectivities marginal to the market economy engage in.[50] We could enlarge that term to include the nonhuman collectivity and speak of a moral ecology. Environmentalists (and others) have highlighted the dependence of ecological ethnicities on their environment. What has not as often been pointed out is that the dependence is mutual. Without what has been referred to as the practices of restraint and prudence toward "natural resources" on the part of human collectivities, the continuity of life of the nonhuman collectivity is threatened as well. We prefer a different language to that of "restraint" and "natural resources" and speak of the actions by both human and nonhuman collectivities that bring about the articulation between both collectivities' regenerative cycles. We prefer to speak of a moral ecology characterized by regenerative cyclicity and mutuality.

As Donald Worster has so eloquently shown, the amorality of the market economy undergirded by an amoral knowledge, namely, science,[51] has fostered the extractive and exploitative activities that are exhausting the nonhuman world, exterminating numberless plant and animal species.[52] However, the morality evinced by ecological ethnicities is not an individualistic one but a collective one. As Addelson has pointed out:

> philosophers' moral theories have to do with *individual agency*. The theories of individual agency *presuppose* a social order or a way of life in which the individual makes decisions and choices. He justifies choices to himself and others in a world that is already named. Theories of individual agency rarely pay attention to how the way of life comes to be named as it is.[53]

The theory has to be one of *collective agency* and *collective responsibility*.[54] This is what one finds among the peasants, artisans, and fisherfolk of the area around Bali Haracandi and among ecological ethnicities in general. One of us has described in detail the communal organization of the villagers, how decisions are taken communally and agency and responsibility is collective.[55] Guha and Gadgil, as well as one of us, have advocated the renewal of local level communal organizations and governance to bring a turnabout in the continued environmental degradation of the subcontinent.[56]

2. In the context of the subcontinent (and probably most other contexts as well), any one religion cannot become the rallying point for an ecological reorientation, given the political conflicts escalating between the various religious communities and the spread of fundamentalism.

Instead, we propose the notion of moral ecology that captures the practices of many—especially, although by no means exclusively, those we have called the ecological ethnicities—who belong to many varied religious traditions. What is required is a moral responsibility toward the nonhuman collectivity. Mahatma Gandhi's selective denial of the British romantic tradition might be instructive here:

> Gandhi, despite his student days in London and his extensive reading of English works, was hardly indebted for his environmentalist views to Wordsworthian romanticism; and indeed, for all his attachment to the Ruskin of Unto This Last, he seems not to have been affected at all by the Ruskin who reveled in nature and rhapsodized over misty mountains and gushing streams: Gandhi's ashram at Sevagram could not have been further, in an intellectual and in a scenic sense, from Ruskin's lakeside home at Brantwood. Rather, Gandhi's environmentalism had its roots in a deep antipathy to urban civilization and a belief

> in self-sufficiency, in self-abnegation and denial rather than wasteful consumption. Gandhi was not going back to nature but to the village and to the peasantry as the heart and soul of India, to rural asceticism and harmony as against urban bustle and industrial life.[57]

For Gandhi, there was no "nature" separate from the daily activities of humans pursuing their livelihood; no nature to be adulated and exalted. Gandhi passionately rejected the amorality of the market economy and of science. His vision of village republics was saturated with a deep sense of morality, of restraint in the expression and fulfillment of human desires. Self-sufficiency, self-abnegation, denial, and rural asceticism take on their meaning in opposition to the doctrine of *homo economicus* and the infinity of his wants. Urban life for Gandhi is modern, industrial life where the logic of the market rules. Restraint or self-abnegation may also be viewed in terms of the logic of sacrifice understood as a reciprocity for the gifts of the nonhuman collectivity. Gandhi's approach to all of life is emphatically a moral one: restraint, ethics, the curbing of desire and acquisitiveness are not to be located elsewhere, in the sacred precinct of wildlife preserves; they are to inform daily activity, here and now. This is how a great majority of smallholder peasants, fisherfolk, and other members of what we have called ecological ethnicities live today. Like Gandhi, we see in these practices a viable alternative to modernity and environmental destruction.

Our major source of inspiration is the collectivity of ecological ethnicities we continue to be engaged with. We are inspired by how ecological ethnicities everywhere act ethically and in concert both with the human and the nonhuman collectivities. We also want to emphasize that the precise scope and nature of these actions is yet to unfold, especially in the more politicized movements. We dare not predict nor can we know the precise nature of the outcomes of the actions of ecological ethnicities.

The system of parks, biodiversity preserves, and sanctuaries is the only other alternative so far suggested to stem the catastrophic decline of bio- and geodiversity. Such a solution does not challenge our behavior while we pursue a livelihood; it does not challenge our daily practices. It does not challenge either our worldview or our ontology; quite to the contrary, it is deeply embedded in it. If, as we have argued, the ontology of the religion of modernity, namely, science, is one that leads to—and perhaps necessitates—the mastery and domination of nature, then it is inextricably implicated in the current ecological crisis. Furthermore, this system is one modeled on the modern West, its economic system and its science. The governments and elites of the world today, like those of seventeenth-century England and France, want the power and wealth that such a system delivers. But at the close of the twentieth century, the social and environmental price to be paid for such power is not affordable (if it ever was). This, as we have argued in this paper, is no alternative at all. What we propose as possible alternatives that avoid the pitfalls inhering in both religion and science/modernity are the notions of *ecological ethnicities, moral ecology,* and *regenerative cyclicity and mutuality.* We propose these notions that we hope capture, in some respect, the collective actions of both human and nonhuman collectivities whose livelihoods depend entirely on the regenerative cycles of nature and whose ethnic makeup cannot be captured by the usual communal/religious labels. In our naming, we have tried not only to be intellectually responsible but to be ethically responsible to the people with whom we have worked and with whom we collaborate as well as to the places where they live.

Notes

1. Donald Worster, *The Wealth of Nature: Environmental History and the Ecological Imagination* (New York: Oxford University Press, 1995), 219.

2. Richard Grove, *Green Imperialism* (Cambridge: Cambridge University Press, 1995); Sivaramakrishnan, "Colonialism and Forestry in India: Imagining the Past in Present Politics," *Comparative Study of Society and History* 37, no. 1 (1995): 3–40.

3. Madhav Gadgil and Ramachandra Guha, *Ecology and Equity: The Use and Abuse of Nature in Contemporary India* (London: Routledge, 1995), 91 passim.

4. Ibid., 91.

5. Madhav Gadgil and Ramachandra Guha, *This Fissured Land: An Ecological History of India* (Delhi: Oxford University Press, 1992), 88.

6. Pramod Parajuli, "Tortured Bodies and Altered Earth: Ecological Ethnicities in the Regime of Globalization" (unpublished manuscript).

7. Madhav Gadgil and Subash Chandran, "Sacred Groves," *Indian International Center Quarterly* 19, no. 1-2 (1992): 185.

8. Gadgil and Guha, *This Fissured Land,* 88–89.

9. The last ruler known to have supported Buddhist religious establishments was Gajapati Mukundadeva in the sixteenth century, according to H. von Stietencron, who adds that even thereafter Buddhism was never completely extinguished. H. von Stietencron, "The Advent of Visnuism in Orissa: An Outline of Its History According to Archeological and Epigraphical Sources from the Gupta Period up to 1135 AD," in *The Cult of Jagannath and the Regional Tradition of Orissa,* ed. A. Eschmann, H. Kulke, and G. C. Tripathi (Delhi: Manohar, 1978), 4.

10. N. K. Sahu, *Buddhism in Orissa* (Bhubaneswar: Utkal University Press, 1958), 176 passim.

11. Communities of Ikat weavers in Nuapatna, northwest of Bhubaneswar, consider themselves Buddhist; Stephen A. Marglin, "Losing Touch: The Cultural Conditions for Worker Accommodation and Resistance," in *Dominating Knowledge: Development, Culture and Resistance,* ed. Frédérique Apffel-Marglin and Stephen A. Marglin (Clarendon: Oxford University Press, 1990), 256 passim. Many peculiarities of Jagannātha Temple, such as the contra-caste-rule custom of allowing, even encouraging, Brahmin and low caste to eat out of the same bowl of *mahāpraśād* (consecrated food), is said by many locals as well as a few scholars to be a Buddhist remnant.

12. Frédérique Apffel-Marglin, *Wives of the God-King: The Rituals of the Devadasis of Puri* (Delhi: Oxford University Press, 1985), 243 passim; Anncharlott Eschmann, "Prototypes of the Navakalevara Ritual and Their Relation to the Jagannath Cult," in *The Cult of Jagannath,* ed. Eschmann, Kulke, and Tripathi, 265 passim; Hermann Kulke, "Tribal Deities at Princely Courts: The Feudatory Rajas of Central Orissa and Their Tutelary Deities [Ishtadevata]," in *The Realm of the Sacred,* ed. Sitakant Mahapatra (Bombay: Oxford University Press, 1992), 56 passim.

13. Another famous site for the celebration of Raja Parba is close to Bali Haracandi, nearer the eastern shore of Chilika Lake, and is sacred to the goddess Ugra Tārā. There are many Tārā goddesses in Orissa evincing Buddhist iconographical styles, a clear indication of the strength of Buddhism in the region.

14. Savyasaachi, "An Alternative System of Knowledge: Fields and Forests in Abujhmarh," in *Who Will Save the Forests? Knowledge, Power, and Environmental*

Destruction, ed. Tariq Banuri and Frédérique Apffel-Marglin (London: Zed Books, 1993), 64–65.

15. Parajuli, "Tortured Bodies and Altered Earth."

16. Frédérique Apffel-Marglin, "Of Pirs and Pandits: Tradition of Hindu-Muslim Cultural Commonalities in Orissa," *Manushi* 91 (1995): 17–26.

17. Sheik Imabul and Mr. Fariduddin were not isolated voices. With my long-time field and research assistant Purna Chandra Mishra, we interviewed Muslims in many locations. These statements have been chosen for their representativeness. Frédérique Apffel-Marglin with P. C. Mishra, "Sacred Groves: Regenerating the Body, the Land, the Community," in *Global Ecology" A New Arena of Political Conflicts,* ed. W. Sachs (London: Zed Books, 1993), 197–207.

18. Parajuli, "Tortured Bodies and Altered Earth."

19. The term "ecosystem people" is taken from Gadgil and Guha, *Ecology and Equity.*

20. However, one of us is of a cultivating Brahmin caste of Nepal and a living testimony to the fact that Brahmins are not always unwilling to till the land.

21. Parajuli, "Tortured Bodies and Altered Earth."

22. Apffel-Marglin, *Wives of the God-King,* 245.

23. Frédérique Apffel-Marglin, "Rationality, the Body, and the World: From Production to Regeneration," in *Decolonizing Knowledge: From Development to Dialogue,* ed. Frédérique Apffel-Marglin and Stephen A. Marglin (Clarendon: Oxford University Press, 1996), 142–81.

24. Here, the work of philosopher Kathryn Pyne Addelson is extremely helpful. She speaks of collective action as moral passages. It is only with hindsight that these actions are named both by participants and by outside observers. Religious literati have the cognitive authority to name what has happened but this cannot be confused with the actions themselves. Kathryn Pyne Addelson, *Moral Passages: Toward a Collectivist Moral Theory* (New York: Routledge, 1994); and "Collective Agency and Responsibility," keynote lecture of the Society for Cultural Anthropology, Washington, D.C., November 1997.

25. Addelson, *Moral Passages.*

26. Frédérique Apffel-Marglin, "Rationality and the World," introduction to *Decolonizing Knowledge: From Development to Dialogue,* ed. Frédérique Apffel-Marglin and Stephen A. Marglin (Clarendon: Oxford University Press, 1996), 1–39; and "Rationality, the Body, and the World."

27. Apffel-Marglin, "Rationality, the Body, and the World"; Ramachandra Guha, "Two Phases of American Environmentalism: A Critical History," in *Decolonizing Knowledge: From Development to Dialogue,* ed. Frédérique Apffel-Marglin and Stephen A. Marglin (Clarendon: Oxford University Press, 1996).

28. Anthony Giddens, *The Consequences of Modernity* (Princeton: Princeton University Press, 1990); Karl Polanyi, *The Great Transformation* (New York: Holt, Rinehart, and Winston, 1944).

29. Morris Berman, *The Reenchantment of the World* (Ithaca, N.Y.: Cornell University Press, 1981).

30. Bruno Latour, *We Have Never Been Modern* (Cambridge, Mass.: Harvard University Press, 1993); Stephen Toulmin, *Cosmopolis: The Hidden Agenda of Modernity* (New York: Free Press, 1990); Berman, *The Reenchantment of the World;*

Steven Shapin and Simon Schaffer, *Leviathan and the Air-Pump: Hobbes, Boyle, and the Experimental Life* (Princeton: Princeton University Press, 1985).

31. Latour, *We Have Never Been Modern,* 27, 29.

32. Ibid., 17.

33. Bruno Latour, *The Pasteurization of France* (Cambridge, Mass.: Harvard University Press, 1988), and *We Have Never Been Modern.* See also the work of many other social studies of science, such as Berman, *The Reenchantment of the World; Mapping the Dynamics of Science and Technology,* ed. Michel Callon, John Law, and Arie Rip (London: MacMillan, 1986); Donna Haraway, *Primate Visions: Gender, Race, and Nature in the World* (London: Routledge and Kegan Paul, 1989), and *Simians, Cyborgs, and Women: The Reinvention of Nature* (New York: Chapman and Hall, 1991); Ruth Hubbard, *The Politics of Women's Biology* (Boston: Beacon Press, 1990); Thomas Hughes, *Networks of Power: An Evolutionary Account of the Social and Conceptual Development of Science* (Chicago: Chicago University Press, 1983); Bruno Latour and Steve Woolgar, *Laboratory Life: The Construction of Scientific Facts,* 2d ed., with a new postword (Princeton: Princeton University Press, 1986); Donald A. MacKenzie, *Inventing Accuracy: A Historical Sociology of Nuclear Missile Guidance Systems* (Cambridge, Mass.: MIT Press, 1990); Shiv Visvanathan, *A Carnival of Science* (Delhi: Oxford University Press, 1997).

34. Berman, *The Reenchantment of the World.*

35. Toulmin, *Cosmopolis.*

36. Latour, *The Pasteurization of France,* and *We Have Never Been Modern.* We also draw on the work of a group of Andean intellectuals belonging to the grassroots organization PRATEC, with whom one of us collaborates on a regular basis. They have written extensively about the Andean world as being inhabited by a multiplicity of beings—humans and nonhumans—who collectively continuously make and remake that world. PRATEC, *Desarrollo o Descolonizacion?* (Lima, Peru: PRATEC, 1993); and *Crianza Andina de la Chacra* (Lima, Peru: PRATEC, 1994).

37. Ilmar Altvater, "Ecological and Economic Modalities of Time and Space," in *Is Capitalism Sustainable? Political Economy and the Politics of Ecology,* ed. Martin O'Connor (New York: Guilford Press, 1994), 76–90.

38. Lillian Silburn, *Instant et Cause: Le discontinu dans la pensée philosophique de l'Inde* (Paris: Librairie Philosophique, J. Vrin, 1955).

39. Emile Benveniste, *Indo-European Language and Society* (Coral Gables, Fla.: University of Miami Press, 1973), 380.

40. As Emily Martin has shown, menses for women of all classes and ethnic background in the United States is something one has; the woman's body is passively subjected to this biological phenomenon. Emily Martin, *The Woman in the Body* (Boston: Beacon Press, 1987).

41. The myth about the creation of this grove tells of a young beautiful woman approaching a passing cloth merchant in this forest and demanding that he give her a particularly fine and expensive sari. Under her spell he agreed and then asked how he should be remunerated. She answered, "Go to the king and he will pay you." When the merchant got to the king, the king had a dream in which the goddess appeared and told him that he should not only pay for the sari but establish a worship place for her in the forest. She was that young woman. The association of the forest, the goddess, and cloth is a theme echoed in many other places in the subcontinent. It is a major

theme in the *Mahābhārata* (Alf Hiltebeitel, "Draupadi's Garment," *Indo-Iranian Journal* 22 [1980]: 98–112) and for the Jharkhand Adivasis, who call the forest the cloth or shawl of the earth (Parajuli, "Tortured Bodies and Altered Earth").

42. Apffel-Marglin, "Rationality, the Body, and the World."

43. Frédérique Apffel-Marglin, in her many years of research surrounding the festival of the menses, never heard an Oriya term equivalent to the Adivasi use of *sarna*. The grove is simply referred to as the place of Haracandi of the Sand (Bali Haracandi), since it is located on the sandy strip by the sea between Puri town and Chilika Lake. Everyone knows that the trees are never cut and everyone respects such an injunction.

44. William Cronon, "The Trouble with Wilderness, or Getting Back to the Wrong Nature?" in *Uncommon Ground: Toward Reinventing Nature,* ed. William Cronon (New York: W. W. Norton and Co., 1995), 72. See also Guha, "Two Phases of American Environmentalism"; and Donald Worster, *Nature's Economy: A History of Ecological Ideas* (Cambridge: Cambridge University Press, 1977), 98 passim.

45. Cronon, "The Trouble with Wilderness," 79.

46. Gadgil and Guha, *Ecology and Equity,* 92–93.

47. It is necessary to point out that the term is also used among historians of religion to refer to such ancient phenomena as the sacred groves of the druids or of the pre-Israelites, for example. It would be interesting to know when the use of the phrase "sacred grove" became current among them.

48. Gadgil and Guha, *Ecology and Equity;* B. D. Sharma, *Globalization: The Tribal Encounter* (New Delhi: Anand Publications, 1996); Parajuli, "Tortured Bodies and Altered Earth."

49. Vijaya Nagarajan, "Rituals of Embedded Ecologies: Drawing *Kōlams*, Marrying Trees, and Generating Auspiciousness," and Madhu Khanna, "The Ritual Capsule of Durgā Pūjā: An Ecological Perspective," both in this volume.

50. James Scott, *The Moral Economy of the Peasant: Rebellion and Subsistence in South East Asia* (New Haven: Yale University Press, 1976).

51. The amorality of science has been succinctly expressed by Einstein as follows: "The concepts which [the scientific way of thinking] uses to build up its coherent systems do not express emotions. For the scientist, there is only "being", but no wishing, no valuing, no good, no evil—in short no goal"; cited in N. Maxwell, *From Knowledge to Wisdom: A Revolution in the Aims and the Methods of Science* (Oxford: Basil Blackwell, 1984), 131.

52. Worster, *The Wealth of Nature,* 211.

53. Addelson, "Collective Agency and Responsibility," 2.

54. See also Worster, *The Wealth of Nature,* 110.

55. Frédérique Apffel-Marglin, "Gender and the Unitary Self: Looking for the Subaltern in Coastal Orissa," *South Asian Research* 15, no. 1 (1995): 78–130; and "Rationality, the Body, and the World."

56. Gadgil and Guha, *Ecology and Equity;* Parajuli, "Tortured Bodies and Altered Earth."

57. *Nature, Culture, and Imperialism: Essays on the Environmental History of South Asia,* ed. David Arnold and Ramachandra Guha (Delhi: Oxford University Press, 1995), 19.

"If You Cut a Branch You Cut My Finger": Court, Forest, and Environmental Ethics in Rajasthan

ANN GRODZINS GOLD

Vansh Pradip Singh ruled the twenty-seven-village kingdom of Sawar in North India from 1914 to 1947, for over thirty years. Accounts elicited from his former subjects—their memories still vivid today—impress that he deliberately embodied, and thereby persuasively enforced, certain strong conservationist values. My focus in this chapter is squarely on the king: his motivations for protecting nature and the remarkable efficacy of his rule in this regard. I approach the king's persona largely through memories elicited from his former subjects, the people of Sawar. These memories supply the substance of my story—a story about the ways royal identification with the well-being of nature acted in the past to protect and secure ecological balance for a small piece of the earth.

Environmental scenarios are evidently to be understood within other power plays. Age, gender, and status divisions are salient. The king, the king's agents, farmers, herders, laborers, and housewives all have divergent motives and perspectives—not only in acting but in representing actions, their own and others. Another matter central to my inquiry here is how to locate these speakers' motivating ideas about human interactions with the natural world in the broad context of "Hindu traditions," and the far narrower setting of the waning years of princely power and colonial rule.[1]

I begin with a highly condensed account of the research context. I then paint a portrait of my protagonist in the words of elderly persons who lived under his regime. After these intimate recollections, I expand my scope considerably in quest of cultural resources—pan-Hindu and regional—that may have inspired this king, and other rulers, to such commitment to sustaining woods and wildlife. While multiple influences may be traced in this local context, one particular concept—*zimmedārī* (responsible authority, or responsibility based in identity)—stands out as repeatedly evoked by interviewees. In concluding, then, I highlight responsibility, as a source not only of past efficacy but of prospective regeneration.

Questions to be addressed (but hardly answered) include: Why did Vansh Pradip Singh care about his trees and proclaim his own physical identification with them? What was the popular perception of his motivations? How did his strong stance impact on the common good? Why does his death mark, in people's minds, the end of an era for "nature" itself? Or, put another way, why did the regimes that followed his—and that also possessed a mandate to protect trees—fail where he had succeeded? Most tenuously, how, in the present crisis, can these failures be redressed?

Doing Environmental History in Sawar

I first landed in the village Ghatiyali in the kingdom of Sawar in 1979. I have often returned during intervening years, but my research topics and my village networks have altered thoroughly, more than once. About ten years ago I began to focus on relationships between religion and the environment. I looked first at sacred groves, but they were by definition highly limited in space and thus in impact.[2] I therefore went back to Rajasthan in 1993 intending to explore religious constructions of nature in ordinary territories: fields and hillsides. Unexpectedly to me (but hindsight tells me inevitably), my inquiries opened up vistas through time.

Although tracing its origins and its lineage of rulers to the premier princely state of Mewar, with its capital in Udaipur, Sawar was for many years a tributary of Ajmer—a city 132 kilometers southwest of the state capital of Jaipur. From the seventeenth through the early nineteenth centuries, Ajmer was a prize that frequently shifted hands, ruled in turns by Mogul, Rajput, and Sindhia of Gwalior.[3] Ajmer district was ceded to the British in 1818 to became part of the dual province they called Ajmer-Merwara: 2,711 square miles, "isolated" as they saw it, in that it was the only part of Rajputana under direct British rule. Although local rulers in Ajmer-Merwara undoubtedly acted under pressures from above (whether Ajmer, Delhi, Gwalior, or London), for villagers these royal persons—with whatever displays of pomp they could muster—controlled conditions of daily life.

Sawar was enough of a kingdom in the early twentieth century to possess three elephants and (at the highest estimate) eighty horses. Its rulers at an earlier era were enough patrons of the fine arts that art historians identify and characterize a Sawar style of painting.[4] Even in today's greatly reduced circumstances, the royal family retains a strong sense of their own importance and status. Vansh Pradip Singh, whose presence dominates Sawar's remembered past, was a proud man with a strong personality. In our interviews, he is usually referred to as the *darbār*—literally, "the Court"—a term of reference commonly applied to rulers of large and small kingdoms in Rajasthan. I will follow this practice.[5]

In 1993, working in close collaboration with Bhoju Ram Gujar, I began to track environmental change in oral narratives of the past.[6] The local environment around Ghatiyali, as in much of Rajasthan, has changed quite radically within living memories, and so have the ways that humans interact with it. There have been no cataclysmic events, no huge dams or displacements.[7] Nonetheless, deforestation and quiet revolutions in agricultural and other technologies have transformed the conditions of living. Moreover, as the visible landscape and the built environment are altered, so—people tell me—are interior realms of conscience and morality; interpersonal relations of affection and courtesy; religious experiences both ritual and emotional.[8]

When Bhoju and I began tentatively to explore the past of nature, our first exploratory questions evoked what were for me unexpectedly vehement and dramatic descriptions of how Vansh Pradip Singh, who was the last *darbār* before independence, had cared for wildlife. Indeed, everyone old enough to remember him told, with varied proportions of bitterness, astonishment, and admiration, how he had scattered popcorn for the wild boars who destroyed his subjects' crops and how he had threatened persons who harmed any forest animals, or cut reserved trees, with ignominious beatings and heavy penalty fines.

While I at first assumed that such sanctions were regularly leveled, more extensive probings suggested that Vansh Pradip Singh never actually had beatings administered. Even fines were often mitigated or completely forgiven with stern, paternalistic warnings along the lines of "Never do it again or else. . . ." It becomes all the more interesting a puzzle that the court's rules were complied with as well as they appear to have been. This compliance was far from slavish. Resistance, both bold and sly, certainly did take place; we gathered a few heroic poaching tales as well as indications that women quietly gathered illicit wood, giving the king's guards the slip. Nevertheless, the overall efficacy of the court's protection is unquestionable: during his rule the jungle was dense and wild animals were populous.

After the death of the court (and here I use the word with both biological and institutional referents), everything changed: politics, economy, landscape. By far the most dramatic visible evidence of radical environmental transformation is deforestation. Ghatiyali, whose name derives from its valley location, is surrounded by hills, once densely wooded but now largely denuded. Each hill formation has a name, and some of these evoke the vanished woodlands. For example, one is called "Wild Pig Hill" and another, "Thieves Pass"—each having once afforded shelter in its dense foliage to marauders, whether animal or human. Hunting towers, formerly used to spot game over the treetops, now stick up from scrubby slopes and afford goatherds and shepherds the only shade around.

A schematic, unnuanced synopsis of why this happened—according to over one hundred oral history interviews—goes as follows: Vansh Pradip Singh died without progeny in 1947, resulting in a succession dispute to the Sawar throne that lasted until 1951 when his widow's adopted "son," Vrij Raj Singh—who was of the same generation and lineage as Vansh Pradip and already ruled nearby Chausala —was enthroned. During this four-year interregnum—coinciding of course with a tumultuous period in India's history—the British Court of Wards in Ajmer officially controlled Sawar. Presumably, theirs was a lame-duck regime at this point. It was then that the rapid decimation of trees, and the animals that sheltered in them, commenced.

This free-for-all attack on the wooded commons did not cease with Vrij Raj Singh's ascension to the Sawar royal seat. Land reform that would disempower the former princes was well underway. There was no longer an incentive for the court to protect the jungle. In fact, it appears there was considerable incentive to exploit. A crude "let's get ours while the getting's good" sentiment prevailed—perhaps exacerbated in Sawar where the ruler was an "outsider" and spreading infectiously from nobility to opportunist mercenaries to ordinary folks who stockpiled firewood. Within a few decades the densely wooded hills where tigers, deer, antelope, jackals, and wild boars once found sustenance and shelter were stripped almost bare.

Archival research reveals that deforestation was a concern of the British collectorate in Ajmer-Merwara long before 1947. But I do not doubt vivid, multiple local accounts of highly accelerated denudation in the post-independence decades.[9] This doubtless overshadowed public awareness of earlier and more gradual depletion. However, some persons we interviewed asserted that Vansh Pradip Singh not only protected an existing forest, but caused both tree cover and wildlife to increase significantly. In Ghatiyali, people strongly associate in their minds the former ruling kings, old growths of indigenous trees, and wild animals. These three came to an end together.[10] And, I hasten to add, almost all but the princes welcomed democracy, the "rule of votes" (*voṭ kā rāj*). At the same time, regret for the loss of natural abundance, nostalgia for the lost woods, is pervasive.

One summary explanation for the ecologically disastrous course of events is that after the *darbār*'s passing, there remained no "responsible authority" (*zimmedārī*). The *darbār* had embodied *zimmedārī*—in his rootedness, his vigilance, his personal care. Whatever injustices may have prevailed in his time, however self-indulgent were his own behaviors, his twenty-seven villages meant the world to him (to use a cliché literally), and his interest in the trees and animals was lifelong. After him, politics are appraised not only as corrupt but as deracinated.[11]

Tree-Culture Passion

A fragment from the archives helps us begin to contextualize Vansh Pradip Singh's intimate concern for the jungle, his sense that trees are extensions of his royal body. W. J. Lupton, settlement officer of Ajmer-Merwara, writes in a document titled "Improvement of Agriculture" that he was "struck forcibly by the dearth of trees" in Ajmer-Merwara, as compared with the "well-wooded tracts in many districts of U.P." Tree culture, he continues, is "not an in-grained passion" among farmers in this region: ". . . no tree which has leaves on it that cattle will eat is sacred from the devastating knife of the cattle-grazer and the tender of sheep and goats."[12] This was penned in 1908 when Vansh Pradip Singh was a young prince. It suggests (as do many other documents in the British records) that the colonial regime was sharply critical of local practices vis-à-vis trees. As elsewhere in India, the British in Ajmer-Merwara wished—for their own self-interested reasons—to assume the role of protectors of the forests.[13] Persons who lived under Vansh Pradip Singh's rule recall his distinctive passion for tree culture. At times we see it set in stubborn opposition to the herders and their voracious charges.

Interviews with persons at various levels of society vividly evoke the *darbār*'s relationship with certain aspects of the natural world. To portray this with the full nuances available in oral testimony, I shall present passages from six different interview texts with persons based in different locations and situated at varying levels of economic, political, and ritual hierarchies.

In a conversation with Ram Singh Shaktavat, a Rajput whose family is closely connected to the royal house, we heard of Vansh Pradip Singh in his youth, decisively bent on fostering the forest, even before he ascended the throne:

> Ram Singh: So Ummed Singhji [father of Vansh Pradip Singh] gave these two villages to Vansh Pradip Singh, but he [the young Vansh Pradip Singh] said, "I don't need villages, I need jungle!"
> Bhoju [startled]: What? At this time there was no jungle?
> Ram Singh: There were hills but less trees, not so many. So he set his special men—he had five men when he was prince—he set them as watchmen to protect the forest; he started to foster and protect it. . . .
>
> And after that he brought animals here, from Bhinai, and from the neighboring jungles.[14] He had sweet potatoes and corn dropped on the road and they came here, eating, eating. And so, little by little, the animals came here.
>
> So Vansh Pradip Singh made this jungle very good, and he created a Forest Protection Department [which didn't exist previously in Sawar]. . . .
> Bhoju: Is there an example of someone who tried to harm the jungle, and the *darbār* punished him?
> Ram Singh: No, because the horses [guards] wandered in the jungle and people were afraid of them.
>
> Suppose you have an acacia in your field, you can't even cut that; so

> how would you cut from the jungle? Suppose somebody complained [tattled] to him, "Bhoju cut a tree," then the court would say, "*my hand hurt all night, I feel pain in my hand, don't do it again.*"

An interesting implication here—beyond the king's claim that he physically suffered the tree's pain—is that such pain would in itself be sufficient punishment and deterrent against future offenses.

A Brahmin from the town of Sawar itself, whose family had been domestic priests to the court (*rāj purohit*), told us:

> The court paid a huge amount of attention to the jungle and whenever anyone cut a branch he would say, "*it is just as if my finger were cut.*" He allowed people to take the dry wood from any jungle, but he didn't let anyone take any green wood. If someone took green wood he became very angry, and he would say, "O you good-for-nothing![15] Why did you do such a thing?"

Here, in a less subtle mode, the court actively chastises the offender. Nonetheless, the chief sanction remains nothing but shame, based on the ruler's personal distress and accusatory chiding. We gather not only that the king identifies with his trees, but that the people's affective bonds with their ruler will influence their behavior.

Mul Chand Joshi, a Brahmin astrologer from a smaller village, Kalera, gave us this version of the court's interest in trees, while answering a query about what projects Vansh Pradip Singh had pursued with passionate interest (*śauk*).

> Mul Chand Joshi: He dug wells and step-wells and planted gardens, and he planted the shade trees by the side of the road between Ghatiyali and Sawar.
>
> He said, "*if you cut the smallest branch of a tree it is just as if you cut my finger.*"
>
> Bhoju: So what was his motive for planting trees?
>
> Mul Chand Joshi: He wanted travelers to have shade. The road looked beautiful, there was shade, it was the beauty of the village, and he himself would wander in the gardens, and he could take any officer to wander in the garden.

Mul Chand's focus is upon the roadside trees and cultivated gardens, rather than the wild jungle growth. Although the court protects both, he is particularly attached to the shade trees he himself caused to be planted. Here Mul Chand stresses the values of beauty, shade, and potential hospitality—all of which are traditionally associated with the religiously valued species of trees that once lined the Sawar-Ghatiyali road.[16] The eventual cutting and selling of these same trees was often evoked by persons interviewed as an index of the community's moral decline.[17]

We heard many stories highlighting the court's extremely vigorous, direct personal interventions in order to protect trees and animals. This one, narrated by Ram Narayan Daroga, is part of the teller's own family history.[18]

> Ram Narayan: Vansh Pradip Singh planted them [the roadside trees]. I myself, when I was young, I used to make the little walls around the *nīm* trees [to protect them]. And all the trees met together over the road, to make a canopy of shade.[19]
>
> Well, once Sukh Dev Char[20]—he had 40–50 goats, and one time, one of them ate one of the *nīm* trees, a branch of a sapling *nīm*. At that time my father was the watchman [*chaukidār*] of the *nīm*. And he got upset and said to Sukh Dev Char, "Uh oh! The Court will make our skin fly off—yours and mine!"
>
> At that time there was another guard, Chandra Singh, who was a bad man. So he went to the Court and said, "Your man, Ukar Daroga [Ram Narayan's father], knowingly failed to protect the trees, and a goat has eaten a *nīm*. Then the Court came to see the tree, and he said, "Where is Ukar? Summon him here."
>
> Well, my father was scared shitless [literally, "he had to open his loincloth"]. But he approached the Court. The Court was upset [and demanded], "What are you doing, eating the trees? or what?" [That is, are you destroying the trees or protecting them?]
>
> My father said, "It was a lizard (*khaṇgeṭyā*)[21] that ate it."
>
> So the Court said, "find it!" And my father's luck was good, so he found a lizard, and at that time it just happened to be eating another plant.
>
> So the Court said, "Call the Bhils, and kill all the lizards on the whole road!"[22] And after that they killed all the lizards between Ghatiyali and Sawar. And the Court rebuked Chandra Singh, "You are a liar! It was a lizard, not a goat; but you deliberately provoked my anger, you good-for-nothing [*nālāyak*]!"

This anecdote reveals, significantly for us, that the court's protective attitude toward wildlife is neither indiscriminate nor necessarily ecological.

Vansh Pradip Singh demands wholesale slaughter of a species of lizard in order to protect trees. Moreover, he commands this violence mistakenly, as he is duped by his subordinates. The trick, in their minds, appears well justified. The court's rage over the loss of a single sapling branch is portrayed as excessive and almost childish. Through deceit the guard is able to deflect this rage from himself, the goatherd, and the goats and onto lizards—about which, presumably, nobody much cares. Yet this tale also serves as a parable of the court's righteous protection when contrasted with the post-independence destruction of the same trees which the court had looked after with such relentless attentiveness.

The court adored wild pigs, and many described his affection and care for them. While his protection of trees may sometimes be opposed, as we just saw, to the interests of herders, his protection of pigs is always opposed to the interests of farmers. Wild pigs cause enormous trouble to farmers and their crops.[23] A 1993 interview with Pratap Regar, a leatherworker, vividly encapsulates this conflict of interests:

> Pratap Regar: At that time there was much grass, and many wild animals, and people had to stay in their fields from the minute they planted seeds, because there were wild pigs and the court fostered them.
> Bhoju: Did he let people kill them?
> Pratap Regar: No no no! And in the hot season he built water tanks for them and made popcorn for them to eat where there was no appropriate food for them; I myself have come many times to scatter popcorn for them. And the court used to sit on the balcony in his hunting tower and talk about them, "Oh, there's a brown one! Look, there's a little one!" He was very happy to watch them.
> Bhoju: If there were so many what happened to them?
> Pratap Regar: Enough! When the court died their number was up. People came from all around and picked off every one of them.

Notice Pratap's explicit linking of the death of the court to the death of the pigs. Notice also that here, too, the court is made to seem somewhat obsessed and immature.[24]

Now and then our interviews give glimpses of a broader religious context for the court's protective stance toward nature. In one conversation in Kosaita, a large village not far from Sawar, we were interviewing a wealthy merchant. Banna Nath, a religious teacher of the Nath community, had been listening and only occasionally adding his own opinions or comments. When we came to the subject of protecting wildlife, however, Banna Nath as religious expert became chief respondent. He told us about a particular water reservoir in Kosaita where hunting is forbidden, behind which lies a story involving the court—but not the usual story.

Banna Nath's tale reveals an occasion—unusual but not the sole example in our recordings—in which lessons about the need to protect nature move from village to court rather than vice versa. Of course it is a miraculous manifestation that teaches the court what the villagers already know.

> Bhoju: Did the court only love pigs or all animals?
> Merchant: He loved all animals
> Banna Nath: In the area around the water reservoir, no one was allowed to shoot. He himself didn't shoot and he didn't allow anyone else to shoot.
>
> Well, once Vansh Pradip Singh tried to fire at ducks there; but when he was aiming his gun his dog ran in front of it and he stopped; then he aimed again and once more the dog ran in front; so he asked the villagers, "What's going on?"
>
> The villagers said, "This is Dharma's Pond [*dharam kā nāḍī,* its name even today] and no one can shoot at this tank." And so the court didn't shoot, and today also no one shoots here.
> Bhoju [seeking confirmation]: Do villagers still pay attention to its protection?
> Banna Nath: Yes.

I could easily continue with detailed, personalized, contextualized examples of ways Vansh Pradip Singh interacted with his subjects in connection with trees and wildlife—recollections stretching back fifty years and more into the past through fluid paths of memory. However, with this much evidence amassed, it is time to turn to the central question: What inspirations and ideals motivated Vansh Pradip Singh's behavior?

To answer this, I can do no more than speculate. Certainly, the court must have made an individual choice to hold his particular attitudes so steadfastly, if theatrically at times. Informing our understandings of the ruler's character and predilections are various sets of possible, converging influences, including pan-Indian and regional Hindu traditions; colonial forestry agendas; and local theories of good government as they have played out through history.

The Court's Finger Contextualized: Speculative Sources for Vansh Pradip Singh's Environmental Care

I begin by juxtaposing the most pervasive image from our interviews —the court's finger, and its shared pain with trees—with the broadest context: Hindu textual prescriptions for moral action. There are powerful antecedents in Sanskrit scriptures, echoing through centuries of South Asian literatures, for a physical identification between human beings and trees. If we go by the gesture that each person who quoted the court made when saying, "If you cut a branch . . .," Vansh Pradip Singh would raise and wag his right forefinger when making this statement. The right hand is highly valued, pure, and indispensable. This gesture suggests that the royal person is coextensive with the royal jungle. It would likely be a forbidding trope—if not a perfect barricade—when wood-hungry peasants contemplated infractions of well-known rules.

Christopher Chapple, in his selective exploration of Hindu scriptural resources for environmentalism, cites an Upaniṣadic passage physically linking trees and people; he comments that such "images of the human body establish a strong kinship between the self and the world of nature, indicating a world view that holds inherent respect for the non-human realms of existence."[25] In the *dharmaśāstra*s (treatises on moral action) there is much to confirm the sentience of trees, as well as statements on the special relationships between trees and kings. V. P. Kane's digest cites texts that teach that ". . . trees have life since they feel pain and pleasure and grow though cut." Moreover, ". . . the king should award . . . [fines] against those who wrongfully cut a tree. . . ."[26]

Regarding the *dharma* of kingship, O. P. Dwivedi also finds kings held responsible for environmental care:

> Environmental ethics, as propounded by ancient Hindu scriptures and seers, was practiced not only by common persons, but even by rulers and kings. They observed these fundamentals sometimes as religious duties, often as rules of administration or obligation for law and order, but either way these principles were properly knitted with the Hindu way of life.[27]

Closer to Sawar in time and space, if not immediately local, would be Rajasthan's famous if small Bishnoi community. Bishnoi religion included a particular reverence for trees. As one contemporary Bishnoi author puts it, "In trees is the breath of life. . . . Hundreds of years earlier, Guru Maharaj [the sixteenth-century founder of Bishnoi religion, Guru Jambheshvar]—knowing the importance of trees—established the religious practice of not cutting green trees, which today's world governments are only now beginning to establish."[28] This passage highlights a nexus of religion and politics in tree protection. In Rajasthan, such convergence is pervasive and reflects, I believe, both the broader *dharmaśāstra* basis for royal tree protection discussed above and the ways this has played out locally within Rajasthan's particular climate and ecology, where trees have enormous value.

I have no evidence from Sawar that Bishnoi ideas about trees were common currency fifty years ago. However, the now internationally famous story of the Bishnoi heroine who led the first tree-huggers in fatal protest against Rajput lumber operations was well known when I began researching environmental folklore in the early 1980s.[29] In Bishnoi lore, the rural population put their bodies on the line to protect trees *against* a ruler's wanton exploitation—quite the opposite of the situation in Sawar. R. J. Fisher suggests that Bishnoi tree protection is "a symbol of political resistance to the [ruling] Rajputs," but he also points to local nobility's recent attempts to co-opt an identification with nature and establish themselves as allied with Bishnoi environmentalist values, which are globally admired.[30]

Vansh Pradip Singh doubtless absorbed many other culturally available teachings in history and religion from his childhood. Even in today's diminished era, the kingdom of Sawar is a place rich in revealed religious knowledge transmitted partially through texts but even more vividly through oral traditions, performance, and ritual. Certain tree species, for example, figure frequently in Sawar folklore and festivals, most especially the pan-Indian religiously emphasized species of *nīm, pīpal,* and banyan (*baṛ;* or, in Sanskrit, *vata*).[31] These culturally valorized trees do not grow in the jungle but are cultivated with care by humans and located near houses, temples, meeting places, water tanks, and wells in the fields. None of these species were numerous enough to be major economic resources or to play major parts in local ecology, but all are praised for shade and beauty and are employed religiously and medicinally.[32]

The *pīpal* tree, for example, is worshiped on the day of "*Pīpal* Mother Pūjā" as a form of the goddess.[33] Both *pīpal* and banyan trees offer shelter and sustenance to wayfaring children in numerous folktales.[34] Gujars, the major herding community in this area, have special duties to respect and protect *nīm* trees, which are associated in several ways with the Gujar divinity Dev Narayan and his epic tale.[35] Sometimes Gujar goatherds lop *nīm* leaves to feed their charges, but they do so almost always with pangs of conscience.

In my own earlier work on environmental concepts in Rajasthani folklore and collective rituals, I observed a number of strong patterns or messages. Foremost among these was that "the environment is responsive to and participates in human identities, states of mind, and moral enterprises."[36] Such observations are pervasive in anthropological studies of South Asian environmental attitudes. Tariq Banuri and Frédérique Apffel-Marglin cite the words of a Hill Maria elder, who says, "The sun, the moon, the air, the trees are signs of my continuity. . . ." They explicate: "The fate of the elder is indissolubly tied to the fate of the group and of their world—'nature', the forest, are the very conditions of their existence."[37] Pramod Parajuli speaks of a worldview that assumes identity of social and natural and human worlds: "nature does not stand apart from who we are."[38] All these sources argue that such ways of being in the world contrast strongly with the separation and mastery paradigm easily observable in Western development schemes, in scientific forestry, in green revolution rhetoric.

Yet it would be unrealistic to ignore British influence as another possible source of Vansh Pradip Singh's principles—even though that influence was arguably based in an entirely different set of ideas about people and nature. Vansh Pradip Singh by reputation held the British in contempt as impure barbarians. He did all he could to keep them out of Sawar, refusing to entertain them personally, to let them in his palace, to reveal to them his favorite hunting grounds, or to shake their hands. Nonetheless, he attended the British-established Mayo College in his youth; and, of course, his treasury was regularly subject to heavy taxation by the colonial power.

At least one interviewee, a member of the Charan, or bardic, caste from Rajpura, whose relations with the royal house historically were strained, suggested non-indigenous motives for Vansh Pradip Singh's tree passion, as he reminisced about former wildlife:

> Chalak Dan Charan: There were so many lions living here, at least two hundred; and rabbits, uncountable. But these days there is nothing. And Vansh Pradip Singh planted the trees on both sides of the road.
> Bhoju: What was his reason for planting the trees, his motive?

> Chalak Dan Charan: He said that he wanted there to be shade, and they were beautiful to see, and the English had planted trees by the side of the road from Devli to Ajmer, and from Kota all the way to Ajmer—*nīm* and *baṛ*—so the court got the idea from seeing that.

Vansh Pradip Singh might well have been inspired by British example to plant roadside trees. But I would argue that the court's public discourse about trees and tree protection did not draw on a British rationale, but rather on a distinctively Hindu identification of humans with plants—merging sensations and consciousness.

All these observations leave us, however, with a glaring paradox: If there is such an evident array of cultural resources to explain both Vansh Pradip Singh's determination to protect the forest and his marked success at the job, what happened after he died? Why did everything change so fast? Why did the same community, which today admiringly recalls the court's identification with trees, participate in their destruction? Why did the next king abandon all pretense of royal *dharma,* or even human conscience, in his relations with nature? Strong values held in common do not usually change so suddenly.

O. P. Dwivedi gives us some help, arguing simply and bluntly that "religious values which acted as sanctions against environmental destruction do not retain a high priority because people have to worry about their very survival and freedom; hence, respect for nature gets displaced by economic factors."[39] In Sawar, while the situation was not desperate, economically, it would seem that a conjunction of historical and political circumstances with economic pressures combined to override inhabitants' traditions of self-restraint and reverence—not only toward nature but toward governments. The major explanation offered within the community lies in qualitative attributes of the changed regime and crystallizes in the concept of responsible authority.

Responsibility and Its Opposite

The term which I translate as responsible authority is *zimmedārī*.[40] *Zimmedārī* is of Urdu/Persian origins, but it has been well assimilated into Rajasthani Hindus' discourse. It is associated with involvement and identity, with protection and care, with genuine, effective power based in a union of person and place. When Bhoju Ram Gujar and I were researching sacred groves, people told us that the gods had *zimmedārī* for their places. In our earlier essay, Gujar wrote, "Every god loves nature and loves natural things within his own boundaries. He protects them, maintaining his responsible authority (*zimmedārī*). He wants to be surrounded by green trees and plants and water; he wants animals and birds to receive shelter." [41]

In 1993, when we asked Dayal Gujar to describe the difference between the old days and the present, here is how he expressed it:

> Just as the *rājā mahārājā* are now finished the jungle too is finished. . . . The kings were finished, their time was finished, then people started to cut the trees because there was no responsible authority (*zimmedārī*).

In thinking about this situation we should keep pragmatic factors and material circumstances well in mind. People needed wood in the past as they need it now. People took forbidden wood in the old days, too, but they took it in moderation. One source of that moderation was the presence of the court and his vulnerable, powerful finger.[42] The pain he claimed afflicted that finger when citizens cut trees epitomized his *zimmedārī* for the land, a responsibility that is personal, immediate, and physically realized.

An old Gujar woman, Kesar, whose voice was emotionally rich and volatile, evoked for me the difference between then and now, between *zimmedārī* and its absence, in a particularly poignant fashion. This was a three-way conversation including a much younger woman, Bali Gujar.

> Ann: So, during the rule of the great kings, at that time you didn't cut the trees?
> Kesar Gujar: We cut them, we cut a lot of them! They forbid us, the guard patrolled, and used to forbid us.
> Bali: So when the guard came, what did you do?
> Kesar: We joined our hands and sometimes he gave us a fine. But at that time there was much jungle, and now there is none. . . .
> Ann: What happened, how was the jungle finished, what happened to the trees?
> Kesar [in a weeping voice]: Alas, people took them, cutting, cutting, you took them, I took them, he took them, she took them, in this way everyone. . . . After the king died, even in the night they took them. People were free. Who was the one who would put an obstacle in their

> path? Were it not so, how many *nīm* trees there were on the Sawar road! On the road you got shade from these *nīm* trees, they met over the road. And now nothing is left. Here in Ghatiyali and in nearby villages people with saws cut in the night so no one could hear. Who was sitting there?
> Ann: This is sad.
> Bali: Under the rule of the great kings, people feared, but of today's government there is no fear at all.
> Kesar: The government doesn't sit here for twenty-four hours. If he grabs us, we are the government's thieves, and otherwise we are the government's grandpa.
> Bali: There is a government, but then again, there really isn't [*sarkār hai aur hai bhī nahīṃ*].

Kesar does not speak directly of *zimmedārī,* nor does she acknowledge the court's pain. But she does tell us plainly that in the past there was an "obstacle" in the path of greed. This never prevented people from collecting a little firewood when they needed it—whether this was done furtively, defiantly, or "with joined hands" (subserviently). But it did prevent uninhibited grabbing, such as followed the court's death. Today, Kesar and Bali agree, people's relationship with the government is unstable and contingent; much of value has thereby been lost.

In the concept of *zimmedārī*—uniting personal dominion, accountability, rootedness in place, and raw-nerved passion—we may have a clue, not only to what was lost at the end of an era, but also to grounds for recovery. It strikes me that the scattered but sometimes sensational successes in India of joint forest-management projects, of "our rule in our villages" movements, of Chipko, and of many other examples of small scale ecoactivism might reflect this principle, transformed in a world beyond autocracy. Today, under the rule of votes, effective responsible authority can no longer reside in a royal person. But it may take new incarnation in a mobilized, educated community sharing work and benefits alike. In such situations, fostered as "user committees" or other engineered collectivities, the social body suffers selfish depredations as the king's body did in times past.[43] Psychological and physical identification merged with rooted, effective, responsible authority retains, I believe, persisting potential and potency for conjoined social and ecological regeneration.

Notes

This research was supported in 1993 by a senior research Fulbright fellowship and in 1997 by a senior short-term fellowship from the American Institute of Indian Studies funded by the National Endowment for the Humanities. In 1993 and 1997, respectively, my advisors in India, to whom I offer heartfelt thanks, were Professor T. N. Madan of the Institute for Economic Growth of Delhi University and Dr. Rajendra Joshi of the Institute of Rajasthan Studies, Jaipur. In 1997–98, while composing this chapter, I had the wonderful support of a Fellowship for University Teachers from the National Endowment for the Humanities, an independent federal agency. I owe an enormous debt of gratitude as always to Bhoju Ram Gujar and his family and to the many people of the *Sāvar Sattāīs* who gave us their time and words. Colleagues' comments have enriched this work; special thanks to Chris Chapple, Ron Herring, Frank Korom, and Mary McGee. Much appreciation but no fault is due to these institutions and persons.

1. For an excellent and thorough discussion of Hindu textual resources for environmental action carefully juxtaposed with a range of modern behaviors, see Vasudha Narayanan, "'One Tree Is Equal to Ten Sons': Hindu Responses to the Problems of Ecology, Population, and Consumption," *Journal of the American Academy of Religion* 65 (1997): 291–332.

2. See Ann Grodzins Gold and Bhoju Ram Gujar, "Of Gods, Trees and Boundaries," *Asian Folklore Studies* 48 (1989): 211–29.

3. Har Bilas Sarda, *Ajmer: Historical and Descriptive* (Ajmer: Scottish Mission Industries Company, 1911): 160–75.

4. Linda York Leach, *Indian Miniature Paintings and Drawings,* vol. 1 of *The Cleveland Museum of Art Catalogue of Oriental Art* (Cleveland: Cleveland Museum of Art, 1986).

5. I note here that persons whom the British regularly label "princes" or perhaps "petty princes" Rajasthanis even today refer to as kings (*rājā log*) or great kings (*rājā-mahārājā*). Moreover, common terms of address for local rulers are identical with terms of address to deities, such as *ṭhākur* and *annadātā*—roughly, "Lord." By contrast, *banā*—the literal word for prince in Rajasthani—is used for an actual ruler's sons; or as a term of respect for any young man who is a member of the Rajput caste; or, ceremonially, for bridegrooms of any caste. Although it confers honor, "prince" does not bear the masterful and divine connotations of king, lord, or court.

6. Bhoju Ram Gujar was my research assistant for many years; more recently he has become collaborator and often coauthor. We have worked together since 1979 and it is difficult to disentangle our respective contributions to this research, culminating in Ann Grodzins Gold and Bhoju Ram Gujar, "The Rule of the Shoe: Fragments of History in a Rajasthani Kingdom" (unpublished manuscript, 1998).

7. Not surprisingly, much work on environmental narratives is about more dramatic losses and highly conflicted sites. See, for example, on the ongoing Narmada conflict, Amita Baviskar, *In the Belly of the River* (Delhi: Oxford University Press, 1995); and *Toward Sustainable Development? Struggling Over India's Narmada River,* ed. William F. Fisher (Armonk, N.Y.: M. E. Sharpe, 1995).

8. See Ann Grodzins Gold with Bhoju Ram Gujar, "Wild Pigs and Kings:

Remembered Landscapes in Rajasthan," *American Anthropologist* 99, no. 1 (1997): 70–84; Ann Grodzins Gold, "Sin and Rain: Moral Ecology in Rural North India," in *Purifying the Earthly Body of God: Religion and Ecology in Hindu India,* ed. Lance E. Nelson (Albany: State University of New York Press, 1998), 165–95; Ann Grodzins Gold, "Abandoned Rituals: Knowledge, Time, and Rhetorics of Modernity in Rural India," in *Religion, Ritual, and Royalty,* ed. N. K. Singhi and Rajendra Joshi (Jaipur: Rawat Publications, 1998), 295–308.

9. For the big picture of post-independence deforestation in North India, see John F. Richards, Edward S. Haynes, and James R. Hagen, "Changes in the Land and Human Productivity in Northern India, 1870–1970," *Agricultural History* 59, no. 4 (1985): 523–48.

10. Gold with Gujar, "Wild Pigs and Kings."

11. At the same time, taxes are far lower for everyone; unpaid labor is no longer conscripted; former disadvantaged communities receive government benefits. I estimate that 95 percent of our interviewees would not wish to return to the time of the kings.

12. W. J. Lupton, "Improvement of Agriculture" (Ajmer: Rajasthan State Archives, Ajmer Branch, 1908, unpublished document).

13. For a helpful discussion of colonial forest policy and its self-interested appraisals of indigenous practice, see K. Sivaramakrishnan, "Colonialism and Forestry in India: Imagining the Past in Present Politics," *Comparative Studies in Society and History* 37, no. 1 (1995): 3–40. For a detailed and incisive account of late-nineteenth- and early-twentieth-century interactions among colonial government, princely states, and environment in Rajputana's Alwar and Bharatpur states, see Shail Mayaram, *Resisting Regimes: Myth, Memory, and the Shaping of a Muslim Identity* (Delhi: Oxford University Press, 1997): 75–82.

14. Bhinai was a larger kingdom than Sawar, also under the British in Ajmer.

15. *nālāyak;* actually a mild enough word, "good-for-nothing" was reported by many to have been the *darbār*'s strongest insult—an insult feared by his subjects. Given the rich and earthy imagery of everyday insult language in the village, this reflects the court's gentility.

16. These are *nīm, pīpal,* and *baṛ*.

17. Gold, "Sin and Rain."

18. Darogas were attached as high-ranking servants to royal families; Daroga women were often concubines to Rajput rulers; their offspring, within three generations, may be granted Rajput status.

19. This man's memories seem to merge the time when the shade trees were newly planted and the time when they were fully grown.

20. Char is a Gujar lineage; Gujars are a herding community.

21. *khaṇgeṭyā* is local language and I have not located it in any dictionary. According to my collaborator, Bhoju Ram Gujar, *khaṇgeṭyā* is the Rajasthani word for *giragiṭ*—Hindi for chameleon. Chameleons, however, have a diet of insects, not greenery. Possibly the lizards were chewing or clawing the seedlings, not actually consuming them. (Thanks to Ron Herring for pointing out the zoological inconsistency here.)

22. Bhils are a "tribal" group; they are sometimes hunters, and in Sawar were bonded laborers to the court.

23. See, for example, B. Seshadri, *India's Wildlife and Wildlife Reserves* (New Delhi: Sterling Publishers, 1986), 76.

24. Wildlife protection is not necessarily the moral high ground in India in this era. Gandhi himself wrote in 1946, "If I wish to be an agriculturist and stay in the jungle, I will have to use the minimum unavoidable violence in order to protect my fields. I will have to kill monkeys, birds and insects which eat up my crops. . . . To allow crops to be eaten up by animals in the name of ahimsa, while there is a famine in the land, is certainly a sin." Mahatma Gandhi, *Daridra-Narayana: Our Duty in Food and Cloth Crisis* (Karachi: Anand T. Hingorani, 1946), 60.

25. Christopher Key Chapple, "Hindu Environmentalism: Traditional and Contemporary Resources," in *Worldviews and Ecology*, ed. Mary Evelyn Tucker and John A. Grim (Maryknoll, N.Y.: Orbis Books, 1994), 115. For folkloric and literary visions of the slippage between human and tree identity, see, respectively, A. K. Ramanujan, "A Flowering Tree," in *A Flowering Tree and Other Oral Tales from India* (Berkeley and Los Angeles: University of California Press, 1997), 217–26; Sumatheendra Nadig, "The Man Who Grew Roots," in *Masumatti and Other Stories*, ed. Kadambi Hayagrivachar (Bangalore: Premasai Prakashana, 1992), 116–23.

26. P. V. Kane, *History of Dharmaśāstra* (Poona: Bhandarkar Oriental Research Institute, 1974), vol. 2, pt. 2, 895. See Mary McGee, "State Responsibility for Environmental Management: Perspectives from Hindu Texts on Polity," in this volume, for an expanded and lucid discussion of Sanskrit texts on kingship and environmental responsibility; she directly compares the perspectives available in these texts with the case of Vansh Pradip Singh in a fashion that illuminates the material presented here. For strong associations between kingship and environmental well-being in precolonial Tamil Nadu, see also David Ludden, "Productive Power in Agriculture: A Survey of Work on the Local History of British India," in *Agrarian Power and Agricultural Productivity in South Asia*, ed. Meghnad Desai, Susanne Hoeber Rudolph, and Ashok Rudra (Delhi: Oxford University Press, 1984), 51–99; David Ludden, "Archaic Formations of Agricultural Knowledge in South India," in *Meanings of Agriculture in South Asia*, ed. Peter Robb (Delhi: Oxford University Press, 1995), 35–73.

27. O. P. Dwivedi, "Satyagraha for Conservation: Awaking the Spirit of Hinduism," in *This Sacred Earth: Religion, Nature, Environment*, ed. Roger S. Gottlieb (New York: Routledge, 1996), 162.

28. Shri Krishna Bishnoi, *Bishnoī Dharm-Sanskār* (Bikaner: Dhok Dhora Prakashan, 1991), 22 (my translation).

29. Gold and Gujar, "Of Gods, Trees and Boundaries," 218–19.

30. R. J. Fisher, *If Rain Doesn't Come: An Anthropological Study of Drought and Human Ecology in Western Rajasthan* (Delhi: Manohar, 1997), 64–70.

31. For more on these trees, their meanings, uses, and associated mythology, see Maneka Gandhi, *Brahma's Hair: On the Mythology of Indian Plants* (Calcutta: Rupa and Company, 1991); H. Santapau, *Common Trees* (New Delhi: National Book Trust, 1966); Prabhakar Joshi, *Ethnobotany of the Primitive Tribes in Rajasthan* (Jaipur: Printwell, 1995).

32. A number of references to these trees' uses and lore found their way into my earlier writings; see, for example, Ann Grodzins Gold, *Fruitful Journeys: The Ways of Rajasthani Pilgrims* (Berkeley and Los Angeles: University of California Press, 1988), 134–35, 163–64; 248–50.

33. Ann Grodzins Gold, "From Demon Aunt to Gorgeous Bride: Women Portray Female Power in a North Indian Festival Cycle," in *Gender, Religion, and Social Definition*, ed. Julia Leslie (Delhi: Oxford University Press, forthcoming). Madhu Khanna discusses the integration of certain tree species with goddess worship in Bengal; see her chapter, "The Ritual Capsule of Durgā Pūjā: An Ecological Perspective," in this volume.

34. For two examples see Ann Grodzins Gold and Bhoju Ram Gujar, "Drawing Pictures in the Dust: Rajasthani Children's Landscapes," *Childhood* 2 (1994): 73–91; and Ann Grodzins Gold, "The 'Jungli Rani' and Other Troubled Wives in Rajasthani Oral Traditions," in *From the Margins of Hindu Marriage: Essays on Gender, Religion, and Culture,* ed. Lindsey Harlan and Paul Courtright (New York: Oxford University Press, 1995), 119–36.

35. See Joseph C. Miller, Jr., *The Twenty-Four Brothers and Lord Devnārāyaṇ: The Story and Performance of a Folk Epic of Rajasthan, India* (Ann Arbor: University Microfilms, 1994).

36. Ann Grodzins Gold, "Cultural Constructions of the Natural Environment in and Beyond Rajasthan" (paper presented at the conference "Common Property, Collective Action and Ecology," Joint Committee on South Asia of the Social Science Research Council and American Council of Learned Societies and the Center for Ecological Sciences, Indian Institute of Science, Bangalore, 1991).

37. Tariq Banuri and Frédérique Apffel-Marglin, "A Systems-of-Knowledge Analysis of Deforestation," in *Who Will Save the Forests?* ed. Tariq Banuri and Frédérique Apffel-Marglin (London: Zed, 1993), 16.

38. Pramod Parajuli, "No Nature Apart: Adivasi Cosmovision and Ecological Discourses in Jharkhand, India," in *Sacred Landscapes and Cultural Politics,* ed. Philip Arnold and Ann Grodzins Gold (Aldershot, Hampshire, Great Britain: Ashgate Publishing, forthcoming).

39. Dwivedi, "Satyagraha for Conservation," 161.

40. For a more extensive discussion of this concept, see Ann Grodzins Gold, "Rooted Responsibility: Locating Moral Authority in North India," in *The Responsibility Reader,* ed. Winston Davis (Charlottesville: University Press of Virginia, in press).

41. Gold and Gujar, "Of Gods, Trees and Boundaries," 215. For deities relating to sacred groves as kings do to their "pleasure gardens," see Rich Freeman, "Folk-Models of the Forest Environment in Highland Malabar," in *The Social Construction of Indian Forests,* ed. Roger Jeffery (New Delhi: Manohar, 1998), 55–78.

42. Certainly, increased population pressure is another factor today, recognized by many. Few, however, would attribute the whole blame to increased necessity; most insist that increased greed and lack of external control are more important.

43. There is a large literature on joint forest management initiatives in India; see, for example, *Village Voices, Forest Choices: Joint Forest Management in India,* ed. Mark Poffenberger and Betsy McGean (Delhi: Oxford University Press, 1996).

Flowing Sacrality and Risking Profanity: The Yamunā, Gaṅgā, and Narmada Rivers

River of Love in an Age of Pollution

DAVID L. HABERMAN

> But is it possible to reach all the way to heaven simply by going along the river? The same heaven where the gods live? Yes. Rivers are all created in the land of King Himail (Himalaya), the land that unites this world and the heaven.[1]

In the beginning was Puruṣa with a thousand heads, a thousand eyes, and a thousand feet. He pervaded the earth on all sides and stretched beyond it by ten fingers. Puruṣa is this entire universe, all that has been and all that is to be. Three quarters of Puruṣa rose upward to become the transcendent immortal that is in heaven, but one quarter remained behind, and from this came the sum total of all that we experience with our senses. So says the well-known *Ṛgveda* hymn 10.90. Creation myths tell us much about a religious tradition; indeed, here is one of the earliest theological expressions of what was to become rather common in Hindu scriptures: a conception of divinity that is simultaneously transcendent and immanent. Although the transcendent dimension of divinity is clearly important, I want to focus on the radically immanent conception of divinity found in this and other Hindu texts. For the world itself is often considered to be a direct manifestation of divinity. The Upaniṣads continue this theme, describing how the world came into being as the result of a division of the one *ātman,* or *brahman,* into a male and female.[2] The sexual interaction of these two produced the entire universe of concrete forms. Later theistic traditions within Hinduism carry on this trend. The

Devī Māhātmya, for example, identifies the Goddess with the world, and the poems of the twelfth-century saint Akka Mahādevī recognize the world of nature as Śiva:

> You are the forest
> You are all the great trees in the forest
> You are bird and beast playing in and out of all the trees
> O lord white as jasmine, filling and filled by all.[3]

The identification of the world of nature with divinity is pervasive in much Hindu theology, but perhaps no scripture states this more directly than the *Bhāgavata Purāṇa*. Therein we read that the entire visible world is the body of God, here known as Kṛṣṇa. The trees, for example, are the hair on his body, the mountains his bones, and the rivers his veins and arteries.[4] As a playful youth, God also appears in this text as an enjoyer of his own body, delighting in the beauty of nature.[5] Specifically, the *Bhāgavata* tells us that Kṛṣṇa was particularly delighted by the sight of the forest, Vṛndāvana, the mountain, Govardhana, and the river, Yamunā.[6] These three represent special forms of divinity in the sacred realm of Braj, which is depicted as a vibrant world of nature.

But what are the implications of these textual ideals for lived religious culture? Moving from scripture to ethnographic texts in Braj—a distinctive area in northern India greatly influenced by the *Bhāgavata*—we find similar ideas being expressed. It is widely believed that the land of Braj is itself divine. Here, natural forms, such as ponds, rocks, trees, and rivers, are worshiped as concrete manifestations of divinity. One of the *mahāvākya*s, or "great sayings," of Braj is that "Kṛṣṇa is Braj; Braj is Kṛṣṇa."[7] These ideas are borne out in much of the local religious activity. The forest of Vṛndāvana, for example, is worshiped in this region as the goddess Vṛndā Devī, often represented by a potted *tulasī* plant. Even more prominently, the mountain Govardhana is worshiped as a natural form of Kṛṣṇa. Rocks from this mountain are bathed, decorated, fed, and worshiped as natural forms of the body of Kṛṣṇa.

But it is the third feature of the divine landscape which delighted Kṛṣṇa that I would like to focus on in this essay: the Yamunā River. Worshiped for centuries in this region as a natural form of divinity, the river is considered to be a tangible form of the goddess Yamunā Devī. I will proceed by constructing a portrait of the Yamunā as a goddess, drawing primarily from an important poem written in this region, and then I will go on to examine how the religious perspective reflected in this poem is in increasing conflict with a utilitarian perspective that views the Yamunā as a sewer to carry away human and industrial waste. I will conclude with a brief consideration of the implications of the collision of these differing perspectives.

The Yamunā River has been recognized as a goddess for centuries, as is evident in much temple sculpture[8] and religious literature. Yamunā first appears in Hindu scriptures as Yamī in *Ṛgveda* 10.10.[9] There, the story is told of Yamī's excessive love for her twin bother Yama, who later became the god of death. We learn little about Yamī, or Yamunā, as a goddess from this passage; Yama rejects his sister's advances, expressing the wish that she find another man to embrace her. Yamunā appears in much Purāṇic literature as a powerful river capable of removing sins from anyone who bathes in her waters and enabling such a person to achieve heaven. The *Padma Purāṇa* (chapter 30), for example, narrates the story of a pious man's decadent son who reached heaven by inadvertently bathing in the Yamunā. The *Varāha Purāṇa* (chapters 150–51) describes the powers of Yamunā and claims that one is freed from sins and achieves Viṣṇu's heaven simply by bathing in her water. And the *Bhāgavata Purāṇa* (Book 10, chapter 22) tells that the cowherd women of Braj achieved Kṛṣṇa as their lover by bathing in the Yamunā. The loving characteristics and purifying powers of Yamunā are greatly expanded in later theological developments.

Theological thinking about Yamunā, however, seems to have reached its zenith in Braj during the sixteenth century, the place and time of a remarkable cultural efflorescence. I have only just begun work toward the construction of a comprehensive portrait of Yamunā Devī, which I plan to complete in future research. Such a portrait necessarily involves examination of a vast amount of literature, but for the purposes of this essay I have chosen to limit my comments to an important work composed by the sixteenth-century saint Vallabhācāryā, founder of the Vaiṣṇava *sampradāya* known as the Puṣṭi Mārga. This work has had a profound effect on the way the Yamunā has been viewed in the religious culture of Braj. The work in question is a nine-verse Sanskrit *stotram,* or hymn, entitled the *Yamunāṣṭakam.*[10] This short text is considered to be the first among the Ṣoḍaśa Grantha, or "Sixteen Works," of Vallabhācāryā that are given special significance within the Puṣṭi Mārga. Numerous commentaries have been written on this short work that amount to advanced theological thinking about Yamunā Devī. Perhaps most important among them are the two authored by Vallabha's son Viṭṭhalanātha and by the seventeenth-century writer Harirāya, grandson of Viṭṭhalanātha's second son Govindarāya.

The *Yamunāṣṭakam* is considered first among Vallabha's works since it reveals the true nature of Yamunā Devī and lays the foundation for loving devotion to Kṛṣṇa. After translating the poem, I will move through it verse by verse fairly quickly, examining briefly relevant points in both the original poem and the single commentary of Harirāya, which expands the earlier commentary of Viṭṭhalanātha. Yamunā Devī shares in a generic South Asian goddess theology that is typically associated with creative life forces.[11] As manifestations of *śakti,* all Hindu goddesses are possessors of power, but it is important to realize that each of the goddesses in the Hindu pantheon has a very distinctive nature. Harirāya's commentary on Vallabha's *Yamunāṣṭaka* gives us clear indication of the specific nature of Yamunā's powers, as he has interpreted each of the eight main verses of the hymn as signifying one of the eight unique powers of Yamunā Devī. Of interest here is the shift back and forth between transcendent and immanent references to the goddess.

Vallabhācāryā's *Yamunāṣṭakam*

1) I bow joyfully to Yamunā, the source of all spiritual abilities.
You are richly endowed with innumerable sands glistening from contact with the lotus-feet of Kṛṣṇa.[12]
Your water is delightfully scented with fragrant flowers from the fresh forests that flourish on your banks.
You bear the beauty of Kṛṣṇa, Cupid's father, who is worshiped by both the gods and demons.[13]

2) You rush down from Kalinda Mountain, your waters bright with white foam.
Anxious for love you gush onward, rising and falling through the boulders.
Your excited, undulating motions create melodious songs, and it appears that you are mounted on a swaying palanquin of love.
Glory be to Yamunā, Daughter of the Sun, who increases love for Kṛṣṇa.

3) You have descended to purify the earth.
   Parrots, peacocks, swans, and other birds serve you with their various songs, as if they were your dear friends.
   Your waves appear as braceleted arms, and your banks as beautiful hips decorated with sands that look like pearl-studded ornaments.
   I bow to you, fourth beloved of Kṛṣṇa.

4) You are adorned with countless qualities, and are praised by Śiva, Brahmā, and other gods.
   Your complexion is that of a dark rain cloud ever ready to shower grace upon devotees such as Druva and Parāśara.
   The holy city of Mathurā resides on your banks, and you are surrounded by all the *gopa*s and *gopī*s (male friends and female lovers of Kṛṣṇa the cowherd).
   You are in union with Kṛṣṇa, the Ocean of Grace.
   May you bring happiness to my heart!

5) By merging with you Gaṅgā, who was born of the lotus-feet of Kṛṣṇa, became a beloved of Kṛṣṇa and became capable of granting all spiritual abilities.
   If anyone can compare to you it is Lakṣmī, since she is a co-wife.
   May you Yamunā, Beloved of Kṛṣṇa, always remain in my mind!

6) Forever I bow to you, O Yamunā. Your behavior is extremely amazing.
   One is spared from the destruction of Yama,[14] god of death, from drinking your milky water.
   For how could even Yama strike his own sister's children, even if they are wicked?
   By serving you, one becomes a beloved of Kṛṣṇa, just like the *gopī*s.

7) May I achieve your presence and thereby attain a new body that makes loving Kṛṣṇa easy, O Beloved of Kṛṣṇa.
   In this way, may you be lovingly nurturing.
   By merging with you Gaṅgā, River of the Gods, became greatly renowned in the world.
   Therefore, devotees established in the path of grace never worship Gaṅgā without you.

8) Who is capable of praising you, Beloved Co-wife of Lakṣmī?
   By serving Lakṣmī along with Kṛṣṇa one achieves the happiness of liberation.

But your greatness is superior to this, since drops of perspiration that arise from love play with Kṛṣṇa on the bodies of all the *gopī*s mingle in your waters.

9) O Daughter of the Sun, all sins are destroyed for the one who regularly and joyfully recites this poem, and this person certainly develops love for Kṛṣṇa.
   All spiritual abilities are acquired and Kṛṣṇa himself becomes pleased. One's nature is thereby entirely transformed. So says Vallabha, beloved of Śrī Kṛṣṇa.

The first verse opens by praising Yamunā—a beautiful river lined with bountiful forests and fragrant flowers—as the "source of all spiritual abilities (*siddhi*s)."[15] Indeed, Harirāya identifies this characteristic as the first of Yamunā's eight powers enumerated in his commentary. Harirāya is quick to point out, however, that in this context the word *siddhi* does not refer to the traditional spiritual abilities listed in yogic texts, such as becoming minutely small or being in two places at the same time, but rather to special abilities relevant to Kṛṣṇa-*bhakti,* such as acquiring a body appropriate for direct service to Kṛṣṇa, the ability to see directly the *līlā*s of Kṛṣṇa, the experience of divine joy (*rasa*), and the perception of everything as Kṛṣṇa (*sarvātma-bhāva*). Also of interest is the fact that Kṛṣṇa is referred to in the first verse by the name Murāri, the "Enemy of Mura," a demon very much associated with water pollution. Kṛṣṇa is here identified as the one who removes pollution from the water, particularly from the Yamunā, so that his companions can freely enjoy its delights and beauty. Contemporary Kṛṣṇa theologians engaged in environmental reflection, such as Shrivatsa Goswami of Vṛndāvana (Vrindaban), make much of this facet of Kṛṣṇa and link it to the episode told in the *Bhāgavata Purāṇa* in which Kṛṣṇa subdues the poisonous serpent Kāliya and removes him from the Yamunā so that the cowherds can once again enjoy swimming in the river.[16]

The second verse provides a description of Yamunā as she rushes anxiously from her Himalayan source near Kalinda Mountain to meet her beloved Kṛṣṇa in the beautiful forests of Braj located on the plains below. Since she is herself a lover of Kṛṣṇa, Harirāya comments that the power indicated in this verse is Yamunā Devī's "ability to increase the devotee's love for Kṛṣṇa."[17] She is also identified in this verse as the daughter of the sun, a traditional position that reaches back to

Vedic literature. The Yamunā theology of Braj has it that, whereas Gaṅgā flows from the feet of Nārāyaṇa, Yamunā flows from the more intimate realm of Nārāyaṇa's heart.

Descriptions of Yamunā's physical beauty continue in verse three, which reveals that Yamunā came to Earth not only to meet with her beloved Kṛṣṇa, but also to "purify the world."[18] Harirāya elaborates on this disclosure, explaining the specific nature of her purifying activity to be preparing the soul for an encounter with Kṛṣṇa by removing any obstacles to the divine relationship. This verse also identifies Yamunā as the "fourth beloved of Kṛṣṇa."[19]

In verse four Yamunā is portrayed as a compassionate granter of the wishes of devotees. Devotees such as Dhruva, who bathed in the Yamunā daily and meditated on her banks to achieve a vision of Kṛṣṇa, are mentioned as examples. Harirāya informs us that this indicates Yamunā's power "easily to help one to establish a relationship with Kṛṣṇa."[20] Also of interest, whereas some earlier Purāṇas recount that Yamunā's dark color comes from the fact that Śiva fell into her waters while in a state of madness caused by the death of his consort Satī,[21] her dark color is accounted for here by the fact that she is in union with her dark lover, Kṛṣṇa.

Verse five announces that Gaṅgā also became a beloved of Kṛṣṇa with the ability to grant the special powers of love by merging with Yamunā. By herself, Gaṅgā can grant only liberation (*mukti*), not love; in contrast with Yamunā, who is filled with love and compassion, Gaṅgā is associated more with knowledge and asceticism. There is a short rhyme spoken in Braj: *Gaṅgā snān; Yamunā pān* ("Gaṅgā-bath, Yamunā-drink"). This is explained as meaning that Gaṅgā prepares one for liberation through bathing in her waters, whereas Yamunā prepares one for devotional love through sipping her waters. The verse ends by addressing Yamunā as the one who, according to Harirāya, "removes the faults of the present age of Kaliyuga for those dear to Kṛṣṇa."[22] This comprises the fifth of Yamunā's powers enumerated by Harirāya.

The sixth verse describes the effects of drinking Yamunā water. It declares that by sipping her water, comparable to a mother's milk, one is spared the agony of death, since Yama, Lord of Death, is her older brother. The poem asks, how could Yama punish the children of his younger sister, even if they are naughty? Mythology informing us that Yamunā and Yama are sister and brother is quite old; but according to Braj Vaiṣṇavism, this implies that the worship of Yamunā Devī involves a powerful protection from death.[23] There is a Braj saying: *Jahā̃ Yamunā, tahā̃ Yama na* ("Where Yamunā is, there Death is not"). We are told at the end of this sixth verse, however, that worship of Yamunā Devī more importantly results in becoming a beloved of Kṛṣṇa. Harirāya comments that the sixth specific power of Yamunā is her "ability to make the devotee great and attractive to Kṛṣṇa."[24] Far from being a jealous consort (as Lakṣmī is said to be), Yamunā delights Kṛṣṇa by preparing souls for an encounter with him.

Verse seven consists of a request for a new body that is appropriate for serving Kṛṣṇa. Harirāya marks her "ability to grant a new body"[25] as Yamunā's seventh power. Braj lore tells how the mighty ascetic Śiva was meditating high atop his Himalayan retreat of Kailāsa when he heard the call of Kṛṣṇa's flute and developed a desire to join in the latter's love play in Vṛndāvana. Śiva, however, could not get across the Yamunā, which functioned as a protective barrier to the love forests. In order to accomplish this he first had to bathe in the waters of the Yamunā and thereby acquire a new *gopī*-body that would allow him/her to approach his/her beloved Kṛṣṇa in the forests of Vṛndāvana.[26] A dip in the Yamunā, therefore, has been considered in many religious contexts a prerequite to any true entrance into the love forests of Braj.

The eighth verse compares Yamunā with Lakṣmī. Yamunā is held to be superior. Because she thinks only of pleasing Kṛṣṇa, her love is selfless, whereas Lakṣmī's love is selfish, since she is jealous and wishes for her own pleasure. Likewise, Yamunā is approached for love, whereas Lakṣmī is approached for material wealth. Therefore, it is declared that it is better to worship Yamunā, since her selfless love leads her to cause other souls to become lovers of Kṛṣṇa, as she did in the case of the *gopī*s whose love juices flow in her water. Harirāya identifies the eighth power of Yamunā to be her "ability to make one a beloved of Kṛṣṇa."[27]

The ninth verse indicates no additional powers of Yamunā, but rather the results of encountering Yamunā through the recitation of the poem itself. These include the destruction of all one's sins, the development of a love for Kṛṣṇa, the aquisition of all spiritual abilities, pleasing Kṛṣṇa, and the transformation of one's entire nature into a form that is appropriate for a loving relationship with Kṛṣṇa. The poem closes with the signature line of Vallabha, a beloved of Kṛṣṇa, who had the important vision of Kṛṣṇa on the bank of the Yamunā in Braj that informed him how to save souls in the present age. Yamunā appeared and assisted Vallabha in this pivotal episode, and the Puṣṭi Mārgīya tradition has it that his *Yamunāṣṭakam* was written in response to his vision of Yamunā Devī on this occasion.[28]

Although much is disclosed in this poem, the main point is that Yamunā, perhaps more than any other goddess in the Hindu pantheon, is a goddess of and for divine love. She is first and foremost that loving and compassionate aspect of divinity. Her primary role may be summarized in the following way: Yamunā Devī initiates into the world of divine love the souls of those who bathe in and drink her waters, and unites them with Kṛṣṇa. It is with great irony, then, that I turn to the final portion of this paper, for today the story of the Yamunā must unfortunately include an account of severe environmental degradation.

If you were to journey to Yamunotri, the source of the Yamunā, and follow the river through the mountains to the town of Dakpatthar, where it spills out onto the plains, you could not help being struck by the amazing beauty and purity of the Yamunā as it winds through the mountains.[29] The source of the river at Yamunotri is twofold: a magnificent waterfall tumbles down from the surrounding high snowy peaks; but the source that is worshiped by most pilgrims is a hot spring that bubbles boiling water out of a small crevice in a massive rock. The hot water runs into a stone tank that functions as a natural hot tub in which pilgrims bathe to commune with the goddess. It is a great pleasure to relax in this extremely hot water while gazing at the surrounding twenty-thousand-foot, snow-covered peaks.[30] There is a temple next to the hot spring that houses an image of the goddess Yamunā Devī, but it is clear that the natural form of the river itself is more important than the temple image. From here, the water descends, roaring clear aqua-blue in color as it rushes over large boulders and sweeps past exquisite white-sand beaches, cutting an impressive gorge through the hillsides covered with tall pine trees and small oaks. It is not difficult to understand why this river has been declared a goddess of nature.

But at Dakpatthar all this changes. The transformation of the majestic mountain river as it meets human civilization on the plains is shocking. In many ways Dakpatthar marks a boundary—between the mountains and the plains, between the natural and the industrial, between river worship and river management, between Yamunā as a majestically wild river and Yamunā as a greatly reduced stream—for at Dakpatthar a huge barrage has been built across the entire river and the great majority of the water is removed from the riverbed to be channeled off into various irrigation canals.[31] Only ten percent of the water reaches Delhi, which, historically, was built on the bank of the Yamunā because of its bounty and beauty.

By the time the Yamunā reaches Delhi it has become seriously polluted. But from there it becomes even worse. By many accounts Delhi is second in the world in having the most polluted air. During the mid-1990s, around five hundred cars were being added to the streets every day, and the economic changes of recent years have allowed new industries to spring up with little pollution control. Two thousand tons of pollutants are being released into Delhi's air every day, causing it to be referred to as a "virtual gas chamber."[32] Lung cancer is sharply on the rise, and it has been reported that one out of every five people living in Delhi has a serious respiratory disorder. These life-threatening environmental problems extend also into the diminished waters of the Yamunā. Barrages, dams, and canals have drastically reduced the original flow of water. Much of the year none of the original water is allowed to pass downstream of Delhi; from here, the water is only industrial and human waste. At the beginning of 1997, *India Today* commissioned the Shriram Institute for Industrial Research to test the Yamunā waters before and after it entered Delhi. Beyond the frightening statistics the researchers gathered, their report concluded generally that in Delhi "the living river becomes a sewer."[33] Pictures accompanying the study show a beaker of light brown water that had been taken out of the river as it enters Delhi and a near oil-black beaker of water taken out of the river downstream from the center of the city. Recent years of population growth, rapid industrial development, and expansion of consumer commodities in India have given rise to dramatic levels of environmental degradation. The waste from over 50,000 industries and over eight million people flows largely untreated into the river. Today, the Yamunā water that leaves Delhi contains dangerous amounts of arsenic, cyanide, lead, mercury, and other industrial pollutants, as well as considerable amounts of human excrement. People get chemical burns and skin diseases from bathing in the water, and the government has declared all crops grown on the banks of the Yamunā unfit for human consumption. All of this is increasingly becoming part of public discourse.[34]

As a student of the religious cultures of India, I am interested in investigating the effects the current environmental degradation is having on the traditional religious culture and natural theology associated with the Yamunā and in learning how this traditional theology is being used by Indian environmental activists to resist environmental degradation. What I say about both of these must necessarily at this time be somewhat speculative, since this not only represents a new area of research for me, but—I would add—is also a new area of human and cultural experience for India. There is probably no better site to examine these two developments than the pilgrimage town of Vṛndāvana, which is located one hundred miles downstream from Delhi, in the heart of the cultural region of Braj, an area closely associated with the worship of Kṛṣṇa and in which Yamunā theology has been most fully developed. Yamunā water has for centuries been used in this area in a variety of temple rituals. For example, the temple deities are bathed in Yamunā water, and this water is then traditionally drunk by worshipers as consecrated water, or *caraṇāmṛta*. It is still a fairly common sight to see the residents and pilgrims of Vṛndāvana and other areas of Braj worshiping the Yamunā River, offering milk, sweets, and incense to the aquatic form of the goddess. In years past it was common to see pilgrims approach the river, gather a few drops of Yamunā water on their fingers, and sprinkle them into their mouths. These practices continue, but the social and religious life that centers on the Yamunā is changing rapidly. I spent the year of 1981 in Vṛndāvana, conducting research for my doctoral dissertation. I survived the hot days of June and July by parking myself before a swamp cooler and burying myself in Sanskrit texts. Around four or five o'clock in the afternoon, I would venture out of my room to join many of the residents of Vṛndāvana on the bank of the Yamunā at a site known as Keśī Ghāṭ, where sandstone steps and meditation platforms made out of huge slabs of sandstone lead down to the water. There I was met with the wonderfully boisterous scene of dozens of people swimming in the river for the extent of the remaining hours before sunset. I have delightful memories of those days. Today, however, everything has changed. Nowadays I personally would not even think of getting into the water at the same site where before I swam without concern, for today the water flowing past Keśī Ghāṭ is black and foul-smelling. Raw sewage from the town now pours directly into the river just upstream from this site. I have seen almost no one swimming there for a number of years, and—needless to say—no one is drinking the water either. Moreover, initial conversations lead me to believe that fewer and fewer people are using Yamunā water to bathe their deities, as they have in this region for centuries. The maharāja in charge of the main temple in Gokul, for example, has stopped using Yamunā water in the temple service because of its polluted condition.[35]

Many experiences become unrealizable when physical means of access are blocked. Particular realms of worship are lost when the doors of the temple are slammed shut or, even worse, when the temple is razed. A unique kind of rapture disappears when the forest on the mountain is cut down. It is clear from the stories told in Braj that access to the higher realms represented by Yamunā Devī is achieved through the physical river. In other words, the transcendent dimensions of the goddess are reached through the immanent dimension of the river. In another important work Vallabhācāryā discusses a river—in this case the Gaṅgā—in the context of exploring philosophical ideas central to his theology. In a Sanskrit work entitled the *Siddhāntamuktāvalī,* he identifies three dimensions of anything within the world. The first dimension is the physical form, in this case the river itself, the water which comprises the earthly form of the Gaṅgā. This dimension is called the *ādhibhautika* by Puṣṭi Mārgīs. For Vallabha, the worldly form is fully divine. The second dimension of the Gaṅgā is its spiritual form, called by Puṣṭi Mārgīs the *ādhyātmika*. This is what allows the Gaṅgā to function as a holy pilgrimage site that will cleanse away sins. Vallabha uses this example to illustrate the spiritual or formless form of Kṛṣṇa, known as *akṣara* Brahman. The third and highest dimension of the Gaṅgā is her divine form of the goddess. This divine form is called by Puṣṭi Mārgīs the *ādhidaivika,* and it is said to be the form which yields the highest joy, *ānanda*.

Although within the nondual framework of Vallabhācāryā's philosophy of *śuddhādvaita* all three forms are accepted as Kṛṣṇa, they are ranked hierarchically. The second form encompasses the first, and the third encompasses the other two. In this sense, the third form is considered to be higher and more complete, but, importantly, it is not unrelated to the other two. In fact, it is important to realize the significance of the first form for any higher achievement. Vallabha indicates that the *ādhidaivika* dimension is available only to one who understands the nondifference between the *ādhibhautika* form and the others; that is to say, one can arrive at the highest, transcendent *ādhidaivika* level only through the immanent *ādhibhautika*. One modern teacher of the

Puṣṭi Mārga explained this to me by using the metaphor of a television. He said that the *ādhibhautika* is like the material substance of the television, the glass, wires, and plastic; the *ādhyātmika* is the electrical power; and the *ādhidaivika* is the resulting visual picture. The point is that one cannot see the picture without the material substance of the television; likewise, one cannot arrive at the highest level of the *ādhidaivika* by abandoning the world of concrete forms, but rather by going through it, while understanding that the world itself is divine. Related to the Yamunā, this means that the physical form of the river is essential to achieving the higher realms of loving devotion. What, then, happens when the physical river ceases to exist as a purifying reality? What happens when the physical form of the goddess disappears in a world of severe pollution?

Amidst the depressing reality of severe environmental degradation, there are a few signs of hope, and many of these are coming from religious communities. In a previous work I wrote of the efforts of residents of this region to combat environmental destruction, motivated by a religious perspective that viewed the land of Braj as a natural form of Kṛṣṇa.[36] For example, one Vaiṣṇava *baba* I met, named Mādhavadās Baba, accomplished remarkable feats of environmental protection motivated by his conviction that the landscape of Braj was divine. He restored ponds, saved tracts of trees, and forced the government to pass legislation to protect several sacred hills in Braj that had been auctioned off to be crushed and used as road-building material. Just before he died I asked him what the most important accomplishment of his life had been. He replied: "I planted trees at the foot of Mount Govardhana." Other organizations, such as Rakesh Jaiswal's Eco-Friends of Kanpur, Veerbhadra Misra's Svach Ganga of Benares, Sunderlal Bahuguna's Save the Himalayas in Uttarkashi, and the Friends of Vrindaban, are all engaged in serious environmental thinking motivated by a religious understanding that the Yamunā and Gaṅgā are sacred goddesses.

There is an obvious conflict between those who see the river as a powerful and loving goddess essential to the devotional life and those who view it as a utilitarian sewer to carry away human and industrial waste. The severe pollution I have just referred to poses a serious religious challenge for those who view the world of nature as divine; clearly, a great deal of cultural damage accompanies environmental damage. Questions arise: Will environmental degradation force those who see the world of nature in terms of immanent divinity to abandon this view for a much more transcendent view of divinity? Will the environmental degradation lead to a "death of god" movement within Hinduism? Will the pollution transform Yamunā Devī into a malignant goddess (she is, after all, the sister of Death)? Or will traditional ways that view natural phenomena like the Yamunā as divine be marshaled effectively to resist the environmental degradation facing India today? Only time will tell.

In the meantime, Yamunā Devī continues to be worshiped. I was in Vṛndāvana during the winter of 1998 and happened to be present at the Yamunā temple located at Keśī Ghāṭ during the evening *pūjā*. After waving the *āratī* lamps before the decorated image of the goddess Yamunā in the small temple, the priest turned and waved the lamps before the river itself, honoring what he considers to be by far the most important form of Yamunā Devī. The crowd gathered before the small shrine worshiped enthusiastically, the image of Yamunā in the temple was adorned beautifully, and, illuminated by the light of sunset, the river looked exquisite.

I close with a poem well-known in Braj that is attributed to the famous sixteenth-century Braj poet Sūrdās. These verses indicate many of the important themes in Yamunā theology:

> Sight of you brings me great delight, O Yamunājī.
> You flow past the blessed land of Gokul, bringing the beauty of your waves.
> You give all joys and take away all sorrows for those who rise early in the morning and bathe in your waters.
> O Yamunājī, you are the dear beloved of Kṛṣṇa, Enchanter of the God of Love.
> Therefore, you are called his favorite consort.
> You make love happen in the forests of Vṛndāvana, where Kṛṣṇa plays the flute.
> Sūrdās says: your purity and glory are sung about in all the holy scriptures.[37]

Notes

1. From Advaita Malla Barman, *A River Called Titash* (Harmondsworth: Penguin Books, 1992), 179.

2. See, for example, *Bṛhadāranyaka Upaniṣad* 1.4.

3. *Speaking of Siva,* trans., A. K. Ramanujan, (Baltimore: Penguin Books, 1973), 122.

4. *Bhāgavata Purāṇa* 2.1.32–33.

5. See, for example, ibid., 10.15.

6. Ibid., 10.11.36.

7. David L. Haberman, *Journey through the Twelve Forests* (New York: Oxford University Press, 1994), 125–26.

8. See, for example, Heinrich von Stietencron, *Gaṅgā und Yamunā: Zur symbolischen Bedeutung der Flussgottinnen an indischen Tempeln* (Wiesbaden: Otto Harrassowitz, 1972).

9. A translation of this text can be found in Wendy Doniger O'Flaherty, *The Rig Veda* (Harmondsworth: Penguin Books, 1981), 247–50.

10. Although an "aṣṭakam" is literally an eight-versed poem, a ninth verse is added that identifies the results of chanting the poem. The copy of the text I discuss is *Śrī Yamunāṣṭakam,* ed. Kedārnātha Miśra (Vārāṇasī Ānanda Prakāśana Saṃsthāna, 1980).

11. This goddess theology is often associated with the three Sanskrit terms *prakṛti* (nature), *māyā* (creative process), and *śakti* (energy or power).

12. Although the word used here is Murāri, I have translated Kṛṣṇa's various names as Kṛṣṇa throughout the poem.

13. The poem shifts between addressing Yamunā in the third person and the second person. For effect and consistency, I have decided to translate the entire poem in the second person. Later commentaries identify the gods and demons referred to in this verse as humble and assertive lovers, respectively.

14. Yama is the elder brother of Yamunā.

15. *sakalasiddhihetutvam.*

16. See *Bhāgavata Purāṇa* 10.16. Recent performances of this episode in the traditional *rāsa-līlā* dramas staged at Jai Singh Ghera, residence of Shrivatsa Goswami, have emphasized an interpretation that links Kāliya to current pollution of the Yamunā.

17. *bhagavad-bhāva-varddhakatvam.*

18. *bhuvanapāvanītvam.*

19. The fourth here means *nirguṇa*. The other three special beloveds of Kṛṣṇa are Rādhā, Candrāvalī, and the Kumārikās.

20. *anāyāsena tatsambandhasampādakatvam.*

21. See *Vāmana Purāṇa,* chapter 6.

22. *bhagavatpriyakalinivārakatvam.*

23. A festival known as Yama Dvitīyā is celebrated in northern India. It consists of a ritual bath in the Yamunā on the day following Diwali and is performed together by brothers and sisters for protection against Death.

24. *bhagavadīyotkarṣādhāyakatvam.*

25. *tanu-navatva-sādhakatvam.* This "new body" is often glossed as a body appropriate for service: *sevopayogideha.* In the Gauḍīya tradition it is called the "perfected body," *siddha-deha.* See my *Acting as a Way of Salvation: A Study of Rāgānugā Bhakti Sādhana* (New York: Oxford University Press, 1988).

26. See my *Journey through the Twelve Forests,* 19–29.

27. *bhagavat-priyatva-sampādakatvam.*

28. This event is told as the first story in the *Caurāsī Baiṭhaka Caritra,* a text written in Braj Bhāṣa in the seventeenth century by Gokulnāth, a grandson of Vallabhācāryā. The edition I refer to is one edited by Niranjandev Sharma (Mathura: Sri Govardhan Granthamala Karyalaya, 1967), 1–4.

29. I made such a journey in October 1996 with a group of international bicycle riders organized by the Friends of Vrindaban.

30. In many ways the Braj theologies of the two rivers Gaṅgā and Yamunā, which connects the first river with the asceticism leading to liberation (*mokṣa*) and the second with the devotionalism leading to love (*prema*), are reflected in the natural surroundings at their source. The upper Gaṅgā valley is bare of vegetation and windswept, and the bath at Gomukh is quite austere as it takes place in an icy glacial pool, whereas the upper Yamunā valley is lush with vegetation, and the bath at Yamunotri is pleasurable as it takes place in a warm and soothing hot spring.

31. Several residents of Dakpatthar told me in conversation that they believed Yamunā was only a goddess upstream from the Dakpatthar dam. At Dakpatthar, the Yamunā as goddess ended. Although this accounts for the seriously reduced state of the Yamunā downstream from Dakpatthar, this viewpoint is denied completely by the residents of Braj, who claim that the Yamunā is most sacred when she arrives in Braj and unites with Kṛṣṇa.

32. "Gasping for Life," *India Today,* 15 December 1996, 38–47.

33. "The Rivers of Death," *India Today,* 15 January 1997, 102–7, especially 103.

34. An article entitled "An Eyesore in the Backyard of Taj" that ran in the 29 July 1997 edition of *The Hindu* newspaper, for example, laments the "pathetic condition" of the Yamunā that now flows past the Taj Mahal.

35. Yamunā water, however, is still used daily in the Śrī Nāthajī temple of Nathdwara, Rajasthan, perhaps the most popular of all Kṛṣṇa temples in India. The water is delivered to Nathdwara by bus from Mathurā.

36. Haberman, *Journey through the Twelve Forests,* chap. 3, esp. 128–29.

37. For my translation of a poem that appears in many collections of Braj poetry, I used the edition *Yamunāṣṭakam,* ed. Kedārnātha Miśra, appendix 2, poem 64, p. 12.

Separate Domains: Hinduism, Politics, and Environmental Pollution

KELLY D. ALLEY

Religion and politics are areas of dynamic interchange in India. In fact, the nexus between religion and politics has proved so convincing that analysts of Indian politics, society, and religion have focused exclusively on the successes of these interchanges. By doing so, they have neglected to mention areas of conflict and debate in which religion and politics are deliberately held apart. While doing field research in northern India in 1995–1996, I began to think about why religious and political leaders were not coming together over a particular issue in which I had great interest, the pollution of the river Gaṅgā. These leaders were failing to create any social, moral, or political valence for the call to "save" or "protect" their sacred river Gaṅgā from material pollution. Likewise, the Government of India, after spending millions of rupees in its first environmental program to "clean" the river, had not motivated the public to participate in cleaning activities.

Given the dynamic interchanges associated with the struggle to control sacred spaces in northern India, this paper asks: why do many Hindu religious leaders avoid connecting their spiritual or organizational agendas to the ideology of combating river pollution? This chapter begins with a case from the colonial period in which Indian nationalists and Hindu religious leaders agitated to prevent government exploitation of the river Gaṅgā. I use this example to show how the historical precedent of connecting political and religious ideologies was a short-lived one that has not inspired momentum in this direction in the contemporary period. Then I outline discussions with contemporary religious leaders and ritual specialists to explain why they are not concerned with the problem of environmental pollution, and river pollution in particular.

Made conspicuous by public rallies and auspicious events, the connections between religion and politics dramatize the transformations taking place in Indian society today. The religious symbolism infusing political contests casts new meaning on the identity of the nation-state and the population groups that occupy it. In the politicization of religious events, secular narratives and political rhetoric are used to give legitimacy to religious claims and the meaning of sacred space. Through this political symbolism is reflected the rapprochement of religious and political leaders—leaders who, in recent struggles to define and possess sacred territory, have joined hands to expand their respective audiences.[1]

The religious organization Visva Hindu Parisad (VHP), or World Hindu Organization, stands out in recent interchanges. Picking up on an old conflict, the VHP has called upon Hindus in India and abroad to support a movement to repossess several sacred sites colonized by Muslims in the precolonial period. In 1984, they vowed to destroy three places of Muslim worship: Babri Masjid in Ayodhya, Gyanvapi Mosque in Varanasi (Banaras), and Shahi Idgah in Mathura. On those spots, they pledged to build Hindu temples. During the 1990s, the VHP joined hands with the Bharatiya Janata Party (BJP), a political party that had little strength in Parliament and state governments at the time. With the support of VHP members, the BJP increased its representation in state and central governments from 1991 through 1998, and assumed control of the central government for thirteen days in 1996. Again in 1998, they formed the ruling party at the center.[2] Combining their networks to some extent, the VHP and the BJP have championed the cause of *Hiṅdutva,* or Hindu nationalism.

During the early 1990s, as the VHP and BJP stepped up their conflict in Ayodhya, political and cultural reporting on South Asia voiced concern for the potentially dangerous nexus developing between religion and politics. Consequently, scholars were drawn into debates on the nature of secularism and communalism.[3] As the scholastic focus intensified, analysts of Indian politics, society, and religion grew more interested in what brings religious and political leaders together in conflict than in the processes that keep them apart. While doing field research in New Delhi in 1995–1996, I began to think about the differences between this religious-political movement to claim sacred space and the environmental movement to save or clean the river Gaṅgā. Since both involved a revisioning of Hindu faith and ritual to some degree, I wondered why religious and political leaders were not riding the waves of the fledgling environmental movement. I thought about the differences between these movements rather nervously as I watched the VHP step up its campaign to destroy the other two Muslim places of worship on its agenda (they had already destroyed the mosque in Ayodhya).[4] I saw that issues defining river pollution in India had an unmined potency that both religious and political leaders could appropriate for their own ends. For instance, an opposition political party might refer to the failures of the Ganga Action Plan as a way to attack the political legitimacy of the Congress Party. Or, religious leaders could mobilize their followers by pointing an accusing finger at Muslim tanners who dump industrial waste into the river. Yet neither these nor other kinds of appropriations had been made.

The symbolic dissonance between religious and environmentalist worldviews is a complex one in India and is partly responsible for the distance between religious and political leaders in this field. I wrote a detailed article in 1998 on the semantic complexity of the terms purity and pollution to outline this symbolic dissonance.[5] Taking a specific group of religious practitioners, I argued that the pilgrim priests, or *paṇḍā*s, of Daśāśvamedha Ghāṭ in Banaras were making a distinction between purity and cleanness when considering the impact that wastewater flows was having on their sacred river Gaṅgā. They explained that Mother Gaṅgā could unfortunately become unclean (*asvaccha* or *gandā*), but that she could never be impure (*aśuddha* or *apavitra*). For these *paṇḍā*s, the river Gaṅgā is a goddess who possesses the power to absorb and absolve human and worldly impurities. In a fashion cited in the *Viṣṇu Purāṇa,* she is able to wash over the impurities of the world like she washed over the foot of Lord Viṣṇu on her descent to earth (*Bhāgavata Purāṇa* 5.17.1). Using this transcendent power in the contemporary context, Gaṅgā can stave off the degeneration of contemporary society without defiling herself. Remaining pure and fertile, she is worshiped by devotees and pilgrims through ablutions, meditation, and worship (*snān, dhyān,* and *pūjā*).[6]

Over a period of three years, I was able to witness the reactions of pilgrim priests of Daśāśvamedha to government pollution prevention schemes. Whenever government officials made public claims about pollution in the river, the priests would look askance, muttering that government officials actually create the "pollution" they claim to control. Openly opposed to government bureaucracy, these Banaras *paṇḍā*s associated scientific techniques for "cleaning" the river with the corruption and inefficiency of government and distrusted both the theories and the officials who espoused them. Yet, while noticeably antistate, their criticisms seldom extended to industrialists, such as tanners or other large company executives. Focusing only occasionally on the irritations produced by local fish markets and slaughter houses, whose immorality and unpleasantness they deplored, they had more vigor in their denunciation of government officials and bureaucrats.[7] This was reflected in their view that industrial effluent, though not distinguished in terms of toxins or heavy metals, was considered only a bit more dangerous (*hānikārak*) than household and city waste. Household, city, and industrial wastewater together created the dirtiness, or *gandagī,* that Gaṅgā was forced to carry away. These forms were just as worrisome as *pān* (betel nut) spit and uncremated bodies; in fact, the latter two were more commonly cited as problematic for everyday life.[8]

Despite their disdain for government, fish markets, and slaughter houses, pilgrim priests at Daśāśvamedha have shown little interest in "cleaning" the sacred river. Certainly, the notion of environmental pollution has no "fit" in their Hindu worldview and little currency in local political debates.[9] Moreover, the ideology of environmental cleanliness is potentially subversive to the ideology of purity upon which a pilgrim priest's own position of authority rests. But this does not mean, as others have suggested, that Banaras residents are simply blind to waste problems.[10] They are, in fact, acutely aware of the wastewater situation in their own neighborhood and highly critical of pollution prevention schemes. The absence of purity rhetoric, therefore, does not reflect a lack of concern for the suffering brought about by dirty drains, solid waste, and general uncleanliness. Rather, the disjunction points to an occupational predicament: while depending upon a vital divine Gaṅgā, they cannot actively engage in "saving" her material form. In other words, while encouraging Hindu pilgrims to respect Gaṅgā's grace, they can neither ask them to consider her fall from grace nor take up the higher power to control, or "clean," her.

In 1995, while researching interpretations of environmental pollution in Uttar Pradesh, I met an environmental activist who wanted to make connections with VHP leaders in New Delhi, hoping they would adopt a call to save the sacred river Gaṅgā from material pollution. The Kanpur activist was successful at getting VHP leaders in New Delhi to include this message as a theme of a pilgrimage rally held in 1995. However, during the actual processions, the message was never publicized. In effect a nonstarter, the issue failed to attract the attention of leaders of the VHP or members of other religious organizations. There was no apparent political valence latent in or created by the call to "save" or "protect" Gaṅgā from pollution.

The conceptual dissonance between the Hindu worldview and the view of contemporary environmentalism and the absence of instrumental appropriations or revisions of the concept of Gaṅgā's sacred purity might, therefore, lead to the conclusion that in modern history religious and political leaders have never simultaneously exploited the symbolism of Gaṅgā's purity. However, a case from the colonial period demonstrates that religious and political leaders once joined hands to defend rhetorically Gaṅgā's purity. Almost a century ago, Indian nationalists opposed the colonial government's plan to arrest and redirect the Gaṅgā near the sacred city of Haridwar. Arguing that Gaṅgā's essence and purifying power would be altered by any kind of canal engineering, they rallied around the demand to ensure her unobstructed flow past the sacred spots of Haridwar.

Given the explosive conflicts over sacred spaces today, what makes Gaṅgā's purity different from the sacrality of sacred space in the eyes of Hindu religious leaders and ritual specialists such as *paṇḍa*s? To provide a historical contrast, I would like to outline how the issue of Gaṅgā's purity was coopted by the political movement to resist colonial rule. Data for this historical analysis are drawn from documents and records of the Irrigation Department of the colonial government of India and from colonial gazetteers. Following this, I examine data collected from interviews with religious leaders, ritual specialists, and members of religious organizations during field visits from 1994 through 1996. These interviews, held during important religious festivals, such as the Māgha Melā in Allahabad, Kārtik Pūrṇimā in Banaras, and during *yajña, pūjā,* and *āratī* on the banks of the Gaṅgā, contain references that shed light on the contemporary distance between religious and political leaders in this topical field.

Purity and Unobstructed Flow

I begin with colonial politics. By 1916, a conflict between colonial officials of the Irrigation Department of the United Provinces and leaders of the Indian independence movement had reached a crescendo. Led by Pandit Mohan Malaviya, a leader of the movement for self rule, this conflict drew the support of Hindu religious leaders, nationalists, princes, and intellectuals across India, taking place in the less visible corridors of the Irrigation Department and in correspondence and meetings between Indian leaders and British irrigation officers. Indian leaders were objecting to the colonial government's attempts to modify a project that partially diverted the river Gaṅgā to agricultural fields in the Doab, the alluvial tableland lying between the valleys of the Gaṅgā and Yamuna Rivers in northern India. Arguing that Gaṅgā's sacred purity was associated with the power of her flow, Indian leaders opposed the government's plans, using rhetoric that demanded her unobstructed flow.[11]

The site of this conflict was Haridwar. For Hindus, Haridwar is the gateway to the earth for Mother Gaṅgā. It is where she descended from heaven on the locks of Lord Śiva to relieve human suffering and purify souls. Earlier names for this place, such as Kapilā, suggest that this site has been connected with Gaṅgā's descent from heaven for centuries. The early Muslim chroniclers Abu Rihan and Rashid-ud-din of the Moghul period and Hiuen Tsang, a seventh-century traveler, also mentioned Gaṅgā-dwāra, or the gateway to Gaṅgā.[12] Historically, both Śaivites (followers of Śiva) and Vaiṣṇavites (followers of Viṣṇu) have worshiped at Haridwar and have developed different meanings of the city and its sacred spaces over time. Śaivites spell the city Hardwar and consider it the abode of Śiva (or Har), while Vaiṣṇavites call it Haridwar, the home of Viṣṇu (or Hari) This paper will use Haridwar, the official name of the city today.

Har-kī-paurī Ghāṭ, situated on the western bank of the Gaṅgā, lies within the center of the sacred space of Haridwar.[13] At the base of the steps that form this *ghāṭ* is Brahmā Kuṇḍ, the spot where Brahmā arranged the descent of mother Gaṅgā to Earth. Brahmā Kuṇḍ is filled with the nectar of immortality, the nectar that fell from the primordial jug the gods wrestled away from the demons.[14] The *ghāṭ* leading to this sacred *kuṇḍ* is known today as Har-kī-paurī, Hari-kī-pairī, or Hari-kā-caraṇ. Alternatively privileging Śiva and Viṣṇu, the first name means Śiva's door, while the latter two mean Viṣṇu's foot.

Against the backdrop of this eternal sacredness, we find that the material shape of Har-ki-pauri Ghāṭ has changed significantly over the past two centuries. In 1828, Walter Hamilton commented that the *ghāṭ* was a narrow passage carved out of a steep mountain with "room for only four persons to pass abreast."[15] In this period, several fatal incidents occurred in the narrow passage. In 1819, 430 persons, including British sepoys, were crushed to death as thousands pushed to bathe in the Gaṅgā on the most auspicious day of the Kuṃbha Melā. After the stampede, the British Government expanded the access way to one hundred feet and increased the *ghāṭ* to sixty steps.[16] When the Śrī Gaṅgā Sabha was established in 1916, this organization took control of maintaining the *ghāṭ* for pilgrims and enforced rules for public behavior. Shoes, cameras, and cooking equipment were not allowed on the *ghāṭ,* and soap or oil could not be used while bathing in the stream. Later, the Municipal Committee constructed a cement platform, connected to Har-ki-pauri Ghāṭ by two foot bridges, to allow pilgrims to congregate on the other side of Brahmā Kuṅḍ.

Hamilton's nineteenth-century description of the river landscape at Haridwar gives some detail about the nature of Gaṅgā's flow past this sacred site. At that time, the Gaṅgā flowed through several channels that lay a mile across an open gorge. One channel, departing from the main stream two and one-fourth miles above Haridwar, flowed past Haridwar and the pilgrimage places of Mayapur and Kankhal before rejoining the parent river.

On this branch of the Gaṅgā, the Government of India harnessed the river's flow and created the Upper Ganges Canal. Modeled on the Eastern Jumna Canal built in the 1830s, Colonel Proby Cautley followed a path demarcated over two centuries earlier by Ali Mardan Khan, a minister of Shah Jahan, and charted out the route for the canal on horseback. The canal was to provide relief to a dry tableland which, lying at an elevation of nine hundred feet above sea level, was prone to drought.

After completing the survey work, Cautley was appointed by the Government of the United Provinces to supervise the construction of the canal from headquarters in Roorkee. Cautley, an engineer and military officer, opened the canal for irrigation in 1855 and then ordered the construction of offshoots and distributors to widen its reach. The colonial government, however, did not begin to reap the rewards until over a decade later, well after Cautley had been charged with overspending. By the turn of the twentieth century, the Upper Ganges Canal had become the lifeline for the Doab.[17]

Negotiating the Canal Works

After the opening of the Ganges canal in 1855, several engineering problems developed, requiring officials to close the plant every year for maintenance work. Because the canal head at Mayapur stood at a higher ground level than the main stream of the Gaṅgā, the flow in the sacred stream (the supply channel) was slower (see figure 1). This caused boulders and shingle to settle and block the headworks downstream. At the end of every rainy season in September and October, the Irrigation Department had to clear the boulders and shingle by diverting the entire flow of the river away from the branch stream and the irrigation canal. Without river water to fill Brahmā Kuṅḍ in front of Har-kī-paurī Ghāṭ, pilgrims could not perform the ablutions that were central to their pilgrimage. Closing the canal also proved problematic for farmers who required water for standing *kharīf* (autumnal) crops and sowing winter crops.[18]

In 1909, engineers of the Irrigation Department of the United Provinces developed plans to create permanent works at Bhimgoda, above Har-kī-paurī Ghāṭ, to regulate the flow of water into the supply channel. Without informing the public of their plans, officials assured the local Brahman priests, or *paṇḍā*s, that the work would ensure an increased flow of the Gaṅgā past Har-kī-paurī Ghāṭ year round. While the priests were convinced and raised no opposition, some resistance was raised by a small organization for Hindu nationalist unity, the All-India Hindu Sabha. Their concerns, however, had no effect on the government's plans.[19]

Work began in 1912, but two years later government officials radically modified the plan to create a canal headworks at another site. In the new plan, the government proposed building a masonry dam across the main stream of the Gaṅgā and constructing a regulator with gates above that dam to divert water to the supply channel. To do this, they proposed digging a new channel across Laljiwala Island, a stretch of land lying between the supply channel and the main stream of the Gaṅgā (see figure 1). Officials of the Irrigation Department speculated that this new regulator would control the flow of the river into

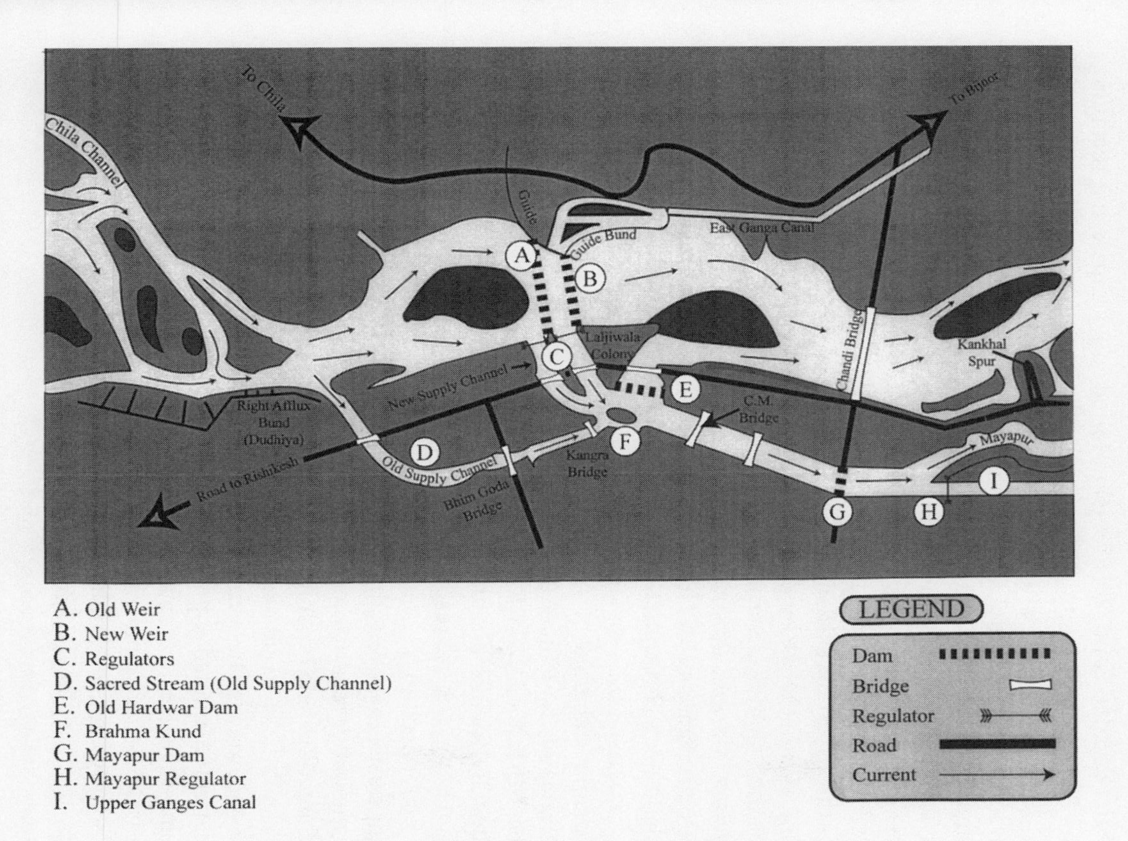

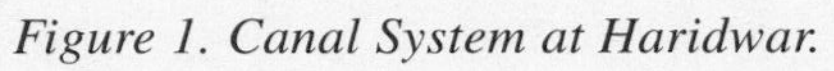

Figure 1. Canal System at Haridwar.

the original supply channel and help prevent high floods. It would also reduce shoaling and ensure an adequate supply of water for irrigation in the months of September and October.

Officials advocating this revision failed to anticipate the reaction of the Hindu community. Unimpressed by the projected benefits of canal irrigation, many Hindu leaders opposed the government's new plan. In a public meeting at Haridwar on 5 June 1914, the local Hindu community expressed alarm at the projected works and laid down their demand for an uninterrupted flow of the Gaṅgā past Har-kī-paurī Ghāṭ.[20] But, in spite of their demands, the Government of India moved ahead and sanctioned the expenditure for the revised plan.

As the government dismissed the protests of the Hindu community, the agitation, fueled by demands for self rule, gained momentum. When the colonial government could ignore their demands no longer, Sir James Meston, the lieutenant governor of the United Provinces, convened a meeting of Hindu leaders to discuss the issue. The meeting, held on 5 November 1914, was attended by public figures from various parts of the vast subcontinent. The maharaja of Jaipur, a minister of the maharaja of Alwar, the maharaja of Darbhanga, Justice Chatterjee of the Calcutta High Court, Lala Sukbir Sinha (secretary of the All-India Hindu Sabha), Mahant Lakshman Das of the Sikh Gurudwara, Dehradun, and several engineers were present.[21] During this meeting, the representatives made several decisions that were later announced in a communique.

Appearing to waver under pressure from the Hindu community, the government agreed to keep the opening to the supply channel free from any blockade or sluice, even though some shoaling would result and would have to be cleared annually. They also agreed to keep a "free opening" in the weir over the main stream of the Gaṅgā to provide an unobstructed supply of water for bathers. Responding to their demand that the new supply channel not take on "the name or the appearance of a canal," colonial officials assured the public that the new supply channel would be cut "naturally" out of Laljiwala Island. To do this, engineers agreed to refrain from lining the canal with masonry.

The commitments spelled out in the communique were vague in several respects and avoided mentioning that work designed under the original plan was, in fact, already underway. Suspicious about the government's assurances, Pandit Malaviya insisted that the maharaja of Jaipur write a letter to Meston suggesting that plans to build both the regulator and the new feeder channel be shelved. Then Malaviya and the maharaja of Jaipur called other ruling princes into the agitation. Another public meeting was held in Haridwar on 4 September 1916. At that time participants expressed their regret that the Irrigation Department's assurances were not upheld. Copies of the resolutions from this meeting were sent to the maharajas of Udaipur, Jaipur, Mysore, Travancore, Indore, Bikaner, Alwar, Gwalior, Jammu and Kashmir, and Darbhanga. The All-India Hindu Sabha was also notified.

At this time, the *paṇḍā*s of Har-kī-paurī created an organization to promote their interests. This organization, the Sri Ganga Hindu Sabha, also submitted its regret to the Government of India. In a separate correspondence, Malaviya and other maharajas, rajas, former judges, and *śaṅkarācārya*s convinced Meston to communicate their concerns to the Government of India. When Meston wrote his letter of 5 February 1917, he warned the government against violating Hindu sentiment:

> It is necessary to face the fact that Hindu sentiment was not consulted before these works were determined on, that it is extraordinarily powerful in all matters concerning the sacred stream at Hardwar and that the agitation—engineered and dishonest though much of it is—has penetrated and if unchecked will further penetrate vast depths of Hindu sentiment, which it would be most inexpedient to array against Government at any time and more especially at present.[22]

Through Meston, the Hindu community pushed the Government of India to abandon the construction of the regulator and the other canal headworks leading to the new supply channel above Har-kī-paurī. They also demanded that an opening of at least thirty feet be kept in the weir across the main stream of the Gaṅgā. Calling the protests irrational, officials argued that Hindu demands ran counter to the needs for flood protection and irrigation. "Unobstructed flow," as Sir James Meston put it, "was stretched to mean much that would be inconvenient."[23]

Nevertheless, Meston decided to accommodate the opposition by convening another conference in Haridwar on 18 and 19 December 1916. He invited the All-India Hindu Sabha, the Ganga Hindu Sabha of Haridwar, and those who had attended earlier meetings on the issue. The maharaja of Jaipur brought along his own contingent of ruling princes and their technical adviser, a retired British chief engineer of the Punjab. After taking a tour of the canal works, Meston sought the opinions of those present and recorded them. Malaviya, who was the most articulate of the group, claimed that participants of the 1914 conference had misunderstood the government's decisions. He argued that they had believed the new channel running from the regulator to the original supply channel would not bear the appearance of a canal. They had also expected the weir across the main stream to have an opening without shutters to allow the free and unfettered flow of the Gaṅgā. Since the government had disregarded both these points by lining the canal and constructing a shutter in the weir, he graciously submitted that perhaps the Hindu community had misinterpreted the government's intentions. The Hindu community expected a larger opening in the weir, to permit enough water to maintain the river stream all the way to Allahabad. Malaviya pleaded:

> From the discussion which has appeared in the papers it is clear that people want a sufficient opening so that they can bathe. It must be wide enough to let a sufficient stream through. Thousands of people bathe along the course of the river. The opening must be so wide that all places below it will receive pure water. People seem to be dissatisfied with the Regulator. Possibly there was a misunderstanding both on the part of those who were present in 1914 and the outside public. It is important to remove this dissatisfaction in view of the holiness of the river. Even if some extra cost is incurred the feeling of the people should be soothed. It is said that the agriculturists will suffer if the volume of water that passes into the canal is reduced, but no Hindu would place his material prosperity above the dictates of his conscience and his religion. In his opinion a five feet opening was not enough. Five to ten lakhs of bathers come to tiraths. They come from great distances and undergo great discomforts. But they will stand any trouble because it is a matter of faith. Even if the cost were one lakh or two lakhs, that should not matter when it was a question of belief with the people. This must be borne in mind. They believe that the Ganges makes people pure and removes sin.[24]

The government strongly disagreed with the option of maintaining a permanent opening in either weir or regulator, citing the problem of shoaling and the danger of flooding. Nevertheless, Meston was eager to forge a compromise and adopted a plan to keep a bay permanently open on the regulator. He also promised to maintain a free opening in the weir that would allow for a minimum discharge of four hundred cusecs in the main stream during the dry season. In addition, the Mayapur regulator would have a free opening to provide a flow of two hundred cusecs to the Kankhal *ghāṭ*s. A minimum supply of one thousand cusecs would be kept in the supply channel to feed the Haridwar *ghāṭ*s.

While Malaviya was satisfied with the settlement, the Government of the United Provinces and engineers of the Irrigation Branch felt that they had been forced to make concessions to a public ignorant of the facts of the situation. Not surprisingly, in early 1917, the government retracted its accommodations while Meston was away attending the Imperial Conference in London. Altering the plan, they proposed to divert the new channel running from the regulator to a point on the original supply channel downstream from Har-kī-paurī Ghāṭ (see figure 1). By redirecting the route of this new diversion, they could argue that the new channel would not mix with the sacred stream running past Har-kī-paurī. In March, the Government of India authorized the Uttar Pradesh government to inform the All-India Hindu Sabha that an alternative arrangement was being considered. But the Uttar Pradesh government did not make the announcement and the Irrigation Department proceeded, without public consent, to demarcate a line for the new channel on Laljiwala Island.

Noticing the work on Laljiwala Island, the Hindu community protested to Meston, who had by that time returned from London. Meston wrote to the Government of India to condemn the government's departure from the December 1916 agreement. The Sri Ganga Hindu Sabha of Haridwar followed with a similar plea. Surprisingly, on 20 September 1917, the Government of India agreed to return to the 1916 resolution.

That resolution put in place a standing order that the Irrigation Department has since used to guide its engineering and public works projects. In 1995, the executor engineer of the Northern Division Ganga Canal in Roorkee pointed to the resolution in Appendix 1 of the *Standing Manual for River Ganga and Ganges Canal* and claimed to be following the order. The minutes of the 1916 meeting printed in the appendix included the guarantee of an uninterrupted flow of the Gaṅgā through Har-kī-paurī. Above the *ghāṭ,* where the supply channel branches off from the main stream, the Irrigation Department is permitted to construct a *temporary* dam to close off the stream during the cleaning and repairing of the canal. Second, the head of the new supply channel cannot be fitted with gates. Several bays in the regulator should remain permanently open to allow the "free" flow of Gaṅgā. In the weir on the main stream, a free gap would extend down to floor level and allow a minimum discharge of four hundred cusecs during the cold weather when the river is low. They also agreed to maintain a pool of four feet in Brahmā Kuṅḍ.[25]

At the sacred bathing *ghāṭ* in Haridwar today, strict rules regulating public behavior are enforced by the Sri Ganga Sabha. The Sri Ganga Sabha, whose origin can be traced to that crucial December meeting in 1916, is a trust which maintains facilities for pilgrims and represents the interests of the *paṇḍā*s of Har-kī-paurī Ghāṭ. Originally established to protect the sacrality of Brahmā Kuṅḍ, today the Sri Ganga Sabha assumes exclusive control over the management of the material affairs of Har-kī-paurī. The trust employs a staff of over one hundred fifty to clean the *ghāṭ* daily and enforces rules for public use of the sacred space. Shoes and sandals cannot be worn on the *ghāṭ,* and photography, smoking, shaving, cooking, using oil or soap, and fishing are prohibited. The sabha runs a free first aid center and ambulance service for pilgrims, provides cold drinking water during the summer months, and has volunteers on hand to assist the elderly and women with children. The sabha is financed by donations given during *āratī,* a fire lamp offering to Gaṅgā made daily at sunrise and sunset.

The Transcendent

For nationalists opposing colonial rule, the demand for Gaṅgā's unobstructed flow was a stratagem that revised a key cultural symbol to oppose political rule. This demand was not part of an environmental movement; rather, it invoked religious symbolism to bring religious leaders and nationalists together on a common platform. In the postcolonial present, by contrast, my discussions with religious leaders and ritual specialists have shown that they associate the notion of pollution with government attempts to interfere in the domain of religious affairs. By extension, they use the notion of Gaṅgā's sacred purity to defend their religious ideology against scientific ideology and interventions by government bureaucracy. This gives their interpretations of purity a defensive tone. Environmentalists and foreign observers interpret this position as general disinterest in "preventing pollution" and getting involved with secular or civic cleanliness.

This does not mean that political leaders, on the other hand, have stepped in to exploit the ideological field. Although they have taken up the issue of "cleaning" or "saving" the river on several occasions, they have allowed this message to sink into a bureaucratic maze that works to dispossess the public of a role in civic cleanliness. Government officials have made references to the spiritual significance of Gaṅgā when explaining the importance of their cleaning programs; however, like religious leaders, they have failed to make the problem of river pollution a significant public issue.

In the mid-1980s, Rajiv Gandhi created the Ganga Action Plan, the first government pollution prevention scheme to rid the sacred river of industrial and domestic sewage. In Varanasi, at the inauguration of this central government plan, Rajiv Gandhi pledged to revitalize the Gaṅgā's sacred place in Indian society by setting to work to "clean" the river. Since 1986, programs under the Ganga Action Plan have installed sewage disposal and treatment systems in many large cities bordering the river. During this time, many residents watched this implementation and had the opportunity to hear and read about the views of Ganga Action Plan officials.[26] Several nongovernmental organizations were also established during this period to oppose what they saw as the consolidation of government power in waste management.[27] Meanwhile, residents grew vocal in their skepticism of the government's touted successes in cleaning the river.

In 1995, religious leaders briefly publicized the issue of pollution prevention only to avoid following through with any plan of public action. In 1995, the VHP organized a pilgrimage they called the Ekatmata Yatra, and agreed to incorporate the pollution message suggested by the Kanpur environmental activist introduced earlier. The Ekatmata Yatra, or "All-India Harmony Expedition," originated in several different parts of the country and converged in Nagpur, Madhya Pradesh, for a final rally. Branches of the pilgrimage started at Rameshwaram and Kannyakumari in Tamil Nadu, Pashupatinath in Nepal, Gangasagar in West Bengal, Somnath in Gujarat, Parashuram Kund in Arunachal Pradesh, Amritsar in Punjab, Mahavirji in Rajasthan, and Haridwar in Uttar Pradesh.

The aim of the rally was to teach Indians about the importance of Mother India, Mother Cow, and Mother Gaṅgā. In their mandate, the religious leaders claimed that the pilgrimage would bring unity to a

Hindu community that had "fallen asleep" and remind Hindus of the importance of the cow for agriculture and Gaṅgā for the advent of Hindu civilization. Taking this mandate further, they linked the goals to their 1984 promise to destroy the mosques in Ayodhya and Kashi and the idgah in Mathura. Dressing these goals in more symbolic language, the VHP associated each sacred site with the sacred river it bordered: the Sarayū with Ayodhya, the Yamunā with Mathura, and the Gaṅgā with Kashi (Varanasi). Gaṅgā, the archetype of all rivers in India, was a key symbol in their tripartite message. Their published mandate for the event stated:

> The presiding deities of all these yatras will be the three Mothers, Ganga Mata, Bharatmata and Gowmata, whose images will be adorning the motorised chariots. In these Yatras, there will be chariot programmes, worship of Ganga kalash (pot), worship of cow-along-with-her-calf [sic], and worship of the sacred soil of India (sacred soils of Sri Ram Janma Bhumi, Ayodhya, Sri Krishna Janmasthan, Mathura, and Sri Kashi Vishwanath, Gyanvaapi, Varanasi). These worships will be conducted in 300,000 villages. There is the possibility of this number increasing.[28]

Among the objectives of the *yātrā* was the aim to: "build up assertive public awareness to ensure maintenance of sanctity and purity of the Ganga, and also awareness against environmental pollution." Yet, this had little to do with the dominant concern of the mandate, which clearly set out the Hindu community's enemy and used this identification with a common hatred to fuel the *yātrā*.[29] What is significant here is that this common Muslim enemy, while cast as the other in the struggle for sacred space, fades away when the focus shifts to the sacred purity of the Gaṅgā. Hindus have not cast Muslims as threats to Gaṅgā's purity. Moreover, Muslims are not enemies of the river. They also bathe in the Gaṅgā and offer religious donations *(dān)* in the form of fish, what they call *machlīdān*. Although Hindus in Varanasi may complain that Muslims consider Gaṅgā a mere river (by calling her *nadiyā*), they do not focus on a hatred of Muslims to defend Gaṅgā's purity. So any environmental message, howsoever it might revision notions of sacred purity and divine power, cannot build upon the sentiment wrapped up in the VHP's opposition to the Muslim other.

In Haridwar, at the beginning of one part of the Ekatmata Yatra, leaders of the VHP gathered with swamis, *śaṅkarācārya*s and members of parliament on Har-kī-paurī Ghāṭ to inaugurate the event.[30] Afterwards, pilgrims and tourists who had gathered to watch the *pūjā* took *darśan* (auspicious sight) of the spiritual leaders and registered their moral support. In the events that followed, as many predicted, the *pūjā*s spoke more to the goal of altering the landscape of Islam than saving Gaṅgā.

Despite the public's general awareness of the Ganga Action Plan and of secular calls for pollution prevention, by 1995 many environmentalists complained that a curious kind of apathy plagued the public. This "apathy" for the problems of material pollution, they argued, was most noticeable among religious leaders and pilgrim priests who promote pilgrimage to the banks of the Gaṅgā. Early in the year after the Ekatmata Yatra, I followed up this claim by interviewing religious leaders and pilgrims who had gathered in the month of Māgha (January–February) for religious observance and spiritual healing on the banks of the Gaṅgā in Allahabad. At the confluence of the Gaṅgā and Yamunā Rivers, *śaṅkarācārya*s, swamis from large missions and ashrams, and *brahmacārī*s and *sādhu*s gathered to live by the Gaṅgā, cook with *gaṅgājala,* do daily ablutions in the Gaṅgā, and perform *yajña*s. Through these rituals, they sought the grace of Gaṅgā and the power to bring about the purification of mind and soul.

Discussions with religious leaders at the Māgha Melā in 1995 helped to explain the conceptual dissonance between religious and environmental worldviews, but they also pointed toward other obstacles, both in ideology and praxis, that prevented an instrumental use of rhetoric about Gaṅgā's purity. I asked religious leaders to respond to questions regarding the extent to which they thought material pollution impacted the sacred purity of the river. When responding, several respected leaders refrained from mixing what they saw as two unrelated subjects: Gaṅgā's transcendent purity and the disintegration of the tangible or temporal world (represented here by "environmental pollution"). Like VHP leaders, these religious specialists also claimed distance from government agencies and the political process at large when they refused to link the protection of Gaṅgā's purity to their own organizational or occupational activities. While many leaders acknowledged that industrial pollution and domestic waste were problems that civic authorities needed to deal with, they did not consider it their responsibility to assist officials in cleaning or protecting Gaṅgā from material pollution. Their role was to oversee the spiritual worship of Mother Gaṅgā.

On the other hand, many alluded to the need for moral regeneration in contemporary society, a regeneration that could have positive effects for civic cleanliness. They stressed the importance of becoming "workers" in the quest for Gaṅgā's grace. Becoming workers involved casting away desire and practicing nonviolence and celibacy (*brahmacāryā*). These commitments could bring one closer to Gaṅgā's grace and to the regeneration of the moral order. A swami from the Rama Krishna Mission explained:

> We should become workers to obtain the ever-flowing grace of Gaṅgā. There should be religious conduct giving birth to truth; birth should be given over to nonviolence, and when one comes for residence (*kalpvās*) on the banks of Gaṅgā one should become a full *brahmācārya*. Stay away from enjoyments or unnecessary things. Do this so your conscience can be pure, so you can be a worker, so you can get Gaṅgā's grace, free from illusion. Like Lord Śaṅkarācārya said, Who is a worker? Who is a worker for grace? They say there are four forms of worship. So what is the first? The first is being full of understanding, the second is having the benefit of sons and service to the cow; the third is the eternal peace (*samādhi*) of saints; and the fourth is salvation. Why? Because we should get salvation (*mokṣa*) from the grace of Ma Gaṅgā.[31]

The meaning of Gaṅgā's purity developed by Hindu nationalists in the colonial period involved the significance of unobstructed flow. By contrast, while many residents of Varanasi and Allahabad commented that Gaṅgā's flow accounted for part of her purificatory power, they made no mention of dams or other obstructions. Rather, Gaṅgā's flow was generally valorized for its self-cleaning power, its mixing or hydraulic effect. One *paṇḍā* who lives and works at Daśāśvamedha remarked, "Gaṅgā cleans herself during the monsoon" (*Gaṅgā khud hī sāph kartī hai*). During the monsoon, as many told me, Gaṅgā climbs up the ghat steps and takes away the silt and *gandagī*. A merchant explained, " From the scientific view, there is pollution, but I do not understand this. Gaṅgā takes the pollution away in floods." Leaders at the Māgha Melā, likewise did not mention the connection between purity and unobstructed flow. Instead, they argued that Gaṅgā's purity was distinctly removed from the transformations taking place in the material or transitory world. By extension, her purity cannot be described using the scientific or materialist notion of pollution put out by government agencies and some environmental organizations. The transcendent cannot be collapsed into the temporal or material. The *śaṅkarācārya* of Jaganath Puri explained:

> *gaṅgājala* gives unparalleled contribution to spiritual and bodily purification. Even scientists give a high position to *gaṅgājala* when comparing it with water from other rivers. Water from the faucet or a channel spoils very quickly compared to *gaṅgājala*. *Gaṅgājala* is never spoiled. *Gaṅgājala* is not impacted by the process of degeneration or defilement (*vikritiyo ko dūr kar*); there is a capacity in *gaṅgājala* for mixing. So, one should reside at the places where *gaṅgājala* has always been pure, and perform religious service (*sevam karnā*) as it should be done according to the śāstras. Gaṅgā must also be pure on the outside. When taking *darśan,* a devotee in his heart cannot see defilement. This is part of the responsibility of having devout sentiment. Also, from the scientific perspective, scientists welcome the importance of *gaṅgājala*. So the outer form of Gaṅgā should also be kept clean. They should not be opposed to purity (*śuddhtā*). When one is opposed to cleanness (*svacchtā*), then one's faith in purity falters. This means cleanness (*svacchtā*) is one thing and purity (*śuddhtā*) is another. We make soap from the oil of animals. Some people use it to bathe, to clean clothes. This is the way to make your life clean, but not to make it pure. Where *gaṅgājala* is pure (*śuddha*), there, on the outside, one's faith is secured. The eyes are convinced. One's faith is also formed by drinking *jala*. Therefore it is important that we pay attention to her cleanness. In whatever place (*sthān*) Gaṅgā is important, and wherever Yamunā is important in her place, and wherever the hidden form of Sarasvatī is important, and where all three meet at a confluence (*saṅgam*), this conception is manifested in the Vedas. Therefore, from the spiritual perspective, we worship the boundless importance of the *saṅgam*.[32]

The *śaṅkarācārya* was responding to my question about the distinction between purity (*śuddhtā*) and cleanness (*svacchtā*). To this, the *śaṅkarācārya* responded by arguing that purity (*śuddhtā*) is associated with Gaṅgā's transcendent power and cleanness (*svacchtā*) with her outer form (*bāhar se*). Gaṅgā's transcendent purity cannot be collapsed into her outward form; rather, the two are necessary complements to each other. The outer form is important to a pilgrim's faith in Gaṅgā's grace because it convinces one of truth through visualization. The outer form figures into Gaṅgā's purity, making cleanness (*svacchtā*) complementary but not conditional to purity (*śuddhtā*). While this distinction is multivalent both linguistically and culturally, a point I dealt with in my article "Idioms of Degeneracy: Assessing Ganga's Purity and Pollution," it is important here as well because it allows the *śaṅkarācārya* to take the idea of separation further into distinguishing between the domains of religion and science.[33] First, he continued, religion is separate from but not opposed to science:

> But about the matter of science, our religion is not opposed to it nor uncongenial towards it. We support scientific research. Like the matchbox. . . . But in our active moral work (*kriyā karm)* we will respect our own *dharmaśāstra*s. In worldly pursuits to the point that science is helpful, I welcome the usefulness of science. Scientists have created electricity; they have given computers, rockets. We don't oppose this. Opposition occurs when there is conflict (*takrār*). They are not one subject but separate subjects. The Vedas gave special importance to one subject and scientists did not give importance to this. Our *śāstra*s dealt with the subject of behavior and our *śāstra*s have freed us from bodily suffering and worldly affairs. Because in the end, in plight or prosperity, these are the two benefits of our faith (*dharm*).

Therefore, separate expertise is required to understand each realm. He continued:

> For example, we can ask, do you believe in your eyes or your ears? Do you understand? To see the essence of form, do you believe in your eyes or your ears? What answer will we give? On our eyes! Because the understanding of form is based on the eyes. The eyes are supreme in relation to form. So someone may say that in regard to sensation (*sparś*), some worship with the eyes and others do not. If knowledge of sensation is possible in other ways than through the eyes, then isn't the subject of science separate? The subjects of religion (*dharm*) and the *śāstra*s are separate (*alag*) from science. When both subjects are separated (*alag alag*), then the contributions of both in their own fields can be appreciated. The lawyer is important in the court; the astrologer is important for making horoscopes. Both are important in their own right. But when we say that an astrologer should do the lawyer's work or if we tell a lawyer to become an astrologer, it will not work.

By advocating the separation of religion and science, this religious leader claimed that practitioners of each profession have their own rights and duties that must be appreciated and protected; one cannot do the work of the other. Without directly referring to the problem of power, he implied that the integrity of each expertise should be based on respect for each sphere of influence. This sense of separation was also echoed in the casual complaints of many ritual specialists on Daśāśvamedha Ghāṭ when they argued that cleaning was not their work but the work of the municipality and state agencies such as the Jal Nigam. These distinctions between occupational specializations can be taken as metaphorical extensions of the specialization of duties within the caste hierarchy. The principle of non-overlapping functions that Frédérique Apffel-Marglin and Robert Lingat described might be applied to the domains of religious leaders/ritual specialists and government officials.[34] In Apffel-Marglin's discussion, separation and wholeness are relevant in the ascending order of caste; when inferiors cross caste boundaries, their behavior invokes pollution, chaos, or confusion. Thus, high caste groups in particular have symbolic interest in keeping duties, food, articles, and other personal effects separated from the reach of lower caste groups. This separation allows them to appear autonomous and independent from other groups. On the other hand, these separations may be taken much less abstractly, as a rejection by upper caste Hindus of contact with lower caste occupations, such as sweeping and cleaning, that deal with impure substances.[35] What these data show, however, is that while these two reasons are tenable, they do not completely explain the other tensions that characterize relations between citizens and the government.

In a broader sense, the *śaṅkarācārya*'s idea of separation is congenial with the notion of separation advocated by Swami Chidananda, leader of the Divine Life Society. In 1995, Swami Chidananda gave a lecture to devotees and environmental activists on Gaṅgā's birthday (Gaṅgā Daśahrā) in the Himalayan town of Gangotri. In his talk, he advocated separation between the people and government by making a more direct reference to power and struggle. He said:

> People say they come to destroy mosques and temples. At that time [of the conflict in Ayodhya], there was so much passion generated. In one day, from the Himalayas to Kanya Kumari, from Arunachal and Meghalaya to the Punjab, they carried this passion. Why don't they have this much passion for saving the Gaṅgā and for the movement to save other rivers for the next generations? Why not? I have said that passion should be applied because humanity takes direction from love or hatred. This is the petrol and diesel used to move men. There in Ayodhya the passion is directed against Muslims. They are engrossed in anti-Muslim sentiment. It is a political thing. So everyone works for

> this passion. But here there is not this kind of passion because if opposition (*takrar*) occurs, it must occur with the government. But at every step we need the grace of government: for a permit we need the government, for information, we need them, for everything. We don't want to need the government! It's like that. There is a *concept* that we don't need any pomp and show from government. Go ahead and take what you want from government but I am ready for a hunger strike. I don't want anything from *sarkar* [government]! Then why should I care for *sarkar?* When we want something from them then our internal structure becomes weak. When we don't need anything we are strong. So without desire, with connection to God, we can sacrifice our own lives.[36]

Swami Chidananda alluded to the pervasiveness of Indian bureaucracy and the embeddedness of bureaucratic power in everyday life. While rejecting government assistance, he proposed that activism proceed outside the web of bureaucracy. Deriving his vision of resistance from the emergent model provided by Mahatma Gandhi, he explained that Gandhi had a soul force (*ātma bhāv*) that life workers should also generate. This soul force, the swami explained, comes from direct communication with God, a communication Gandhi achieved by reading the *Bhagavadgītā*. Connection with the "cosmic, infinite, inlimitable [sic], boundless, immeasurable divine power that you call God (*adiśakti bhāv, parāśakti bhāv, mahāśakti bhāv*)," he proposed, "will give you the power to persevere and the power to succeed." Since a "blind and noncomprehending attitude in the minds of the bureaucracy" was preventing real communication and undermining Hindu life and culture, *bhaktī-bhāv* and *visvās* were needed to create work with moral vision.

Environmental activists argue that this kind of opposition to bureaucracy, while explicitly advocating moral action, has led the public to pursue a kind of passive resistance to government power. For example, the *paṇḍā*s of Daśāśvamedha often blamed the government for failing to clean the river under the Ganga Action Plan and for creating institutions that remain unresponsive to citizens' needs.[37] This allocation of blame was a kind of everyday resistance for them.[38] They complained that in a field controlled by a government that blocks citizen action, they can do very little. While Swami Chidananda agreed that the web of bureaucracy is pervasive and oppressive, unlike the *paṇḍā*s, he advocated an active reform, taking assistance from divine power and proceeding independently from official government.

However, the tradition of religious mobilization has never exhibited efforts to engage in any form of environmental activism. As I suggested earlier, the symbolic dissonance between religious and environmental worldviews has tended to discourage such an interchange. Eulogies to Gaṅgā and worship of *gaṅgājala* are central sacred symbols in pilgrimages and in *pūjā,* but they do not describe *gaṅgājala* as a finite resource whose contours are shaped by a larger ecosystem.[39] Moreover, religious rituals cannot be confused with civic ethics. *Pūjā* and pilgrimage, according to their pristine meanings, are not undertaken to generate social effects. Rather, they are the means to communicate with and seek grace and power from a divine force, a presence that predetermines any social effect that may arise from ritual.

On the other hand, we find historical evidence for the political appropriation of the symbolism of Gaṅgā's purity, an appropriation that served the purpose of mobilizing religious and political leaders under a common political cause. But this hardly serves as a historical precedent for political use of purity rhetoric, because the symbol of Gaṅgā's purity has not played a consistent role in religious mobilization during colonial or precolonial periods. The appropriation of symbols of purity and unobstructed flow by Hindu nationalists in the colonial period was a strategy that, while serving the purposes of a broader political movement, ended in an agreement between government, scientific, and religious communities. In the current period, movements for reclaiming sacred spaces reproduce models of religious mobilization that successfully use religious symbols for political effect, yet leaders have not transferred the dynamism of these movements into environmental programs to save or clean Gaṅgā.

Shifting to ecosystem factors, the relative absence of political rhetoric about Gaṅgā's purity during precolonial and colonial periods may have reflected the vast powers of the river ecosystem.[40] Akbar's praise of *gaṅgājala* and his commitment to storing and drinking the river water spoke to the munificence of the river system. Before this, eulogies to Gaṅgā in the Purāṇas celebrated her bountiful virtues and Gupta sculpture depicted a voluptuous, motherly Gaṅgā.[41] The image was that her flow was abundant and powerful, providing for all without strain or limit. Today, some Hindus suggest limits to Gaṅgā's purificatory power when they refer to her "capacity" (*ksamtā*) to purify or take away worldly dirtiness and impurity. Patterns of resource use in India put all rivers in a new predicament and force residents to acknowledge that waste is having some kind of impact on the Gaṅgā. Part of the way religious leaders deal with ecological change is to outline a separation between the domains of Hinduism and science (and with science, environmental pollution) and Hinduism and politics. This does not simply reproduce an opposition between tradition and modernity because, at another level, modernity is already incorporated into their position.[42] By defining Hinduism and science and Hinduism and politics as separate areas of specialization, the contradictions between their cultural logics can be momentarily elided, minimizing ideological conflict at a time when resources are in great demand. The problem for environmental activists is that this position of relative remove from active civic clean-up work aggravates the ecological health of the river.

Religious leaders and ritual specialists alike acknowledge that pollution prevention, like struggles to protect Gaṅgā's purity, involves competing claims to resources. While categories in the literature on natural resource management put the Gaṅgā somewhere along the continuum between an open-access resource and common property, a closer look reveals that Gaṅgā defies such a categorization.[43] This is because there are explicit and implicit (or sanctioned and subversive) institutions that govern use of and access to the river. At the explicit level are the departments under the Government of India that deal with resource management and the judiciary that enforces their policies. These are the dominant institutions. Religious institutions—sectarian organizations, temple committees, trusts, and public service agencies—are the implicit players, not recognized by the government as legitimate resource managers but respected by the public as moral authorities.

The government regulates uses of the river at the legal level, by restricting navigation, fishing, the building of dams or other obstructions, and the extraction of water in specified zones. Moreover, officials can exact penalties for dumping waste and effluent under the laws of the Pollution Control Boards.[44] However, these legal restrictions do not limit religious uses of the river and do not provide the only codes of "resource management" understood and respected by the citizenry. Procedures for performing ablutions in the sacred texts (*śāstra*s), for example, were rules that in some measure set out to regulate use of a common resource. Rules encoded in textual traditions and folk and oral narratives have taught devotees how to perform ablutions, how to think and act while on pilgrimage to Gaṅgā, how to enter sacred space, how to transport and store *gaṅgājala,* and how to use it in *pūjā* and other rituals. Priests at places of pilgrimage today loosely "enforce" these prescriptions by preaching a similar set of instructions to pilgrims. They do not punish those who make minor transgressions, but teach and direct them. In addition to these cultural traditions, competing definitions of what waste is today make it difficult for officials to persuade the public to follow government policies or judicial orders. Religious leaders do not consider that ritual materials immersed in Gaṅgā—the statues (*mūrti*) of gods and goddesses and the saris and religious books offered to her—contribute to the waste problem in any significant way.

Yet, these freedoms from government control do mean that, despite their relative autonomy, the one set of claims is tied up with the other.[45] Contenders of the political-legal right to regulate behavior by punishing transgressions of the law and those upholding the religious-moral right to teach and direct public behavior in the practice of ritual must measure their respective powers through a kind of imagined compromise of power with the other. By proposing such a coexistence, however, the one set of rights becomes divorced from the other. The result is that the government that holds the legal mandate to enforce a code of conduct has little moral or social power. Government officials buckle under the pressure of mass movements such as pilgrimage and are unable to exact the punishments they are, in theory, empowered to exact. Religious leaders, while teaching and directing behavior, have no power to enforce conformity to the code they uphold and are reluctant to regulate the hand that feeds them. Religious and political leaders who avoid coming together in the name of preventing pollution share limitations to their power. Seeing Gaṅgā through different lenses, the respective groups keep their distance and disregard and discredit each other's traditions and bureaucracies.

There are, however, religious leaders and ritual specialists who defy acknowledgment of the impact of industrial modes of resource use and refuse to accept the possibility of a capacity to Gaṅgā's purifying power. Pollution is, for them, an altogether erroneous notion. Gaṅgā's powerful flow and motherliness attest to its fallacy. Eternal power (*acyuta viṣay*) cannot be understood through the eyes that affirm science, as the *śaṅkarācārya* said. Because this knowledge, as

Swami Chidananda put it, comes from direct communication with God, an understanding of Gaṅgā cannot be collapsed into knowledge of material transformations. It cannot be approached in the way one washes dirt off the body with soap. It must be done through a holistic commitment to *dharma*. Scientific treatments to "clean" the river are only as good as soap; they cannot reach or transform divine power. Cleaning the river and gaining knowledge (*jñāna*) from Gaṅgā are separate pursuits and should remain in their place.

At the end of the interview with the *śaṅkarācārya* from Jagannath Puri, he stressed that the vectors of space and time (*deś aur kāl*) intersect with any manifestation of sacredness. This is the visual truth (*darśnik tathya*) that underlies Gaṅgā's importance. Gaṅgā's power is especially apparent at Prayāg, during the month of Māgha, on the last day of the dark half of the month (*amāvasyā*) and on the full moon day (*pūrṇimā*). Her power intersects with radiant place (*ujjval deś*) and radiant time (*ujjval kāl*). This does not mean that Gaṅgā is sacred only at times of cosmic juncture that produce auspiciousness, but that her power can be accessed most readily by mortals when rituals are performed in radiant place and time. This does not reflect Gaṅgā's intermittent sacrality, but reflects a tradition that uses time and place to organize access to sacrality.[46] Gaṅgā occupies a fixed sacrality beyond space and time (*acyut*). She is therefore imperishable in both space and time.

Many scholars have argued that Western notions of sacred and secular cannot be applied to Indian religious and political circumstances.[47] Yet, despite its inappropriateness, the foreign model of separation has been adopted by some modern Indians. Government and its secular ideology, by negating religion or attempting to remove religious power from public life, have made little headway in India. Rather, secular ideology remains "encompassed" by the religious.[48] Other scholars, however, have explained the cultural roots of this encompassment to raise a warning flag. Indeed, religious institutions, with their capacity to teach and direct, are making the far more significant strides in mobilizing the public, while government powers to regulate public behavior remain largely ineffective. But in defense of Gaṅgā's purity, religious ideology does not encompass, but pulls away from, secular and scientific ideologies to achieve a kind of territorial integrity. Religious leaders pull away from political leaders to prevent government intervention in religious affairs.[49] They do not appropriate the political process to achieve a movement in religious consciousness. To remain powerful in their own right, religious leaders maintain claims to Gaṅgā in distinctly transcendent terms, distinguishing politics, science, and environmental pollution as pursuits of a separate and more mundane order.

Prognosis

Possibilities for collaboration among religious and political leaders in the area of environmental activism might arise if the distinction between purity and cleanness were used in a sensitive manner to mobilize Hindu sentiment. However, the boundedness of the domains of Hinduism, politics, and science/environmental pollution that many religious leaders articulate will likely act as a formidable obstacle to any kind of coordinated coalition building. In the colonial period, nationalist leaders and religious leaders worked together to resist the canal projects engineered by a common enemy, the British government official. An opposition to a common enemy assisted in the process of unification. In the postcolonial period, political and religious leaders no longer have a common enemy to oppose when they discuss the importance of the river Gaṅgā. The Muslim "other" staged in struggles over sacred space fades away in discussions about Gaṅgā's sacrality. Moreover, contemporary political leaders, unlike nationalist leaders, negotiate a dynamic field of power. Their grasp on political positions is often tentative and short-lived. Therefore, ideological struggles must produce quick results and, to do so, must draw upon a strong base of public sentiment. Since the resistance to canal modifications at the turn of the century, no other concerted efforts to protect Gaṅgā's purity and sacrality have evolved in the modern historical record. The particular nationalist resistance I described did not move other leaders toward the creation of a robust field of public activity and sentiment equal to the struggles over sacred spaces that have moved political leaders in colonial and postcolonial periods. I suggest that if environmental activism continues to develop in India, it will likely expand out of coalitions between nongovernmental activists and public interest lawyers long before it does so at the crossroads of religious and political movements.

Notes

1. See Peter van der Veer, "God Must Be Liberated! A Hindu Liberation Movement in Ayodhya," *Modern Asian Studies* 21, no. 2 (1987): 283–301; van der Veer, *Religious Nationalism: Hindus and Muslims in India* (Berkeley and Los Angeles: University of California Press, 1994); and van der Veer, "The Ruined Center: Religion and Mass Politics in India," *Journal of International Affairs* 50, no. 1 (1996): 254–77.

2. The Bharatiya Janata Party (BJP) rule in state governments is as follows. In Madya Pradesh they ruled from 1990 to 1992. In Uttar Pradesh they held the government from 1991 to 1992. Again in Uttar Pradesh in 1995, the Bahujan Samaj Party (BSP) split with the Samajwadi Party (SP) with BJP support, but it was a BSP government. In 1996, the BJP formed another coalition after elections with the BSP, but then split from the BSP to head the government. In Rajasthan, they held power from 1990 to 1992 and from 1993 to 1998. In Maharashtra, they were in power from 1995 through the writing of this article in early 1999. The governments that ended abruptly in 1992 were ends imposed by the central government after the demolition of the Babri Masjid.

3. The events in Ayodhya in 1992 brought a great deal of attention to bear on the issues of religious nationalism, secularism, and communalism. For discussions of these issues, see: Ajit Kumar Jha, "Hindutva Swadeshi: Waves of Economic Nationalism," *Times of India* (New Delhi), 6 September 1995; T. N. Madan, "Whither Indian Secularism?" *Modern Asian Studies* 27, no. 3 (1993): 667–97; Barbara Stoler Miller, "Presidential Address: Contending Narratives—The Political Life of the Indian Epics," *Journal of Asian Studies* 50, no. 4 (1991): 783–92; Subrata Kumar Mitra, "Desecularising the State: Religion and Politics in India after Independence," *Comparative Study of Society and History* 33, no. 4 (1991): 775–77; Manju Parikh, "The Debacle at Ayodhya," *Asian Survey* 33, no. 7 (1993): 673–84; Vidya Subrahmaniam, "Redefining Secularism: Gap between Theory and Practice," *Times of India* (New Delhi), 22 September 1995; Ramesh Thakur, "Ayodhya and the Politics of India's Secularism," *Asian Survey* 33 (July 1993): 645–64; Prakash Chandra Upadhyaya, "The Politics of Indian Secularism," *Modern Asian Studies* 26, no. 4 (1992): 815–53; and van der Veer, "The Ruined Center: Religion and Mass Politics in India." However, comments by T. N. Madan in his article "Secularism in Its Place," *Journal of Asian Studies* 46, no. 4 (1987): 747–59, several years earlier, make it clear that secularism has been a conundrum for scholars and nationalists since the independence of India and Nehru's first designs for national policy.

4. See the *Statesman* (11 August 1995, 12 August 1995, 14 August 1995, 15 August 1995, and 18 August 1995, p. 1) and the *Times of India* (14 August 1995, p. 1, and 24 August 1995) for news reports about the VHP's agenda in Mathura around the time it planned and executed a *māhāyajña* on Janamāṣṭamī (Kṛṣna's birthday).

5. See Kelly D. Alley, "Idioms of Degeneracy: Assessing Ganga's Purity and Pollution," in *Purifying the Earthly Body of God: Religion and Ecology in Hindu India,* ed. Lance Nelson (Albany: State University of New York Press, 1998).

6. Pilgrim priests on Daśāśvamedha, as well as religious leaders I introduce later in the text, tended to focus discussion of Gaṅgā's power on sacred purity. They did

not elaborate on Gaṅgā's fertility or auspiciousness or the impact waste might have on these conditions. This contrasts markedly with the statements of informants in Maharashtra gathered by Anne Feldhaus in *Water and Womanhood: Religious Meanings of Rivers in Maharashtra* (New York: Oxford University Press, 1995). Her informants have associated sacred rivers with fertility and their powers with auspiciousness. While I have not looked at this contrast systematically yet, I would venture the argument that the significant male bias in the discourses I represent here contributes to the emphasis on purity over fertility.

7. The fish market just behind Daśāśvamedha Ghāṭ is the biggest culprit. Vendors apparently let their waste roll through open drains down the embankment into the Gaṅgā. This attract flies and other insects to the area.

8. See Alley, "Idioms of Degeneracy," for a discussion of *gandagi*.

9. By environmental pollution, I am referring to the modern scientific notion and not in any way to the anthropological notion of ritual impurity or pollution. Ritual pollution can, however, occur in conditions of material or environmental pollution and when doing so can make the distinction between the two harder to define.

10. See F. Fitzjames, *Preliminary Report of the Sewerage and Water Supply of the City of Benaras,* Government of NW Provinces and Oudh, Public Works Department (Allahabad, 1888); quoted in Kelly D. Alley, "Ganga and Gandagi: Interpretations of Pollution and Waste in Banares," *Ethnology* 33, no. 2 (1994): 132. See also Nita Kumar, *Friends, Brothers, and Informants* (Berkeley and Los Angeles: University of California Press, 1992).

11. M. A. Parmanand, in his book *Mahamana Madan Mohan Malaviya: An Historical Biography* (Varanasi: Malaviya Adhyayan Sansthan, 1985), 244, summarized the Hindu sentiment toward canal engineering in the following passage:

> The reverence which the Hindu feels for the Ganga and its water is equalled only by his veneration of the Brahman and the cow. He believes Ganga water to be so holy that, let alone a bath or ablution, even a sip, touch or sight of it, washes away sin. He cannot however ascribe the same virtue to water drawn from the Ganga into a canal which, in his eyes, does not possess the same virtue.

12. W. W. Hunter, *The Imperial Gazetteer of India,* vol. 4 (London: Trubner and Company, 1888), 1–5.

13. A *ghāṭ* is a flight of steps that provides access to the river. It is also considered a place of sacred power.

14. This is the site of the Kumbha Melā, a festival that takes place once every twelve years in Haridwar, Allahabad, Nasik, and Ujjain.

15. Walter Hamilton, *East-India Gazetteer,* 2d ed., vol. 1 (Delhi: Low Price Publishers, 1993 [1828]), 667-669.

16. Hunter, *Imperial Gazetteer of India,* 4, and Dangli Prasad Varun, *Gazetteer of India—Uttar Pradesh, District of Saharanpur* (Lucknow: Government of Uttar Pradesh, Department of District Gazetteers, 1981), 336.

17. Over the last quarter of the nineteenth century, irrigation canals in Roorkee District have effectively combated drought (see Varun, *Gazetteer of India,* 112). In

contrast to the drought of 1868–69, when almost all the autumn crops were lost except those in the irrigated region, in 1870, 162,317 acres in Saharanpur District were irrigated, out of which 84,404, or more than half, were watered from canals alone (see Hunter, *Imperial Gazetteer of India,* 104). In 1884–85, canals supplied water to 71,916 acres (29,100 hectares). This increased to 80,724 acres (32,668 hectares) in the following decade and to 1,21,550 acres (49,190 hectares) annually thereafter.

18. Parmanand, *Mahamana Madan Mohan Malaviya,* 254.

19. Ibid., 245.

20. Ibid., 247.

21. Ibid., 247–48.

22. Chief Secretary, Government of Uttar Pradesh, to Home Department of India, 5 February 1917; cited in Parmanand, *Mahamana Madan Mohan Malaviya,* 250).

23. Parmanand, *Mahamana Madan Mohan Malaviya*, 252.

24. Ibid., 254–55.

25. The appendix reads:

> I. To guarantee an uninterupted flow of the Ganges through Har-ki-Pairi and past the other ghats of Hardwar, Katchha (temporary) bund will be made at the head of channel no. 1 when necessary a minimum supply of 1000 cusecs in that channel being guaranteed except at periods when clearing of the shoaling in channel no. 1 is in progress; the Irrigation Branch undertaking that this work will be carried through as expeditiously as possible in order to ensure a flow from this channel into the Har-ki-Pairi.
>
> II. The opening to supply channel no. 1 will be left for the present exactly as it is. Should experience show that this is dangerous, it may be necessary to curtail the width of the present opening and to take measures to prevent the retrogression of the bed. But no steps beyond these will be taken without prior consultation with the Hindu Community.
>
> III. The head of new supply channel will not be fitted with gates.
>
> Some bays contiguous to each other will be completely closed up with masonry and earth banks, some bays will be completely open. All bays whether open or closed will have a foot bridge. The floor on the Hardwar side will be level with the sell and with the bed of the supply channel. The Irrigation Branch reserves to itself the right of keeping open or closed by masonry as many of the bays as experience from time to time shows, may be necessary for the purpose of feeding the canal which starts at Mayapur and also in the interest of the safety of the town of Hardwar and the existing canal works but such bays as are kept open will be completely open as described above and whatever bays are closed, will be kept completely closed with masonry and banks as mentioned above.
>
> The existing grooves will be left in what is now called the head of the supply channel. These grooves will never be used except in case of some impending calamity of accident to Hardwar or the existing canal works, when it is of vital importance to close completely the openings by means of wooden planks for a short period.
>
> IV. A free gap will be left in the weir which will go down to floor level. The

openings will be so constructed that it will give according to calculations of Irrigation Branch a minimum discharge of 400 cusecs at the cold weather low level of the river. For this purpose a record of gauges will be kept by the Irrigation Branch.

N. At Mayapur regulator a free opening going down to the upstream bed level will be made calculated to provide a permanent floor of 200 cusecs for the service of the Kankhal ghats, which after leaving Kankhal will ultimately flow into the Ganges.

N.B. The free gaps at Bhimgoda weir and Mayapur Dam may be closed by the Irrigation Department when necessary for the purpose of floor repairs or other emergencies but only for a short time and on informing the Hindu Sabha.

V. A depth of 4 ft. of water in the Har-ki-Pairi pool will be guaranteed by and at the cost of Irrigation Department.

26. Officials with this secular worldview work in the Ministry of Environment and Forests and in the Uttar Pradesh Jal Nigam.

27. See Alley, "Ganga and Gandagi," and "Idioms of Degeneracy," for discussions of the environmental activism of the Sankat Mochan Foundation of Banaras. See also Kelly D. Alley, "Urban Institutions at the Crossroads: Kanpur under the Ganga Action Plan," *Urban Anthropology* 25, no. 4 (1996), for a discussion of the work of Eco-Friends of Kanpur.

28. The memorandum later stated that the aim was: "to get people [to] take pledge to construct stately temples symbolizing our national heritage and moorings at Sri Ram Janma Bhumi, Ayodhya, Sri Krishna Janmasthan, Mathura, and Sri Gyaanvaapi Vishwanath spot, Varanasi" (*Ekatmata Yatra,* October/November 1995, 2).

29. See also Dipankar Gupta, "The Search for Martyrs: Hindutva's New Logic of Imagination," *Times of India* (New Delhi), 2 September 1997.

30. The minister of parliament at that time from Madhya Pradesh, Swami Chinmayananda, Ashok Singhal, Vishnu Hari Dalmia, Śaṅkarācārya Rambhadracarya of Chitrakoot, Śaṅkarācārya Jagadguru Ramanandacarya Ji (the identity of both as *śaṅkarācārya*s is disputed), Haryacarje Maharaj of Ayodhya, Swami Prakashanand Giri Mahasnandedeshwar, and Jagadguru Vasudevanand Saraswati were present.

31. These statements, provided by a swami of the Rama Krishna Mission, were tape-recorded during an interview at the Melā, January 1996.

32. From a tape-recorded interview with Śaṅkarācārya Sri Nischalanand of Saraswati Goverdhan Muth in Jagannath Puri.

33. See Alley, "Idioms of Degeneracy."

34. See Frédérique Apffel-Marglin's discussion of Robert Lingat in "Power, Purity and Pollution: Aspects of the Caste System Reconsidered," *Contributions to Indian Sociology* 11, no. 2 (1977): 252.

35. Frank Korom, "On the Ethics and Aesthetics of Recycling in India," in *Purifying the Earthly Body of God: Religion and Ecology in Hindu India,* ed. Lance Nelson (Albany: State University of New York Press, 1998).

36. Excerpt from a tape recording of his talk. The talk was given in Hindi, but the phrases in bold mark where he spoke in English.

37. Alley, "Ganga and Gandagi," 138–39.

38. I am using the term in the sense James C. Scott describes in his book *Weapons of the Weak: Everyday Forms of Peasant Resistance* (New Haven: Yale University Press 1985).

39. See Feldhaus, *Water and Womanhood: Religious Meanings of Rivers in Maharashtra,* and Peter van der Veer, *Religious Nationalism: Hindus and Muslims in India* (Berkeley and Los Angeles: University of California Press, 1994).

40. See Madhav Gadgil and Ramachandra Guha, *This Fissured Land: An Ecological History of India* (Berkeley and Los Angeles: University of California Press, 1992), for a discussion of broadly generalized modes of resource use in precolonial and colonial periods.

41. See Savitri V. Kumar, *The Pauranic Lore of Holy Water-Places* (New Delhi: Munshiram Manoharlal, 1983).

42. See, for example, the *śaṅkarācārya*'s reference to the possibility for a science that is congenial with the moral work.

43. For discussions of these categories, see Harry W. Blair, "Democracy, Equity and Common Property Resource Management in the Indian Subcontinent," *Development and Change* 27 (1996): 475–99; *Making the Commons Work: Theory, Practice, and Policy,* ed. Daniel W. Bromley et al. (San Francisco: Institute for Contemporary Studies, 1992); David Feeny, Fikret Berkes, Bonnie McCay, and James Acheson, "The Tragedy of the Commons: Twenty-Two Years Later," *Human Ecology* 18, no. 1 (1992): 1–19; *The Question of the Commons: the Culture and Ecology of Communal Resources,* ed. Bonnie McCay and James M. Acheson (Tucson: University of Arizona Press, 1987).

44. See Alley, "Urban Institutions at the Crossroads," for a discussion of environmental laws and the duties of the Pollution Control Boards.

45. These sets of claims do not fall into the hierarchical model that Louis Dumont provided in *Homo Hierarchicus* (Chicago: University of Chicago Press, 1970), a model in which the sacred principle of pure and impure encompasses the "secular" power of Kṣatriyas. Putting to rest Dumont's version of the sacred-secular opposition, this connection between claims must be understood in terms of the specific legal and moral codes that seek to define Gaṅgā's power and uses.

46. See Vijaya Rettakudi Nagarajan, "The Earth as Goddess Bhūdevī: Toward a Theory of 'Embedded Ecologies' in Folk Hinduism," in *Purifying the Earthly Body of God: Religion and Ecology in Hindu India,* ed. Lance Nelson (Albany: State University of New York Press, 1998), for a discussion of how sacrality moves around with rises and lapses. She uses this notion to argue that the *kōlam* ritual art tradition practiced by South Indian women locates the sacrality of Mother Earth (*Bhūdevī*) only temporarily in the space where the *kōlam* is created.

47. See, for example, Madan, "Secularism in Its Place," 755.

48. Ibid., 753. T. N. Madan's categories of sacred and secular are not to be conflated with Louis Dumont's: Madan is referring to contemporary relations between institutions, whereas Dumont was concerned with relations between caste groups. Still, by using Dumont's original phraseology of encompassment, the risk is that his sense of the secular might be applied. Frédérique Apffel-Marglin's clear critique of Dumont's opposition, in her "Power, Purity and Pollution," enables one to move beyond the artificial separation of status and power to consider Madan's more meaningful

references to the institutional tensions between secular ideology and what he refers to as "the religious."

49. I do admit a few exceptions to this general rule. For example, Dr. Veer Bhadra Misra, *mahant* of the Sankat Mochan Temple in Varanasi, has worked diligently for years to challenge the government's sewage management plans in that sacred city and to formulate alternate proposals. I have known Dr. Misra (Mahantji) for several years and write about his work with the Sankat Mochan Foundation in Alley, "Ganga and Gandagi," and Alley, "Idioms of Degeneracy." In a book manuscript I have just completed I also outline how Swami Chinmayananda, a *sādhu-sant* and BJP politician, has issued calls to protect Gaṅgā's sacrality against environmental pollution. But, in that case, again, the calls have proved to be short-lived political appropriations. For more on *sādhu-sant*s in the BJP, see Virginia Vandyke, "Sadhus, Sants and Politics: Religious Mobilization and Communalism in India" (Ph.D. diss., University of Washington, 1999).

The Narmada: Circumambulation of a Sacred Landscape

CHRIS DEEGAN

She cleanses the soul by simple *darśan*.

Space is fundamental to all systems of thought, and in the Hindu pilgrimage tradition it plays a central role in defining perceptions of the past as well as landscapes of the present. Usually the focus of a pilgrimage is a particular place, or a series of otherwise unconnected places. These centers are most often specific idols or representations of the sacred. Sometimes boundaries extend to include whole towns or cities. Quite often there is a direct connection with the immediate physical environment as the playground of the gods. Regardless of the structure of any center, they all become *tīrtha*s, or crossings to another realm. The Narmada River is one such center. As a river, it is also an idol (*mūrti*). As a deity, it is subject to and giver of *darśan* (auspicious sight). As a temple, it is circumambulated. As a *tīrtha,* it is a sacred crossing. These core determinants—*mūrti, darśan, parikrama, tīrthasthan*—are what shape the spectacular *līlā* (divine "play") of the Narmada and define its sacred space. In this sense, the Narmada and its environs are a spiritual domain defined by its pilgrimage behavior, occasionally controlled and manipulated by priests and politicians. As a sacred space it has developed from collective memories, which through the centuries have adjusted their meanings, intents, and roles.

Pilgrimage behavior on the Narmada has evolved and survived over hundreds of years by relying on both natural and political landscapes. This tradition has also long been dependent on networks and centers beyond the object of pilgrimage. Yet, somehow, a very unique circumambulation of the Narmada has survived at least since the sixth century because of a special relationship pilgrims have with their environment: walking, barefoot and with an alms bowl, in a clockwise direction around the entire sixteen hundred miles from the source at Amarkantak to the Arabian Sea at Broach, and back again. Because the circumambulation is not linked to man-made space, it has not been greatly influenced by political expediency or affected by sacred specialists, and this has resulted in an almost total historical absence of Narmada temples.

The Narmada pilgrim patterns have many of the same pieces that other circulations have: its population is made up of ascetics, wanderers, and the ordinary pilgrim with a purpose. Their reasons for being on the footpaths are not unusual: spiritual liberation, relief from sickness, a more fertile body, release from economic encumbrance. There remains, as well, the cultural leadership of sacred text and spiritual leadership of the *sannyāsi,* yogi, or ascetic, and, as in centuries before, sacred specialists and common cultural practice continue to reinforce the value of gift exchange.

Culture and History

As a religiocultural landscape, the Narmada region is defined most clearly by the behavioral choices made by people who live there. These choices are made up of sets of beliefs, including sacred myth and symbol, regarding the relationship—the history—of people to their past in the context of their geographical setting. The results of manipulation by priests, potters, politicians, and pilgrims of behaviors and symbols have given it a sense of place and created multiple and contested mental maps.

Culture is made up of both metropolitan and folk traditions. Even with common roots, there is a distinct differentiation between the two, particularly the ecology of place, where the folk aspect of culture merges legend with history. Whether in the action of pilgrimage or the ideational systems of folk belief, both maintain an authority for expressing elementary human needs. Folk traditions give order to human feelings and provide knowledge of how people organize the expression of those feelings. More than any other cultural activity, they reveal, through literature, music and song, dance, and art a society's fundamental development and scope over time. Particularly in the Hindu context, and in consideration of its political, economic, and social hierarchies, pilgrimage is a specific folk tradition in which a consistency of cultural activity can be clearly seen.

A great challenge for the study of culture and history in India is the interface of legend, myth, and spiritual fantasy, and "fact" as documented by Western traditional research methodologies. There is ongoing discussion about the location of Laṅkā as described in the *Rāmāyaṇa*. Some scholars place it in the region of the Narmada. Many Narmada valley residents would agree.

The study of culture as history is more than the recording of a sequence of past events and their effects, both of which are largely confined to the metropolitan tradition. Rather, we should look behind the politicians, the philosophers, the economists, and the generals and try to understand the supporting cast. *Itihāsa,* "thus it was," is the modern Hindi term for history and in some ways deals with explanations of the who, what, when, where, and why of the past. This does not necessarily fulfill the requirements for a pragmatic approach, which includes "scientific" evidence with the assumption of myth as false and history as true. But both draw equally from cultural models of interpretation.

Explanations of much of ancient and medieval Indian history by Indian recorders was done by "applied" historians—those who recounted and interpreted past events because of their relevance to the present. This was, and is, predominantly a metropolitan or high culture exercise. Folk history, however, relies on the retelling of events—also because of their relevance to the present—but in partnership with other diagnostic tools, including religion and the environment. The cultural historian, on the other hand, must identify and synthesize "submerged" history. This approach looks for signs and symbols which are repeated in the detail of everyday life: year after year; decade after decade; century after century. These, then, become the active (versus static) links between a people and their environment. In the Indian context the relationship is constantly changing and the role each plays is interchangeable and reinterpretable. The perception of the relationship between people and environment can move to that of people and deities. People may be performers and deities listeners. The reverse is also true and is seen most commonly in classical texts like the Purāṇas and their derivatives: *māhātmya*s (glorifications of a deity) and *kaṭha*s (religious stories). Important to identify as the critical piece is not who the actors are but rather the relationship between the two. For it is these relationships between the people (the folk) and the environment (the deities/cosmos) which also provides a framework for rational reasoning and, therefore, a functioning *itihāsa*.

The Physical Landscape

The Narmada River is one of a very few which flow east to west on the Indian subcontinent. Although geographers would rightly argue that the Narmada is a catchment area of hundreds of spring- and rain-fed streams, as one of the seven sacred rivers of India, it has a "source," and it is at Amarkantak in the eastern Vindhya Mountains of Madhya Pradesh. The river is unnavigable except for thirty miles inland from the Arabian Sea by flat bottom boat. During the summer months, the flow of water decreases significantly and in places the river can be crossed by foot. Along its banks are temples to both local deities and Śiva and Viṣṇu of the greater Hindu pantheon. Some temples, particularly at Amarkantak, Omkar, and Barwani, are destinations for pan-Indian pilgrims.

Human activity along the banks of the Narmada is dated variously from several thousand to several tens of thousands of years before the common era. The most ancient human remain yet found on the Indian subcontinent was discovered on the southern bank of the central Narmada River and may date back as far as 150,000 years.[1] Information about early agriculturists along the Narmada's banks is more definitive. There is solid archaeological evidence, for example, of habitats at Navdatoli on the south bank as far back as 1600 B.C.E. Excavations at the site have shown that the building layouts were circular or rectangular, with wooden posts of *babbula* to support roof and walls of wattle and daub construction. Floors were of tamped earth, *kaṅka* (pebble), and straw. These continue to be the basic building materials and tools today for poor rural agriculturists of this area.

Navdatoli is of particular interest also because of earthen storage jars found intact. Most had remains of cottonseed, wheat, rice, barley, and lentils—all continuing staples of the area today. One jar, however, remains a curiosity. It shows "figures of a female and lizard with what looks like a shrine between them."[2] There are four such representations on each side of the jar. Interpreting such artifacts is hazardous, but if the female figure is taken as a worshiper, a *yakṣi* or mother goddess, and if the lizard is a less developed representation of the *makara* (alligator)—both having historical iconic and thematic association with the Narmada—then this may be an early symbolic display of river-cum-mother goddess worship in this region.

Integration of life-style, worldview, and physical surroundings results in the expressive behavior of the participants. For rock shelter inhabitants of the Narmada valley, the motif of their physical and spiritual environment was clearly displayed in their rock wall painting. For these early settlers of the Vindhya and Satpura ranges, a close association with their physical environment defined their "universal" center as both workplace and spiritual theater. Expressed in another level of symbology are the later contexts of kings who styled themselves as *cakravartin,* designing a world mandala with the king not only "at" the center but "as" the center. For the *cakravartin,* centers relied on social hierarchies and networks and were expressed in the political format of rock edicts and copperplates. Whatever their design, spiritual and political centers in the Hindu world have long been a defining element of that culture.

The Narmada has traditionally been regarded as the north/south dividing line of peninsular India. Initially, this most probably developed because of the topographical boundary of the Vindhya and Satpura ranges which cradle the river and its valley. The definition was compounded by a vast forest land and all its unknown and undiscovered elements. In its beginnings, the region was a boundary, a frontier, and a transition zone. Cosmological metaphor gave definition to this environment:

> The Vindhyas function as a kind of earthly equivalent of the Lokaloka mountain: they separate this real, Aryan, northern world from that unreal, non-Aryan, southern world. In particular, when demons or demonic deities are inadvertently created, and there is need for a place "out of the system" in which to deposit them, they are usually consigned to the Vindhyas.[3]

Earliest habitation and life-style is generally guesswork and relies on fossil evidence. One exception is a nearly complete skull found on the banks of the Narmada in 1982. It is, as mentioned earlier in this essay, the oldest human remnant so far recovered on the Indian subcontinent and may date back 150,000 years.[4] If confirmed, it would cast the Narmada valley as one of the world's most important early habitats.

In the sixth century B.C.E., the Narmada was referred to as "the middle line of the earth."[5] This was also a place recorded by early writers as having an extensive forest, one which covered most of central India; a place where only *muni*s and *ṛṣi*s resided; but also *adharma-deśa*, a place for demons or other liminal characters to live. It was a dense ecological and poetic cultural zone defined through metaphor. One thousand years later, in the sixth century C.E., these attributes were described in detail:

> Listen now to the rising of the sage Agastya, who stemmed the Vindhyan mountain;
> which threatened to obstruct the course of the Sun's carriage by raising its peaks;
> which was beautified by flags hoisted aloft in the form of clothes hanging and waving from the bodies of the Vindhya damsels;
> which was scratching the firmament with its raised peaks that had the sonorous sound of the humming of swarms of intoxicated bees;
> which peaks contained hyenas, bears, tigers, and monkeys;
> which is embraced closely by the river Reva (Narmada) having the mandana trees growing on her banks;
> whose gardens are occupied by the gods;
> and where dwell sages who live on water, roots, and nothing.[6]

The Figurative Landscape

The creation story of the Narmada begins with the all-devastating flood which occurred at the end of the Satyuga period.[7] Śaṅkara was alone and doing *tapasya* on the Amarkantak plateau to bring back all the gods. Through his *pasīna* (sweat), Narmada was reborn. With her came the gifts (*bardān*) she had earned for her own ten thousand years of *tapasya* (ascetic practice):

> to be *amara* (forever);
> to be holy;

to have her *bhakti*'s sins washed away by bathing in her waters;
to be the southern Gaṅgā;
to offer whatever rewards one gets from other rivers;
to enable those who do *tapasya* on her banks to find a place with Śaṅkara;
to have those who are given to her at death reach Śaṅkara;
to be forever known as a sin cleanser;
to have those who live and die on her north bank go to *Amarpuri* (where the gods are);
to have those who live and die on her south bank go to *Pitṛloka* (where their ancestors are).

Connecting specific sites (*tīrtha*s) to the *tapasya* of Śaṅkara and others is the central part of the *Revā Khand* of the *Skanda Purāṇa*. In communities along the river, this text is known as the *Narmadā Purāṇa*. In it, the cosmography of the Hindu world is transposed to the specifics of the Narmada River. To establish her central place in the deity hierarchy, early in the *Revā Khand* she is described as equal to and/or the same as the Sarasvatī and Gaṅgā Rivers. In further deconstruction, she is, in fact, more important because she has been granted her *bardan* and been the sole survivor of three destructions of the world.

The name "Narmada" has not been constant. Like a child in a joint family, she has many names depending on who is calling for her attention. The earliest literary reference is in the fourth-century work *Raghuvamsa,* where she is called "Reva" (meandering/flowing river). She is also referenced to Śiva with her name "Jaṭā Śaṅkari," and to her origination from the moon with "Somodbhava," and "Induja." She is also the Ganges with her names, "Maheśvari Gaṅgā," "Kāśī Nāḍī," "Dakṣin Gaṅgā," and "Pūrva Gaṅgā."

But the most expressive names for Narmada are those which describe human attributes or point to her sacrality. They are found in the *Revā Khand* and among the names are: "Mahati" (great, excellent); "Surāsa" (flavorful, sweet); "Kṛpa" (of graceful attitude); "Vipāśā" (one who absorbs but does not release sin); "Karabha" (one who radiates happiness); and "Vāyuvāhinī" (carrier to heaven).

By the thirteenth century, the Narmada valley region was being recognized as an administrative unit with specific geographic labeling, for example, *Narmadā-tata-maṇḍala* (land on the banks of the Narmada) and *Narmadā-deśa* (Narmada valley).[8] As its population grew and diversified, the region also changed ecologically. But the river maintained its identity as a spiritual center.

Parikrama/Pilgrimage

All along her banks, *tīrtha*s are represented by actual *ghāṭ,* temple, or village sites, or are the generic name for an area. Most, as might be expected, are connected directly to Narmada's specific history: *liṅgatīrtha*s for Śiva, *kuṅḍ*s, or reservoirs, for bathing, and areas where ascetics gathered for *tapasya*. There are *tīrtha*s at the spot where Viṣṇu is believed to have washed his *sudarśan* (discus); where the goose did *tapasya* to become the *vahana* (vehicle) for Brahmā; where Indra, Agni, and Varuṇa did *tapasya*. There are two *tīrtha*s for Gaṅgā: one where she came to do *tapasya* and another where she came in the form of a cow to bathe. On the south bank is Jyotismatipuri, the place where Rāma and Lakṣmaṇa did *tapasya,* and Yojaneshwar, where they bathed before going off to kill Rāvaṇa. There is Suklatirtha, the place to which all other *tīrtha*s do pilgrimage and the oldest by modern historical text dating. It was well established as a *tīrtha* site at the time of Chandragupta Maurya who came here to bathe to be cleansed of the sin of killing his brothers in the fourth-century B.C.E.[9] And Omkareshwar, still today the most important destination for Narmada pilgrims and the place of one of the twelve all-India *jyotirliṅga*s.

Although both pilgrims and academics might argue that the very act of this pilgrimage puts one in what Victor Turner has called a "liminal state," to undertake Narmada *parikrama* is to no longer remain *alug,* or separate from the world. This pilgrimage, like others, is a ritual reinforcing of shared values of community. Even for those not able to undertake a *parikrama,* there are alternatives for the true devotee. Reading the *Narmadā Purāṇa,* or bathing in her waters, or catching sight of her every day are enough for sinners to cleanse themselves. Doing a *parikrama* of Amarkantak (place of the Narmada's source), or of a cow, is to circle the world. Pilgrimage of this dimension—the barefoot circumambulation of an eight-hundred-mile-long river—is an extraordinary event and there is no evidence to suggest that this is duplicated on any other river of the subcontinent. Movements through sacred space require both guidelines and rules. In addition to the essential behaviors of going barefoot and not cutting any hair, a

Narmada *parikrama* requires a pilgrim to: eat only once a day; bathe in the Narmada once a day, thereby also completing the necessities of *darśan;* abstain from sexual activity; sleep on the ground; not lie; collect food through alms; refrain from malice or hostility; suppress desire; speak from love. Following the templates of the *Revā Khand,* the contemporary pilgrim starts the *parikrama* at any place along the river and on either the south or north bank. The overall itinerary is to walk with the river always on the right side, never crossing the river except at its mouth at Broach. At the start, there is a simple ceremony where the pilgrim's hair and both finger and toe nails are cut. As the only things produced by a pilgrim—and therefore essentially "of" the pilgrim—they are offered to Narmada and placed in the river. During the *parikrama,* no hair is cut and men do not shave. A container of Narmada *jala* (water) is filled at the place of departure. At both the source at Amarkantak and where the Narmada meets the sea at Broach, half the water is taken out and then replaced. On completion of the circulation, the pilgrim goes to Mandhatta and the island of Omkar in the middle of the Narmada. At the temple of Omkareshwar on the south side of the island, the pilgrim pours half of their Narmada *jala* over the *liṅgam* in a final offering and refills the container again at the *ghāṭ*. Once home, the pilgrim sprinkles this Narmada *jala* around the house as a holy water to protect all who live there.

Engaging in a level of religious activity like the Narmada *parikrama,* one is almost certain to be saying prayers for others—family members, others in your *biradri* (community), friends. For this there are two sets of rewards: one to the person it is being done for (e.g., the old and sick mother), and one for the person remembering them in prayer and action (the pilgrim). Narmada *parikrama* can also be undertaken by someone else on behalf of another person, although with some parameters and various formulae for determining amounts of *phal* (benefits). Ideally, the surrogate pilgrim should be from the same *biradri* and have equal *jñāna* (knowledge/wisdom). The rewards for this arrangement are highest for both if a husband or wife do it for the other. In this case, there is an equal distribution of *phal.* In descending order of receipt of spiritual *dān,* it is most important for pilgrims to undertake Narmada *parikrama* for their *chela* (student), their own son or parents, their brother's son, their sister's son, and then for all other relatives.

Shifting Landscapes

Economic, political, and demographic interests are now, more frequently, main determinants of the interrelationship between human and natural landscapes. In the Narmada valley, the sacred space of its *tīrtha*s and the routing of its *parikrama* are being precluded by the pragmatic realities of competition for resource management.

The Narmada has a systemic order which interconnects all its parts. Thinking about and trying to come to some safe analysis of the interaction between cultural history, *itihāsa,* folk tale, and cosmography and how these dovetail with the ecology of this cultural landscape is complex. Ongoing efforts to reconfigure the Narmada River through the completion of a major dam project represents a competing concept of the meaning of "ecology" in this region. For those who live there it is less complicated, for they are headed, without choice, dictated by custom and culture, and necessitated by the rules of engagement between the living worlds of humankind and nature, to the banks of the Narmada in life and in death.

From the source of the Narmada in the sacred grove at Amarkantak, to the Reva Kuṅḍ of Baz Bahadhur and Rupmati on the plateau at Mandu, to the *nābhi* (navel) of the Narmada at Nemawar, to the seaside village of Palri where pilgrims are ferried across the estuary to the north bank with the salutation of *Narmade Har!,* this central Indian river is alive with *itihāsa.*

Notes

1. *Ancestors: The Hard Evidence: Proceedings of the Symposium Held at the American Museum of Natural History, April 6–10, 1984,* ed. Eric Delson (New York: A. R. Liss, 1985).

2. M. K. Dhavalikar, "Chalcolithic Religion," in *Archaeological Congress and Seminar, 1972: Papers Presented at the Sixth Annual Congress of the Indian Archaeological Society and the Seminar on the Late Harappan and other Chalcolithic Cultures of India,* ed. Udai Vir Singh (Kurukshetra: B. N. Chakravarty University, 1976).

3. Randolph W. Kloetzil, "Maps of Time, Mythologies of Descent," *History of Religions* 25, no. 2 (1985): 140, n. 43.

4. *Ancestors: The Hard Evidence,* ed. Delson.

5. Ram Gopal, *India of the Vedic Kalpasutras* (Delhi: Motilal Banarsidass, 1983), 93.

6. *Varahamihira's Bṛhat Saṃhitā,* trans. M. Ramakrishna Bhat (Delhi: Motilal Banarsidass, 1981), 154.

7. *Reva Khand,* 2d ed. (Varanasi: Gyansatra Prakashan Nyas, 1983).

8. D. C. Sircar, *Studies in the Geography of Ancient and Medieval India* (Delhi: Motilal Banarsidass, 1971), 108.

9. *Broach District Gazetteer,* ed. M. R. Palanda (Ahmedabad, 1961).

Sacred Rivers, Sacred Dams: Competing Visions of Social Justice and Sustainable Development along the Narmada

WILLIAM F. FISHER

In the interactions among religion, ecology, and politics, politics usually prevails. This paper explores the evocations of varying visions of the Narmada River by those involved in the struggle over the damming of the Narmada River. Visions of the Narmada as goddess, homeland, or development resource punctuate the debate over the Narmada and often stand in for more complex and conflicting moral arguments about social justice and sustainable development. Through the cryptic use of symbols and simplified dichotomies, developmental, environmental, and moral concerns become politicized and fundamental differences obscured. The concern here is with where, how, and why key symbols of cultural and religious values enter into development discourse and what happens when they do.

This discussion takes heed of and evolves through the consideration of two related aspects of the Narmada debate. The first is that the Narmada controversy involves a highly complex set of issues and a multiplicity of stakeholders (including residents of the Narmada valley, residents in the command area, Gujarat development planners and politicians, various antidam activists in India, and "northern" environmentalists). Two issues here are pertinent: first, that these different stakeholders often have widely divergent perceptions of development, "facts," problems and solutions, and costs and benefits; and, second, that not all stakeholders are equally tolerant of other perspectives.

The second consideration is the degree to which these divergent positions, values, and beliefs are simplified, dichotimized, misrepresented, or otherwise obscured in the struggle over Narmada. This emerges and becomes manifest in a variety of ways. One of the most important is through the use of the same rhetoric and claims to some of the same powerful symbols, a process which serves to obscure a fundamental clash of worldviews. These clashing worldviews each derive from a different set of assumptions about the good life and the relationship of humans to nature. Insofar as the appropriateness of a solution depends in large part on how a problem is defined, the power to shape the definition of the problem—as one of poverty, or as a shortage of resources, or as the need to maximize resource use for modern development—offers the opportunity to designate and describe appropriate "solutions." Sorting out the differing worldviews enmeshed in shared rhetoric and symbols reveals and illustrates the often inconspicuous but nevertheless treacherous power of rhetoric. Development actions and language have a complex relationship as multiple stakeholders jockeying for political advantage constantly recast and redescribe ends and means in response to critical feedback from a range of different influential audiences. In development debates, just as in others, language has the potential either to open up possibilities of thought or to obscure, either to clarify or to gloss over fundamental differences, either to generate new concepts or to coopt ideas and values (and thereby either transform or defang them).

The creative use of language is particularly apparent in highly emotional and deeply politicized debates like that over the damming of the Narmada. Rivers evoke deep and yet varying responses from different constituencies. As in this case, they may be valued as living goddesses, protected as complex habitats, or coveted as a store of resources for the vast quantity of "wasted" water running untapped to the sea. In the Narmada case, these contrasting visions serve a range of conflicting moral arguments for social justice and sustainable development. But not all of these visions have received equal attention in the public debates about Narmada. Least heard are the views and values of local people resident along the stretch of the Narmada valley destined to be inundated by the waters of the Sardar Sarovar reservoir. Instead, idealized versions of *ādivāsi* and tribal beliefs can be found appropriated into both development and environmentalist positions.

Background: What's at Stake?

Narmada has become a familiar case to those concerned with either environmental or South Asia issues.[1] The damming of the Narmada River in western India is an issue where the perceived conflicts between economic development and environmental protection have become the battleground of other conflicts over human rights, decision-making processes, and development objectives. By articulating issues that have a significance far beyond the particular details of the Sardar Sarovar Project, a domestic construction project along a remote sector of an Indian river has become a highly emotional symbol at the center of an international controversy.

Tapping the resources of the Narmada, the largest westward flowing river in India, and one of the most sacred rivers in India, has been the dream of political leaders and development planners for decades. The Narmada River, 1,312 kilometers in length, rises in the state of Madhya Pradesh in central India and passes through the states of Maharashtra and Gujarat on its way to the Gulf of Khambhat. The Narmada drainage basin covers 98,796 square kilometers, with an estimated population of 22 million people. Exploiting the water resources of the river is complicated by the fact that 90 percent of the Narmada's flow occurs during the three months of monsoon rains, from June through September. The river is the subject of the largest river development scheme in the world that would include, if it were to be completed, 30 major, 135 medium, and about 3,000 minor dam projects in the Narmada River valley.[2]

The terminal dam of this project, Sardar Sarovar, stands almost complete. It includes a dam, a riverbed powerhouse and transmission lines, a main canal, a canal powerhouse, and a 75,000-kilometers-long irrigation network which will occupy 80,000 hectares of land. The water is to be collected in a storage reservoir which will resemble a narrow lake extending more than 200 kilometers upstream. At its full reservoir level of 455 feet, the reservoir will submerge 37,000 hectares of land and will adversely affect at least 100,000 people in 245 villages. A larger number of farmers, as many as 140,000, will lose land to the canal and irrigation systems.

Financing to initiate the current project was secured in 1985 when the World Bank entered into credit and loan agreements with the government of India, providing US$ 450 million for the construction of the dam. Construction began in earnest in 1987. Its supporters claim that it is the only viable means of delivering critically needed irrigation water to drought-prone areas of northwestern Gujarat and Rajasthan and electrical power and drinking water to thousands of other rural and urban communities in Gujarat.

From the beginning, the project has been the subject of local, national, and international opposition that has criticized it on environmental, technical, and humanitarian grounds.[3] Critics of the project cite its potentially negative environmental and social impacts, particularly the relocation of tens of thousands of people, the majority of them members of lower socioeconomic communities. In the past decade, social action groups representing rural communities along the Narmada and independent nongovernmental organizations have lobbied prime ministers of India, chief ministers of the Indian states of Gujarat, Maharashtra, and Madhya Pradesh, state and national bureaucrats, the Japanese government, the World Bank, and governments funding the Bank.[4]

On 14 March 1991, in response to growing criticism of the project and under pressure from the global campaign of international NGOs against the project, the president of the World Bank commissioned an unprecedented independent review of the Sardar Sarovar Project. The review of environmental impacts and resettlement and rehabilitation began in September 1991. Its final report, submitted in June 1992, was very critical of the World's Bank involvement in the project. It stated:

> we think the Sardar Sarovar Projects as they stand are flawed, that resettlement and rehabilitation of all those displaced by the Projects is not possible under prevailing circumstances, and that the environmental impacts of the Projects have not been properly considered or adequately addressed. Moreover, we believe that the Bank shares responsibility with the borrower for the situation that has developed.[5]

World Bank funding of the project ended in March 1993, but the government continued to build the dam and opponents continued to oppose their efforts. As of late 1998, despite lawsuits, coordinated lobbying, an unprecedented independent review of a World Bank project, the Indian rejection of World Bank funding for the project, a negative review by an Indian review team, and a high-profile protest movement characterized by marches, rallies, and hunger strikes, "Narmada" remained an unfinished story, its outcome not yet resolved.

When World Bank funding for the project ceased in March 1993, the government of Gujarat proceeded with the project on its own while the opposition pursued its case through the courts. The project continues to be controversial: the dam is near completion but approval to proceed remains stalled in the Supreme Court. The government of Gujarat remains determined to complete the dam as planned and the activists remain just as determined to prevent it.[6]

The Significance of Narmada

There is not one but many Narmada stories. One of the striking aspects of the local, national, and global debates over Narmada is the array of actors with different objectives who nevertheless defend their positions in the same terms, a phenomenon made possible when these terms—like "sustainable development"—are conceptually flexible and vague. While the range of actors and issues involved in the Narmada controversy are remarkably diverse, both those resisting and those defending the Sardar Sarovar Project use the same moral vocabulary of social justice, the same economic rhetoric of sustainable development, and similar evocations of the legacy of Gandhi.

Narmada has captured the imagination of many actors on many stages: depending on the audience, the name evokes the hopeful image of an exploitable and renewable natural resource, refers to a homeland in the river valley, stands for the river itself as a powerful religious symbol, or denotes a historic religious pilgrimage route. Both in India and in the international arena, Narmada has become a symbol for the struggle for local autonomy against forced displacement associated with state-directed and internationally funded development. The struggle over Narmada provides a poignant example of how local people are caught between the threat of destruction of their way of life and the promises of development, while government agencies, NGOs, activists, and academics step forth to speak for them. In this process, idealized, sanitized, or simplified versions of local beliefs are appropriated into developmentalist, Gandhian, and social ecological positions.

The Narmada controversy has had significant ramifications far beyond the river valley. An internal review of the World Bank's performance in the Narmada case is partly responsible for structural changes within the Bank;[7] the transnational alliances of NGOs supporting the Narmada Bachao Andolan (Save the Narmada Movement) helped establish more permanent linkages among dam-affected peoples all over the world and led to the first international meeting of peoples affected by large dams in Curitiba, Brazil, in 1997;[8] and the lessons of Narmada were a primary motivation for the formation in 1998 of a world commission on large dams sponsored by the World Bank and the International Union for the Conservation of Nature and Natural Resources (IUCN).[9]

Deploying Gandhi

Religion creeps into the Narmada issue in many ways, not least because the Narmada is, for Hindus, one of the most sacred rivers in India: its banks are lined with numerous sacred monuments and sites and it is the subject of circumambulation by devout pilgrims. Religious identity is also invoked by the Morse report,[10] in its defense of the indigenous identity of many people affected by the Sardar Sarovar Project, and by critics of the report, who contest the characterization of project-affected people as indigenous and tribal people, arguing instead that they are "backward Hindus."[11]

Even more striking in the Narmada issue, perhaps, is the omnipresence of Gandhi. No matter where one turns in the Narmada issue, one encounters the icon of Mahatma Gandhi: from the nonviolent strategies of the Narmada Bachao Andolan, which opposes the dam as an example of "destructive development"; to the Gandhian groups working to assure adequate resettlement rights for the "oustees"; to the Gandhian groups now working in the command area of the project, who long ago took seriously Gandhi's admonition to settle in remote villages and who now look to the promised irrigation water from the Narmada as a boon that would help further their work to improve the lives of poor Gujarati villagers; to the large statue of Gandhi not far from the massive block building which serves as headquarters to the Sardar Sarovar Narmada Nigam in Gandhinagar.

The image or icon of Gandhi is evoked in the Narmada debate for numerous reasons—and there are variations in the degree to which it is underscored by the evoker—but it is used most frequently to assert a link between one's own actions, on one hand, and the goals of social justice and the needs of the poor and marginalized groups, on the other. Gandhi as an icon for those concerned with the poor and with traditional values contrasts most clearly with the forward-looking, development-oriented industrialization legacy of Nehru.[12] While the contrasts and contradictions between the visions of these two men are sharp, and in many ways represent two opposed views of modernity (one predicated upon industrialization and the other opposed to industrialization),[13] in the Narmada case, many dam proponents argue that they have made a successful and progressive merger of Nehru's modernist vision with a Gandhian concern for the poor. It was Nehru who laid the cornerstone for a dam on the Narmada in 1961, but, ironically, references to Gandhi's concern for the poor are now used to justify it. Dam opponents, on their part, characterize the dominant Nehruvian development vision as one which results in worsening conditions for the poorest of the poor and most marginalized elements of society, and one which results from a decision-making process that fails to consult with the poor and thus fails to take seriously their needs and concerns.

The omnipresence of Gandhi as a legitimizing symbol to both dam builders and dam opponents can be a bit disorienting to everyone. One day in November 1991, as I was conducting interviews in an office of the headquarters of the Sardar Sarovar Narmada Nigam, the Gujarat institution responsible for building the dam and canal system, official after official came by to volunteer their genuine concern for the poor and cited as part of their credentials that in their own way they were followers of Gandhi.[14] In justifying their current activities in the Nigam, these officials made attempts to reconcile their Nehruvian visions of a modernized India with the views of Gandhi, while at the same time dismissing the dam opponents as misguided antidevelopment, antinationalist neo-Luddites. Echoing a widely held view, one official told me that, "the Narmada Project holds all the hopes for the future of Gujarat." Another insisted that "the project is necessary for the poor farmers of northern Gujarat." And as one of my visitors put it, "the masks must be removed from the environmentalists, exposing them as antidevelopment agitators."[15]

"Baba Amte," Nigam officials told me repeatedly, referring to the famed Gandhian activist who resettled along the Narmada to show his support for the opponents of the Narmada development projects, "has been misled by his supporters."[16] Each of them was convinced that the SSP was a project which offered real hope for the alleviation of drought in northern Gujarat and one that had been unfairly mischaracterized and victimized by the protesters.

Later that same day, accompanied by Amar Gargesh, who was at that time in charge of public displays supporting the dam, I left the massive Narmada building in Gandhinagar and headed toward the city of Ahmedabad. Outside, we found ourselves again in the presence of Gandhi in the form of the prominent statue celebrating Gandhi and his spinning wheel that seems to stand guard in the city of his name outside the modernist building that houses the staff responsible for building a major nature-altering technological project.

At the museum in Ahmedabad, Amar and I toured the Nigam's display extolling the technological wonders of the dam and the features of the canal system. The display revealed enthusiasm, even if a bit uninspired, and featured confident assertions like those which are displayed on the billboards that line the road to Kevadia and the dam site:

> "Backbone of the western Indian economy."
>
> "The only remedy against recurring drought."
>
> "A planned ecological harmony among men, water, and vegetation."
>
> "A ray of hope to thirty million people."

Though time was short, the engineer and the museum curator were insistent that I visit the permanent exhibit next door which traces the life of the Indian independence movement and, particularly, the roles played in it by Gujarat's native sons. Throughout this tour, my hosts pointed out, in hushed tones of reverence, the hardships suffered by these men, their remarkable determination and dedication to their cause, and their willingness to go to prison for their principles. In front of a picture of Gandhi and Sardar Patel emerging from prison, Amar took a long pause, following which, in a low voice, almost to himself, as if the thought were occurring to him for the first time, he said quietly and respectfully, "they [Gandhi and Patel] were just like the [Narmada Bachao] Andolan" [the antidam activists].

Competing Visions: Worshiping a Goddess or Taming Nature

While dam advocates and opponents are generally sincere in their advocacy of sustainable development and social justice, and in their evocation of Gandhi, what they mean by the use of these terms and symbols differs profoundly. Sharing the same rhetoric to describe very different goals and means disguises fundamental philosophical differences, maintains confusion in the debate about development, and makes it more difficult to mount an effective challenge to the dominant development paradigm. Underlying the conflicting arguments are visions of the Narmada as goddess, homeland, or development resource. These contrasting visions serve a range of positions on social justice and sustainable development.

For millions of people in India, "Narmada Mai" is a goddess.[17] This is one of the most dramatic and long-standing visions of the river. Along its banks are thousands of temples dedicated to Gaṅgā and Śiva, and each year thousands of pilgrims throng to these and other sacred places. The *parikrama,* or pilgrimage, involves the circumambulation of the river from its source at the spring on the Amarkantak plateau to the mouth of the river and back again along the opposite bank. The *parikrama basi*s (lit., dwellers on the circuit) carry no money or extra clothing and accept food along their journey.

Along their journey these pilgrims pass through a valley which serves as homeland for hundreds of thousands of people. For these residents, too, the river, like many aspects of their environment, is sacred. It is a timeless environment of which they are an integral part and which is consequently to be treated with respect and reverence.

Contrasted with these views of the timeless sacredness of the Narmada geography, the focus on the Narmada as a resource which might supply irrigation water for up to one hundred years seems a narrow, temporal concern. The vision of the relationship between human and nature implied by this view is also dramatically different. The Sardar Sarovar Dam is a vivid example of modernist convictions that one can obtain mastery over nature, and that the failure to do so will mean ruin. It derives from the conviction that as humans we can and must make our own destiny, that human history has been a history of progress, and that we can find technological solutions to all the problems we encounter. It reflects Descartes's conviction that the general good of all humankind could be pursued by the attainment of knowledge that is useful in life so as to make ourselves "the masters and possessors of nature."[18] This perspective on development defines and responds to two aspects of nature. Nature is seen as threatening and dangerous—in need of containment—while, simultaneously, it is viewed as a stockroom of resources for technological advancement. For development planners, both aspects present problems requiring technical solutions. The diverting of the Narmada waters to drought-prone areas of northern Gujarat is promoted as an appropriate technical response to both of these aspects of nature, diverting the "wasted" water of the Narmada to prevent the continued disasters caused by drought.[19]

From this perspective, it is less the river that is sacred than the dam. Hailed as "the lifeline of Gujarat," the dam and its complex canal system become an embodiment of Nehru's modernist vision that high dams would become "the secular temples of modern India."[20] This emergence of the dam as a sacred icon of modernization has the unfortunate consequence of presenting the dam as an end rather than a means of development. And the devotees of this temple of modernization have demonstrated an ardent commitment to their shrine that will not permit them to step back to reconsider its efficacy. The focus remains fixed on completing the dam and refusing to consider the possibility of other creative solutions to the initial problem.

With this unquestioning commitment, the chief minister of Gujarat in 1991 insisted that: "1) a review of the Sardar Sarovar Project will not be accepted under any circumstances; 2) the work will not be stopped for even one day; 3) and the height of the dam will not be lowered by even one inch." While the government has been forced to accept a review (indeed, more than one review), and the work has stopped, the devotion to the dam has not lessened. Six years later, in 1997, another chief minister of Gujarat fervently insisted that he was still committed to the dam, a commitment driven by his concern for "the weaker and downtrodden segments of society," and the chief minister linked his efforts to a concern for "the Adivasis, Harijans, Dalits, and other Backward classes and Scheduled Tribes."[21]

The view of progress and modernization from which projects like Sardar Sarovar emerge has been expressed by Vidyut Joshi:

> we have welcomed change in the name of progress, development or modernization. This being so, why should anyone oppose when tribal culture changes? A culture based on lower level of technology and quality of life is bound to give way to a culture with a superior tech-

> nology and higher quality of life. This is what we call "development." What has happened to us is bound to happen to them because we both are parts of the same society.[22]

This view allows no room for the tolerance of different ways of life and different relationships with the environment, especially where conflict over resources is at issue. Instead, it makes clear the conviction that dominant elements of society are justified and even duty-bound to force the change of marginal populations. Of course, despite Joshi's generous and inclusive use of the first personal plural, the oustees of the Sardar Sarovar Project cannot look forward to the same life that Joshi enjoys.

Three practical consequences that emerge from the attempt to conquer nature are apparent in the Narmada case. One is the transfer or redistribution of resources from low-resource-use populations to high-resource-use populations—a transfer that is done without the consent of the low-use group and justified in terms of both human need and progress. Resources perceived as unused or wasted are taken as part of the manifest destiny of high-use portions of the population. Second, this diversion of natural resources is done in such a way that it entails further alteration and domination of nature. Third, the process allows and even requires that governments consolidate their control over both resources and people.

Drowning Voices

While the Narmada conflict may be a vivid illustration of a paradigm shift in process, a working-out through conflict and struggle of the nature of social justice and sustainable environmental use that concerns communities all over the world, the completion of this paradigm shift is not assured.[23] Nor does the process itself ensure that either the environment or justice will be served. Meanwhile, in many ways, the process itself continues to do violence to the views and lives of local people in the Narmada valley.

Despite the wide array of actors with a stake in the Narmada controversy, the struggle is often oversimplified as a battle between two ardently held positions pitting the people of the Narmada valley against a large development apparatus. The oversimplified division of actors into developers and resisters emerges from the dominant rhetorical exchanges of the controversy between developmentalists and environmentalists, and it imbues the struggle over Narmada with a compelling black-and-white character that allows it to resonate far beyond the valley. Ironically, while everyone steps up to talk about and talk for the local people of the Narmada valley, both the views and values of those in the Narmada valley and the values of groups in the command area of Gujarat are appropriated and oversimplified by this dominant contestation.

It is important to highlight what happens in the politicization of religion and the environment in the Narmada conflict and the violence done to local lives and views. As the struggle between dam builders and dam opponents evolves from a struggle over a specific dam to a clash about the process of development—a clash in which the dam builders and opponents each harden their points of view, and a contestation in which a sacred dam is made to confront a sacred river—the voices of the local people are at risk of being drowned out, their views and lives reduced through overgeneralization to simplistic caricatures.

Critiquing Development

From the massive block building in Gandhinagar with its fleet of chauffeured Ambassador cars, through the network of comfortable Nigam guest houses with their plentiful buffets, one travels a long way to the spare, narrow, cramped third-floor room in Baroda that served as the base for the activists of the Narmada Bachao Andolan. Here, while the terms of sustainable development and social justice and the evocation of Gandhi are familiar, the meanings are profoundly different. For the activists, the dam, far from being sacred, is a sacrilege. The Sardar Sarovar Project is just one more of too many projects said to be in the "national interest," but which in fact undermine the ability of the rural poor to control and use local resources. The positions of environmentalists and activists opposing the Sardar Sarovar Dam are ideologically heterogenous and include a number of hybrid positions blended from Gandhian, Marxist, and "indigenous knowledge" positions.[24] In the view of some of these activists, the struggle of the inhabitants of the Narmada is a living example of a true environmental movement, a challenge by communities who worship nature and use it sustainably.[25] Some activists would go so far as to claim that the beliefs and practices of these local communities contain an implicit critique of development and an alternative vision of the relationship between humans and nature.[26]

This is an attractive and compelling point of view, and use of this compelling image was extremely effective in rallying support against the dam, but there are other stakeholders in the Narmada conflict whose voices are not widely heard and whose views are often oversimplified in the conflict.

Respecting Nature

The journey from Baroda to the villages of the submergence area, while physically more difficult, doesn't seem as far, conceptually, as that from the Nigam headquarters to Baroda. There, six months after my tour of the museum in Ahmedabad, I sat outside one warm evening listening to talk about gods and nature; people spoke about their way of life, their relationship to the earth, the forests, and the river. At first encounter, one is struck by the tranquility of life and the respectful attitude toward nature. Here, the environment is not merely a stockroom of resources, but a living landscape where the natural and the supernatural are intricately intertwined. Spiritual power which resides within trees, rocks, or hills is perceived as intervening actively in people's lives. Virtually all of them emphasized their ties to ancestral land, to the river, to the goddess Narmada, and to the local spiritual world: "our gods cannot move from this place," one said to me; "how can we move without them?"

Are the cultural and economic histories and conditions of the people living in these different landscapes distinct? The answer to that question is essentially political. Identifying or labeling this local set of practices as "Hindu" or "indigenous" does it a great injustice and misrepresents the specificity of local life. Dam proponents have argued that the potential "oustees" of the submergence area are simply "backward Hindus" and do not have a proper "indigenous" tradition. Activists have been careful to counter this image and to emphasize so-called tribal characteristics of their social practices. A great deal is at stake in the way valley residents are represented. Scheduled Tribes and Castes are of course entitled to concessions from the central government.[27] "Indigenous and tribal" people are covered by World

Bank and other international guidelines that require the expenditure of additional caution and money when their way of life is disrupted by development projects.[28]

The hybrid culture characteristic of the valley residents does not fit easily with bureaucratic and academic needs for sharply distinct categories. Active within the daily social lives of valley inhabitants is a multicultural panoply of Hindu and local gods and spirits.[29] In conversations with me, the inhabitants did not repudiate Hinduism. Indeed, aspects of it are obviously part of their life: there are Hindu as well as local specific deities and shrines. Calling them either Hindu or non-Hindu would seem inappropriate and irrelevant, were it not for the political ramifications of these labels.

Material life in the valley is also closely tied to the forests. Homes are built of local materials, and even in so-called degraded forests the inhabitants gather useful fruits and medicines. The environment is vibrant with life: trees and rocks become shrines and the river is seen as the source and support of spiritual life. In this context, what can the dam symbolize but the end of that life?

But these observations too easily slip into an idealized view of what is in fact a complex and messy relationship with a difficult, degraded environment. While it is difficult to desegregate the environment from cosmology, cultural and spiritual values, human life, and identity, it is important to examine this complex relationship closely. Most simply, nature, like gods, may be both threatening and protecting and, like gods, may require propitiation. But nature, like gods, then, can also be trusted to care for itself, to rejuvenate itself when abused and overexploited—an attitude that shares less with the advocates of sustainable development and more than we might want to acknowledge with the development planners of the Sardar Sarovar Dam.

While society in the Narmada valley is respectful of nature, it is not necessarily one still in harmony with it. There is little point in idealizing life in the remote and difficult terrain of the valley. It is a life neither as harmonized with nature as sometimes presented in the West nor as riven with poverty as portrayed by the Nigam. But understanding it as it is, with all of its wisdom or tribulations, is often overwhelmed by the need for immediate political activism.

Even before the difficulties created by the scheme to dam the Narmada, life in the valley was hard and the relationship with the environment was a complex and ambiguous one. That realization must accompany the criticism of simplistic technological solutions which are offered to ease life for some (while in fact making life more difficult for others).

Writing about similar groups elsewhere in India, Christoph von Fürer-Haimendorf asked that we consider how it came about that populations which were self-sufficient for centuries now need to be protected, aided, or rescued by the government.[30] From his own observations, he asserted that these populations enjoyed well-balanced ecologies only one or two generations ago and had a quality of life superior in many ways to that of large sections of the Indian rural population—with adequate food, nonexploitative social structures, and freedom from indebtedness.[31] Contemporary problems, he argued, do not stem from within these societies but derive from the loss of land and resources.

When so much is at stake, when people are faced with displacement from their homeland and the disruption or termination of their way of life, it may become politically expedient in the defense of that way of life to misrepresent it and to emphasize, even exaggerate, the harmony of their relationship with their environment. It is also tempting to find in local practices all that is missing in development ideologies and practices, to find and use indigenous ideas to add legitimacy to our own ecological revisionist views. What is at stake is the idea of different kinds of society, but, inspired by political necessity and ideological hopefulness, the conflict becomes simplified into a view of two contesting views of nature and the environment.

Conclusion

It has been widely recognized in recent development literature that while the impacts of environment and development policy choices are often experienced most acutely at the local level, local communities often have little voice in the policy-making process. As a consequence, when they are able to make themselves heard at all, it is often in resistance to policies already decided elsewhere and implemented locally. In the Narmada case, too, people in the submergence area had no voice in the policy-making process, despite World Bank guidelines that project-affected and relocated people should be the first beneficiaries of a development project.

Environmental and human rights are varyingly described in the back and forth between Sardar Sarovar opponents and proponents. In these debates the right to water is opposed to the right to a low-energy use, low impact way of life; the right to development benefits is opposed to the right to participate in the decision-making processes that determine the reassignment of natural resources; and the development of a high-technology distribution of diverted water resources to (perhaps) needy segments of the population is opposed to the understudied, undervalued damage that will be done to an existing ecology. All of these concerns are framed within and subsumed under the terms sustainable development and social justice.

In the Narmada controversy we find every reference to Gandhi except, perhaps, his devotion to truth. By that comment I mean to strike a cautionary note about the power of rhetoric. Rhetoric is not an irrelevant and easily dismissed by-product of the development process. It may have the power to open up new possibilities, but it may also mystify what is actually happening. Rhetoric mystifies when it suggests consensus where there is none, directs attention away from conflict, and obscures relationships of inequality and power. Rhetoric does all of these things in the Narmada conflict.

The academic practice of solving problems by coining new terms is unlikely to cease anytime soon, but we need to direct our attention to the ease with which these new terms are easily hijacked by policy makers to conduct business as usual. It is so easy to embrace the rhetoric of sustainability and social justice and then to use these terms to defend essentially unchanged actions directed toward essentially unchanged ends. If, in the interactions among religion, ecology, and politics, politics usually prevails, we must resist premature celebrations of changes in rhetoric that are unaccompanied by changes in practice. Dominant discourses—be they developmentalist or environmentalist—have the power to absorb, coopt, and alter the way the views and values of local people are represented. The Narmada controversy is not just a simple disagreement about whether this dam is or is not a viable project. It strikes right to the heart of the philosophical, political, and moral debates about contemporary development efforts. What is called for, then, is not simply a more informed mechanism for deciding the costs and benefits of building the dam, but both a more fundamental transformation of the way development decisions are made and a reexamination of the measures by which difficult development trade-offs should be weighed.

It is clear that within this debate some points of view get more hearing than others. A great deal of violence has been done to the people of the Narmada valley—the great bulk of it by those who wish to flood the valley and uproot their communities, but some by those who misrepresent them in order to save them. In the immediacy of struggles, many "truths" are politicized and many nuances are unfortunately, but perhaps inevitably, overlooked. It is a conflict which, in part, pits those who can afford the luxury of further abusing the environment in order to exploit it against those who can afford the luxury of protecting it, while those who have no choice but to come to terms with life in a difficult environment must struggle on.

Notes

1. Narmada has become almost a cottage industry for journalists and a fertile topic for academics. It has been the subject of numerous books, several international conferences, at least three documentary films, a PBS radio special, and hundreds of magazine articles. Of the books published, see Y. K. Alagh, R. D. Desai, G. S. Guha, and S. P. Kashyap, *Economic Dimensions of the Sardar Sarovar Project* (New Delhi: Har-Anand Publications, 1995); Amita Baviskar, *In the Belly of the River: Tribal Conflicts over Development in the Narmada Valley* (Delhi: Oxford University Press, 1995); *The Dam and the Nation: Displacement and Resettlement in the Narmada Valley,* ed. Jean Drèze, Meera Samson, and Satyajit Singh (Delhi: Oxford University Press, 1997); *Toward Sustainable Development? Struggling over India's Narmada River,* ed. William F. Fisher (Armonk, N.Y.: M. E. Sharpe, 1995); Vidyut Joshi, *Rehabilitation: A Promise to Keep: A Case of the SSP* (Ahmedabad: The Tax Publications, 1991); and Vijay Paranjpye, *High Dams on the Narmada,* Studies in Ecology and Sustainable Development, no. 3 (New Delhi: Indian National Trust for Art and Cultural Heritage, 1990).

2. For sources on the Sardar Sarovar Project, see Alagh et al., *Economic Dimensions of the Sardar Sarovar Project;* Y. K. Alagh and D. T. Buch, "The Sardar Sarovar Project and Sustainable Development," in *Toward Sustainable Development?* ed. Fisher; Thomas A. Blinkhorn and William T. Smith, "India's Narmada: River of Hope," in *Toward Sustainable Development?* ed. Fisher; *The Dam and the Nation,* ed. Drèze, Samson, and Singh; Government of Gujarat, "Comment on the Report of the Independent Review Mission on Sardar Sarovar Project," draft (Gandhinagar, 1992); Bradford Morse and Thomas Berger, *Sardar Sarovar: The Report of the Independent Review* (Ottawa: Resource Futures International, 1992); Paranjpye, *High Dams on the Narmada;* C. C. Patel, "The Sardar Sarovar Project: A Victim of Time," in *Toward Sustainable Development?* ed. Fisher; and World Bank, Staff Appraisal Report: India, Narmada River Development—Gujarat. Supplementary Data Volume (1985).

3. See, for example, Asia Watch, "Before the Deluge: Human Rights Abuses at India's Narmada Dam" (1992); Shripad Dharmadhikary, "Hydropower at Sardar Sarovar: Is it Necessary, Justified, and Affordable?" in *Toward Sustainable Development?* ed. Fisher; William F. Fisher, "Development and Resistance in the Narmada Valley," in *Toward Sustainable Development?* ed. Fisher; Lawyers Committee for Human Rights, "Unacceptable Means: India's Sardar Sarovar Project and Violations of Human Rights," October 1992 through February 1993; *Lokayan Bulletin,* Special Issue on Dams on the River Narmada, 9, no. 4/5 (1991); Kalpavriksh, *Narmada: A Campaign Newsletter* (New Delhi), no. 5 (1990); Morse and Berger, *Sardar Sarovar*; Anil Patel, "What Do the Narmada Valley Tribals Want?" in *Toward Sustainable Development?* ed. Fisher; Medha Patkar, "The Strength of a People's Movement," in *Indigenous Vision: Peoples of India Attitudes to the Environment,* ed. Geeti Sen (New Delhi: Sage Publications, 1992); Medha Patkar, "The Struggle for Participation and Justice: A Historical Narrative," in *Toward Sustainable Development?* ed. Fisher; Rahul N. Ram, "Benefits of the Sardar Sarovar Project: Are the Claims Reliable?" in *Toward Sustainable Development?* ed. Fisher; Ashvin A. Shah, "A Technical Overview of the Flawed Sardar Sarovar Project and a Proposal for a Sustainable Alterna-

tive," in *Toward Sustainable Development?* ed. Fisher; and Lori Udall, "The International Narmada Campaign: A Case of Sustained Advocacy," in *Toward Sustainable Development?* ed. Fisher.

4. Opposition to the project has a long complex history. See Gail Omvedt, *Reinventing Revolution: New Social Movements and the Socialist Tradition in India* (Armonk, N.Y.: M. E. Sharpe, 1993); Anil Patel, "What Do the Narmada Valley Tribals Want?"; Patkar, "The Struggle for Participation and Justice: A Historical Narrative"; and Udall, "The International Narmada Campaign."

5. Morse and Berger, *Sardar Sarovar,* xii.

6. For the moment attention has shifted elsewhere, including upstream to another proposed dam site along the Narmada. See, for example, Narmada Bachao Andolan, "Narmada NAPM Tour," electronic release, 7 March 1996; Narmada Bachao Andolan, "Update," electronic release, 20 February 1998; National Front for Tribal Self Rule, "A Major Victory," press release, 5 March 1996; International Rivers Network, "World Commission on Dams Launched," press release, 16 February 1998; and the "Declaration in Support of the Struggle for the Promised Suspension of Construction on the Maheshwar Dam."

7. See the discussion in Robert Wade, "Greening the Bank: The Struggle over the Environment: 1970–1995," in *The World Bank: Its First Half Century,* ed. Devesh Kapur, John P. Lewis, and Richard Webb (Washington, D.C.: Brookings Institution, 1997).

8. See First International Meeting of People Affected by Large Dams, Declaration of Curitiba, Brazil, 14 March 1997.

9. International Rivers Network, "Independent Commission to Review World's Dams," *World Rivers Review* 12, no. 3 (June 1997); and Patrick McCully, with assistance from Peter Brosshard and Shripad Dharmadhikary, "An NGO Report on the April 1997 World Bank–IUCN Dams Workshop and on the Proposal for an Independent International Dam Review Commission," 4 May 1997.

10. Bradford Morse and Thomas Berger, *Sardar Sarovar: The Report of the Independent Review* (Ottawa: Resource Futures International, 1992).

11. Government of Gujarat, "Comment on the Report of the Independent Review Mission on Sardar Sarovar Project."

12. Gandhi's own words on this subject make this contrast clear. In 1928 Gandhi said: "God forbid that India should ever take to industrialization after the manner of the West. . . . If an entire nation of 300 million took to similar economic exploitation [as that of England], it would strip the world bare like locusts." Mohandas Gandhi, *Collected Work of Mahatma Gandhi,* vol. 38 (Delhi: Publications Division, Ministry of Information, Government of India, 1958), 243–44.

13. One might also compare Sir Mokshagundam Visvesvarayya's exhortation, "industrialize or perish," and Gandhi's response, "industrialize and perish" (cited in Sunil Khilnani, *The Idea of India* [New York: Farrar, Straus Giroux, 1998], 73).

14. Nigam officials often talked to me about their commitment to uplifting the lives of Scheduled Tribes and Castes. Most were sincerely convinced that current resettlement efforts would achieve this upliftment. Some officials were unable to restrain their frustration with the reluctance of project-affected people to agree to resettlement and expressed annoyance with valley inhabitants who were holding up

the completion of the dam. One, frustrated with the slow rate at which inhabitants of the valley had accepted resettlement, suggested that "they will move quickly enough when the waters of the reservoir rise and they are forced to climb to higher ground like monkeys" (personal communication, 12 January 1992).

15. These are but a few representative comments of many made to me by dam planners that echoed these sentiments. In conversations with me, Nigam officials seemed determined not to give in to Malthusian pessimism, or to let the activists cancel their hopes for the future.

16. Baba Amte moved to the banks of the Narmada to show his support for the inhabitants of the valley. In 1990 he accepted the Templeton Prize for Progress in Religion, saying, in part:

> "How long are we to watch passively as all that is our common heritage is destroyed and lost forever in the name of 'development'? Today I have become part of the battle to save the Narmada, one of the most sacred rivers in India, from massive dams which would destroy a whole way of life that depends on the river and its life-sustaining water. The battle is not for the Narmada alone, it has an even larger meaning. The battle is for the whole earth, to stop the immorality of destructive 'development' and replace it with a new vision, a new way of human living.
>
> "We must seek a path of greater kindness, tolerance and respect for all forms of life; a way of living founded on compassion, which seeks sufficiency for all rather than superfluity for some. *Real* development is natural . . . this, the way of nature, is the only basis for real development; it is our solemn responsibility to preserve and enrich our natural heritage for the sake of children yet to be born." Cited in Geoffrey Waring Maw, *Narmada: The Life of a River,* ed. Marjorie Sykes (Hoshangabad, 1991), 82.

17. For a discussion of the sacredness of the Narmada, see Chris Deegan, "The Narmada in Myth and History," in *Toward Sustainable Development?* ed. Fisher. See also Maw, *Narmada: The Life of a River.*

18. René Descartes, *Discourse on Method and the Meditations,* trans. F. E. Sutcliffe (Hammondsworth: Penguin, 1971), 78.

19. C. C. Patel, "The Sardar Sarovar Project"; and Alagh et al., *Economic Dimensions of the Sardar Sarovar Project.*

20. Nehru also observed in 1954, at the site of a high dam: "As I walked round the site I thought that these days the biggest temple and mosque and gurdwara is the place where man works for the good of mankind. Which place can be greater than this, this Bhakra-Nangal?" (Jawaharlal Nehru, *Jawaharlal Nehru's Speeches,* vol. 3, *March 1953–August 1957* (New Delhi: Publications Division, Ministry of Information and Broadcasting, 1958), 3.

The deification of Gandhi in the pantheon of nationalism and the designation of high dams as the temples of modern India seem peculiarly discordant.

21. This later claim serves as the basis for the rather astonishing insistence of the dam proponents that the Sardar Sarovar Project is an example of sustainable development. The dam builders insist that sustainable development is compatible with large-

scale, ambitious, centrally controlled schemes that are capable of mitigating the effects of natural catastrophes and meeting the increased needs of a growing economy for food, water, and energy.

22. Joshi, *Rehabilitation: A Promise to Keep,* 68.

23. William F. Fisher, "Full of Sound and Fury? Struggling Toward Sustainable Development," in *Toward Sustainable Development?* ed. Fisher.

24. Ramachandra Guha has identified three distinct strains (crusading Gandhian, appropriate technology, and ecological Marxism) that he believes contribute to contemporary environmental positions within India: see Ramachandra Guha, "Ideological Trends in Indian Environmentalism," *Economic and Political Weekly* 23 (1988): 29, and "The Environmentalism of the Poor," in *Between Resistance and Revolution: Cultural Politics and Social Protest,* ed. Richard G. Fox and Orin Starn (New Brunswick, N.J.: Rutgers University Press, 1997). The Narmada controversy also awkwardly encompasses the environmentalism of the poor as well as the various ideological environmentalisms of Indian intellectuals and northern NGOs.

25. I base this summary on extensive conversations with Narmada Bachao Andolan activists as well as on their own publications.

26. See, for example, Sen, *Indigenous Vision,* and Pramod Parajuli, "Power and Knowledge in Development Discourse: New Social Movements and the State in India," *International Social Science Journal* 127 (1991): 173–90. Baviskar (*In the Belly of the River,* 47) summarizes this position this way: "The collective resistance of indigenous people is not a rearguard action—'the dying wail of a class about to drop down the trapdoor of history'—but a potent challenge which strikes at the very heart of the process of development."

27. Article 46 of the Indian constitution reads: "The State shall promote with special care the educational and economic interests of the weaker sections of the people, and, in particular, of the Scheduled Castes and Scheduled Tribes, and shall protect them from social injustice and all forms of exploitation.

28. See, for instance, the International Labor Organization Convention no. 107 and the 1987 statement of the World Commission on Environment and Development, both cited in Independent Commission on International Humanitarian Issues, *Indigenous Peoples: A Global Quest for Justice* (London and Atlantic Highlands, N.J.: Zed Books, 1987), and World Bank, Operational Directive 4.01 (1991).

29. For a detailed discussion of the Bhilala community of the Narmada valley, see Amita Baviskar's excellent book, *In the Belly of the River,* particularly chapter 7.

30. Christoph von Fürer-Haimendorf, *Tribes of India: The Struggle for Survival* (Berkeley and Los Angeles: University of California Press, 1982), and *Tribal Populations and Cultures of the Indian Subcontinent* (Leiden: E. J. Brill, 1985).

31. Fürer-Haimendorf, *Tribal Populations and Cultures of the Indian Subcontinent,* 170.

Green and Red, Not Saffron: Gender and the Politics of Resistance in the Narmada Valley

PRATYUSHA BASU and JAEL SILLIMAN

> It was not as though we did not have an ideological framework. This whole alternative paradigm of development which we are propagating and also the ways and means to achieve it stem from that framework. I have always been for democratic socialism as far as political ideology is concerned. But at the same time we feel that both kinds of dogmatic forces . . . have somehow not taken [into account] the man-nature relationship, which is a newly added factor and newly realised issue which would define . . . the lifestyle as a whole. . . . So that factor has to be brought in which might lead us to something like environmental socialism. . . . [T]he combination of green and red values and ideas has constituted our ideological perspective.
>
> —Medha Patkar, leader of the Narmada valley movement, in a 1992 interview

The construction of a series of dams on the Narmada River is being vigorously opposed by the Narmada Bachao Andolan (Save Narmada Movement). The primary concern of the movement is that large-scale human displacement has already occurred and will occur as a consequence of these constructions. The Indian central government and the state governments of Gujarat and Madhya Pradesh claim that these dams will provide electricity, irrigation, drinking water, and flood control. To the Andolan, this emphasis on benefits and discounting of costs signals the need for an alternative development paradigm. This alternative development would be linked to social justice and would serve the interests of the poorest sections of India's population, instead of catering mainly to upper-class and urban segments. This focus on social justice comprises the "red" aspect of the Andolan's ideology.

The environmental destruction that accompanies dam building is also a major concern in the Andolan's resistance. This has enabled the Andolan to draw the attention of and mobilize support for their cause from a number of national and international environmental organizations. This transnational environmental network has, to a large extent, forced the state and multilateral development organizations, like the World Bank, to reconsider their pursuit of development through megadams. This comprises the "green" aspect of the Andolan's strategy of resistance. A combination of green and red ideologies ensures that a focus on environmental protection does not restrict the access of the poor to environmental resources in the name of conservation.[1] The environmentalism pursued by the Andolan is thus one that seeks to balance the interests of human beings and nature, instead of pitting them against each other.

Religious ties to the river are also significant as the Narmada is one of the holiest rivers for Hindus in India. In fact, it is said that even the Gaṅgā, the most important of India's rivers, bathes in the Narmada to purify herself.[2] Hindu religious practices which center on the river are very severely affected by obstructions on the river and by consequent alterations in its course and submergence of areas along its banks. A major Hindu pilgrimage consists of circling the Narmada, beginning from its source, going down one bank, and coming back to the source along the other bank (this circumambulation is known as *parikrama*). As Chris Deegan points out:

> Pilgrimage tradition in India has long been dependent on networks and centers to fulfill the needs and expectations of its sojourners. Essentially geographic behavior, it is both the belief in, and a reaction to, a chosen landscape whose fundamental energy comes from a combination of its sacred geography, religious circulation, and attachment to place.[3]

The submergence of holy sites, which functioned as rest stops for pilgrims at various points along the bank of the river, has disturbed the original route of the Narmada *parikrama.* An obstructed river is in fact no longer as holy. For worshipers who see the Narmada as a divine force, dams represent a human interference that can only bring destruction.

For tribals too, the Narmada has been a source of cultural meanings and attachment. According to Bava Mahalia, a tribal leader speaking on behalf of the tribals who are part of the Andolan, "[w]e [tribals] worship our gods by singing the gayana—the song of the river. . . . The Narmada gives joy to those who live in her belly."[4] In Amita Baviskar's words:

> The gayana, creation myth sung during indal ['indal pooja' is 'the worship of the union of the rain and earth which brings forth grain'], links the origin of the world to the river Narmada. The gayana, perhaps more than any other part of their religious complex, sets the Bhilalas apart from the Hindus to whom the myth is entirely unfamiliar. . . .[5]

Moreover, tribals residing along the Narmada have built economic and social relationships with the pilgrims passing through their villages. A documentary by Anurag Singh and Jharana Jhaveri (*Kaise Jeebo Re* [How Do I Survive, My Friend], 1997) on the effects of submergence and displacement due to the Bargi dam, briefly focuses on the relationship of tribals with pilgrims. In it a tribal man laments the loss of opportunities to meet *sādhu*s (holy men), as the pilgrimage stops are no longer near tribal villages.

Amita Baviskar, however, cautions against a romanticization of the relationship between tribals and the river. She notes that while the tribals "have rituals for the propitiation of almost every natural phenomenon, little or no ritual surrounds the river." A number of pilgrims also complain about the harassment and robbery that they are subject to in tribal villages. According to Baviskar: "The ritual of circumambulation has no meaning for the adivasis [tribals]; pilgrims are fair game to them."[6] Despite these qualifications, the Narmada plays an important role in both tribal and Hindu cultures. The alteration of the river's course, as well as the actual human displacement due to the dams, deprives both men and women of their cultural and spiritual relationships with the river.

Environmental movements within India and transnationally have, in some instances, stressed spiritual connections to nature. The Andolan, in order to avoid links or associations with Hindu fundamentalism, has stressed cultural and religious links to nature only to highlight the distinctiveness of tribal populations. The strident "saffron" ideology of Hindu fundamentalism is thus deliberately eschewed in the Andolan's opposition to the dams. Notwithstanding this reluctance to invoke Hindu symbols, the representation of the Narmada River in Hindu and tribal cultures as *mātā,* or mother, is central to the symbolism of the anti-dam movement. According to Medha Patkar, "The concept of womanhood, of *mātā,* has automatically got connected with the whole movement, . . . the concept of Narmada as *mātā* is very much part of the movement. So if a feminine tone is given—both to the leadership and the participants—then the whole thing comes together."[7]

Our aim is to examine this "feminine" tone of the Andolan, as expressed in women's leadership, women's participation, and the voicing of women's concerns.[8] We show how women's experience of displacement due to dams is crucially shaped, not just by their position in the gender hierarchy, but also by tribal-Hindu differences. The activists of the Andolan have very successfully centered global attention on flaws in the current paradigm of development and the need to rethink development. Building on the powerful possibilities opened up by women's leadership and participation in the struggle against dams on the Narmada, our study focuses on the significance of opposition to gender inequities in the process of formulating a truly radical alternative to development.

Gender and Networks: Examining Tribal Links to the Land

Official development agencies, including the Indian state and the World Bank, treat land as "a fungible commodity without regard for the social and spiritual ties people have to the land."[9] The Narmada movement opposes this narrow view of land as commodity and highlights the impossibility of compensating people for the loss of their cultural ties to the land, river, and forests. The attachment to land is more marked in the case of tribal people and the movement thus foregrounds the tribal when it stresses links to the land. In this vein Bava Mahalia says:

> To you officials and people of the town, our land looks hilly and inhospitable, but we are very satisfied with living in this area on the bank of

> the Narmada with our lands and forests. We have lived here for generations. On this land did our ancestors clear the forest, worship gods, improve the soil, domesticate animals and settle villages. It is that very land that we till now.[10]

The Andolan is one with Bava Mahalia in asking of the state: "What price this land? Our gods, the support of those who are our kin—what price do you have for these? Our adivasi life—what price do you put on it?"[11]

Yet, to focus on links to a specific place is to present only a part of what is being lost in displacement. For more than links to a particular piece of land, what is being destroyed is the particular relationships between places. This loss of specific cultural sites as well as networks within and between places is stressed by the activists when they draw attention to

> the dismantling of production systems, desecration of ancestral sacred zones or graves and temples, scattering of kinship groups and family systems, disorganization of informal social networks that provide mutual support, disruption of trade and market links, etc. Essentially, what is established in the accumulated evidence in the country suggests that . . . forced displacement has resulted in a "spiral of impoverishment" . . . [T]rade links between producers and their customer base (and systems of exchange and barter) are interrupted and local labor markets are disrupted. Additionally, there is also a loss of complex social relationships which provided avenues of representation, mediation and conflict resolution. . . .[12]

The gendered nature of the networks lost due to displacement are for the most part not highlighted in the movement's oppositional discourse, even though Medha Patkar and the other activists of the Andolan are very sensitive to the particular predicaments of women in the valley. For women, networks that connect places are even more crucial, as they provide social support, financial assistance, and shelter in times of distress. Thus, "Women typically depend a great deal on such informal support networks, which they also help to build through daily social interaction, marriage alliances which they are frequently instrumental in arranging, and complex gift exchanges."[13] According to Baviskar, among the Bhilala tribals, "a wedding produces further structures of help, reciprocity and security. Affines are the source of economic aid in times of distress."[14] Such networks are lost and cannot be easily replicated in resettlement sites. For women, displacement leads to a loss of support that differs from and probably exceeds that suffered by men.[15]

Moreover, the widening of the river has specific implications for women, as it impinges on the maintenance of links between villages. In tribal areas, the Narmada River allows for easy travel between women's natal and marital homes, especially since natal and marital homes are not very far from each other in the case of tribal communities. In places where the river has become particularly wide after dam construction, such links between villages are no longer available, and tribal women can no longer obtain their natal kin's support with the same ease. According to Patkar, "The river is converted into a sea, the one small little plank boat of the tribal people cannot even cross the river, and the country boats are also becoming more and more useless even for going from one village to another."[16] Baviskar explores the impact of dam-related submergence on Binda, a woman belonging to the Bhilala tribe:

> if the dam is built . . . an ocean will separate her from her natal village Arda. At present, Arda is across the river [and she goes there on a rough raft built by her husband, Khajan]. . . . When she goes there, she can get a short respite from her responsibilities, get news about all her relatives and spend time with her parents. The children look forward with great excitement to visits to Arda. . . . For Khajan [her husband] too, Arda has been a source of support. . . . And Binda's parents always send their daughter back home with a gift—a bagful of *muhda* flowers, a bamboo basket woven by her father, or some roasted gram. If the dam is built, this network of affection, caring and help, will come to an end and Binda will be marooned on the mountain top.[17]

For a woman, then, "two places define her life, her sense of location, her identity."[18] This relationship between places is severely disrupted by displacement.

The extent to which the geography of female kinship might frame the opposition to displacement is lucidly expressed by Bava Mahalia:

> Our daughters' and sisters' husbands' villages are close by here; our wives' natal homes are also near. When we go away from here, then we will never get to meet our relatives. They will be as dead to us. The women of our village threaten us, "We are willing to leave our husbands; we can always find other men. But we can't get other parents;

> so we will never leave this place. In Gujarat [where the resettlement sites are located], if any sorrow or evil befalls us, to whom can we go to tell of our troubles? You are not going to give us the bus fare and send us back, are you[?]"[19]

Patkar quotes an old woman in Chimalkheda as saying: "'I will not move. I will die here. They may give land and water and whatever. But can they give me my Chimalkheda? That is not possible. So, why should I go?'[20] An understanding of the gendered nature of networks allows us to appreciate what this woman is opposing. It is not just the loss of Chimalkheda, but the facility with which she can access her extended kin due to her placement in Chimalkheda.

Land Ownership and Women: Losing Tribal Rights

Issues of resettlement and compensation for the displaced have been at the core of the Narmada struggle. More recently, the movement has decided to oppose any construction of dams on the Narmada, and hence has moved away from focusing on the particulars of an ideal resettlement and rehabilitation package. In spite of this, the Andolan's previous efforts have very successfully ensured that tribals are not deprived of their land without proper compensation. The lack of legally recognized title to land in tribal societies no longer stands in the way of tribal rights to be compensated for loss of land. This inclusion of tribal ownership has proceeded alongside the exclusion of women's rights to land. Thus, tribal women in the Narmada valley have been disempowered as the investment of ownership rights in men has served to further strengthen the position of men in tribal communities.

A particular conception of the household and of women's proper position within it drives the Indian government's view of compensation for displacement. The Award of the Narmada Water Disputes Tribunal defines a "family" as the "husband, wife and minor children and other persons dependent upon the head of the family, e.g. 'widowed mother.'"[21] In keeping with this award, the various state governments promise land to each family and give this land right to the head of the household, nominally assumed to be male. Furthermore, the tribunal treats every major son as a separate family, which is seen as a radical improvement on previous compensation packages. Giving in to protests by those seeking more equitable resettlement packages, the state governments decided to recognize the rights of widows and unmarried daughters to land. In 1990 the Gujarat government decided to provide all women who became widows after 1980 with rights to land, and in 1993 the Maharashtra government included adult daughters (unmarried as of 1 January 1987). All other women are considered to be completely dependent on their husbands.

The assumption that drives national resettlement policy is that land rights for men will ensure the well-being of women in their households. Central to this view is the notion that the household is a unitary category and that there is a shared harmony of interests among household members. Feminist scholarship has challenged these gender-blind assumptions by documenting persistent intrafamily inequalities in access to and use of resources.[22] Gender inequalities also exist in allotment of household tasks among family members and in the ability of household members to participate in financial decision-making. Gender differences in perceptions of family welfare, and hence in household priorities, have also been underlined in feminist analyses of the household. By investing land rights primarily in men, the tribunal award ignores the inequalities that exist within households.

More significantly, the tribunal award ignores female-headed households and the fact that not all women can depend on men for financial support. The presence in rural India of deserted, separated, and divorced women testifies against the norm of the male-headed household.[23] According to Brinda Rao, "Informal estimates of deserted women made by voluntary organisations indicate that a third of women [in rural areas] between the ages of eighteen and thirty years are either deserted by their husbands, or choose to leave them."[24] Often, due to social pressures, these deserted women pretend to be married to men who live and work in the cities.[25] Government officials remain very unwilling to consider land rights for such women in their resettlement policies. According to Priya Kurian, standard responses include:

> "In this country, there is no custom of women holding land."
>
> "You don't come across [deserted or unmarried] women. You will always find women with parents or a brother or someone to look after them."
>
> "[I]f a woman is single, she will have to have witnesses to establish that she is virtuous and of noble character."[26]

In this framework, only those women that fit the dominant patriarchal norm—daughters, wives, and widows—are bestowed with some, though unequal, rights to land. In contrast, single, divorced, or deserted women, who do not adhere to these norms or violate them, have their rights to land forfeited.

What is occurring in the Narmada valley is a process whereby a patriarchal notion of land ownership is being imposed by the state on tribal societies. It is significant that the official viewpoint on land ownership replicates women's rights to land under Hindu law. Under Hindu law, women's right to property is solely their "right of maintenance as incoming wives (including as widows) and as unmarried daughters: when she [is] married, a daughter [is] . . . also entitled to marriage expenses and associated gifts."[27] Further, only "chaste" widows can inherit land in the absence of sons and, "Unlike sons, married daughters (even if facing marital harassment) have no residence rights in the ancestral home."[28] While the situation for women in tribal societies is not ideal, in some ways it is less restrictive than under Hindu law. For example, Govind Kelkar and Dev Nathan, in documenting tribal women's rights to land in Jharkhand, show that widows have land rights equal to that of their late husbands.[29] This right is the one that is most under attack currently in the Jharkhand region. There, too, tribal widows are being relegated to a state similar to that of Hindu widows. However, there are still some striking differences between tribal and Hindu widows due to:

> (a) the continued participation of such adivasi [tribal] widows in labor, both in the field and forest as well as the home, as a result of which widows too can have their own income from the collection and sale of various kinds of forest produce; and (b) the right which widows still have, in the event of maltreatment by their relatives, to demand the allotment of a separate plot of land for their maintenance.[30]

The differences between tribal and Hindu land ownership patterns are not being seriously investigated. Consequently, the lack of land rights for tribal women in resettlement policies mirrors the legal status of Hindu women.

According to Bina Agarwal, "the gender gap in the ownership and control of property is the single most critical contributor to the gender gap in economic well-being, social status, and empowerment. In primarily rural economies such as those of South Asia the most important property in question is arable land."[31] Thus, the empowerment of women depends crucially on their access to land ownership, an issue that has not been raised in connection with displacement. The Narmada movement has set precedents in its attention to environmental and tribal issues. It is probable that a sensitivity to gender inequities here would have far-reaching consequences in terms of an alternative development that would be truly inclusive. As Agarwal points out, the issue of land rights for women is not raised by social movements for pragmatic reasons: "acknowledging the varying and sometimes conflicting preference and interests and access to resources among family members, means admitting new contenders for a share in this scarce and highly valuable resource . . . [and bringing the conflict over land] into the family's innermost courtyard."[32] But as Kelkar and Nathan insist, "The demand of land for women cannot be taken up in isolation, nor can it be won by women alone. . . . It is also necessary that the men realize that the demand is not such as will disrupt and destroy whatever is left of the community."[33] The issue of land rights for women in the Narmada is crucial to the participation of women as equals in tribal and Hindu communities. Thus, by emphasizing tribal rights to land without questioning the investment of this right in men, the Andolan might be inadvertently contributing to a weakening in the position of women in tribal societies and aiding the uncritical assimilation of tribals into Hindu law.

Women, Work, and Environment: Domestic Duties and Knowledge of Place

For women, the place-based experience of the Narmada valley is intrinsically linked with their primary responsibility for domestic work, which requires them to collect fuelwood, water, and forest produce for consumption at home. The much-debated links between women and the environment should thus be "understood as rooted in their material reality, in their specific forms of interaction with the environment."[34] Thus, both men and women depend on the forests and the rivers, but their gender-specific use of and relationship to the surrounding environment plays a major role in determining their particular environmental knowledge. According to Ratna Kapur, "tribal women are dependent on nature—the forests and the rivers form the cultural parameters in which their lives are laid out."[35] To be uprooted from this environment is thus to lose familiar patterns of work and be subjected to disempowering uncertainties.

Andolan activists have focused on the extent to which tribal women's workloads increase considerably in resettlement sites. Thus, while livestock and cattle could obtain water directly from the river in the original villages, in resettlement sites women have to pump water for hours in order to obtain enough to provide for animals as well as household needs. So, even as development measures would present the provisioning of water through tubewells as an indicator of progress, it is access to the river that works out to be advantageous for tribal populations. Besides water, the fish caught in the river are an important source of protein in tribal diets, a food item that is sorely missed in resettlement sites. The right to fish has actually been taken away even from the villagers who have stayed behind. The government has sold rights to fishing in the new lakes created by the dams to private parties who refuse to allow tribals access to the fish for personal consumption. The contempt of the fish contractors for the tribals is vividly portrayed in the 1997 documentary film *Kaise Jeebo Re*. Similarly, the absence of nearby forests in resettlement sites deprives tribals of supplementary sources of food and fuel. Labeled "minor forest produce" by the government, these forest products in fact constitute a significant part of tribal livelihood. Further, medicinal products obtained from forests provide tribals with affordable cures for everyday ailments. Knowledge, often among women, of these forest-based cures becomes useless with the loss of the forest. The loss of an ecological environment that can be put to a variety of uses impoverishes women in terms of their ability to provide for their households. Women in resettlement sites thus become dependent on the market or have to travel long distances to fulfill their domestic duties.

Yet, raising the issue of gender-specific impacts only within the context of women's work and displacement leaves the traditional division of labor in tribal societies unchallenged. For Kapur, "the situation of tribal women must not be glorified. The culturally accepted division of labor within the family leaves the collection of household needs like fuel, fodder and water to women," and this is both time-consuming and arduous.[36] Moreover, it must not be assumed that tribal women's interactions with the environment occur in the absence of power relationships based on gender hierarchies. As Kelkar and

Nathan show in the case of Jharkhand, "Though women do play a major role in the labour of gathering of forest produce they do not play a commensurate role in the management of forests, which today is the province of either the *panchayat* or the village community."[37] Thus, relationships with the environment and the gender division of labor are shaped by patriarchal power, and environmental experiences are not freely chosen. The Andolan's limited focus on gender in the context of the increased difficulty associated with fulfilling domestic duties in resettlement sites points to the movement's strategically motivated unwillingness to address gender inequalities in tribal societies. The lack of a critical examination of the gendered structure of work in the tribal context and a too easy celebration of tribal links to the environment glosses over the very crucial gender inequalities which maintain such workloads and environmental linkages.

New Employments: Tribals in an Unfamiliar Economy

Tribal women and men are, often, not equipped to perform the types of jobs that might be available around resettlement sites, due to illiteracy and unfamiliarity with the dominant culture. Supporters of the dam argue that the Sardar Sarovar Dam will generate employment. However, the government of Gujarat does not specify the skills required to access these jobs or the wages provided by these employment opportunities. It also does not address the differences in employment opportunities for men and women. Global employment data for the 1980s compiled by the International Labor Organization (ILO) has found that "women made the most significant gains in the areas that extend their traditional domestic roles (i.e., community, social and personal services)."[38] According to Brett O'Bannon, this "suggests that female participation at Sardar [Sarovar] will be largely in unskilled or semi-skilled positions which are poorly remunerated."[39] Drawing upon a 1993 study conducted by Bombay's Tata Institute of Social Sciences, O'Bannon states that:

> Although Narmada valley tribal women traditionally engaged in agricultural production on a relatively equal footing with men, [displaced] women in Maharashtra have experienced wage disparities in their new economic roles. Increasing their wage labor activity in response to growing cash requirements in the new economy, women in Maharashtra are

> earning only Rs. 7 per day, compared to Rs. 10 per day earned by the male counterparts.[40]

Since women's demands have not been identified, there are no policies or programs designed to give women the skills and training they would need to improve their employment prospects.

The lack of access to forests has made resettled tribals dependent on the market, which makes the earning of cash wages urgent. Besides this, the land provided to them is in many cases waterlogged, saline, or generally inferior to their original landholdings, which makes nonagricultural wages even more crucial.[41] Of course, not all the displaced are agriculturists. As Vasudha Dhagamwar, Enakshi Thukral, and Mridula Singh point out, "The planners of the Gujarat policy seemed to have forgotten that people belonging to other categories, such as artisans, fisherfolk, barbers, potters, blacksmiths, and so forth, also reside in the submergence zone."[42] New employment is necessary for survival; yet, new opportunities are unlikely to match the existing skills of the displaced.

Also overlooked by the planners, in their emphasis on employment and urban facilities, is that the cash economy is likely to be unfamiliar terrain for tribals. Typically, tribals are excluded from the benefits of the market. As Kapur notes, "the advantages of urban life accrue much more to the rich than to the poor."[43] According to Dhagamwar et al.:

> Cash in the hands of the poor, especially the tribals who have little or no exposure to the outside world, has very little meaning. Unaccustomed as they were to so much money, the tribals fell prey to the petty businessmen selling colorful trinkets and consumer items like transistors and watches, or they lost the money gambling and by spending it on liquor.[44]

Bava Mahalia, too, makes a similar point regarding the insignificance of money in tribal society: "We [tribals] grow so many different kinds of food, but all from our own effort. We have no use for money."[45] About the forest, Bava says: "We have lived in the forest for generations. The forest is our moneylender and banker. In hard times we go to the forest."[46] Of course, the tribal economy is not completely nonmonetized and Baviskar's study of Bhilala society reveals the many ways in which tribals are integrated into the surrounding market economy.[47] Yet, it is also true that the surrounding forests provide the tribals with a certain amount of "free goods" and tribal agriculture ensures their subsistence. Also, tribals are integrated into the market economy, not as equals, but via a series of exploitative relationships. In the resettlement sites the fulfillment of subsistence needs is threatened by the complete dependence of tribals on these exploitative relationships. For women, especially tribal women, limited rights in land and agricultural produce are further eroded through displacement that provides remuneration to men. In this context, women's dependence on new employment opportunities further disempowers them.

Tribal Women Encounter Hindu Customs: The Burden of Difference

The differences between tribal and Hindu cultures are especially striking in the area of gender relationships.[48] In the Nimad plains, tribal women have greater mobility as compared to many Hindu women who are subjected to domestic seclusion. Further, tribal women "address all men by their names and speak to outsiders far more easily than the women from the plains."[49] For tribals, the choice of a marriage partner is one that, for the most part, lies equally with men and women, and adolescents are allowed to mix freely even before marriage. Divorce, too, is relatively easier. Women will have a place in their natal homes in the event of a breakdown of marital relations, and widow remarriage is possible. Baviskar details some aspects of gender relations within Bhilala society:

> The festival of Bhagoria in February-March, when adivasis [tribals] gather together in the haats [markets] to select their partners celebrates that freedom of choice. Marriages are sealed by the payment of brideprice to the woman's father; they are broken when she leaves her husband, who then tries to get his money back. . . . Finally, the notion of the gift of a virgin appears ludicrous in a society that does not place a high value on the chastity of unmarried women. Sexual liaisons are easily acknowledged and people live together before wedding rites are performed. All of which is quite contrary to the Hindu norm.[50]

Hindu women in these locales have relatively little choice in terms of marriage partners. They have no rights in their natal home and are particularly powerless when widowed. Remarriage is possible only in the rarest of circumstances.

In the Narmada valley tribal customs are being undermined in the process of Hinduization. Hindu caste society labels tribal life-styles "deviant," "immoral," and "backward," especially due to the "freedoms" extended to tribal women. Development planners, who seek to resettle tribals, do not consider the dissonances between Hindus and tribals as potentially dangerous for tribal men or women living alongside nontribals in resettlement sites. In the view of officials, "development" for tribals quite naturally requires them to give up tribal ways of living. An especial point of attack is the way tribals clothe their bodies. Kurian quotes a member of a pro-dam organization as saying:

> The Bhils were always in loincloth, the women were bare-breasted. But when they came here, the men were wearing trousers and all were fully clothed. They are learning to type and learn new skills. But is that wrong? The element of coercion is minimized. . . . No, the project is not destroying the people.[51]

This narrow focus on clothing, which is then taken to be an expression of widespread "immorality," avoids a focus on tribal culture as a way of living that may have much to commend itself.

In fact, displaced tribals are thrust into an alien context which "others" them and seeks to absorb them completely into the lowest levels of the society and economy. As S. Srinivasan shows, the reasons given by the displaced for returning to their original homes include "hostility of nearby villages," "lack of safety for women," and "cultural alienation."[52] Tribal women, especially, bear the brunt of sexual attacks and harassment. These are justified by the attackers, who see tribals as "loose." Since women's wages are crucial for displaced populations, tribal women most likely face sexual advances in their everyday work. In fact, the lack of employment opportunities combined with social stigma has led many tribal women to prostitution. According to the *Lokayan Bulletin,* "The documented case of Ukai [dam] where tribal women displaced by the projects were regularly soliciting truck drivers on the national highway from Baroda to Ahmedabad is not an exception."[53]

The freedoms of tribal society and the constraints placed on women by Hinduization should not be taken to imply that gender inequalities do not exist within tribal communities. Besides the relative inequality in terms of land rights and political participation, witch hunting is also an established practice among tribals, and the violence against women sanctioned through this practice is, as Kelkar and Nathan point out in the case of Jharkhand, "not the suppression of a rebellion, but the process of establishing the authority of men, the process of establishing patriarchy."[54] The Andolan to a large extent recognizes the evil of witch hunting, the practice of *dakan,* and seeks to persuade tribal men and women to desist from this practice.[55] Yet, witch hunting has, for the most part, been raised only as a peripheral concern by the Andolan.

Rao points out that violence against women increases in times of stress.[56] It is highly likely that the stress of displacement, and the consequent inability of men to negotiate alien economic contexts, is resulting in increased alcoholism and wife-beating. The issue of violence against women has never been taken up by the movement, whether in the context of tribal societies or in the context of displacement. For tribal women, then, and women in general, displacement is an experience that reduces their power at home and outside.

Women's Participation in the Movement and the Use of Gandhian Tactics

Women have been very actively involved in the struggle against dams on the Narmada. The leader of the movement, Medha Patkar, is herself a woman, and so are a large number of the urban and rural activists who play central roles in the struggle.

Before considering women's participation in resistance, it is useful to consider the way in which women are represented in the state's pro-dam publicity exercises. This representation gives a sense of the position that the state seeks to accord women in its project of development. We have already talked about the way in which the state has ignored women in its compensation package. Yet, this has not prevented the state from using tribal women's images to symbolize the "success" of its resettlement programs. At the official dam authority guesthouse, a caption beneath a colorful photograph of women in tribal attire represents them as "happy" and contented women. This conceals the way in which tribal women have been further disempowered and impoverished by resettlement. They have neither been provided for nor consulted in official planning. What is also erased in this image is that tribal women are usually called upon to perform such dances for outside viewers, typically government officials and their guests, to demonstrate the success of the project. Performed for outside visitors, such dances lose a major part of their cultural meaning. These isolated performances, far removed from the forests and rivers that are the bedrock of tribal culture, no longer express tribal festivity; instead, they become pro-resettlement performances that hide the trauma of displacement. According to Kelkar and Nathan, "the practice of community dancing among the tribals is looked down on as a sign of 'backwardness.'"[57] It seems, then, that the government is not averse to appropriating "backward" cultural expressions when it suits the pro-dam purpose.

Against this blatant manipulation of tribal women's images by state officials, the Andolan provides women with opportunities to be active participants in the struggle against the dam. According to Baviskar:

> The Andolan has . . . enabled the women of Nimad, traditionally jealously cloistered, to come out of their homes and take to the streets, demonstrating in front of Project authorities' offices, raising slogans, challenging the police and taunting bureaucrats and politicians. This revolutionary change has been brought about by Medha Patkar and other women activists of the Andolan. They inspire women with a vision of what they can safely accomplish outside the home; their presence reassures men that women are 'safe'. This sensitivity of the Andolan to the gendered nature of its constituency is one of its greatest achievements.[58]

Patkar sees the inclusion of women as something that is best achieved gradually over time. In her words:

> When I started working in tribal villages, tribal women were not at the forefront. I have never insisted that women be dragged into it from the beginning. I have been criticized for this—why women are never seen at my rallies, etc. . . . I think it doesn't make sense to insist that women must attend our rallies, only for the sake of appearance, included in decision-making structures. . . . Women got some exposure as they began to attend our big rallies. At the village level, we tried to include women as much as we could. Things have changed a lot since those early days. We were conscious of this but we were never for somehow or the other enforcing it. . . . Women now know about the fight and gradually their participation is increasing.[59]

Women's active role in the Andolan includes participating in long marches, in demonstrations, and in hunger strikes. One instance of women's participation is provided by Patkar:

> Take Manibeli for example. On 3rd August when all the men went to jail, the women immediately took over. . . . The women used to take out rallies three times a day. . . . If a policeman picked up a lathi, women would turn their bare backs and say *maro, maro*. . . .[60]

The presence of women in the Andolan's demonstrations is very palpable in the films made on the struggle in the valley. Ali Kazimi's 1996 documentary, *Narmada: A Valley Rises,* records very severe attacks on women participants by the police in the course of the Andolan's peaceful march to the Sardar Sarovar Dam site. Baviskar also describes attacks on women and links these to the Andolan's strategic placing of women at the forefront of demonstrations. Thus, in the course of these encounters between the Andolan and the police, women's "clothes are ripped off in public, they are dragged along by the hair—in one incident, a pregnant woman was repeatedly hit on her stomach with a rifle butt."[61] Patkar stresses women's ability to endure personal hardship:

> We were really shocked and surprised to see the extent woman can go—whether it is a 70 year old woman who walks 100 km for a *padayatra* [a march] or a woman who can snatch a form from an *adhikari* [an official]. . . . Women have tremendous force. . . . Men, too, are capable of an intense emotional attachment to an issue but it is more so in women.[62]

In many ways, and especially in this invocation of women's natural ability to endure suffering, the Narmada valley struggle follows the model for women's participation provided by the nationalist struggle, and especially by Gandhi. This is not surprising, as Gandhi has become a symbol that is repeatedly evoked by social justice movements in India, especially since Gandhi serves as a foil to Jawaharlal Nehru, who is seen as primarily responsible for the centrality of dams in India's development. Regarding women, Patkar stresses that

> As the strength of interior, traditional, tribal societies lies in their homogeneity or integrity, similarly the strength of women stems from the fact that they are not exposed and, therefore, not *kalankit* [tainted]

> or *bhrasht* [corrupted] by outside influences. Therefore, women have conserved the strength that comes from their honesty or integrity or perseverence.[63]

Richard Fox writes about Gandhi's perceptions of women in very similar fashion: "Gandhi believed that women were essentially stronger than men because they could bear suffering better, especially better than cowardly men. The ability to suffer, as Gandhi's non-violent methods required, came from an inner strength and spiritual discipline."[64] Gandhi expanded the spaces available to women by extending the "family" to encompass the "family of the nation." Women's moral strength was seen by him as crucial for a strong national movement, and women were revered for their selfless devotion as mothers of the nation. Similar to women's strategic placement at the head of demonstrations in the valley, Fox describes how women in the nationalist movement formed protective cordons between men and the police, and thus ran the risk of being beaten.[65] In the nationalist movement this stress on feminine strength mainly served to bind women even more to their domestic roles and to their position as mother. Women's emancipation was seen by nationalist leaders as a spontaneous outcome of the national movement; yet, as Saraswati Raju points out, independence has not naturally guaranteed gender equality.[66]

However, Fox emphasizes some positive outcomes of Gandhi's essentialization of women: "Women's public participation in nationalist meetings and reformist associations developed very quickly and almost entirely from Gandhi's legitimization after 1920. The Gandhian Constructive Programme became a training ground for their further participation."[67] Gandhi's essentialism, in providing legitimacy to women's political participation, ultimately led to women's empowerment and participation in struggles beyond the protection of the (patriarchal) home. Thus, it "had provoked and legitimated women's active involvement in nationalist protest to the extent that they could break with the restraints he would have placed on them. . . . His essentialism, although still affirmative, could not any longer contain the modern consciousness of Indian women, even though it had once been a major inspiration for that consciousness.[68] Similarly, Patkar believes that the "rally in Ferkua was a new phenomenon in . . . [women's] lives. You just have to ignite a spark and then leave it and it will burn on its own."[69] Yet, against this optimism, Joni Seager's words need to be invoked as a caution: "The weight of history suggests that if women's issues are not taken up in the midst of social change, indeed as central to the agenda for social change, then they will not receive a hearing in the flush of whatever new social order replaces the old."[70]

As the experiences of the nationalist movement show, unless the very bases of patriarchal power are explicitly questioned, there is a high likelihood that participation in resistance will do very little to change women's subordinate position.[71] In Kapur's words, "A perspective that does not include the 'voice-consciousness' of women remains as much a patriarchal project as the development project of the World Bank or the Indian government."[72] Thus, participation may be doing relatively little to empower women if it proceeds without specific attention to gender.

For instance, one of the Andolan's major demands has been that the government provide the displaced with complete information about the dams and their impacts. Given that so far the government has carried out its planning exercise without even informing the people who will be most affected by the implementation of the plan, the right to information is especially crucial in the context of the Narmada dams, and more broadly in the context of development planning in India. Yet, this information, when it reaches the community, may not flow equally to men and to women. Providing women with access to information requires an understanding of prevailing customs and social norms, including women's daily work patterns, as well as of the places that are suitable for women to exchange and receive information. A part of a conversation recorded by Baviskar is of significance in this context of gender, work, and participation in the movement. Baviskar is recording the words of Khajan and Binda, members of the Bhilala tribe:

> Khajan complains, "Other villagers don't appreciate what I do as an activist." He tells Binda, "Even you don't understand why I have to travel and be away from home." Binda jokingly tells him, "Get yourself a second wife to help with the housework and then I will roam around with you."[73]

Changes in work patterns imposed by displacement or submergence of surrounding land in fact might have very significant repercussions on women's participation in resistance. In Rao's words:

> dwindling ecological resources such as water, fuel, land, forests and fodder not only makes them [women] work harder—walk more miles, get less nourishment, endure more hardships, etc.—to acquire these resources, but also affects their life-views, self-images, coping strategies, choices and consciousness and, consequently, their ability, potential and willingness to organize as a group and struggle for social change.[74]

It is lack of time that Patkar presents as defense against the lack of attention to gender issues. Thus:

> organising people, carrying out dialogues, finding documents, establishing a support network, explaining things from one point of view to the human rights activists, and from another point of view to the environmentalists as well as worrying about the forest dwellers. . . . In the process, there was literally not a single moment, to arrange separate meetings with the village women. If you want to deal with both men and women, you need special manpower for that, special approach, different strategies. We didn't have the time to do any of this.[75]

Though the magnitude of the task undertaken by her is very apparent, her presentation of the women's struggle makes it seem separate from the resistance against the dam, instead of integral to the formulation of an alternative and socially just development. According to Baviskar, "Activists feel uncomfortable with the systematic way in which adivasi society downgrades its women members, yet their inherent belief in cultural plurality keeps them from protesting."[76] What might also be pertinent here is that since men are the most powerful members in any given community the Andolan prefers not to alienate them.[77]

Despite the subordinate position of women in the communities in the valley, urban and rural female activists have very successfully organized the movement with very little male resentment. In part, this is the result of the remarkable way in which Medha Patkar combines charisma and absolute dedication to what is a very daunting struggle. Yet, it also arises in part from the perception of her as a mother and a goddess. According to Baviskar, "To the people of the valley who are dedicated to the Andolan, Medhajījī (sister) as she is called, is not just a leader but a little bit of a goddess, in whose power they have faith."[78] Similarly, Rashmi Shrivastava quotes an "agitator" as saying: "For us Medha is Mother Narmada. If she leaves, we will be motherless."[79]

This invocation of the leader as mother or goddess may serve to distance her from "ordinary" women. Yet, Medha Patkar does serve as an inspiration for women in the valley. Baviskar mentions how "in the space of a few years, the man who learnt about 'social service' from Medha Patkar, wants his daughters to do the same thing when they grow up."[80] Perhaps what this points to is the pragmatism of the urban activists who have left the gender status quo relatively undisturbed, and consequently have become more acceptable to the communities they seek to organize.

Separating Religion from Resistance: Steering Clear of the Hindu Right in the Narmada Valley

The specific strategies of resistance adopted by the Andolan must be evaluated within the current political context in India. Thus, the rise of the right-wing Bharatiya Janata Party (BJP, Indian People's Party) and the consequent strengthening of fundamentalist Hindu forces has expressed itself in attacks on non-Hindus and on their institutions. Against this, the Narmada movement has provided a community identity that derives from links to the land and the river and participation in resistance, as opposed to divisive party or caste identities.[81] Thus, Baviskar writes: "For young men, the Andolan has widened the political spectrum: involvement in the Andolan is a welcome alternative to the fundamentalist politics of the Bharatiya Janata Party or the traditional Congress."[82] Moreover, the new "Andolan identity" is based on an attempt to weave together the various communities residing in the valley.

As Patkar emphasizes, "The extremely different scenarios posed a major challenge from the point of view of using different strategies, skills, media, and idioms on the one hand, and bringing them together in a situation where the upper-caste landlords were used to using derogatory language when referring to the Adivasis [tribals]."[83] Thus, during the initial organizing of the Andolan in the Nimad, meetings used to be held in *mandir*s (temples) into which Harijans (lower-caste untouchables) were denied entry. The Andolan steadily worked against this exclusionary attitude and succeeded in making place for all castes at the meetings.[84] Further, the Andolan organized celebrations on Ambedkar Jayanti (Ambedkar being a famous nationalist leader who was also an untouchable), which were attended by upper-caste farmers. As a consequence, according to Patkar, "Lately we have found that the people here are not particularly interested in the Mandir-Masjid [temple-mosque] issue."[85]

In an attempt to knit various castes together, the Andolan foregrounds tribal cultures in its struggle against the dam. In some ways this foregrounding takes the shape of a romanticization of tribal communities. In Patkar's words:

> But let me tell you, there was a potential which was already there in these tribal communities. So it was comparatively easier to organize them to rally around this. Also, they are homogenous communities and somewhat removed from modern influences which would lead them towards individualisation and a selfish kind of attitude with the focus on individual rather than the community and obligations towards community. Their direct interaction with nature is something which is very crucial for the life they know, so snatching away their land and forest is bound to have repercussions.[86]

The Andolan categorically denies that its evocation of the tribal seeks to relegate them to living in the past. According to Kurian:

> The NBA [Narmada Bachao Andolan] rejects the charge that it seeks to preserve the tribals as "museum pieces." Those interviewed said repeatedly that they support the provision of roads, schools, and access to medical facilities, as well as the introduction of agricultural facilities that serve the purposes of the tribals. But all these, they say, the government ought to provide without making them a bargaining tool in the effort to get the tribals to resettle.[87]

The particular way in which tribals are positioned in resistance can be seen in Simantini Dhuru and Anand Patwardhan's 1994 documentary *A Narmada Diary*. In a particularly moving scene in the film, drums beat solemnly and plaintively as tribal men adorn themselves in ceremonial attire to perform a ritual dance on their sacred, ancestral land. The scene is filled with foreboding as the viewers are made acutely aware of the fact that this might be the last time this dance is performed on this land. Such an evocation of tribals is in keeping with the Andolan's stress on the cultural losses of displacement. Similarly, Baviskar describes a gathering of the Andolan in the town of Hapeshwar in 1989, this being one of the sites scheduled to be submerged by the

Sardar Sarovar dam. Here, "thousands of hill adivasis [tribals] gathered in a stirring ceremony and held aloft their bows and arrows and, with the waters of the Narmada cupped in their palms, pledged to fight the dam until their dying day."[88] Baviskar notes that it is not just the urban activists of the Andolan who arrange the tribals in ways that tend toward an exoticization and romanticization of their culture. It is likely that tribals themselves participate in such evocations in order to marshal all the resources available to them to oppose the dams. The victimized tribal being passively manipulated by the Andolan is as much of a myth as the pristine tribal who opposes the dam in order to remain untouched by "modernity." The context of resistance within which particular notions of the "tribal" are evoked has to be critically examined before such notions are dismissed as urban and naïve impositions on the tribal.

Part of the reason for this particular use of the "tribal" is, paradoxically, to attack the accepted notion of tribals as backward and uncivilized. For the Andolan the "tribal" actually embodies the knowledge that is required to rescue the world from its current descent into environmental destruction. Thus, Patkar stresses the nonconsumerist life-style of tribals and their harmony with the environment.[89] The stress on tribal culture is also to allow for a distinction between tribals and Hindus and to work against the notion of tribals as "backward Hindus," whose destiny therefore is to be assimilated into the Hindu "mainstream."[90] The Andolan's knitting together of tribals and Hindus stands in marked contrast to the communal unrest occurring in the rest of the state of Gujarat and in other parts of India. For instance, Stany Pinto shows how tribals in southern Gujarat have become increasingly anti-Muslim and anti-Christian and have attacked non-Hindu populations living around them:

> the sectarian movements have provided fertile ground for the spread of communal consciousness among the tribals and at the same time their incorporation into various religious sects and active promotion of 'Bhajan Mandalies' supposedly to promote or raise the so-called 'ethical standard' of the tribals seem to have blunted the class consciousness in the area.[91]

In the Narmada valley, upper-caste Hindu landlords profess (at least outwardly) that they stand united with their tribal brethren, especially, but not solely, because of the exigencies posed by dam construction.

The task of bridging tribal-Hindu differences and opposing caste injustices is crucial in contemporary India. The Andolan has done a commendable job in this area. However, by sidestepping the issue of gender inequalities, it maintains injustices against women in the valley. This lack of focus on gender thus almost seems to affirm the immutability of gender identities in Hindu and tribal cultures.

Conclusion

The Narmada movement brilliantly combines social justice with environmental concerns, allowing the needs of the poor to be addressed alongside environmental issues. The Andolan mobilizes red and green ideologies against the saffron wave of Hindu fundamentalism that is sweeping India. However, its lack of attention to gender inequities compromises its efforts to challenge mainstream development policies and construct an alternative development based on social and environmental justice for both men and women. Though tribal subordination has been questioned, the subordination of women in tribal and Hindu cultures has not been made a political issue or priority. Thus, the success of the Narmada movement might lead to very little gains in terms of the transformation of gender inequalities in the valley. As this paper has shown, incorporating gender into its critique of development will enable the Andolan to ensure that its alternative development is truly inclusive. It will also ensure that tribal-Hindu encounters do not further oppress tribal women, who currently bear the brunt of cultural displacement.

Notes

1. Medha Patkar, "The Strength of a People's Movement," in *Indigenous Vision: Peoples of India, Attitudes to the Environment,* ed. Geeti Sen (Delhi: Sage Publications, 1992), 282.

2. Amita Baviskar, *In the Belly of the River: Tribal Conflicts over Development in the Narmada Valley* (Delhi: Oxford University Press, 1995), 91.

3. Chris Deegan, "The Narmada in Myth and History," in *Toward Sustainable Development? Struggling over India's Narmada River,* ed. William F. Fisher (Armonk, N.Y.: M. E. Sharpe, 1995), 65.

4. S. Santhi, *Sardar Sarovar Project: The Issue of Developing River Narmada* (Thiruvananthapuram, Kerala: INTACH, 1994), 37.

5. Baviskar, *In the Belly of the River,* 91.

6. Ibid.

7. Patkar, "The Strength of a People's Movement," 294.

8. This study, for the most part, focuses on the Andolan's resistance to the Sardar Sarovar Dam, the largest dam sought to be built on the Narmada River.

9. William F. Fisher, "Development and Resistance in the Narmada Valley," in *Toward Sustainable Development? Struggling over India's Narmada River,* ed. William F. Fisher (Armonk, N.Y.: M. E. Sharpe), 3.

10. Santhi, *Sardar Sarovar Project,* 34.

11. Ibid., 39.

12. *Lokayan Bulletin* 11, no. 5 (1995): 12.

13. Bina Agarwal, "The Gender and Environment Debate: Lessons from India," *Feminist Studies* 18, no. 1 (spring 1992): 142.

14. Baviskar, *In the Belly of the River,* 120.

15. Vasudha Dhagamwar, Enakshi Ganguly Thukral, and Mridula Singh, "The Sardar Sarovar Project: A Study in Sustainable Development?" in *Toward Sustainable Development? Struggling over India's Narmada River,* ed. William F. Fisher (Armonk, N.Y.: M. E. Sharpe, 1995), 278.

16. Medha Patkar, "The Struggle for Participation and Justice: A Historical Narrative," in *Toward Sustainable Development? Struggling over India's Narmada River,* ed. William F. Fisher (Armonk, N.Y.: M. E. Sharpe, 1995), 177.

17. Baviskar, *In the Belly of the River,* 111–12.

18. Ibid., 112.

19. Santhi, *Sardar Sarovar Project,* 38.

20. Patkar, "The Struggle for Participation and Justice," 158.

21. Dhagamwar et al., "The Sardar Sarovar Project," 272.

22. Naila Kabeer, *Reversed Realities: Gender Hierarchies in Development Thought* (Delhi: Kali for Women, 1995).

23. Priya A. Kurian, "Environmental Impact Assessment in Practice: A Gender Critique," *Environmental Professional* 17, no. 2 (June 1995):167–78; and Brinda Rao, *Dry Wells and "Deserted" Women: Gender, Ecology, and Agency in Rural India* (New Delhi: Indian Social Institute, 1996).

24. Rao, *Dry Wells and "Deserted" Women,* 7.

25. Ibid., 8.

26. Kurian, "Environmental Impact Assessment in Practice," 174.

27. Bina Agarwal, "Gender and Legal Rights in Agricultural Land in India," *Economic and Political Weekly* 30, no. 12 (25 March 1995): A-40.

28. Ibid., A-40, A-43.

29. Govind Kelkar and Dev Nathan, *Gender and Tribe: Women, Land, and Forests in Jharkhand* (Delhi: Kali for Women, 1991).

30. Ibid., 90–91.

31. Bina Agarwal, "Gender and Command over Property: A Critical Gap in Economic Analysis and Policy in South Asia," *World Development* 22, no. 10 (October 1994): 1455.

32. Bina Agarwal, *A Field of One's Own: Gender and Land Rights in South Asia* (Cambridge: Cambridge University Press, 1994), 3.

33. Kelkar and Nathan, *Gender and Tribe,* 99.

34. Agarwal, "The Gender and Environment Debate," 126.

35. Ratna Kapur, "Damming Women's Rights: Gender and the Narmada Valley Projects," *Canadian Women's Studies* 13, no. 3 (spring 1993): 62.

36. Ibid.

37. Kelkar and Nathan, *Gender and Tribe,* 118.

38. Brett O'Bannon, "The Narmada River Project: Toward a Feminist Model of Women in Development," *Policy Sciences* 27, no. 2-3 (1994): 258.

39. Ibid.

40. Ibid., 262.

41. S. Srinivasan, "Disillusionment of the Displaced," *Economic and Political Weekly* 30, no. 26 (1 July 1995): 1555.

42. Dhagamwar et al., "The Sardar Sarovar Project," 273.

43. Kapur, "Damming Women's Rights," 62.

44. Dhagamwar et al., "The Sardar Sarovar Project," 275.

45. Santhi, *Sardar Sarovar Project,* 34.

46. Ibid., 36.

47. Baviskar, *In the Belly of the River.*

48. Kelkar and Nathan, *Gender and Tribe,* 113–44.

49. Kapur, "Damming Women's Rights," 62.

50. Baviskar, *In the Belly of the River,* 96.

51. Kurian, "Environmental Impact Assessment in Practice," 172.

52. Srinivasan, "Disillusionment of the Displaced," 1555.

53. *Lokayan Bulletin* 11, no. 5 (1995): 11.

54. Kelkar and Nathan, *Gender and Tribe,* 99.

55. Patkar, "The Strength of a People's Movement," 285.

56. Rao, *Dry Wells and "Deserted" Women,* 65.

57. Kelkar and Nathan, *Gender and Tribe,* 114.

58. Baviskar, *In the Belly of the River,* 216–17.

59. Patkar, "The Strength of a People's Movement," 289.

60. Ibid.

61. Baviskar, *In the Belly of the River,* 209.

62. Patkar, "The Strength of a People's Movement," 293.

63. Ibid.

64. Richard Fox, "Gandhi and Feminized Nationalism in India," in *Women Out of Place: The Gender of Agency and the Race of Nationality,* ed. Brackette F. Williams (New York: Routledge, 1996), 42.

65. Ibid., 46.

66. Saraswati Raju, "The Issues at Stake: An Overview of Gender Concerns in Post-Independence India," *Environment and Planning A* 29, no. 12 (December 1997): 2191–206.

67. Fox, "Gandhi and Feminized Nationalism in India," 44.

68. Ibid., 47.

69. Patkar, "The Strength of a People's Movement," 293.

70. Joni Seager, *Earth Follies: Coming to Feminist Terms with the Global Environmental Crisis* (New York: Routledge, 1993), 171.

71. Raju, "The Issues at Stake."

72. Kapur, "Damming Women's Rights," 63.

73. Baviskar, *In the Belly of the River,* 113.

74. Rao, *Dry Wells and "Deserted" Women,* 7.

75. Patkar, "The Strength of a People's Movement," 290.

76. Amita Baviskar, "Tribal Politics and Discourses of Environmentalism," *Contributions to Indian Sociology* 31, no. 2 (July-December 1997): 214.

77. Kurian, "Environmental Impact Assessment in Practice," 175.

78. Baviskar, *In the Belly of the River,* 113.

79. Rashmi Shrivastava, "Don't Give a Dam! Woman Leads Tribals against Indian Dam," *Women and Environments* 39-40 (summer 1996): 11.

80. Baviskar, *In the Belly of the River,* 214.

81. Ibid., 216–17.

82. Ibid., 217.

83. Patkar, "The Struggle for Participation and Justice," 163.

84. Patkar, "The Strength of a People's Movement," 286.

85. Ibid., 287.

86. Ibid., 275.

87. Kurian, "Environmental Impact Assessment in Practice," 172–73.

88. Baviskar, *In the Belly of the River,* 207.

89. Patkar, "The Strength of a People's Movement," 285.

90. Stany Pinto, "Communalisation of Tribals in South Gujarat," *Economic and Political Weekly* 30, no. 39 (30 September 1995): 2416–19.

91. Ibid., 2419.

Can Hindu Text and Ritual Practice Help Develop Environmental Conscience?

Rituals of Embedded Ecologies: Drawing *Kōlams*, Marrying Trees, and Generating Auspiciousness

VIJAYA NAGARAJAN

> tāṇam, n. . . . dāna. 1. Gift in charity, donation, grant, as a meritorious deed; 2. (Buddh.) Liberality, munificence, bounty, . . . 3. Gifts, as a political expedient, . . . 5. House-holder's life. . . .
>
> Tōsam, n. . . . dōsa. 1. Fault; 2. Sin, offence, transgression, heinous, crime, guilt; 3. Defect, blemish, deficiency, lack; 4. Disorder of the Humours of the body, defect in the functions of the bile, phlegm, or wind; 5. Convulsion, often fatal and always dangerous; . . . 8. Illness believed to be due to the evil eye, etc.[1]

Toward a Theory of Embedded Ecologies

In the village of Ammangudi along the Kāveri River, Saroja, an elderly Tamil woman shared with me her concerns and insights about the natural world, what she referred to as "dense fog," illustrating how ritual works to bind social relationships between natural and cultural worlds:

> It is about the *mūṭipani* [dense fog], isn't it? The *kōlam,* too, is disappearing. Do you see the connection? We are losing the ability to give to each other and to give to God. We are forgetting that we need to

> practice giving constantly. I understand why the dense fog is disappearing. We have forgotten to be generous. We used to give, our people. We used to give to each other. We used to give to the gods and goddesses. The *kōlam* is about giving, do you know? I keep hearing about the *mūṭipani* disappearing above us in the sky. I have the solution to the problem of the *mūṭipani* disappearing. We have to learn to give again. We have to marry trees again.[2]

When I first heard Saroja use the word *mūṭipani,* I was puzzled. *Mūṭi* means "closed" and *pani* usually refers to fog or snow. When she went on to speak about the disappearance of the *mūṭipani* in the sky, I realized that she must be referring to the depletion of the ozone layer. I gradually understood that Saroja was alluding to the illness of the entire planet—sickly, suffocating, infertile, and increasingly inauspicious. In other words, the "dense fog," or ozone layer, is collapsing because of our declining levels of generosity toward one another. Her solution of "marrying trees" represents a way of restoring the earth by establishing relationships with the powerful, creative, and auspicious force of trees.

The theory of embedded ecologies emerges through an exploration of subtle and complex relationships between the cultural and natural worlds.[3] Cultures often provide the lens through which the perception of the natural world is understood and acted upon. This essay is a further reflection on how the natural world is perceived, articulated, and framed through cultural and religious knowledge in Tamil Nadu in southeastern India. Based on conversations with Tamil women and men conducted while I was researching the *kōlam*, a women's ritual art, I probe specifically the significance of reciprocal generosity between nature and culture in three ritual acts: the daily creation of the *kōlam* on thresholds, the performance of marriages with trees to restore relationships, and the regenerative qualities of particular plants which, in turn, cultivate and transfer auspiciousness from the natural world to humans. Each of these ritual acts illustrates the interplay between the natural and cultural worlds as social exchanges. Similar to human relationships that are based on deference, respect, admiration, and, above all, ritual exchange, the people of Tamil Nadu have equally complex relationships with objects and phenomena in the natural world. The theory of embedded ecologies characterizes the multiple ways in which culture frames and reveals the complex relationships between nature and culture.[4]

Drawing *Kōlams*

Throughout India, one sees bright vermillion and white-grey marks on surfaces of things, a splash of red and white color dabbed on foreheads and bodies, stones and temples, thresholds, doorways, and houses, which, a few hours later, is worn to a light tincture. The designs range from thin, red U-shaped lines, to three parallel horizontal lines, to a smudge of red on the forehead, stone, or lance. Designs made of white rice flour are drawn on thresholds of houses and temples. Complex lines loop around a matrix of dots. These ritual markings are some of the most ubiquitous signs of sacredness in Hinduism. The red dot is known, depending on the specific Indian language, as *kumkum, bindi,* or *pottu,* among others. The threshold designs, too, are called by various names: *kōlam, rangōli, māndana,* and *alpana,* among others.

Like the *bindi, pottu,* and other ephemeral marks of *pūjā* (worship), the *kōlam* is one of many daily rituals that evoke, host, and dehost the divine as guests.[5] The daily creation of the *kōlam* requires that Tamil Hindu women rise at dawn every day and draw or paint a rice flour design on the thresholds of their homes, on temples, and at the base of trees. The *kōlam* is designed to invite, host, and maintain close relationships with the goddesses Lakṣmī and Bhūdevī, who will in turn prevent harm, illness, and laziness from entering the household. When a woman draws a *kōlam* in front of a coconut or banana tree, she is making a *vaḷipāṭu* (an adoring welcome) to the *deyvam* (divinity) in that tree.

The *kōlam* is gradually carried away by the feet of passersby and serves as a feeder of sorts for small animals, including birds, ants, and insects; thus, the *kōlam* disappears a few hours after it is made. One of the purposes of the *kōlam* is to fulfill the dharmic code, to feed one thousand souls every day. As one woman, Saraswathi, told me: "For those of us who are not wealthy by birth or acquisition, who cannot afford to feed a thousand people-souls a day, the *kōlam* is our way to feed a thousand souls. It has the same ritual karmic benefit to feed a thousand animal souls as to feed a thousand human souls!" Its gradual disappearance over the course of the day signifies the notion of reciprocal generosity: the household has "fed a thousand souls," that is, the small birds, ants, and insects that have come to the threshold of the house and grazed on the *kōlam,* and, in turn, the gods and goddesses may protect the household. A few hours after a *kōlam* is made, bits of it disappear as if it had been bitten into, strange breathing

Figure 1. A fourteen-year-old girl drawing a kōlam in the early morning. (Photograph by author)

holes appear in the surface of the dirt, where burrowing holes of tiny creatures appear. As the *kōlam* is dispersed, it also signifies the departure of the gods or goddesses who had been hosted there. The *kōlam* exemplifies the ephemerality of "spirit" in Hindu ideology, and the continual need to attract the attention of the divine through various acts of ritual generosity reflects one aspect of embedded ecologies.

"We Have to Marry Trees Again"

When Saroja had mentioned that "We have to marry trees again," I remembered when I had seen a young man married to a tree in the semi-rural, semi-urban outskirts of the old city of Madurai in southern India in the spring of 1990. I had heard about this wedding during one of my interviews on the *kōlam*. The young woman had said suddenly, bursting into our smoothly running conversation, "The *kōlam* removes suffering, it is a carrier of generosity, *dānam,* just like the wedding of a tree." I exclaimed quickly, "A wedding to a tree! What is that? When does that happen? Does it happen often?" She slowly explained to me as if I were a blind woman who could not see what was clearly in front of her eyes:

> We suffer. There are times when suffering comes at us suddenly and we do not know how to handle the enormity of the suffering. And then we marry trees. Usually it is when someone cannot get married; and there are a lot of obstacles in their path. Each man or woman they see, it does not work out for them. So, then, we know that there is something about the life path of the person that is preventing them from marrying a human person. So, we arrange the marriage of that person to a tree, and then we pray that the tree will take on the burdens of that human being and therefore release that person from their suffering. Then, the human person is free to marry someone else. Usually it works out that way. Do you understand? There is someone who is getting married to a tree tomorrow. Do you want to come?

She looked over at her mother and sister, as if expecting agreement, which she got, with their broad smiles. I nodded quickly, "Yes, I would like to come." I describe the scene below:

> *The next day I arrived there under the tree she had pointed to. The ritual objects were beginning to be organized and placed and, in an*

> *hour, formal aspects of the ritual began. A crowd had gathered, about fifty guests, around the large tree. It was an* araca maram, *a* pīpal *tree, a kind of Indian fig tree, gaily decorated, with marigold garlands and giant* kōlam*s at its base.*
>
> *The man had been unmarried for a long time, causing stress to his family and kin. There was food cooking outdoors for a small feast after the ritual. The priest sat near the tree, dark black hair framing his face, with a beard thickly flowing from his chin. His eyes were focused and serious, intent on the ritual being completed properly. He was setting up the fire ritual, with bricks outlining an enclosure, about two feet by two feet.*
>
> *The finely adorned groom was called to sit on the* kōlam *near the tree. The family sat near him, handing the priest the appropriate ritual objects as he asked for them, a new brass bowl, flowers, and some kindling sticks for keeping the fire aflame. And then, the priest started chanting. It lasted some hours. Meanwhile, wedding guests had gathered around the tree. And, as he chanted, flames licked around the large pieces of wood and the heavy smoke rose in front of all of our eyes*
>
> *We, the audience, witnessed the exchange of garlands between the groom and the tree bride. The priest took on the role of the bride and guided the garland back and forth. Some hours later, the tree bride and the groom were designated married and were then feted. A plate was laid out in front of the tree and a plate in front of the seated groom. Everyone then sat on the ground and ate a feast of a meal. The wedding was over an hour or so later, as everyone was fed and then drifted away.*

At the time, I was puzzled by this ritual and only years later did I begin to unravel some of its many meanings.

Marrying trees is part of the repertoire of solutions offered to families in the throes of suffering. With infertility and illness, or in the case of a daughter or son whose suitability as a mate does not fit the suitors who come, one of the solutions is to organize and conduct a marriage to a tree. Since trees, like humans, are seen as male or female, the tree selected may be of the opposite gender from the afflicted individual. If a man or woman has been unable to marry by the age of thirty, or a married couple finds themselves infertile, it is suspected that he or she is under some kind of *dōśam* (hardship), the effect being that the spouse-to-be could die early from the combustive combination of their essences. The word *dōśam* is commonly used to explain a string of personal or familial disasters, forming obstacles to the fulfillment of one's desires, and reflecting a diminishing of literal and metaphoric auspiciousness in the family, a kind of temporary or long-term marring of one's destiny, akin to an eclipse of the sun.

Suffering within a community is sometimes attributed to the deterioration or souring of relationships between the natural and cultural worlds. In fact, it is believed that ignoring these relationships is the very cause of inauspicious states such as illness, poverty, or death. In order for the individual or community to become whole, healthy, or prosperous, the atrophied relationships with the natural world must be repaired. The ritual of a person marrying a tree is, then, both a symbolic and literal reminder of our "kinship" with the natural world and is integral to healing an unfulfilled life.

Whenever there is a sudden decrease of auspiciousness in the family, an intimate alliance with a tree is said to increase the chances of alleviating the suffering. In general, the ritual of arranged marriages in India is used to cement the bonds between separate families. Establishing relationships with the natural world is as important for the family's survival as the marriage between humans. It is believed that trees have an enormous capacity to absorb suffering, since they have an abundance of auspiciousness, goodwill, and generosity. As part of the greater natural world, their sacredness is inherently more encompassing than that of humans. Therefore, if the marriage to a tree is arranged first, the tree will bear the burden of human suffering and, in a sense, transform the suffering and inauspiciousness into auspiciousness. If the problem is infertility, for example, the tree's natural fecundity will be partly transferred to the couple. After this marriage has been conducted, the person can be married a second time to his or her chosen human partner. Including the tree in a form of kinship—a familiar category with expectations of particular responses—is another manifestation of embedded ecologies.

Generating Auspiciousness through Plants

The third example of embedded ecologies is the generation and transfer of auspiciousness between the plant world and the human world. The connection between auspiciousness and the plant world was explained to me by a Tamil man named Aiyaswamy. Aiyaswamy lived in Madurai and served multiple roles in his community as a well-known astrologer and an actor in a drama troupe. In the middle of a conversation about the *kōlam,* what began as a simple inquiry into the relationship between a temple and a *kōlam* evolved into a series of reflections that linked the *kōlam* ritual with generating and reflecting auspiciousness, divine residency in trees, and reproductive metaphors inherent in plants. Below, I reproduce to the extent that I can, the full conversation, to set the context in which botanical knowledge is embedded within ritual contexts.

Vijaya: What is the relationship between a temple and the *kōlam?*

Aiyaswamy: The *kōlam* has the ability to "indicate" what *viśēśam* [celebration, festivity] is inside the temple.[6] Now, ordinarily women will be putting the *kōlam* on the house. But when festivals come, that very same *kōlam* becomes a huge "size." Then, today you know that there is something or other—a festivity—because you have put a huge decorative *kōlam* on your house. On the day of *Thai Pongal* [an annual festival], the whole street will have *kōlam*s made on it. It is the idea of *viśēśam* [celebration]. That is the key idea. It is just like the mango leaf garland we put on the doorway to indicate the idea of the celebration. And just like the banana tree on the threshold of a house, you know something big is happening in that house—a wedding or birth. Or the coconut palmyra leaves we put on the threshold. Or the use of coconuts. The same coconut tree, that coconut, or *mattai,* is used for death.

Vijaya: For death—how?

Aiyaswamy: On the tenth day, they use the coconut *mattai* for death. That coconut tree has borne a seed; that milk is there, isn't it? They open it and then, they put . . . the *āvi,* the spirit of [*he coughs to indicate that he does not want to say "the dead man"*] into it. Below that is a special building made with coconut frond, with only one opening constructed for the soul.

Vijaya: Then the same coconut is used for showing states of *mangalam* [auspiciousness] and *amangalam* [inauspiciousness]. Why do we [use] auspicious objects for indicating inauspicious events? That does not make sense, does it?

Aiyaswamy: Why do you say this—aren't you listening? [*He scolds me and continues, emphasizing for extra measure.*] I told you, "We use the coconut *mattai* for death and for weddings."

Vijaya: [*I persist*] But why?

Aiyaswamy: Because we use it in a different form. It is the same thing in different form. When we use it to indicate *mangalam* [auspicious] objects, then its *alangāram* [decoration] is quite different, so it has a different set of meanings. For *amangalam* [inauspicious] things it is, again, a different *alangāram*. That is why *alangāram* is so important. That is how you know what is happening around you. How else would you know?

Vijaya: What is the *mattai?*

Aiyaswamy: The *mattai* of the coconut means that it is the space that emerges, where the raw coconut grows. It is the closed stamen of the coconut, the husk that used to be used for fuel. We use the braid of coconut fiber. They take the branches, like shucking the corn, to separate the outside skin; they would put it in water in the villages; after the death, they take the soul into the coconut frame with mantras. That is what we have done for *paramparāya* [generations]. You see, our *aitīkam* [tradition] here in this land is that the coconut tree is the *kaṛpavirus̀am* [wish-fulfilling tree]—a tree that fulfills all your desires. The history is that the *kaṛpavirus̀am* is from the world of the gods. We believe that all the gods live or take up *kudi* [residence] there in the coconut tree and the banana tree. For example, if we take a marriage—if someone has not been able to get married and they are already thirty—then we think that there is a *dōs̀am* [transgression, disorder, illness, sin]. And we will go to the astrologer. But, second, we think that they should not get married, because if someone does get married then their husband's life . . . [*He is reluctant to say more and falls silent to make sure I understand; then he continues softly*]. The same thing can happen to a man as well.

If there is such a belief, then they will arrange a marriage with the banana tree for the first time. The banana tree is made into a woman and the *tāli* [wedding pendant] is tied onto her to take the *dōs̀am* away.[7] Then, the man can be married to the woman. Tying the *tāli* on the woman or the man, making that tree the man or woman, and afterwards that tree is cut down, so that it has died instead of the man or woman that is supposed to die. Then they conduct the second wedding. The power that is in a female is in the banana tree. The small containers, or *thonnai*s, are made [of banana leaves] to hold the ghee during a *hōmam* or *yāham* [fire ceremony]. Have you seen those?

Vijaya: [*I nod*] Why the banana leaf?

Aiyaswamy: Because only in the banana leaf are there no divisions. The other leaves have parts or sections. It does not have any *kulai*s [divisions]. We can only use plants without divisions for such auspicious occasions. When the leaves separate, it is like becoming a family. Father, then son, then younger brother, then younger sister; then the family becomes separated into individuals. At the same time, if there is only one child, then people will say that, like a banana tree, he has only one leaf—only one son. The coconut leaf has divisions. *Vēppa maram* [margosa tree] survives on its own. If you chop down a *vēppa* tree, then it will grow again on its own. A coconut tree is not like that. In the same way a banana tree is also like that. The banana tree only produces in one batch. It bears only one *kulai* [litter]. And bamboo, too. But the coconut seeds are not many, only one seed.

Here, in the conversation with Aiyaswamy, I became aware of the subtlety with which the morphological natures of plant species are perceived through the lens of a religious culture, the ways that distinctions in plant forms are embedded with a religious morphology of auspiciousness: how banana trees and coconut trees and margosa trees are seen as auspicious at times and as representative of inauspiciousness at other times. Reproduction itself is cast as metaphor embedded in plant morphologies. The theme of this conversation seems to be that the reproductive characteristics of trees represent auspicious or potentially reproductive states in humans. Banana and coconut trees are believed to come from heaven and to be inhabited by various gods and goddesses. They are considered abundant because they reproduce their fruits in *kulai*s (clusters, bunches), rather than one at a time, as humans do. The coconut frond is also considered auspicious, because it is an individually fragmented or divided leaf and, therefore, related to the abundance of reproduction. The type of auspiciousness in these trees is directly related to their discernible reproductive characteristics. For example, observing the abundant production of clusters of bananas and coconuts leads to the use of these trees as transferrers of auspiciousness, to bringing, in turn, reproductive qualities to humans. Leaves, whether single-planed like the banana leaf or divided like the coconut frond, are further distinguished by the types of auspiciousness one wants to be associated with, for example, either with many seeds or with one seed at a time.

In a South Asian worldview, fertility in plants is believed to encourage fertility in humans, just as the generosity or auspiciousness embedded in the *kōlam*s evokes generosity in the household and the community at large. Using trees in multiple ritual contexts for their reproductive qualities is believed to bring about the fruition of one's desires. It is as if the trees themselves personify the reproduction of their own desires by their own physical flowering of their fertility; establishing a close relationship with trees then expands your own potential to reproduce your own desires, whether for reproduction or marriage. An ability to reproduce is linked with the ability to give. The ability to reproduce conveys the presence of the divine and is a sign of the divinity's visible generosity.

Mutual Generosity, Auspiciousness, and Fertility

Establishing relationships with the natural world is a necessary component of relationships in the social and cultural worlds. Whether a ritual involves creating the *kōlam,* invoking the goddesses Lakṣmī, Bhūdevī, or Gaṅgā, sacralizing a tree by undertaking vows toward it, or inviting divinities to take up residence (*kudi*) in a tree, a sacred exchange is established.[8] The *kōlam* and its many counterparts belong to a category that I call "rituals of generosity." Through the form of *pūjā,* or ritual offerings, one enacts the hope for a particular type of relationship with the divinities. In evoking a generous heart, rituals of generosity circumscribe human relationships in both cultural and natural contexts. According to the classical Tamil text, the *Tirukkuṟaḷ,* giving hospitality from the household to those who ask is one of the ways of attracting the goddess of prosperity. A proverb under the section of "An Open House" goes like this: "The Goddess of Prosperity will be gladdened in heart and linger in the house of the man whose smiling face welcomes those who seek hospitality" (Akan amarnthu seyyāl uraiyum mukan amarnthu nalvirunthu ōmpuvān il; Rajagopalachari, 10). *Dānam,* the notion of giving, is signified by the relationship between the giver and taker.

We have seen that the *kōlam* establishes a particular type of sacred relationship within cultural areas known as thresholds. And we have seen how divine residency establishes relationships of generosity with trees. These examples suggest an interesting congruence between ritual form and intention. As in any social exchange, these relationships are predicated on the expectation of mutual ritual generosity—that is, one gives freely and receives gifts in return. Just as we can have generous relationships with human beings, so humans can have generous relationships with trees, rocks, rivers, and other aspects of the natural world.

Figure 2. Kōlam*s during the rice festival of Pongal, in Tamil Nadu. (Photograph by author)*

Positive Intentionalities

The *kōlam* is a ritual act that physically embodies blessings, or "positive intentionalities," reducing the accumulation of negative intentionalities such as jealousy, envy, or greed. With the *kōlam,* women's sphere of auspicious power moves in two directions: outward to the world beyond the threshold, and inward to the home, where it contributes to the stability of the household. Making the *kōlam* is like raising the sail on a ship for that day, enabling it to leap forward on its course with a strong tailwind. The positive intentionalities travel from the women's hands, through the *kōlam,* through the feet of those passing through its energy field, and into their bodies. In this way, both the tangible *kōlam* and the auspicious effects of its creation are carried into the larger world. In an environment of scarcity, where the next meal often seems just out of reach, where the world appears capricious and disorderly and death strikes frequently and brutally, where infertility is a source of private and public grief and child mortality rates are high, the notion of reality takes on a different cast. Life is seen as something precious and rare, like the unanticipated arrival of good fortune. The cruelty and heartlessness in oneself and others must be faced, since immoral behavior is seen as the cause of poverty, misfortune, and other forms of inauspiciouness.

In this context, acts of sacredness and faith in divine relationships signify courage and the hope of regeneration from the brutalities inflicted by nature and other humans. In making the *kōlam,* the hope is that the auspicious power embedded within the rice flour and invoked by the ritual will be transfered to the community, generating abundance and goodwill.

Calling upon the divine to inhabit trees or thresholds is part of the larger social fabric of "positive intentionalities" in Hindu popular life. Such acts of generosity in the face of suffering and scarcity are imbued with spirit and energy, and acts of goodness and piety become "weapons of the weak."[9] Positive intentionalities can be seen as manifestations of creativity within the larger social arena.

Similar to the *kōlam,* the motivation for conducting marriages with trees is the desire to create relationships with auspicious elements in the natural world. Auspiciousness can be linked with the ritual principle of productivity and reproductivity, or generation and regeneration. "Rituals of generosity" are established with auspicious sites, such as trees and rivers, with auspicious objects, such as books and tools, and with fertile or potentially fertile states, such as virginity, marriage, and pregnancy. Regardless of whether the association is with the "natural" or "cultural" realm, the auspicious forces are transferred to the individual through the act of rituals of generosity. These rituals of generosity are embedded in a constant relationship between the natural and cultural realms. This mutually reinforcing relationship is best conveyed through the description of a Balinese shaman, who, according to David Abrahms:

> acts as an intermediary between the human community and the larger ecological field, ensuring that there is an appropriate flow of nourishment, not just from the landscape to the human inhabitants, but from the human community back to the local earth. By his constant rituals, trances, ecstasies, and "journeys," he ensures that the relation between human society and the larger society of beings is balanced and reciprocal, and that the village never takes more from the living land than it returns to it—not just materially but with prayers, propitiations, and praise. The scale of a harvest or the size of a hunt are always negotiated between the tribal community and the natural world that it inhabits. To some extent every adult in the community is engaged in this process of listening and attuning to the other presences that surround and influence daily life.[10]

The Tamil woman, too, in a similar way, negotiates for the Tamil community the webs of generosity between the natural and cultural realms through the practice of making the *kōlam*.

A Story to End With: The Coming of the Refrigerator

I sat one day on my balcony, observing the street I lived on in the outskirts of Madurai, a large city south of Madras in southern India. It was not unusual to see both men and women traveling frequently throughout the neighborhood streets and asking for food at each house. Sometimes, one could hear in the distance, a voice raised in entreaty, "Amma, Amma [Mother, Mother]," as a figure approached a doorway. Sometimes, these men and women were dressed in saffron clothes, reflecting a religious order but, just as often, they were dressed in tattered clothes, elderly, widows, or just plainly in need of a meal.

In most households, meals were still prepared fresh; it was an enormous amount of work for the women of the house. From morning to late in the evening, the sounds of the making of a meal could be heard. Pounding, shredding, slicing, rolling, boiling. And, according to custom, the household usually cooked an extra meal's worth beyond what the family needed.

When a beggar would come to the door, then, it was the custom to feed the beggar, to give some rice, lentils, vegetables, sometimes cooked and sometimes raw. There was even a concerted effort to find a beggar, in order not to waste the leftover cooked food. From my upper flat balcony, I would hear voices of mothers calling their daughters to go down the street so that they could bring back a beggar to feed, so the food would not go to waste. Beggars were in demand, morally and physically, soaking up the excess food supplies. If you wasted food, it was as if you were wasting god. The concept was offensive. The goddess Lakṣmī is said to be embedded in every grain of rice.

A refrigerator had arrived some weeks before in my neighbor's house. It was a big moment, when the neighbors from several houses gathered together and shared in the festivities and joys of the first refrigerator coming into the neighborhood.

Some weeks later I observed the following scene. A beggar arrived at the door of the neighbor who had just bought the refrigerator. An onomatopoeic sound issued from the doorway: "Shoo, shoo, shoo, go away, go away, why are you people always coming around here?" I listened to the tone of voice.

A refrigerator-owning woman has, perhaps, a different relationship to leftovers than the non-refrigerator-owning woman. The freedom gained for the woman who no longer has to cook everyday can be juxtaposed against the loss of access to perishable food from the point of view of a wandering beggar. The very concept of waste as it relates to food undergoes a dramatic change with the coming of a refrigerator into a community. Generosity takes on a different character. Rituals of generosity, therefore, also change. Holding onto surplus food and storing it versus letting it move through the household and beyond the individual household becomes more prevalent.

This story returns us to the beginning of the essay, when Saroja was bemoaning the loss of generosity in human beings and the resulting decrease in the protective shield of the ozone layer. How ironic that the very refrigerator which tears a hole in the social fabric of generosity that had existed is the same tool which produces chemicals that destroy the ozone in our atmosphere, reducing our ability to protect ourselves from skin cancer and other ailments.

Notes

1. I follow the system of transliteration of the University of Madras Tamil Lexicon (Madras, 1982).

2. My fieldwork in Tamil Nadu, in southern India, was supported by a Fulbright-Hays Dissertation Research Fellowship, 1992–1994, and I am very grateful to K. Krishnaswamy, my primary research assistant in Madurai, and to S. Chitra, my primary research assistant in Thanjavur, for their invaluable help during this research period. I alone am responsible for errors in fact or judgement. I wish to thank Lee Swenson, Christopher Chapple, Mary Evelyn Tucker, Lance Nelson, Ann Gold, and Jennifer Beckman for their support and encouragement during the writing of this paper.

3. For an earlier step in my understanding of this theory of embedded ecologies, see Vijaya Nagarajan, "The Earth as Goddess Bhūdevī: Toward a Theory of 'Embedded Ecologies' in Folk Hinduism," in *Purifying the Earthly Body of God: Religion and Ecology in Hindu India,* ed. Lance Nelson (Albany: State University of New York Press, 1998), 269–96.

4. For a fascinating introduction to the field of environmental history in South Asia, see *Nature and the Orient: The Environmental History of South and Southeast Asia,* ed. Richard Grove, Vinita Damodaran, and Satpal Sangwan (Delhi: Oxford University Press, 1998); and Ramachandra Guha, *The Unquiet Woods: Ecological Change and Peasant Resistance in the Himalaya* (Berkeley and Los Angeles: University of California Press, 1989).

5. For a fuller elaboration of the *kōlam* as a women's ritual art, see Vijaya Nagarajan, "Hosting the Divine: The Kōlam in Tamil Nadu," in *Mud, Mirror, and Thread: Folk Traditions of Rural India,* ed. Nora Fisher (Albuquerque: Museum of New Mexico Press, 1993).

6. Stringing mango leaves on the outer doorways is a type of "threshold ritual," indicating an auspicious event inside the house. Another threshold ritual entails placing banana leaves on each side of the doorway to indicate an auspicious event.

7. The *tāli* is usually a gold necklace with a pendant that is given to the wife upon the completion of the marriage rite. The pendant's ornate symbols depict each household god or goddess, symbolizing the unification of the two families.

8. In everyday Tamil discourse, the word *kudi* often implies a kind of rented situation. For example, *kudi* literally refers to subjects, citizens, family, house, and residence. In addition, it means "to occupy, to take possession of the mind as a deity." A frequent referent to the residency of a divinity, *kudi* is often used in the context of *kudi irukkuruthu,* or "being housed."

9. James Scott, *Weapons of the Weak: Everyday Forms of Peasant Resistance* (New Haven: Yale University Press, 1985).

10. David Abram, *The Spell of the Sensuous: Perception and Language in a More-Than-Human World* (New York: Pantheon Books, 1996).

The Ritual Capsule of Durgā Pūjā: An Ecological Perspective

MADHU KHANNA

The goddess Durgā is one of the most formidable personifications of the Hindu pantheon. Revered by millions in India as the embodiment of the divine cosmic energy, Śakti, her primary mythological function is to maintain the balance of cosmic order by vanquishing the demons, who represent the antidivine forces of the cosmos. In her role as a demon-slaying battle queen, she symbolizes victory over evil, falsehood, and disorder. However, the goddess is not merely a powerful transcendent force, referred to as *śakti, māyā,* or *prakṛti* of the cosmos; she is a grace-bestowing compassionate Mother who presides over everyday concerns of human life. One of the most popular festivals of the goddess, known as Navarātrī or simply as Durgā Pūjā, falls in the autumn when she is propitiated for nine days.

In the past, Durgā Pūjā was carried out by Hindu kings and their successors under the Moguls and British and later by the zamindars and aristocratic families. The worship was traditionally performed by a *yajamāna,* or patron who conducts the ceremony for the benefit of the family, the household, and the subjects. Gradually, this worship has evolved into a *sarvajanin pūjā,* organized by festival committees and temple trusts. Today an increased democratization of worship is visible, as the *pūjā* far transcends its Bengali associations and is celebrated both in India and abroad by all Hindus. Popular belief holds that the nine-day autumnal festival was introduced in Bengal in the middle of the nineteenth century by one Candracūḍa Tarka Cūḍāmaṇi of the court of Giriśacandra, who was the great grandson of Maharaja

Kṛṣṇcandra of Nadia. However, historical sources place this worship as far back as the fifteenth century.[1] It is a fairly old festival and is mentioned in *smṛti* digests of Bengal.

The most celebrated legend linked to the goddess is described in the *Devī Māhātmya* (hereafter DM), which is unquestionably the first comprehensive statement about the function and significance of the goddess. The myth in the DM recounts that Durgā was created at a time of critical cosmic crisis, brought about by a demon who the divine patriarchy was unable to subdue. Obviously, the gods were running out of steam. The great demon (Mahiṣāsura) was a monster threatening to undo the world. So powerful was he that he had driven the gods from their celestial paradise. The vanquished gods approached Viṣṇu and Śiva for aid. Then, the gods in heaven emitted a flood of energy, a substance spewed forth like streams of flames, which combined into a cloud that grew larger and larger until it condensed into the form of a woman. The three-eyed goddess, adorned with the crescent moon, emerged with eighteen arms holding auspicious weapons and emblems, jewels and ornaments, garlands and rosaries of beads, all offered by the gods. Thereupon, a fierce battle raged between goddess and demons, narrated in three episodes in which each confrontation is more terrible than the last. Durgā displays an unbounded martial strength, slaying one demon after another. In the closing scene the battle queen is hailed as a peacemaker. The gods acclaim that she is the highest principle of existence. In this exalted role she is known as the great goddess, the perennial energy, the power of the cosmos.

There is a second version of the myth[2] known from the Bengali version of the *Rāmāyaṇa* ascribed to Kṛttivāsa. Rāma sought out the goddess Durgā to help enable him to vanquish Rāvaṇa and free Sītā. He performed her worship. Pleased, the goddess blessed him and assured him victory over his adversary. Thereupon, Rāma embarked for battle on the day of Daśahrā and succeeded in vanquishing Rāvaṇa. In addition to these two myths, there is yet another legend that is popular in Bengal. Quite apart from the transcendental image of the goddess in the DM, the folk imagination of rural Bengal has given her another identity. Far away from the resplendent figure of deliverence, the goddess is portrayed leading a householder's life with her husband Śiva and four children on the snowy peaks of Mount Kailāsa. Once a year, during *pūjā,* she returns home to rest from the grind of a comfortless life with her husband. She comes home just for three days. In this role, she is treated as a daughter of the house and enjoys special love and attention. The *Agomani* songs[3] (songs that celebrate the arrival of the goddess) of Bengal record the wide range of emotions experienced by the mother waiting in hope for her daughter's return or suffering pangs of separation at her departure. Contemporary calendar pictures in Bengal often illustrate the arrival of the goddess alighting from a boat with her children.

The Primal Forms of the Goddess

It is generally understood that Durgā Pūjā is primarily performed to commemorate Durgā's victory over the buffalo demon Mahiṣāsura in

Figure 1. A contemporary poster image of the goddess Durgā. (Photograph by author)

order to make devotees conscious of the victory of good over evil and to recall the mercy of the compassionate Mother of the Universe, who bestows fortune and prosperity and restores the balance of life. During the festival, the central image of Durgā represents the goddess as a resplendent ten-armed female figure, in the act of piercing the heart of the demon with a trident. The image is adorned with a sari and jewels. The image is flanked by her children, Gaṇeśa and Kārttikeya, and by Lakṣmī and Sarasvatī. These images stand on a pedestal, arranged symmetrically against a semicircular wooden frame, on which figures of Śiva with his family are painted. The worship of Durgā is not performed on these images but on the ceremonial *ghaṭa,* an earthen ritual-pot, on the wood apple tree (*bel-gācha*), and on a bundle of

Figure 2. The goddess Durgā in the form of barley sprouts. Matrika Ashram, New Delhi. (Photograph by author)

plant incarnations of Durgā, known as *nabapatrikā*. The worship of nine leaves is an old survival and affirms Durgā's intimate connection with the land, fields, forests, and groves. It also reflects the confluence of oral and written traditions. What is worth noting is that the central object of adoration consists of a bundle of plants placed on the *ghaṭa*. In private worship, an earthen pot with freshly sprouted barley seeds functions as a vegetal icon of Durgā. The ceremonial pot is filled with the generating waters of life and a variety of ritual ingredients. The pot is dotted with vermilion and decorated with mango leaves and coconut and serves as a symbol of the goddess who represents the primal womb of the universe. The ceremonial *ghaṭa,* holding the cosmic waters of creation, is set on moist dough in which barley seeds

Figure 3. The bundle of nabapatrikā, *the nine plant forms of Durgā, ready for worship. Matrika Ashram, New Delhi. (Photograph by author)*

Figure 4. The worship of nabapatrikā.
Matrika Ashram, New Delhi. (Photograph by author)

have been planted. The fresh sprouts are the microcosmic image of the fertility of the soil and vegetative life. The offerings reach the goddess through the womb vessel (*ghaṭa*), the wood apple (*bel*) tree and the bundle of nine leaves (*nabapatrikā*), the nine vegetal icons of the goddess. The three symbols of the goddess—the soil, water, and vegetation represented by the tree, the water in the womb vessel placed on the fertile soil, and the plants—are the most fundamental elements of the natural world and are in a constant symbiotic relationship. They form a primary unit in the complex series of a biotic web. Together, they represent the culturally constructed notion of a biotic web that functions as an undivided organism in the ritual. Over the centuries, the primal link between seasonal celebrations and ecology has gotten blurred, overlaid by the overwhelming mythic image of

Durgā. Today, we recognize the great goddess as a demon slayer, but not as one who is responsible for preserving the natural balance of our environment. This chapter attempts to decode the Nabapatrikā Pūjā and to show that it is seminal to the goddess's identity and is one of the principle means of ecological sustainability. This chapter, then, explores the subtle connection between ritual and nature. Nature is sanctified in the embrace of ritual, and ritual offers a traditional strategy for celebrating fertility and the creativity of nature. This inextricable connection between ritual and nature cannot be set aside, but must be seen as a central element of goddess worship. Modern studies on Durgā concentrate either on the textual traditions of the DM[4] or on the sociological aspects of goddess worship.[5] The impact of the text of the DM is so strong that it seems that Durgā worship is an invention of the "śāstric" tradition. A closer look at the festival performance reveals that the roots of this worship lie in the nature-oriented, village-based agricultural traditions of India, which are intimately bound to seasonal rhythms and crop cycles.

For millions in India, the goddess Durgā lives in freshly sprung paddy saplings or in the tender shoots of barley; in golden spikelets of harvest grains; in deep forest groves hidden among clusters of green shrubs, trees, and creepers; in the spices and roots used in the daily diet; in the *bilvā,* or wood apple tree, and its fortune-bestowing fruit, *śrīphala;* and in the rich produce of the harvest season. These nature personifications of the goddess represent the fecund power of the earth with which the goddess Durgā is identified. The unity of agricultural productivity and festive activity is integral to all the traditional societies, and Durgā worship presents but one "ecocosmic" model of this worship. Let me probe deeper into these connections. The chapter is divided into two parts. The first outlines the *pūjā* sequence; and the second explores its ecological dimension.

The Plant Forms of Durgā

During the annual autumn festival in Bengal, in addition to the customary worship of her earthen icon, nine aspects of Durgā are propitiated in her plant forms, collectively referred to as the *nabapatrikā*[6] worship of Durgā. The goddess has been described as one who dwells in nine plants (*navapatrikāvāsinī*),[7] and the nine plants that receive worship are:

1. Banana (*Musa paradisiaca; rambhā*), identified with Brahmāṇī or Brahmā's Śakti. The banana is an auspicious plant in Bengal and is used in a variety of ways. It is cooked when ripe, and the flowers are used as delicacies in curries. It is part of a normal diet. Its leaves serve as plates, its bark is used by craftspersons for decoration. The plant as a whole is used in life-cycle rituals.
2. *Kacchavī/kacu* (*Colocasia antiaquorum*) is an edible root and has a meaty flavor when cooked. It is invariably used as a supplement to a rice diet. It represents the goddess Kālikā.
3. *Haridra,* or turmeric (*Curuma longa*), is an edible root and one of the most widely known spices. It provides a basic seasoning in cooking. It has many medicinal properties and is fervently used as a home remedy. It is the plant form of Durgā herself.
4. *Jayantī* (*Clerodendrum phlomidis*) is associated with the goddess Kārtikī, the energy of Kārtika, Durgā's son. Its leaves are also used as a home medicine.
5. *Bilvā/ bel,* or wood apple tree (*Aegle marmelos*), is one of the most revered trees in *śākta* worship. The goddess is invoked in this tree. A single leaf can serve as a substitute to the elaborate offerings of sixteen or more substances. The fruit of the tree is cooling and useful for stomach ailments. The goddess of this tree is the energy of Śiva, or Śakti herself.
6. *Dāḍima,* or the pomegranate fruit (*Punica granatum*), is a popular offering in goddess worship. The goddess of this tree is Raktadantikā.
7. *Aśoka* (*Saraca indica*) is another sacred tree of India associated with women and is said to promote fertility. The goddess of this tree is Śokarahitā, or "One who is free from sorrow."
8. *Mānkacu,* or the arum plant (*Alocasia indica*), is presided over by the terrific aspects of the goddess Cāmuṇḍā who slew the demons Caṇḍa and Muṇḍa.
9. *Dhān,* or paddy (*Oryza sativa*), the unhusked, harvested rice is the most sacred plant, for it is the basis of life and so essential for *pūjā*. It is the goddess Lakṣmī (Lakkhi, *dhan,* or wealth), whose most common representation in Bengal is a bowl of unhusked rice. We learn from the *Śrisūkta* of the *Ṛgveda* (2.6.19), that the goddess Lakṣmī was adored as Rice Mother

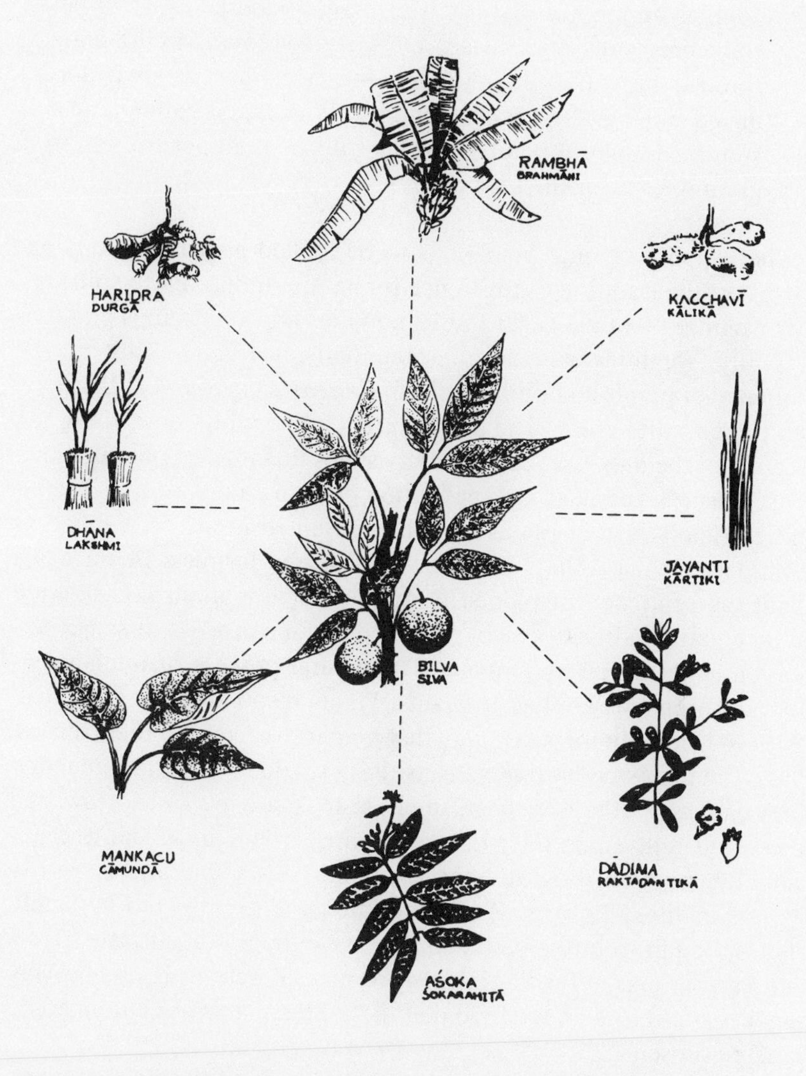

Figure 5. The plant incarnations of Durgā. Center: A branch of the wood apple tree representing the goddess Śivā. Clockwise: banana as the goddess Bramāṇī; kacu *as Kālikā;* jayantī *as Kārtikī; pomegranate as Raktadantikā;* aśoka *as Śokarahitā; the arum plant as Cāmuṇḍā; rice as Lakṣmī; and turmeric root as Durgā. Matrika Ashram, New Delhi. (Courtesy Tantra Foundation, New Delhi)*

> or grain goddess. In Bengal and in all rice-growing areas of India, the grain of rice, raw, unhusked, and cooked, is looked upon as an animated spirit. The grain goddess is said to become pregnant when the spikelets of rice bloom. As the grain sprouts, her offspring—that is, the rice—has its own animated spirit. At harvest, the child spirit of rice is gathered and kept in storage. When the crop is sold, a small portion of the grain is kept for the next crop.

The worship of nine plants has to be looked at within the larger context of the tradition of reverence for nature. India's earth spirituality recognizes that the earth and its bounties are a conscious and living entity, breathing with life. A potentially instructive view on the sentiency of plants and biological life (*jangama*) is found in the *smṛti* and the epic literature. The *Manusmṛti* (1.49), while describing the creation of the universe, categorically states that plants "possess inner consciousness and have the realization of both pain and joy." The latent consciousness in plants is again explained in the *Mahābhārata* (*Śāntiparvan,* chap. 184), where the sage Bhṛgu instructs Bharadavāja about the sentiency of plants. Bhṛgu's most important contribution lay in creating a link between plant life and the process of sense perception experienced by humans. The natural process of human sensory experience is applied to plants. Plants respond to sound, touch, taste, and smell in the same way that humans respond to their senses. Plants are endowed with great sensitivity to touch, heat, and thunder. They can, as it were, see, hear, smell, taste, share joy and sorrow, and repair and rejuvenate their damaged parts. The skin of the tree has tactual perception; it is affected by thunderous sound and is said to possess auditory powers. Plants are known to be affected by smell. They seek sap from the earth, so they have the power of taste.[8] How can humans ever infer that plants are not animate in the same way humans are, if it is understood that these nine plants are animated by a sacred presence?

It is obligatory that the Nabapatrikā Pūjā be held on two occasions: during the annual autumnal festival of Navarātrī, and during the spring worship (Vasantī Pūjā) on the ninth and tenth days of the bright half of the advent of the Indian new year. Although preparations for the *pūjā* may start a month before, the chief observances of Durgā Pūjā[9] begin on the sixth day. Prior to this, the goddess is worshiped every day from the first to the sixth with a minimum of sixteen traditional articles of worship (*ṣoḍasopcāra*); in some temples, officiating priests recite the *Devī Māhātmya* from the *Mārkaṇḍeya Purāṇa,* while a separate priest is appointed to recite the nine-syllabled mantra of the goddess. In private worship, on the day before Durgā Pūjā commences, women of the household sow barley seeds in an earthen pot, and the goddess is invoked in a ritual-pot, filled with water. This is worshiped for nine days.

The *pūjā* program commences with a *saṃkalpa,* a pledge declaring the intent to worship, followed by a series of purifications, annointings in which the deity is invited to reside in the plants. Next is an elaborate worship with special ritual offerings, fire oblations, ending with the worship of virgin maidens (*kumārī pūjā*) and distribution of ritual food. In my presentation of the ritual, I shall mainly draw upon some of the principal mantras of invocation and the ceremonial offerings that are used in this worhship.

Ṣaṣṭhi: Bodhana

Before the commencement of the main worship on the sixth day (*ṣaṣṭhi*), the goddess is first welcomed in the wood apple (*bel-gācha,* or *bilvā*), which is identified with her. The goddess is aroused (*bodhana*) from her dormant state on the sixth day of the light half of the full moon of Aśvin (October–November) with the making of an inner resolution (*saṃkalpa*). She is then addressed, in the tree, with the following mantra:

> Oṃ Bilvā tree come hither, Oṃ, this flower to the Bilvā tree.

Then, touching the northeastern twig of the tree, the worshiper says:

> *Aiṃ,* for destruction of Rāvaṇa and
> success of Rāma, Brahmā had in earlier
> days, awakened thee out of season, I, too, on the sixth lunar
> day do arouse thee. Indra having so
> roused thee gained dominion and
> the heavens, therefore, do I rouse thee
> to superhuman dominion and
> transcendental power: as the ten-faced one
> was destroyed by Rāma, may I have the
> might to destroy my enemies.[10]

The goddess Durgā is then invited (*āmantraṇa*) to take up residence in the tree, and the tree receives offerings of sandal paste, oil, turmeric, rice, and incense. The priest anoints the tree with turmeric and oil and humbly begs forgiveness for causing pain to the goddess who stands before him in the form of a tree:

> *Oṃ Śrīphala* tree thou art born on the mountain Mandara, Meru, Kailāsa and on the top of the Himāvat thou art always a favorite of Śivā. Born on the top of hill, Śrīphala, thou art the abode of prosperity, I take thee away to worship thee as Durgā herself.
>
> I worship Durgā, having taken a branch of yours, oh lord, forgive the pain generated by the separation of the branch, for it is said the gods have worshiped Durgā on your branch, I bow to the *Bilvā* tree born on the Himāvat, favorite to Pārvati and embraced by Śiva.

Then, with a sharp sword, the priests sever the branch, uttering these words:

> *Oṃ phaṭ,* sever, sever, *phaṭ hrīṃ svāhā, Oṃ* for the increase of progeny, life, and wealth. I take thee away, oh beloved of Umā.[11]

The severed branch from the *bilvā* tree and the other plants are tied together with the *aprājitā* creeper. The nine plants are anointed with the oil and turmeric powder. *Nabapatrikā*s are thrice touched with the following auspicious articles: sandal, soil, pebble, paddy, Durvā grass, flower, nutmeg, curds, ghee, the swāstikā symbol, vermilion, conch shell, collyrium, *gorācana* (an orpiment prepared from the bile of cattle), mustard, gold, silver, copper, a fly whisk, a mirror, a light, and a plate. This ceremony completes the *adhivāsa,* or the fumigation of the goddess with incense.

Saptamī: Annointing of the *Nabapatrikā*

On the seventh day (*saptamī*), the ceremonial bathing of the nine plants takes place. For this, an auspicious diagram called the *sarvatobhadra-maṇḍala* is made with colored rice paste. An earthen vessel (*purṇaghaṭa*) filled with pure water is placed on it. The nine leaves are anointed with the following chants:

> *Oṃ* water that is purified, superior, warm filled with the force of fire and life, destroyer of sins, I anoint you. . . .

The wood apple twig is placed on the nine plants and the bundle of nine leaves is established on a decorated throne for ceremonial worship. The nine leaves are bathed nine times by invoking waters of the rivers, oceans, lakes, heavenly streams, and, finally, with a jar of Ganges water. Besides the annointing with water, various other ingredients are used, such as marine drugs (infusions prepared in sea water), infusions made from flowers, herbs, medicinal plants, tonics made from the bark of different trees, decoctions made from the *mahā-auṣadhi* (appendix, section V), oils of different kinds, perfumery, fragrances, a confection of honey, five products of the cow, and soils and earth of different kinds (appendix, section IV). In other words, all the bounties from nature enter the ritual cycle and the goddess is worshiped in the image of her own creation.[12]

After these purifications and annointing, the seven leaves are wrapped in a saffron-colored sari and worshiped as nine forms of Durgā, with the traditional sixteen, thirty-two, or 108 offerings, ending with the following prayer:

> *Oṃ* Brahmāṇī, come here, be established here, having summoned the goddess [the priest recites the mantra:] *Oṃ hrīṃ,* reverence to Brahmāṇī, come and favor us with your presence, O goddess in banana form, bestow peace upon me at all times. I bow to thee.
>
> *Oṃ* Raktadantikā (blood-teethed), come here, *Oṃ hrīṃ,* obeisance to Raktadantikā in the form of the Pomegranate tree. Since in ancient times, during the slaying of the demon Raktabīja, you played the part of Umā, hence, bestow boons upon me.
>
> *Oṃ* Lakṣmī, come here, *Oṃ hrīṃ,* obeisance to Lakṣmī, the presiding deity of grains, dear to Umā. Formerly, you were created by Brahmā, for the protection of the living beings of the world, therefore, protect me at all times.
>
> *Oṃ hrīṃ,* Durgā, come here. *Oṃ hrīṃ,* obeisance to Durgā who presides over the turmeric root. O goddess, you are Umā incarnate, good-natured one, for the destruction of obstacles receive my worship and bless me.

Figure 6. Kala Bau, the Banana-maiden, the wife of Gaṇeśa, being dressed in a sari. Matrika Ashram, New Delhi. (Photograph by author)

Oṃ hrīṃ, Cāmuṇḍā, come here. *Oṃ hrīṃ,* obeisance to the goddess who presides over the Māna tree, and on whose leaves she rests, the beloved of Śacī, for the destruction of obstacles, accept my worship as you please.

Oṃ hrīṃ, Kālikā, come here, *Oṃ hrīṃ,* obeisance to Kālī dwelling in the Arum plant. Formerly, in the war with the demon Mahiṣa, you assumed the form of the Arum plant, good-natured one, beloved of Hara, come here and bestow your grace upon me.

Oṃ hrīṃ, energy of Śiva, come here, *Oṃ hrīṃ,* obeisance to goddess Śivā, the goddess dwelling in the Bilvā tree. *Oṃ* Bilvā tree, you are dear to Umā, you bring delight to Mahādeva, and you are always a beloved of Vāsudeva. I bow to thee.

Figure 7. The Banana-maiden dressed as a bride.
Matrika Ashram, New Delhi. (Photograph by author)

> *Oṃ hrīṃ,* the goddess Śokarahitā (Devoid of Sorrow), come here, *Oṃ hrīṃ,* obesiance to the goddess who is devoid of sorrow, dwelling in the Aśoka tree. O Aśoka, since you bring delight to Śiva, destroy sorrows, and are a favorite of Umā, therefore, dispel my sorrows always.
>
> *Oṃ hrīṃ,* Kārtikā, come here. When at the time of battle with demons, along with Indra and other gods you slayed the demon Niśumba, then, O goddess, you were worshiped as Jayantī. Bestow boons upon us.
>
> *Oṃ,* obeisance to the plants which are in the form of nine aspects of goddess Durgā, you bring pleasure to Mahādeva. Accept the entire worship and protect us, Supreme Empress.[13]

After the *nabapatrikā* is annointed, the idols receive the same worship, followed by a *homa,* or rite of oblations into fire. The banana plant receives special attention and is ritually transformed as a bride of Gaṇeśa (*Kala bau*). A fresh banana plant is purified, ritually anointed, and dressed in a sari. The banana maiden is then married to Gaṇeśa and is ceremonially placed near her husband.

Aṣṭamī Pūjā

The program of the seventh day is repeated on the eighth day, or *aṣṭamī pūjā*. The *nabapatrikā*s are worshiped first, then the goddess receives worship, and finally the rite of *homa* and *kumarī pūjā* is performed.

Navamī-Sandhi Pūjā

The aforesaid *pūjā*s are repeated, except that there is a special worship performed on the conjunction of the *aṣṭamī* and *navamī* (eighth and ninth) days, called Sandhī Pūjā. This marks the moment when the goddess in the form of Kālī slew the demon Mahiṣāsura, as well as the period when Rāma vanquished Rāvaṇa.

Daśmī (Daśahrā)

On the last day, after worshiping the deities, the priest formally makes a farewell address to the goddess, saying: "O goddess may you now retire in order to visit me, and dispel all my difficulties. Move in the stream of water and dwell in thy home for my benefit." The bundle of *nabapatrikā* is carried to the riverbank of a running stream and immersed in the waters with the following mantras:

> *Oṃ,* O goddess, dive into the water together with the prosperous *nabapatrikā*s. I immerse thee in the waters for the prosperity of my children, life, wealth. O goddess, having been worshiped, depart, retire to the best of places, thy abode, so that at the end of the year you may return again.[14]

The twenty-five idols of *pūjā* undergo an identical ceremony. Before the idols are carried away, the women offer special betel leaves and circumambulate each idol seven times. A male participant then scatters a plateful of rice, silver, gold coins, and fruit over the heads of the idols, to the beating of drums. The idols are carefully tied onto a bamboo frame and carried to the riverbank. The idols are placed on a boat and, after drifting for a short time, they are immersed in the river while the following mantra is recited:

> Oṃ, this worship is complete to the best of my power. . . . now lie to thy best home.[15]

The images are thus consigned to the river. Various sports and auspicious rites follow. To the beating of drums and the blowing of conch shells, the worship party returns home, chanting, "When again shall the goddess return, oh where is the goddess gone?"[16] The party returns to the hall of worship, which is now lit with a diffused light. The whole scene is solemn. At the end, the priest and other members write the name of Durgā on a banana leaf twelve times, once for each month, with a pen made from the *khagara* twig (*Saccharum fuscum*). The priest sprinkles water on the worshipers, reciting the same mantras with which the Nabapatrikā Pūjā had commenced, and the rite concludes.

What is Ecological about Durgā Pūjā?

In trying to locate the ecological dimensions of Durgā Pūjā, let me first present the most obvious ones. First, the place of Durgā Utsava in the religious year coincides with the harvest season. The celebration is integral to the crop cycles and natural rhythms of the season. This is the original motive and intent of the worship, which, over a period of history, has been superimposed (and perhaps forgotten) by the mythification of Durgā as demon slayer in the DM. The momentous period of harvesting simply transforms into a celebration of Durgā, the power of the fecund earth, mistress of grains, plants, and animals. A Bengali proverb says, "in twelve months, thirteen festivals," but the celebrations that give structure to the yearly cycle are those that are performed during the turning of the seasons, the "rites of spring" and "rites of autumn." Accordingly, Durgā is worshiped in both spring and in autumn. A metaconcept of great significance to Durgā Pūjā is the Indian perception of sacred time. In the Indian ethos, time is regarded as cyclical. Whatever is born will perish and decay and will be reborn again, for time continuously renews itself. The cyclical time consciousness is grounded in the direct observation of the repeated round of seasonal rhythms.

Settled rural communities in India have adopted two circular time models: one is determined by the course of the sun; the other by the moon's waxing and waning cycle. The solar cycle, like every other category in the Hindu universe, reflects a polarity. In its northern course (*uttarāyaṇa*), the sun passes through the Tropic of Capricorn, the period marked for the winter solstice. This is the period when the cycle of abundance and life-giving fertility is set in motion, a time when the earth's creative potencies burst into bloom. A harbinger of spring, it is conceived of as an auspicious period when gods in their wakefulness offer their benign grace to the world and sun's golden light rents asunder the terror of darkness. In contrast, the sun's southern course, or *dakṣināyaṇa,* in the Tropic of Cancer marks the summer solstice. This phase is identified by the descending energies of the earth, when nature's potent creativity is at its lowest ebb. It is a period of barrenness and unproductivity when the soil is desiccated, gods lie still, and the demonic and inauspicious forces make their home in the bosom of the earth.

The winter and summer solstices are vital to cultivators. The agrarian calendar responds to the necessity of organizing the annual operation with precision and in unison with nature's harmonic rhythms. In India, the agrarian cycle and festive calendars of the cultivators is dictated by the solar and lunar rhythms. Despite the great divergence in ecologies from area to area in India, the periods of fundamental importance to agricultural activity are those of sowing, germination, replanting, growth, harvesting, winnowing, sorting of the seeds, and sorting and cooking certain kinds of food. All are dictated by the rhythms of sacred time. Periods of scarcity followed by periods of great abundance mark the ritual aspects of cultivation and rhythmic periodicity of that economic activity. When the land yields its bounties, the people celebrate the natural order, *ṛta*, of seasonal cycles. Active participation in Durgā Pūjā festivities through rituals of natural bonding confirms our awareness of plants and the seasons in the cycle of sacred time.

The goddess Durgā's association with plants, trees, forests, and fertility of the soil is perennial. This becomes evident in several places in her worship. The earliest known hymn, the *Durgā-stava,* appears as an insertion at the end of the *Mahābhārata* and states "that the goddess dwells on the peaks of mountains, by rivers, and in caves, forests, and groves"[17] and is worshiped by the primal forest communities, the Shavaras and the Pulindas. Her association with the forests may have been the basis for Bana Durgā,[18] or the Durgā of the Forests cult popular in rural Bengal, and the survival of the *nabapatrikā* worship.

Durgā's agrarian roots are confirmed in the ancient belief of the tradition of the goddess as an embodiment of the vegetation spirit. Sarat Chandra Mitra has drawn our attention to the survival of the Vana Durgā, "Forest Durgā," or Bana Devi, "Goddess of the Forest" (also known as Burha Thakrani), cult in Mymensingh and Tippera in eastern Bengal, which may have predated Durgā's representations in the *Devī Māhātmya.*[19] Durgā is meant to be the indwelling spirit of the sheora (*Streblus asper*) and karim (*Murayya exotica*) trees indigenous to that area. The forest Durgā (Bana Durgā) is worshiped at the foot of the tree and is often referred to as the indwelling spirit of the tree (Sakotabāsinī). No image is made to represent this goddess, nor is she worshiped in a shrine or a temple. On the contrary, she is invoked at the foot of the tree. Mitra ascribes a non-Aryan origin to this cult. Bengal was covered with primeval forests. Non-Aryan aborgines who lived in the forest believed in the presence of a spirit. Women would create a miniature forest of twigs from the sheora trees and worship it with offerings. Representations of the goddess from the first to the third century A.D. portray her slaying the demon in a buffalo form.[20] In later icons, the demon is depicted as a human, a buffalo, or in a composite form, with the demon head of a buffalo and the body of a human, or vice versa. The evolution of the image reflects the inextricable link of Durgā with a common domestic animal in rural India, that is, the buffalo. Farmers in rural Bengal offer this explanation for the buffalo as demon: buffalos sometimes wreak havoc in the paddy fields and destroy the tender rice saplings. Hence, they have aquired a demonic character in the myth. Another belief prevalent among agrarian communities in Bengal that has survived is that the buffalo demon Mahiṣa, slain by Durgā, represents the demon of drought that so obstructs prosperity.

This association of the goddess with the fecund earth precedes the systematic codification of her legend in the celebrated text of the *Devī*

Māhātmya. She may well have ascended from the incipient powers of the earth to the more exalted position of a transcendent deity. We find an unbroken continuity in the nature personifications of the goddess in the text of the DM. She appears as Śākhambharī, the nourishing mother; as Annapūrṇā, one of plentiful food; and in the grand image of Śatākṣi, one who has one hundred eyes, who brought relief to the famine-stricken populace by overflooding the earth with nourishment.[21]

The ritual capsule of Durgā Pūjā is double-structured: the first structure is well grounded in the seasonal cycles and is in harmony with the ecology of the environment; the second exalts the goddess's supreme power through her narrative. Both these threads interweave to create an intricate tapestry of symbolic action in the cycle of festival ceremonies. The unbroken structure of the ritual cycle may be compared to a *capsule*.[22] The ceremonies continue uninterruptedly for nine days. Throughout this period, the ritual cycle recognizes the interdependence of a web of relationships: biotic, ethical, social, and cosmic. The integrity and dignity of the ceremony lies in the fact that it does not "split" these elements apart but envelopes them together in a capsule.

Though the goddess is eulogized as cosmic energy in the DM, she is "vitalized" in ritual worship through her *svayambhū-mūrti*s, or natural symbols, such as the sacred cosmic waters (placed in the *maṅgala-ghaṭa*), the *bilvā* tree, and the vegetal icons in both temple and private worship. We have seen that the mantras that are recited during the ceremonies function as a very significant category to express the goddess's domain over nature. They also record the ethos and sensibility of a people who share an empathy with nature. The worship of the vegetal icon of the goddess is the primal stratum that, over a period of time, is overcoded by the legend of the DM.[23] While discussing their relationship, one must never lose sight of an essential modality of the goddess tradition in India. The quintessential principle of physical reality, that is, nature-*prakṛti*-earth and its elemental forces, is chosen to be represented by the feminine principle, conceived of as a source of power of regeneration and transformation. Rarely have male figures in the Hindu pantheon been able to play, subvert, or contravene the role of the earth and the elements as *śakti*. While the forms, functions, and roles of gods and goddesses have evolved, changed, and been underscored or overstressed throughout history, the autonomy of the earth mother and elemental goddesses has remained unchanged and has gone unquestioned. It is precisely in these sources that one may find the primal beginnings of theological ecofeminism, of which Durgā Pūjā is but one example.

A more direct link with ecology is reflected in the selection of the *nabapatrikā*s. The nine plants that are propitiated contribute to the life-support system and have a sustenance value. The *pūjā,* while celebrating the seasonal cycles, also preserves the crops that are useful to the community. Plants that are propitiated also fill the community's medicine chest. Making the countless waters, oils, fragrances, herbs, medicinal plants, seeds, roots, barks, fruits, soils, pulses, grains, and vegetables (see the appendix) essential to the *pūjā*'s rituals year after year is the safest and surest method of preserving the genetic and biotic diversity of the environment.

Another association between ecology and ritual can be observed in the rites of *homa. Homa* is performed by making 108 oblations into fire. The *homa* ingredient (*sāmagrī*) consists of a mixture of medicinal herbs, aromatic substances, wood, and a variety of twenty items (appendix, section VIII). The *homa* cuts through air pollution and purifies and cleanses the atmosphere. It provides a natural form of aromatic therapy that creates an immunity to common ailments.

The holy food that is taken as *prasād* (appendix, section IX) is grounded in the primal understanding of the symbiotic relationship between seasonal foods and the biorhythms of the human body. The intimate link between food and seasons is known to people who live in close partnership with nature. In India, special types of sacred food taken as *prasād* have been developed as natural products of a nutritional harvest. Over the course of India's long agricultural history, the knowledge of an "agriculture modeled on nature" has been the preserve of women. The nomenclature of the food chain, from its raw to its cooked state, is an integral part of a woman's life in rural India. Women are primarily responsible for maintaining and integrating links between trees, animals, crops, and forests, and for preserving the ancient knowledge system of the gastronomic arts. During the *pūjā* festival, women cook sacred food in their homes, a skill they acquire from their mothers and grandmothers. Vandana Shiva has pointed out, appropriately, that "the worldwide destruction of the feminine knowledge of agriculture, evolved over four to five thousand years, by a handful of white male scientists in less than two decades has not merely violated women as experts; since their expertise has been related to modelling agriculture on nature's methods of renewability, its destruction has gone hand in hand with the ecological destruction of nature's processes."[24] She continues: "the most nutritious part of the food is turned into waste because the efficiency of the machine for profit generation is the determining factor, not the efficiency of women for the generation of nutrition."[25] Women's expertise and role in food production and processing has been displaced, with the result that we are mercilessly divorced from the seasonal variation in food patterns preserved in our ritual celebrations.

The traditional idols for Durgā Pūjā are made of biodegradable substances such as straw, clay, resin, and wood. They are painted with natural dyes from grains and flowers and coated with varnish. These substances dissolve readily in the flowing river once the idols are ceremoniously immersed. However, today there is a change in the quality and type of materials being used in the production of the idols. Many artisans have begun to use synthetic colors to paint the idols, thus adding to the toxic waste in the rivers in which the idols are immersed.

The innumerable substructures within the Nabapatrikā Pūjā—as, for example, the *bodhana* ceremony where the plants are "awakened" —preserve a norm for the ethics of the environment. We treat nature according to the ways we know and perceive it. A worldview may either encourage or discourage our attitudes and responses to nature. The imbalance in our ecosphere is rooted in the pollution of the psychosphere. The new culture of a cash economy and profit seeking has led to inequities and irreversable ecological inbalances. Markets do not take into account the sustainability of the "gross nature product" of creation. The natural resources of the land are of negative value. Little heed is paid to the economy of creation. Nature is looked upon as an inexhaustible storehouse, to be wasted, controlled, manipulated, and consumed. The representation of nature in the fragmented and mechanistic worldview of modern industrialized societies only aggravates the violence against and dominance and exploitation of nature, whereas models for appreciation of the natural environment, based on holistic worldviews of traditional societies (such as the one described here), inspire an attitude of reverence, reciprocal partnership, and ecological sustainability. In the ritual capsule of Durgā Pūjā, the plants are treated with love and respect, for they embody a sacred world pregnant with life and meaning. This reverence provides a framework for ethical restraints and control, which function as norms against earth abuse; the ritual creates a silent language of geopiety toward biotic life. This, we have lost.

It is worth noting that, whereas some religious institutions have slowly perished under the impact of invasive modernization, others still struggle to survive. The festival of Durgā Pūjā has shown an amazing resilience and ability to adapt to socioeconomic changes. Recent trends show that, far from disappearing, the festival is becoming ever more popular.[26] The ritual, as it exists today, has the potential to inspire urban dwellers to use symbols rooted in the earth for ecological activism.

Primal ecocosmic systems may be viewed as an *ur*-phenomenon, a study of nature, its rhythms and cycles, that predates the sophisticated and specialized study of distinguishable disciplines that describe natural phenomenon. In systems of primal ecology, transitions from ecocosm to cosmocosm, from ontology of nature to life sciences, and from ritual to ecology could be made without distinguishing disciplines. Tradition provides a homogenous system of thought that knits ontology, life sciences, mythology, and ecology into one whole. All these disciplines are engaged in creating the ritual capsule of Durgā Pūjā.

Appendix: Ritual Ingredients Used for Durgā Pūjā[27]

I. Plants Used in the Preparation of Idols for Durgā Pūjā

1. Dhān (*Oryza sativa*)
2. Kārpās (*Gosspium herbaceum*)
3. Bâs (*Bambusa arundinacea*)
4. Keśe (*Sacharum Spontancum*)
5. Cork of Śola (*Querus suber*)
6. Natural dyes from grain, flower buds, turmeric, and burnt paddy

II. Five Types of Plants, Seeds, Twigs, and Powders

	Plants	Seeds	Twigs	Powders
1.	Bilvā (*Aegle marmelos*)	Dhān (*Oryza sativa*)	Ām (*Mangifera indica*)	Wheat grains (*Triticum sativum*)
2.	Tulasī (*Ocimum sanctum*)	Tila (*Sesamum Indicum*)	Aśvattha (*Ficus religiosa*)	Flower buds
3.	Pān (*Piper betle*)	Mung (*Phaseolus aureus*)	Yagna-ḍumur (*Ficus glomerata*)	Haldi (*Curcuma longa*)
4.	Supāri (*Areca catechu*)	Śveta Sarisā (*Brassica hirta*)	Baṭ (*Ficus bengalensis*)	Burnt paddy (*Oryza sativa*)
5.	Pakur (*Ficus infectoria*)	Jau (*Ordeum vulgare*)	Aśoka (*Saraca indica*)	Rice (*Oryza sativa*)

III. Types of Liquids and Waters Used in the Annointing Ceremony

1. Pure water (*śuddhajala*)
2. Conch water (*śaṅkhajala*)
3. Ganges water (*gaṅgājala*)
4. Hot water (*uṣṇodaka*)
5. Scented water (*gandhodaka*)
6. Pure water (*śuddhajala*)
7. Gold water (*suvarṇajala*)
8. Silver water (*raupyajala*)
9. Copper water (*tāmrajala*)
10. Common holy water prepared with camphor and spice (*arghyajala*)
11. Honey water (*madhujala*)
12. Kuśa grass water (*kusájala*)
13. Lotus water (*kamalajala*)

14. Green coconut water (*dab* in *kusájala*)
15. Honey water (*madhujala*)
16. Clarified butter water (*ghṛtajala*)
17. Milk water (*dugdhajala*)
18. Coconut water (*nārikelajala*)
19. Sugrcane juice (*ikṣurasa*)
20. Curd water (*dadhijala*)
21. Tila oil (*ambikā*)
22. Viṣṇu oil (*cāmuṇḍā*)
23. Dew (*śiśira*)

IV. Nine Types of Soil Used for Annointing

24. Earth from the king's door (*rājadvāra*)
25. Earth from a cow shed (*catuṣpada*)
26. Earth from a holy mountain (*vṛsaśṛṅgi*)
27. Earth scratched by the tusk of an elephant (*gajadantikā*)
28. Earth from the confluence of holy rivers (*nadīsaṃgam*)
29. Earth from the door of a courtesan (*veśyādvāra*)
30. Earth from the holy Ganges (Gaṅgā)
31. Earth from anthills (*vālmikī*)
32. Earth from all the holy centers (*sarvatīrtha*)
33. Sugarcane juice mixed with water (*ikṣurasajala*)

V. Medicinal Plants Used for Annointing

Sarva-Auṣadhi

34. Jaṭāmāṃasī (*Nardostachy jatamansi*)
35. Bach (*Zingiber zerumbet*)
36. Kur (*Saussurea lappa*)
37. Haldi (*Curcuma longa*)
38. Dar Haldi (*Berberis aristala*)
39. Śāli (*Curcuma zedoraia*)
40. Campā (*Michelea campaea*)
41. Motha (*Cyperus rotundus*)

Mahā-Auṣadhi

42. Cākulia (*Uraira legopides*)
43. Śayāma-latā (*Technocarpus frutescens*)
44. Bhṛṅarāja (*Eclipta alba*)
45. Śatamūlī (*Asparagus recemosus*)
46. Golañca (*Tinospora cordifolia*)
47. Śvetabirala (*Sida cordifolia*)

VI. Pañca Kaṣāya (Bark of Five Plants)

48. Jāmuna (*Syzygium cumini*)
49. Bakula (*Mimusops elengi*)
50. Simul (*Bombax Malabricum*)
51. Birala (*Sida rhombifolia*)
52. Kula (*Zizyphus jujuba*)
53. Pure water from four jars (*catuṣṭayakalaśjala*)

VII. Types of Water Used for Annointing Accompanied by Rāgas

Water	Musical Notes
54. Ganges water (*gaṅgājala*)	Mālwa Rāga (accompanied by vijaya vādya)
55. Rain water (*vṛṣṭijala*)	Lalita Rāga (accompanied by dundabhi vādya)
56. Sarasvatī water	Vibhāsa Rāga (accompained by dundabhi vādya)
57. Ocean water	Bhairavī Rāga (accompanied by bhīma vādya)
58. Lotus water (*padmarājamiśrikajala*)	Koḍa Rāga (accompanied by indrābhiṣekha)
59. Waterfall water	Aradhi Rāga (accompanied by śaṅkha vādya)
60. Water from all the holy centers	Vasanta Rāga (accompanied by pañcvādyadhola, vīṇa, damru, dhāk, and tablā)
61. Pure water	Dhanaśrī Rāga (accompanied by vijaya vādya)

VIII. *Havan Sāmagrī*

62. Agar (*Aquilaria agollocha*)
63. Tagar (*Valeriana wallichii*)
64. Charila (*Parmalia perleta*)
65. Jaṭāmāṃsī (*Nardostachy jatmansi*)
66. Khas ki Jaḍa (root of *Andropogan muricatus*)
67. Chandana kā Burādā (*Santalum album*)
68. Kamala Gaṭṭa (*Nilembium speciosum* wild seeds)
69. Pañca-Mevā (a preparation of resin, cashew nuts, coconut, almonds, and dried dates)
70. Khāṇḍa (sugar)
71. Ghee (clarified butter)
72. Belpatra (leaves of *Aegle marmelos*)
73. Wood of the Bilvā tree (*Aegle marmelos*)

74. Wood of the Mango tree (*Mangifera indica*)
75. Wood of the Yagna-ḍumura tree (*Ficus glomerata*)

IX. Ingredients for Holy Food (*Prasād*)

76. Nārikel (*Cocos nucifero*)
77. Khirā (cucumber; *Cucumis sativus*)
78. Chola (chickpea; *Cicer arietinum*)
79. Mung (*Phaseolus aureus*)
80. Kamala-nembu (*Citrus reticulata*)
81. Bātābi-nembu (*Citrus marima*)
82. Piyārā (*Psidium guajava*)
83. Mash-Kalai (*Phaseolus mungo*)
84. Pape (*Carica papaya*)
85. Pāṇi-phala (*Trapa bispinosa*)
86. Mutter or But (peas; *Pisum sativum*)
87. Ām (mango; *Mangifera indica*)
88. Kala (*Musa paradisiaca*)
89. Khajur (dates; *Phoenix sylvestris*)

X. Ingredients for *Siddhi* (Tonic with Hemp)

90. Milk
91. Sugar / honey
92. Coconut water (*Cocos nucifera*)
93. Dalchini (cinnamon stick; *Cinnamomum zeylanicum*)
94. Elāchi (cardamom; *Amomum subulatum*)
95. Lavaṅga (clove; *Syzygium aromaticum*)
96. Tejapāt (*Cinnamomum tamala nees*)
97. Ādā (ginger; *Zingiber officinale*)
98. Kesar (saffron; *Crocus sativus*)
99. Vijaya buṭī (hemp; *Cannabis sativa*)

Notes

1. Chintaharan Chakravarti, "Śākta Festivals of Bengal and Their Antiquity," *India Historical Quarterly,* September 1951, 257.

2. See *Kālikā-Purāṇa,* ed. Biśwanārāyan Śāstri, The Rai-krishanadasa Prāchyavīdyā Granthamālā, 5 (Varanasi: Chaukhambha Sanskrit Series Office, 1972), 60.25–33; *The Mahābhāgavata-Purāṇa,* ed. Pushpendra Kumar Sharma (Delhi: Eastern Book Library, 1983), chap. 46ff.; and *Purohita Darpaṇa* (with *Bṛhannandikeśvara Purāṇa* and *Devī Purāṇa,* chap. 5, pp. 757–98), comp. Surendra Mohan Bhattacharya (reprint, Calcutta, 1381 B.S. [1911]), p. 758.

3. For some representative examples of *Agomani* songs, see Sibanarayan Kabiraj, "Significance of Durga Puja," *Folklore* 32, no. 16 (October 1991): 174–76.

4. See, for instance, Thomas B. Coburn, *Devī-Māhātmya: The Crystalization of the Goddess Tradition* (Delhi: Motilal Banarsidass, 1984), 1–69. Coburn examines the various strands that go into creating a composite figure of the goddess.

5. See Akos Ostor, *The Play of the Gods* (Chicago: University of Chicago Press, 1980); James Preston, *Cult of the Goddess: Social and Religious Change in a Hindu Temple* (New Delhi: Vikas Publishing House, 1980); Amita Ray, "The Cult and Ritual of Durgā Pūjā in Bengal," in *Shastric Traditions in Indian Arts,* vol. 1, ed. A. Dallapiccola et al., 131–41.

6. The list of the nine plants is given in the *Śākhambharī Rahasya,* cited in *Śākta Darśana* (reprint, Guwahati: Pandit Chakreswara Bhattacharya, 1970), 168; and *Durgāpūjā-śyāmapūjā-paddhati,* ed. Ramachandra Jha, Mithila Granthamala, 25 (Varanasi: Chaukhambha Vidyabhawan, 1993), 38.

7. *Śākhambharī Rahasya,* cited in *Śākta Darśana,* 168.

8. For the Hindu perception of trees and plants, see *Vṛkṣāyurveda* (excerpt from *Śāraṅgadhara-Saṃhitā*), ed. Ramchandra Rao (reprint, New Delhi: Bhartiya Cattle Resource Development Foundation, 1993), 2–8, 10ff. See also P. Sen Sarma, *Plants in the Indian Puranas: An Ethnobotanical Investigation* (Calcutta: Naya Prokash, 1989). On the liturgical use of plants, see Vinaya S. Ghate, "Plants in Patra-Pooja: Notes on Their Identity and Utilization," *Ethnobotany* 10 (1998): 6–15.

9. The *pūjā* program is elaborate and requires extensive preparation. See *Purohita Darpaṇa*; *Durgāpūjā-śyāmāpūjā-paddhati,* 1–77; and Pratapachandra Ghosha, *Durga Puja: With Notes and Illustrations* (Calcutta: "Hindoo Patriot" Press, 1871), 14ff. Note that there is considerable regional variation in the style of the *pūjā*. Some temples in Delhi, for example, follow the rituals set out in the *Durgārcanā-paddhati,* which gives emphasis to the daily recitation of the text of the *Devī Māhātmya,* along with *homa* and *kumarī pūjā*. On the regional variations of Durgā Pūjā, see James Preston, "Aspects of Change in an Indian Temple: Chandi of Cuttack, Orissa" (Ph.D. diss., Hartford Seminary, 1974). In this paper I have followed the *pūjā* of the Matrika Ashram, New Delhi, where there is a shrine to Durgā and Kālī. The ashram temple follows the traditional Bengali ritual manuals used by temple priests in Bengal and replicates the *pūjā*s performed in Bengal.

10. aiṃ rāvaṇsya vadhārthāya rāmasyānugrahāya ca/
akāle brāhmaṇā bodho devyāstvayi kṛtaḥ purā/
ahamasyāśvine ṣaṣṭhyāṃ sāyāhne bodhayāmyataḥ//
tasmād ahaṃ tvā pratibodhyāmi vibhūtirājyapratipattihetoḥ/

yathaiva rāmeṇa hato daśāsyastathaiva śatrūn vinipātyāmi//
Durgāpūjā-śyāmāpūjā-paddhati, p. 26; cf. *Purohita Darpaṇa*, p. 758

11. Oṃ merumandarakailāśahimavacchikhare girau/
jātaḥ śrīphalavṛkṣa tvmambikāyāḥ sadā priyaḥ//
śrīśailaśikhare jātaḥ śrīphalaḥ śrīniketnaḥ/
netavyosi mayāvṛkṣa pūjyo durgāsvarūpataḥ//
gṛhitvā tava śākhāntu durgāpūjāṃ karomyaham/
śākhācchedodbhavaṃ dukhaṃ na ca kāryaṃ tvayā prabho//
devairhitvā śākhā te pūjitā ceti viśrutiḥ/
putrāyurdhanavṛddhyarthaṃ neṣyāmi tvāmumāpriyāṃ//
Oṃ chindhi chindhi phaṭ hrīṃ svāhā
Durgāpūjā-śyāmāpūjā-paddhati, pp. 26, 28; cf. *Purohita Darpaṇa*, p. 762ff.

12. Although the hibiscus (*Japa*) is the most popular form of flower offered during the *pūjā*, the sources enjoin a rich array of fragrant flowers grown in sacred groves for the goddess's worship:

> Now I shall describe the fruit to be obtained by worshiping the goddess [Durgā] with special flowers. Let the devotee worship her with flowers of the season, and with Mallikā, Jāti, Kuṃkuṃ, red, white, blue and with off-white flowers, with Kiṃśuka, Keśara, Kaṅka, Kaṅkata, Campā, Bakula, Mandāra, Kuṇda, Kuraṇṭaka, Karavīra, Arka, Turviṣa and Aparājitā; Dhatūra, Atimukta, Brahmaka, Agasti, Damana, Sindhuvāra, and Bhīmā Arka, with creepers of Brahmavṛkṣa, Durvā, and tender buds with filaments of Kuśa grass and with beautiful leaves of Bilvā. (*Durgopāsanākalpadruma* 2.57–62)

13. Oṃ brahmā ihāgaccha ihatiṣṭhetyāvāhya, Oṃ hrīṃ brahmāṇyai namaḥ/
Oṃ brahmāṇi tvaṃ samāgaccha sannidhyam iha kalpaya/
rambhāsvarupe me nityaṃ śāntiṃ kuru namastute//
Oṃ raktadantike ihāgaccha Oṃ raktadantikāyai namaḥ/
Oṃ dāḍimi tvaṃ purā yuddhe raktabījasya saṃkṣaye/
umākāryaṃ kṛtam yasmāt tasmān me varadā bhava//
Oṃ Lakṣmi ihāgaccha Oṃ hrīṃ dhānayādhiṣṭhātryai lakṣmyai namaḥ/
jagataḥ prāṇarakṣārthaṃ brahmaṇā nirmitaṃ purā//
umāprītikaraṃ dhānyaṃ tasmāt tvaṃ rakṣa māṃ sadā//
Oṃ hrīṃ durge ihāgaccha Oṃ hrīṃ haridrādhiṣṭhātryai durgāyai namaḥ/
devi umārūpāsi suvrate mama vighnavināśāya pūjāṃ gṛhāṇa prasīda me//
Oṃ hrīṃ cāmuṇḍe ihāgaccha Oṃ hrīṃ mānādhiṣṭhātryai cāmuṇḍāyai namaḥ//
yasya patre vaset devī mānavṛkṣā śacīpriyaḥ/
mama vighnavināśāya pūjāṃ gṛhṇa yathāsukham//
Oṃ hrīṃ kālīkā ihāgaccha kacupatrikādhiṣṭhātryai kālyai namaḥ/
mahiṣāsura yuddhe tvaṃ kacubhūtāsi suvrate/
mama cānugrahārthāya āgatāsi harapriye/
Oṃ hrīṃ śive ihāgaccha Oṃ hriṃ. bilvādhiṣṭhātryai śivāyai namaḥ/
mahadevapriyakaro vāsudevapriyaḥ sadā/
umāprītikaro vṛkṣo bilvavṛkṣa namostute//
Oṃ hrīṃ śokarahite ihāgaccha aśokādhiṣṭhātryai śokarahitāyai namaḥ./
śivaprītikaro devo hyaśokaḥ [śoka] nāsānaḥ/

umāprītikaro yasmād aśokaṃ māṃ sadā kuru//
Oṃ hrīṃ kārtikī ihāgaccha . . .
Oṃ niśumbhamathane [devi] sendrairdevagaṇaiḥ saha/
jayanti pūjitāsi tvamasmākaṃ varadā bhava//
patrike navadurge tvaṃ mahādevamanorame/
pūjāṃ samastāṃ saṃgṛhya rakṣāṃ [kuru] maheśvarī//

Durgāpūjā-śyāmāpūjā-paddhati, pp. 38–40. Emendations to the Sanskrit texts in notes 10, 11, and 13 are by Pandit Satkari Mukhopadhyay. The English translation is my own.

14. *Durgāpūjā-śyāmāpūjā-paddhati*, p. 72.

15. Ghosha, *Durga Puja*, 82.

16. Ibid.

17. *Mahābhārata*, appendix 1, no. 4, in the *Virāta-parvan* of the critical edition; verses 27–29.

18. See Sarat Chandra Mitra, "On the Cult of the Tree-Goddess in Eastern Bengal," *Man in India*, 1922, 228–41. See also M. N. Chaudhury, "The Cult of Vana Durga, a Tree Deity," *Journal of the Royal Asiatic Society* (*Letters*) 11 (1945).

19. Mitra, "On the Cult of the Tree-Goddess in Eastern Bengal," 18ff.

20. Shanti Lal Nagar, *Mahisasuramardini in Indian Art* (New Delhi: Aditya Prakashan, 1988), 156.

21. *Devībhāgavata Purāṇa*, book 7, chap. 28; *Śākhambharī Rahasya*, pp. 166–75; *Śākhambarī śatākṣī sā saiva Durgā prakīrtitā*, *Śākhambharī Rahasya*, p. 173; *DM*, *Mūrtirahasyam*, 15.

22. The word "capsule" in a generic sense means a case, sheath, or envelope enclosing a membranous sac, which does not allow the contents to be dispersed. The *Collins Cobuild English Language Dictionary*, ed. John Sinclair et al. (London: William Collins and Sons, 1987), 204, gives the meaning of the word capsule as "a structure in a plant or animal which does not open unless it is split, for example the seed case of a poppy." The ritual of Durgā Pūjā may be better understood if we look upon the structure and cycle of worship as a seamless whole.

23. Note that both Indian and Western scholars have generally agreed that the goddess Durgā "is a cultural gift of pre-Aryan India"; Daniel H. H. Ingalls, cited in Coburn, *Devī-Māhātmya*, viii. See also N. N. Bhattacharya, *The Indian Mother Goddess* (Calcutta: R. D. Press, 1971).

24. Vandana Shiva, *Staying Alive* (Delhi: Kali for Women, 1989), 105.

25. Ibid., 111.

26. Preston, *Cult of the Goddess*.

27. This list is reconstructed on the basis of fieldwork conducted at the Matrika Ashram, Dakshina Puri, New Delhi, headed by Her Holiness Madhobi Ma. I am grateful to H. H. Madhobi Ma for her support and guidance. The ritual in this temple follows the text of the *Durgā-Pūjā-paddhati* (in Bengali), based on the Devī Purāṇa. There is little variation in the pūjā mantras given in other Sanskrit texts quoted in this chapter. The list of plants used for Durgā Pūjā cited in D. C. Pal, "Plants Associated with the 'Durga Puja' Ceremony in West Bengal," *Journal of the Bengal Natural History Society* 30, no. 1 (January 1970): 61–68, parallels my list. Pal's list omits sections II, III, IV, VII, and VIII found in my appendix.

Ethical and Religious Dimensions of Chipko Resistance

GEORGE A. JAMES

The interest of specialists in environmental philosophy in non-Western ways of valuing the environment is hardly new. It was in 1949 that Aldo Leopold expressed the view that society stands in need of a new environmental ethic.[1] Some twenty years later the idea was expressed that the West should achieve such an ethic by abandoning classical Western forms of ethics and applying those of Hinduism, Buddhism, and the peasant cultures of Asia.[2] In an essay published in 1967, the environmental historian Lynn White argued that the religious traditions of Asia express an attitude toward nature that is completely different from those of the West, one inherently more congenial to the environmental.[3] He was joined in this opinion by scholars such as Roderick Nash, who claimed that, in Eastern thought, nature was to be honored and cared for.[4] The Judeo-Christian tradition, on the other hand, was believed to legitimate the most exploitative and destructive instincts of humanity.[5] This idea has been opposed by such scholars as the Australian philosopher John Passmore, who cast doubt upon the suggestion that our environmental problems are likely to be solved by abandoning the analytical and critical approach the West had so very painfully learned.[6] He claims that the doctrine of stewardship that appears in the West entails a regard for nature that the Eastern traditions completely lack. He holds that such concern is incompatible with the Eastern religious quest for a salvation achieved by freeing oneself from every kind of earthly bondage.[7]

While many distinguished scholars have defended the Hindu religious tradition against Passmore's charge, the negative judgment concerning the support in Hinduism for an environmental ethic remains strong. The issue has arisen again in a recent work by the environmental philosopher J. Baird Callicott that examines a variety of religious traditions for their ecological insight.[8] In this work Callicott envisions a variety of traditional cultural environmental ethics, each resonating with and helping to articulate a scientifically grounded global environmental consciousness that spans national and cultural boundaries.[9] His judgment about the ecological significance of the Hindu religious tradition, however, is hardly more sympathetic than that of Passmore. He holds that because the Upaniṣads understood the undifferentiated and unmanifest Brahman (supreme reality) to be the source and ground of all manifest and differentiated things, the Hindu religious tradition regards manifest and differentiated things as something less than ultimately real or morally significant. The world in which environmental problems are manifest is either a beguiling appearance or an outright illusion. Hindu religious practice seeks to transcend this world, not to improve it. It seeks to unify the self with the ultimate undifferentiated reality beyond the visible world. "From the Hindu perspective," he says, "the empirical world is both unimportant, because it is not ultimately real, and contemptible, because it seduces the soul into crediting appearances, pursuing false ends, and thus earning bad *Karma*."[10] In Callicott's view, Hinduism entails an understanding of reality that is essentially hostile to environmental concerns.

While Callicott does not claim expertise in Sanskrit literature or philosophy, the view that he and Passmore hold concerning Hindu religious thought has support in a recent study of Advaita Vedānta by the Sanskrit specialist Lance Nelson. While Nelson recognizes that the Hindu religious tradition includes more diversity than is acknowledged by either Passmore or Callicott, he points out that the *saṃnyāsin*s (the renouncers) have been the creators of Hindu spirituality and that, among these, the Śaṅkara tradition that preserved the nondualist viewpoint has been a dominant cultural force in India for more than one thousand years. It is also the central viewpoint of the modern Hindu renaissance.[11] Nelson argues that while this tradition is often seen as an alternative to the Western dualism that elevates spirit over nature, Advaita Vedānta encourages attitudes of devaluation and neglect of the natural universe. While he acknowledges that a revalorization of nature can be found in the teachings of Aurobindo, Vivekananda, Ramakrishna, and others, he holds that no such recovery of the world is to be found in classical Advaita Vedānta. He finds in this tradition no sense of community with nature, no reverence for nature, and no basis for an understanding of nature as having intrinsic spiritual worth. On the contrary, he finds that in Advaita Vedānta nature is to be feared, despised, and transcended.

The sentiment of Advaita toward the natural world, in Nelson's view, is reflected in the attitude of the *jīvanmukta,* the saint who has achieved liberation while remaining in the world. The perception of the world by such a person is described in the analogy of a defect of the eye or the liver. A person with double vision may continue to see two moons, but he *knows* that there is only one; the person who has a disorder of the bile *knows* that sugar is sweet even though he continues to experience it as bitter. Likewise, the liberated saint *knows* the natural world to be false, as the content of dreams. He has experience of the natural world but does not believe in it. And in the highest stages, he experiences it no more. Nelson writes:

> It is difficult to see in such modes of thought anything less than an extreme version of the world-negating, transcendental *dualism* that supports environmental neglect. Advaita achieves its brand of "non-duality" not inclusively but exclusively, at great cost: the world of nature is finally cast out of the absolute, out of existence.[12]

At its best, in Nelson's view, classical Advaita Vedānta fosters devaluation and disregard of the world. Thus, even if we view Advaita as a tradition of thought and practice reserved for an esoteric elite, the fact that it came to carry and continues to carry such great prestige in the religious culture of India has serious consequences for India's ecological future.[13]

The difficulty with Callicott's conclusion concerning the Hindu religious tradition is that it leaves him quite unprepared to make sense of what he himself considers one of the most important environmental movements of modern India. He holds that the Chipko movement of the Indian Himalayas positions itself in uncompromising opposition to conventional neocolonial development schemes, in which local resources are ravaged to supply foreign markets. The movement, he says, derives its name from a Hindi word meaning to hug or to embrace.

It is a movement known for its strategy of literally hugging or embracing trees to stand in the way of the ax that is intended to cut them down. It is remarkable, then, that Callicott also claims that the Chipko movement applies "the foundational ideas of Hindu philosophy." He states that Hinduism has both inspired and grounded this most successful environmental uprising.[14] Given what Callicott has said about the Hindu view of nature, it is difficult to understand how any of the actions of the Chipko movement could have their origin in or receive support from Hindu thought. If Callicott's discussion of the foundations of Hindu thought is correct and if the participants of the Chipko movement are true to the Hindu faith, it is hard to understand why they should take the least interest in the preservation of the material world.

Nelson's views are less perplexing than those of Callicott. While Nelson does not discuss environmental activism in India, he does not deny that there may be grounding for an environmental ethic in some varieties of Hindu religious thought. His point is that such a grounding is not to be found in classical Advaita Vedānta. Nevertheless, because he also argues that the influence of Advaita Vedānta presents an obstacle to the environmental consciousness of India, the history of Chipko activism must remain an anomaly to his vision of India as well.

The question thus arises whether Chipko is a completely secular movement without sanction in Hinduism, whether perhaps it is even antithetical to Hindu religion, or whether perhaps it has support in aspects of the Hindu tradition that are not immediately evident to the Western observer, that have not been the focus of the Western tradition of scholarship in which India is the subject of study. To answer this question, I take a tack that differs from those of Passmore, Callicott, and Nelson. Aristotle once argued that while metaphysics deals with the most subtle of ideas, it is the political or the ethical that ordains what should be studied, what each class of citizen should learn, and the extent to which a subject can be pursued.[15] This approach is justified because it remains a question whether an adequate understanding of the values of a society, particularly as it pertains to nature, can be derived from an analysis of texts reflecting the philosophical views of even the most honored of its men. Rather than focus upon the metaphysical views of what might arguably stand as the most prestigious tradition of Indian philosophy, and attempt to infer their ethical implications, I focus upon the values that have informed Indian social life in its relation to the material world. I then examine the environmental activism expressed in the Chipko movement in the light of this tradition of values. I discuss the support for this environmental movement in the religious life of those most affected by environmental degradation.

Traditional Indian Values

It has been pointed out by some Indian scholars that India's values have not received the scholarly attention which other aspects of India's intellectual history have received. One possible reason is that, while discussions of values in the West have frequently been motivated by conditions under which the grounding of all values has been questioned, the values by which Indian culture has been informed have never become a matter of serious doubt. It is generally recognized that the Hindu religious tradition acknowledges the importance of a variety of values. These are usually discussed in the context of the principal ends of human life (*catuṣpuruṣārtha*): *artha* (political and economic value), *kāma* (sensual value), *dharma* (moral value), and *mokṣa* (spiritual value). Yet, Indian thought has not been preoccupied with the philosophical clarification of values as such or with the effort to integrate them all within an embracive philosophy of ethics. The writers of the Sūtras that treat these differing values do not speak unanimously about how these values are related one to the other. Rather, each maintains a claim for the legitimacy of the particular value with which his treatise is principally concerned.[16]

One of the most important discussions of value that we find in the history of Indian thought is the famous *Arthaśāstra* (321–296 B.C.E.), attributed to Kauṭilya, a minister to the first Mauryan emperor. Along with other ancient writers, Kauṭilya acknowledges the trinity of values (*trivarga*) that pertain to temporal existence: *artha, dharma, kāma.* Among these values he thinks *kāma* is to be recommended simply because a life devoid of pleasure is good for nothing. On the other hand, a life that is pursued for pleasure, and without consideration to virtue and wealth, is likely not to be pleasurable for long. A life lived for pleasure, in opposition to other values, is likely to lead to misery. In this discussion it is evident that *kāma,* or sensual pleasure, is understood to be a good in itself. *Dharma* and *artha* stand above it in the sense that if these things are not taken into account, then sensual pleasure itself will be lost. The most lasting of pleasures will be those in which *kāma* is pursued in harmony with *dharma* and *artha*. Of the three, however, it is neither *kāma* nor *dharma* but *artha* that is to be regarded as the highest of values, because, according to Kauṭilya, without *artha* neither pleasure nor moral virtue can prevail. Had Kauṭilya been writing in the style of Saint Paul, he would perhaps have said: And now there abideth pleasure, power, and virtue, these three; but the greatest of these is *power*.

For Kauṭilya, *artha* is not limited to what we today would associate with economic and political strength. What he calls *Arthaṣāstra* is Kauṭilya's account of the views of the ancient teachers concerning the acquisition and maintenance of the earth. By the earth he means both the material source of the life and welfare of the community and the society that is supported by it.[17] The maintenance of the earth imposes responsibilities of various kinds upon the various parts of society. In Kauṭilya's view a large responsibility falls upon government. Some writers have observed that, like Thomas Hobbes in the West, Kauṭilya believes that a competent monarchy is the form of government most likely to protect the weak from the strong and maintain the welfare of the people. For this to work, much depends upon the character and competence of the sovereign. The sovereign must be a person well versed in the sciences, who maintains control over his senses, who keeps company with elderly persons, and is in possession of a host of natural endowments. Kauṭilya lists hundreds of personal and intellectual qualities which he believes an effective ruler must have.

As the guardian of social as well as ecological order, the sovereign supports the arrangement of duties required of his subjects according to the classes (*varṇa*s) and stages of life (*āśrama*s), which Kauṭilya takes on the authority of the Vedas. Yet, Kauṭilya also specifies duties that pertain to persons regardless of their social class or stage of life. These include nonviolence, truthfulness, purity, compassion, and forgiveness. It is significant that, where Kauṭilya specifically addresses the question of *mokṣa,* he points out explicitly that the secret of heaven and *mokṣa* is simply in doing one's own appointed duty (*dharma*) and that the repudiation of one's duty (*dharma*) leads to destruction.[18]

In the *Arthaśāstra,* it is the responsibility of the king to maintain law and order, to protect the state against threats within and without, both through punishment and good government, and to attend to the moral and spiritual advancement of his subjects. The welfare of the king is the welfare of his subjects. His pleasure is not his own good, but his own good is the pleasure of his subjects. Much of what we today would call agricultural administration, disaster management, and environmental policy falls, in Kauṭilya's view, within the purview of the king. Specifically, he is expected to undertake needed agrarian reforms and establish policies for the maintenance of pastures and forests.[19] It is interesting to examine the details of Kauṭilya's specific recommendations for the maintenance of forests and other features of the environment, and to view them in light of environmental issues of current interest in India today. In the *Arthaśāstra*, for instance, specific fines are recommended for such offenses as disposing of dust on roads, urinating or defecating near a well, pond, or temple, and for inappropriately disposing of a dead animal.[20] For the present it is enough to observe that the care and management of forests is important enough to merit Kauṭilya's attention. The present point, however, is that, for Kauṭilya, one of the most important of the ancient writers concerned with values, it is not *mokṣa*, or release from the natural world, but the very earthly value of *artha* that is of supreme importance.

Nor is Kauṭilya alone. If Kauṭilya subordinates *mokṣa* to *dharma,* and finally subordinates *dharma* and *kāma* to *artha,* it is interesting that Vātsyāyaṇa, the putative author of the famous *Kāmasūtra,* also acknowledges the three temporal values of *artha, kāma,* and *dharma* (*trivarga*). Yet, for him, it is *kāma* alone that is of intrinsic value and should be preferred to both wealth and moral virtue, which are its means. Turning to the question of spiritual values, Vātsyāyaṇa states that a happiness based on virtue and wealth is the most desirable, and that the balanced realization of the three leads to happiness both in the present world and the next.[21]

The sacred law proclaimed by Manu (circa 100 B.C.E.) is widely considered the most authoritative of the ancient discussions of values.[22] Manu argues that while some claim that the good lies in *dharma* and *artha,* and others believe that it lies in *dharma* alone, and still others maintain that *artha* is the main goal, it is rather the case that human good lies in the harmony of all three.[23] Nevertheless, in constructing the harmony he has in mind, clearly *dharma,* the principal subject matter of his treatise, must have executive control over *artha* and *kāma*. Wealth and pleasure that transgress the limits of *dharma* are not to be desired.[24] *Dharma,* the conduct sanctioned by the Vedas and their authoritative commentaries, in Manu's view, constitutes a plenum of universal moral obligations. The contribution of Manu to ancient ethical literature represents an important departure from Kauṭilya and Vātsyāyaṇa in his integration of *mokṣa* into the span of human life. Manu holds unequivocally that spiritual liberation is the *ultimate* goal of human life. But he also holds that the harmonious realization of sensual, materialistic, and moral values is a necessary condition of the realization of *mokṣa*. For Manu, the life of moral duty (*dharma*) is the only way and the only preparation for the realization of that final goal.[25]

It would certainly be hard to deny that in the thought of Śaṅkara and of Advaita Vedānta, *kāma, artha,* and even *dharma* are understood ultimately to stand in conflict with the realization of *mokṣa*. Earthly values present an obstacle to the attainment of the highest goal of liberation from the world of death and rebirth. But the world of death and rebirth is also the world of mundane ethical demands, particularly those concerning the maintenance of the world. Nevertheless, if we look at the Advaita tradition in the context of ancient reflections on values, we find that Śaṅkara's view of *mokṣa* as the value that finally excludes others is one among a variety of views. It is hardly more outstanding than other ancient works in which one or another of the traditional values, like one or another of the deities, is praised and exalted above all others. Attention to the diversity of the views expressed in these texts discredits the claims of Passmore and Callicott that Hinduism is simply incompatible with environmental concern.

The Chipko Movement

I suggested above that the difficulty with Callicott's representation of Hinduism is that it fails to make sense of that movement of Indian environmental activism that focused international attention on the ecological crisis in the Indian Himalayas, that it fails to show, in the context of Hindu thought and practice, how such a movement could even be conceived. I suggest that an appreciation for the ethical dimensions of Hinduism provides better preparation for an understanding of the Chipko movement than we find in Callicott's view of the Hindu religious tradition. To understand the relation of Hindu religion to environmentalism, it is appropriate to ask what kind of support for their ethical concerns the participants found in Hindu religious traditions.

The Chipko resistance is generally understood to have begun in the town of Gopeshwar in the Chamoli district of Uttar Pradesh. In early 1973, a cooperative organization concerned with generating local employment made a request to the forest department for an allotment of ash trees to make agricultural implements. The forest department refused their request. Yet it granted a request for ash trees to a company manufacturing sporting goods for the export market. These were to be taken from the Mandal forest only a few miles away. This blatant injustice moved the organization, the Dashauli Gram Swarajya Sangh (or Society for Village Self-Rule), to organize several meetings in Mandal and Gopeshwar to determine what to do. One suggestion was to lie down in front of the logging trucks when they tried to move. Another was to burn the timber and resin depots, as had been done during India's resistance against the British. When neither of these were agreed upon, Chandi Prasad Bhatt, the leading local activist, suddenly thought of the plan of embracing the trees. "Let them know," he said, "they will not fell a single tree without felling one of us first." To this plan the young members committed themselves with signatures in blood. With this resolution, by most accounts, the Chipko (hugging) movement was born.[26]

In his discussion of this movement, Callicott, following the account of Jayanata Bandhyopadhyay and Vandana Shiva, points out that this uprising gave contemporary expression to a method of resistance that has had profound significance in Indian history. The precedent for action of this kind goes back to the protest of the Bishnoi people of Rajasthan who, in a famous incident in 1731, resisted the decision of the maharajah of Jodhpur to cut down their *khejri* trees for use in a lime kiln for a new royal palace. The Bishnois were committed to the protection of all wildlife, but to the *khejri* tree (*Prosopis cineraria*) in particular. This tree provided food and fodder, as well as building materials for fencing and other purposes. For the Bishnois the tree was sacred. On this occasion they embraced the trees in order to prevent the axmen from cutting them down. In the protest, 363 people died of injuries the axmen inflicted.[27]

While the Chipko movement received its name in Gopeshwar in

1973, the Chipko resistance was actually part of a protracted struggle of the people of Uttarakhand against forestry practices inimical to local needs. From the 1950s on, there is a clear record of local indignation about timber sales to outside contractors in violation of the hereditary rights of the local people. Such disaffection was often combined with hostility toward other features of commercialization, like the widespread distillation and sale of liquor that flourished in the face of the evident dissipation of forest resources.[28] Nevertheless, it was in 1970 that this resentment came into clear focus. That year, in a devastating flood in the Alakanandā valley that brought disastrous loss of life and property, the villagers could see the evident connection between massive deforestation and its ecological consequences. In 101 marooned villages, 604 houses and 500 acres of crops were destroyed. Six motor bridges and another sixteen footbridges, were destroyed. Cowsheds and watermills were lost. Fifty-five people and 142 head of cattle died. The affects of the flood were seen as far away as Haridwar, 300 kilometers down stream.[29] It was, in fact, on behalf of villagers who had suffered the greatest losses in the flood of 1970 that the Dashauli Gram Swarajya Sangh (DGSS) came to be organized. Two years before the Chipko movement got its name, the DGSS had demonstrated in Gopeshwar to bring an end to liquor sales and to untouchability. They had also tried to pressure the government to establish a policy for the just distribution of forest resources for the use of local people.

Reflecting on the *Arthaśāstra*, it is clear that from its origin the Chipko movement was an initiative for the maintenance of the earth, both as the source of the well-being of the community and as the social reality it supports, an obligation that in ancient times was the responsibility of the king. In its very idiom of protest, Chipko identified with similar efforts of the past, evoking not only historical precedent but the support of a religious perception of nature, the sacred nature of the trees. Moreover, it undertook protest coherent with moral duties of purity, truthfulness, and nonviolence.

The place in which the Chipko movement received its name is also the place in which it achieved its first visible success. The agents of the Symonds Company, who had been allotted ash trees for the production of sporting goods, were forced to depart from Mandal without felling a single tree. In June of the same year, however, the forest department allotted trees to the Symonds Company in another location, near the village of Phata in the Maṇḍakini valley, en route to the famous Kedarnath shrine. Word of this scheme reached the DGSS, and a massive demonstration was organized. The result was that the company's agents returned to Gopeshwar complaining to the forest department that, even after depositing money in pledge, they were unable to take any of the trees they had been promised. The Chipko movement was underway.

The inception of the Chipko movement, however, did not immediately alter government policy. The annual government auction of forests continued in November 1973. One of the plots marked for sale was the Reṇi forest near Joshimath, in the Alakanandā valley, an area in which the memory of the devastation of 1970 was especially vivid. Here, two thousand trees had been designated for felling. Chandi Prasad Bhatt again recommended the Chipko strategy, and Chipko activists organized a massive demonstration. On this occasion the government resorted to deceit. On the promise of a meeting with the conservator of forests, Chandi Prasad Bhatt remained at Gopeshwar. On the promise of compensation long overdue for lands appropriated by the Indian army after the Chinese invasion of 1962, the men of Reṇi and the surrounding villages journeyed to Chamoli. With the DGSS and the local men out of the way, the contractor's men avoided contact with Reṇi village and proceeded directly to the forest to begin felling the trees. On their way they were seen by a young girl who reported their presence to Gaura Devi, the head of the village women's organization. Gaura Devi mobilized the women of the village, who proceeded to the forest and implored the lumbermen not to begin the felling operations. At first the women were met with threats and abuse. But they remained firm in their purpose. "This forest," they said, "is our mother's home; we will protect it with all our might." When it was clear that the women would not relent, the men eventually departed from the forest.

After the protest at Reṇi, the chief minister of the state set up a committee to investigate the grievances of the people. It concluded that the deforestation of the Alakanandā valley was the major cause of the flood of 1970. As a result of their findings, commercial felling in the upper catchment of the Alakanandā River and its tributaries was banned for a period of ten years.[30] After the Reṇi resistance, the auctioning of forests was successfully opposed in the Chakrata division forest in Dehradun, in the Chamyala forest, near Sunderlal Bahuguna's famous ashram at Silyara, at Loital, and at Amarsar. In October 1977 activists of the Uttarakhand Sangharsh Vahini (USV) organized demonstrations in Kumaun against forest auctions that had proceeded, despite major landslides in the region. In Almora the USV organized Chipko activities where five thousand trees had been marked for felling. Among other things, the USV demanded a ban on the export of all raw materials from the hills. At Chanchridhar, protesters camped in the forest in which six thousand trees had been consigned to be felled. They remained there until forest workers admitted defeat. By similar strategies other scheduled fellings were stalled.

Chipko activists were disturbed not only by the auction of trees to outside contractors but by the tapping of chir pines (*cīr*) for resin. A government committee formed to investigate the practice had found that standard rules concerning length, width, and depth of the cuts from which resin was extracted were rarely followed. The life and health of the trees was endangered. While the chir pine was of little use to the local people, they had come to realize that forest cover was vital for life in the hills, and they moved to protect the trees. On 30 May 1977 a procession of activists entered the Hemvalghati region of Tehri Garhwal and applied mud and sack plasters to the wounds of chir pines damaged by overtapping. As abusive tapping practices continued, villagers began removing the iron blades inserted into the trees to bleed off their resin. Following this action, forestry officials eventually arrived to inspect the area. Observing the effects of abusive tapping and the results of the work of the villagers, one of them remarked that the village people had done precisely what the forest department should have been doing all along. Soon after, the forest department revoked the contracts for all resin-tapping in these forests.

The Religion of Forest Resistance

Among the several accounts of the Chipko movement, there are none that have made the religious dimension of the movement the specific focus of study. Nevertheless, some of the details suggest a tacit religious disposition toward nature reflected in actions and attitudes of the participants. The incident concerning the protection of the trees of the Advani forest is one of the most celebrated in the legacy of the Chipko movement. In October 1977, in the district headquarters in Narendranagar, the government auctioned the 640 trees of the Advani forest and 273 trees of the Salet forest in the Hemvalghati region. On this occasion, the Chipko leader Sunderlal Bahuguna undertook a fast at the town hall at Narendranagar and appealed to forest contractors and district authorities not to carry out their intended mission. The villagers declared that they would hug the trees in order to protect them. The contractors warned a group of village women that if their men entered the forest to hug the trees, they would be cut down along with the timber. In response, the women themselves took to the forests. Hundreds of women took a pledge to save the trees even at the cost of their own lives.[31] The village women tied sacred threads around the trees and for seven days guarded the forest while listening to discourses from ancient texts.[32] It was on this occasion that a forest officer made a speech to convince the women that the felling of trees was scientifically viable and economically indispensable. He ended his speech with what is now a famous proclamation: "You foolish village women! Do you know what the forest bear? Resin, timber, and foreign exchange!" To this remark the women replied with words that have become perhaps the most memorable Chipko slogan: "What do the forests bear? Soil, water, and pure air! Soil, water, and pure air sustains the earth and all she bears"[33]

There are two details in this episode that merit close attention. All of the available sources make the point that the women attached threads to the trees of the Advani forest, and all affirm that sacred narratives were heard. In describing these details, Callicott expresses frustration that his sources offer no insight into the substance or content of the readings that were undertaken on this occasion. Concerning the threads that were applied to the trees, he seems to defer to the observation of Shankar Sen Gupta that the entire corpus of ancient Indian literature abounds in praise for trees and to the view of O. P. Dwivedi concerning the popular belief that every tree has a deity that is worshiped "with prayers and offerings of water, flowers, sweets, and encircled by sacred threads."[34]

In light of the claim that the Hindu religion encourages the neglect of nature, the details of such behavior cannot be ignored. The practice of dendrolatry, the worship of plants and trees, is hardly alien to India. One of the many seals of the Indus valley civilization (2500–1500 B.C.E.) depicts a horned goddess in a *pīpal* tree, receiving worship from a devotee, with a hoofed animal watching the ceremony and a row of seven women in attendance.[35] It has been observed that the

Indus seals often depict a female deity in the branches of a tree, and that trees are featured prominently in the imagery of the Ganges.[36]

Perhaps for Indian writers the meaning of these texts and the nature of these actions are so transparent as hardly to require discussion. For the environmental philosopher they are a matter of considerable importance. A clue to their meaning can be drawn from suggestions offered casually in the various reports. Vandana Shiva states that the women tied sacred threads to the trees "as a token of their vow of protection."[37] Guha states that the women tied *rākhi*s to trees wounded by abusive overtapping.[38] In Thomas Weber's account, "they tied silken threads around the tree trunks, as sisters do to the wrists of their brothers at the festival of Raksha Bandhan, thus signifying that the trees were their respected brothers."[39] To grasp the significance of this simple gesture, it is necessary to understand the meaning of the *rākhi*. The *rākhi* itself is known more formally as the *rākśā bandhan*. *Rākśā* means protection; *bandhan* means bond. At the festival of Rākśā Bandhan, also called Rākśā Pūrṇimā, girls tie a *rākhi,* or amulet, around their brothers' wrists. It signifies a request for a bond of protection, which the brother who receives the *rākhi* is expected to observe. It is significant, however, that its application is not restricted explicitly to sibling relations: "Any lady who wishes to show favour to a friend and adopt him as her brother may tie the Rakhi round his wrist, and he is honour bound to remain her protector without having any other intentions towards her."[40]

On the face of it, the tying of *rākhi*s to the trees seems to have a meaning completely opposite to its traditional meaning. Here, it would seem, the women are pledging themselves in a bond of protection. Yet, as an Indian colleague put it to me, in India neither myths nor rituals are frozen in stone. They are capable of variation and of application to ever new needs and conditions. The flexible but forceful meaning of this gesture became clear in the course of my own research when, after I was invited to plant a tree outside his village, Vishweswar Saklani, the famous Chipko activist who has himself planted over 100,000 trees, honored me by tying a *rākhi* around my wrist and inviting me to tie a *rākhi* around the wrist of his wife. By tying *rākhi*s around the trees, whether to trees wounded by abusive tapping or to trees marked for destruction, the women of the Chipko movement were recognizing their complete dependence upon these trees for their life and well-being and announcing their vow to protect them. At the same time, they were expressing the standing of these trees within their larger familial embrace.

Most of the available sources affirm that during the seven days in which the women of fifteen villages kept watch over the Advani forest religious texts were heard. What was the content of these texts? Guha reports that a reading of the *Bhagavadgītā* was organized. Vandana Shiva states only that "discourses on the role of forests in Indian life from ancient texts went on nonstop."[41] For some, the question may be trivial, but for a number of my own informants it was a matter of considerable importance that ancient Sanskrit sources would have had little if any meaning for these peasant village women. They would neither have been schooled in Hindu philosophy nor able to read the Sanskrit texts. When I raised this question with Sunderlal Bahuguna, he acknowledged that verses from the *Bhagavadgītā* would have been recited, as they would be at the outset of any momentous communal undertaking. However, he stated emphatically that the ancient discourses that were heard over the entire seven days of the demonstration were not from the *Bhagavadgītā,* but from the *Bhagavad Kaṭha*. *Kaṭha*s are the narrative tales of the actions of divine beings from which practical moral lessons are often derived. He explained that the *kàṭha*s that were recited over the seven-day vigil in the Advani forest were the ancient stories of the entire life of Lord Kṛṣṇa. These included the stories of the birth of Kṛṣṇa, of his miraculous escape across the Yamunā River from the evil king Kaṃsa, and of Yaśodā's cosmic vision. They also included the stories of the sons of Kubera who learned humility by being incarnated as trees, eventually liberated by Kṛṣṇa; of Kṛṣṇa's swallowing of a forest fire; and of Kṛṣṇa's raising of Mount Govardhana to shelter the *gopī*s (or milkmaids) and the cattle from the storm. They certainly included the story of the demon Dhenuka who tried to keep Kṛṣṇa and his companions from the palmyra forest and from enjoying the fruits of the palmyra trees. They celebrated the story of Kṛṣṇa's defeat of Dhenuka and of his opening of the forest so that the people could enjoy its fruit and their cattle could graze on the grass in the shade of the trees. And they assuredly included the story of Kṛṣṇa in the forest of Vṛndāvana and of the compelling sound of his flute in the night, and of the enchanted *gopī*s who left the security of the homes of their fathers and husbands to dance with Kṛṣṇa in the moonlight.

While these stories are known to Western scholars chiefly from the

Bhāgavata Purāṇa and other medieval textual sources, Sunderlal indicated to me that the life and ventures of Kṛṣṇa in the forests of Vṛndāvana are known in Garhwali translation and are popular in the villages of the Garhwal hills.[42] It is not difficult to see how the stories of the forest life of Kṛṣṇa and other such tales legitimated the lives of these forest women. It is hardly surprising that in these narrative traditions the village women found support for their struggle to save the forests upon which their lives and the lives of their families depend.

To offer these observations as a response to the question whether Hinduism supports an environmental ethic is likely to provoke the reply that the actions and stories I have observed above bear little relation to the philosophical or ethical underpinnings of the Hindu religious tradition. It is certainly true that religious beliefs and practices that support the Chipko movement seem distant from those religious texts around which the debate about the possibility of a Hindu environmental ethic have usually revolved. We find in the Chipko movement no polemic based on Hindu religious sources for the intrinsic as opposed to the instrumental value of nature, such as might impress the exponents of deep ecology. The Chipko movement does not seem to undertake arguments from philosophical texts. The Chipko movement is a down-to-earth movement that expresses a down-to-earth concern for the maintenance of the earth as the responsibility of good government. At the same time, the Chipko movement aggressively affirms the spiritual value of nature. This value is affirmed, not on the strength of theoretical arguments, but because such ways of valuing cohere with the experience of those whose lives are most affected by the degradation of nature. The spiritual value of nature that is affirmed by such persons is supported not so much in the "great tradition" of religious life and thought familiar to those occupied with Hindu philosophy as in the "little traditions" familiar to local forest people. These traditions are also unquestionably a part of the living reality that we know as the Hindu religious tradition.

Chipko Leadership and the Ascetic Ideal

I pointed out that for Lance Nelson the attitude toward nature in the philosophy of Advaita Vedānta is expressed in the disposition of the *jīvanmukta*. For this saint who has brought himself to the brink of complete detachment to the material world, the perception of the world that remains a part of his experience is understood in the analogy of a defect of the eye or of the liver. He sees two moons but he knows there is really only one. He experiences sugar as bitter, though he knows it to be sweet. For Nelson, it is difficult to see in such modes of thought anything less than a kind of world negation that supports environmental neglect. But as John Dominic Crossan has shown in his recent studies of early Christianity, ascetic world negation is often as much a negation of a social and political world as it is of material reality.[43] The question that arises here is one that is not normally the focus of philosophical studies of the *jīvanmukta*. What is the social meaning of the person who has renounced the world but continues to live, the person who remains in the world although he is no longer of it? What, in particular, does this person mean to the community that supports him?

This question takes on special importance in the light of Guha's discussion of Chipko as a social movement, and particularly in his discussion of Chipko leadership. I do not wish to argue here that Sunderlal Bahuguna or any other Chipko leader qualifies as a *jīvanmukta*. Neither they nor their followers make such a claim.[44] Nevertheless, Guha makes the point that the Chipko movement, like movements in the Garhwal hills that preceded it, has had a strong moral and religious sanction. For Guha, the recitation of religious narratives, folk songs, and poetry extolling the value of the forest, and of the people's place within it, is an expression of the moral and religious content of the Chipko agitation. This moral and religious content, he thinks, is also expressed in the character of its leadership. For Bahuguna, the most visible leader of the Chipko movement, the main exploiters of nature are the forest officials who, in consort with outside contractors, leave not a splinter of wood in the forests for the use of the local people—who care more for their own prosperity than for the life of the forest or the community that depends upon it. He points out that Bahuguna's call to forest officers to abandon their ways and to serve the local communities evoked a positive response from a people exposed to extortion by officials in earlier regimes. As a person who had himself renounced political ambitions, Bahuguna's own style of life stands out as the antithesis of the self-seeking politician. Guha makes the point that the predominantly Hindu peasantry of the hills responded positively to Bahuguna's nonviolence and personal asceticism. His repeated fasts and strenuous foot marches, undertaken to support and spread the Chipko message, distinguished him as a notable ascetic in our own time. In his capacity for suffering and his spirit of self-sacrifice, the villagers read the renunciation of worldly ambition exhorted by the Hindu scriptures. According to Guha, "Sunderlal's remarkable physical endurance and sage like appearance make him a natural leader whose followers look to him to restore a pristine state of harmony and just government." In Guha's view, Sunderlal Bahuguna is a figure in whom the memory of peasant martyrs is revived and reappropriated. In this movement, as in other Indian peasant revolts, asceticism is a decisive ingredient.[45]

The Chipko Movement and the Feminine

The religious dimension of the Chipko movement is by no means limited to the recitation of *kaṭha*s and the ritual of sacred threads or to the ascetic character of Chipko leadership. It is expressed in details of the movement too numerous to treat within the scope of the present study. Nevertheless, the involvement of women in the Chipko movement brings out a critical facet of this religious dimension. It is perhaps in discussions of Chipko as a women's movement that the religious dimension of the movement has received the most attention. In the view of Vandana Shiva, nature and women are intimately related. She points out that in Indian religious traditions nature is symbolized as the embodiment of a feminine principle. Here, the world is understood to be produced and renewed by the dialectical play of creation and destruction, cohesion and disintegration:

> The tension between the opposites from which motion and movement arises is depicted as the first appearance of dynamic energy (Shakti). All existence arises from this primordial energy which is the substance of everything, pervading everything. The manifestation of this power, this energy, is called nature (Prakriti).[46]

In both its animate and inanimate aspects, according to Shiva, nature is an expression of Śakti, the creative principle of the cosmos. This creative principle is feminine. In conjunction with the masculine principle (Puruṣa), Prakṛti creates the world. Shiva notes that Prakṛti is worshiped as Aditi, the primordial vastness, the inexhaustible source of abundance, and as Adi Śakti, the primordial power. Her image, according to Shiva, is on every feature of the natural world.[47] It is interesting that in Callicott's discussion of Shiva's account of the religious support for the Chipko movement, Callicott simply conflates her discussion of Prakṛti with the philosophical idea of Brahman that appears in Vedānta.[48]

Shiva argues that the Indian attitude to nature differs significantly from that of the West, in which the human being is separate from the natural world and dominates it. The attitude of Hinduism toward nature is gracefully expressed in the daily worship of the sacred *tulsi* plant (*Ocimum sanctum*) that occurs in almost every Hindu home. The *tulsi* is sacred, in Shiva's view, not merely because it is a plant with healing powers, but because it is Vṛndāvana, the symbol of the cosmos. "In their daily watering and worship women renew the relationship of the home with the cosmos and with the world process."[49]

If Indian traditions have venerated the feminine in nature as the creative principle of the cosmos, we should not be surprised that in the forest they have found an especially compelling manifestation of this sacred feminine power. In Shiva's view, the diversity, harmony, and self-sustaining vitality of the forest have provided the organizational principles that have guided Indian civilization. As the poet Rabindranath Tagore has suggested, the distinctiveness of Indian culture consists in its having defined life in the forest as the highest form of culture. This culture, according to Shiva, has not been produced from a condition of ignorance and superstition, but from the experience of forest life, rich with vivid symbols. It has nurtured an ecological civilization that has fostered harmony with nature.[50] It is also the substance of the everyday beliefs of tribal and peasant society. She observes that the forest, which is the highest expression of the earth's fertility and productivity, is also symbolized as Vāna Durgā, the tree goddess and earth mother. Following the research of W. C. Bean, she points out that both Durgā and Kālī originated as the vitalizing energy of the forest. She observes that in Bengal Vāna Durgā is associated with the *sal* tree (*Shorea robusta*) and with the *aśvattha*, or *pīpal* (*Ficus religiosa*). She finds this deity in Comilla as Bamani, and in Assam as Rupeśvarī. She notes that among folk and tribal cultures trees and forests are worshiped as Vāna Devatās, or deities of the forest. She also finds a parallel of these beliefs in the *Devī Māhātmya* (90:43–4), in which the earth mother proclaims that she will slay the *asura* (demon) that personifies drought and sustain the whole world with the life-giving vegetables that grow from her body.[51]

For Vandana Shiva, the Chipko movement was a response to the militant assault upon this relationship to nature. That assault came first to India with colonialism but was perpetuated in postcolonial India with patriarchal scientific forestry and the economic relationships to which it was committed. The forestry that has dominated India's forest policy, in her view, is reductionistic in thought and practice. Within its purview, the ideal of harmony with nature is interpreted as a symptom of underdevelopment. It violates the integrity and harmony of humanity in nature and the harmony between men and women. "It ruptures the co-operative unity of masculine and feminine, and places man, shorn of the feminine principle, above nature and women, and separated from both."[52] The present crisis of the environment, she argues, is the manifestation of a process in which the feminine principle has been subjugated in favor of a male-centered idea of development. But the objective of this view of the forest is very far from true development:

> I want to argue that what is currently called development is essentially maldevelopment, based on the introduction or accentuation of the domination of man over nature and women. In it, both are viewed as the 'other', the passive non-self. . . . Nature and women are turned into passive objects, to be used and exploited for the uncontrolled and uncontrollable desires of alienated man. From being the creators and sustainers of life, nature and women are reduced to being 'resources' in the fragmented, anti-life model of maldevelopment.[53]

The Chipko Movement Today

As an effort to protect trees from contract felling, and thus to save the habitat of the forest people, the Chipko movement has had a significant impact on the development of similar movements in India and elsewhere. While the last agitation for the protection of forests in Uttarakhand occurred in 1980, Chipko activists have continued to disseminate their message by means of foot marches and environmental camps that continue to the present day. The spirit of Chipko is especially vivid in the present agitation against the construction of a major hydroelectric dam over the Bhagīrathi River, the northernmost tributary of the Ganges, at Tehri. The religious ideal of harmony with nature remains a crucial feature of this continuing message.

While the Chipko movement began as an effort to maintain the traditional right of peasants to their own forest resources, it gradually expanded to embrace issues of wider ecological concern. In May 1978, at Gaumukh, where the Bhagīrathi River flows from the Gangotri glacier, Sunderlal Bahuguna took a pledge to devote himself to the protection of the Himalayan environment in all its aspects. For him, this meant not simply the protection of trees from powerful outside contractors, but the protection of the forest from destruction by anyone for any purpose. Trees must be available to meet the needs of local people for food, fuel, fodder, fiber, and fertilizer. They could do so without being destroyed. Bahuguna came gradually to hold that forest-based industries were not necessary to improve the economy of the people. Instead, he envisioned a self-sufficient rural community living in a sustainable relationship with its natural surroundings.[54]

Nevertheless, state forestry was by no means the only threat to the Himalayan environment. While Bahuguna's fast of 1981 succeeded in bringing about a fifteen-year moratorium on all commercial felling in the Uttarakhand region of the Himalaya, and eventually an unconditional ban on the felling of all trees above one thousand meters in elevation, commercial penetration continued to endanger the region. The most visible features of this penetration were the development of mining operations, increased sales of alcohol, and the construction of major dams.[55] In recent years it is perhaps in the agitation against the construction of the Tehri dam that the name of Sunderlal Bahuguna has come to be most widely known.

There is a legend, Bahuguna writes, that the waters of the Bhagīrathi can be contained only within the matted locks of Lord Śiva. The story goes that the king Bhagīrathi wanted the goddess Gaṅgā to come down from the heavens to wash off the sins of his forefathers. After much prayer and penance on his part, she agreed to do so; however, she warned the king that when she came down she would have to be contained. "If I am not tied down," she said, "I will not be the life-giving source you expect me to be, but I will cause chaos and destruction on the earth." King Bhagīrathi then sought out someone strong enough to restrain the tempestuous Gaṅgā. He concluded that the only power sufficient for the task was that of the mighty Śiva. He then prayed fervently to lord Śiva to tie up Gaṅgā as she descended to the earth. Śiva finally agreed, and Gaṅgā descended into the matted locks of his hair, from which she comes forth as a life-giving stream to the northern plains of India. The environmentalist Anil Agarwal indicates that the monsoon is a life-giving system for the Indian subcontinent. He suggests that, seen in another way, it is nothing but the descent of the Gaṅgā that happens every year. In Hindu mythology Śiva is identified with the Himalayas. Bahuguna states that the locks of Śiva "are the natural forests of the Himalayas which help contain the water in the soil and protect the land from floods." In these matted locks the waters of the Gaṅgā are entangled every year, turning it into a life-giving source by regulating its flow. But now, the sacred locks of Śiva have been cut, turning the Gaṅgā into the destructive force against which the story warned.[56]

The catchment area of the Bhagīrathi River is one of the worst victims of deforestation brought about by commercial forestry in India. For Bahuguna, the construction of the Tehri dam is a further violation of the Himalayan environment. If completed, the Tehri dam, presently under construction at the confluence of the Bhagīrathi and Bhilangna Rivers, will have the distinction of being one of the six highest dams in the world. The rock filled structure is designed to rise 260.5 meters above the present river level, creating a reservoir 42 square kilometers upstream. It is expected to generate 2,400 megawatts of electricity, promised mostly to the city of Delhi, and to irrigate 270,000 hectares of land in the western districts of Uttar Pradesh. For the Tehri Hydro Development Corporation and the government of Uttar Pradesh, it is a symbol of progress. Yet, the slogans painted on rocks and walls indicate that for the local people it is a symbol of the destruction of the Ganges and the Himalayas. In 1978, when officers of the Uttar Pradesh government arrived to inaugurate construction on the first diversion tunnel for the dam, they were met by thousands of men, women, and children who blocked their way and shouted: "You love electricity, we love our soil."[57]

Upstream from the dam site is a picturesque valley. Ranged on the mountainsides, Bahuguna says, "are clean, low, whitewashed houses, with carved lintels, that identify village Garhwal. They are comfortably spaced out on separate terraces; crime is rare." If completed, the reservoir will fill this valley. Twenty-two Garhwali villages will be submerged. Their green terraced fields, sculpted over centuries of painstaking work, will be consigned to the depths of a manmade sea.[58] The town of Tehri, on the pilgrimage route to Gangotri, is the cradle of an ancient Garhwali culture. It has given birth to eminent poets, scholars, and artists. The Vedāntic saint Swami Ram Tirtha became a *saṃnyāsin* here and lived near Tehri. In the glow of his influence, Tehri became renowned as a place of spiritual meaning. Freedom lovers have been inspired by the historic martyrdom of Sridev Suman following his eighty-four-day fast for civil liberties in the Tehri jail.[59] If the dam is completed, the entire town of Tehri, with its rich cultural heritage, will be submerged. The town has a population of 25,000; a total of about 109,000 people will be uprooted by the construction of the dam.[60]

Besides the disruption of village life and the loss of cultural heritage, there are several objections to the dam that can be developed on technical ecological grounds. It was mainly these that were presented in the written petition that was filed in 1987 with the Supreme Court of India. The proposed dam is to be located on a seismically active site, putting it at risk from possible earthquakes. The probability of earthquakes is increased by the weight of the water retained behind the dam. The possible failure of the dam presents a grave threat to the lives and property of inhabitants of towns on the banks of the Ganges downstream from the dam. In addition, the dam will present a hazard to the security and safety of heavily populated areas of the flood plains of the Ganges during emergency releases from the reservoir at the time of peak floods. Because of heavy deforestation in the hills of the catchment area, high siltation will shorten the expected life of the dam, reducing its economic value. Despite promises of afforestation, the hope of restoring the forest cover to the now barren hills is remote.

The reservoir slopes are also ringed with human settlements. The hills are steep, unstable, and fragile. Enormous depths of water impounded between these slopes makes them vulnerable to failure, putting the human settlements at risk. Many dams of less ambitious size have suffered accidents, and many have totally failed. Such dams engender feelings of subjugation among the people who are dislocated, and the government's record in managing the relocation of the affected people is bleak.[61]

In these objections it is not difficult to see an analogy to the grievance that was the catalyst for the Chipko movement in Gopeshwar. Just as ash trees were allotted to outside contractors to provide sporting goods for a distant market, the government here endorsed a project to provide hydroelectric power to distant Delhi and irrigation to the western districts of Uttar Pradesh. Just as the Forest Department denied forest resources for the needs of the local people of Gopeshwar, the government here imposed the burden of loss of heritage, history, and habitat upon the local people of Garhwal. It is hardly surprising that the dam is seen by local people as the project of corrupt and greedy politicians, bureaucrats, and contractors, and that it has received the support of the government of India on grounds of political expediency. It is easy enough to interpret the opposition to the Tehri dam in terms of economic justice. For Bahuguna and other opponents of the Tehri dam, the justification for resistance against this project is not a matter simply of one interest over another. While environmental and economic justice remains a fundamental feature of their protest, their opposition to the dam is also ethical, ecological, and religious.

At a meeting of the International Alliance against Large Dams, held in Curitiba, Brazil, in 1997, J. P. Raturi, the representative of the Tehri Bandh Virodh Sangharsh (Tehri Dam Protest Committee) made the point that Indian culture sees divinity in nature. To the rulers, he says, the Gaṅgā is megawatts of power and hectares of irrigated land. To the local people, she is a life-giving goddess.[62] Bahuguna states that when the Gaṅgā flows in her natural course she benefits all, irrespective of caste, creed, or color, poverty or wealth. As soon as she is dammed, she looses her social character.[63] The late Virendra Datt Saklani, the father of the Tehri dam resistance movement, made the point that the Tehri dam "will end the sanctity of the River Ganga and the purity of Ganga water." The language is unequivocal. The dam is an interference, the contamination of something valued on religious grounds. In the words of V. D. Saklini:

> Pure Ganga water purifies other water, but when Ganga water is itself polluted, who will purify the water? The cold, pure and nectar like Ganga water has for centuries been glorifying the Indian soil, has been giving life and fertilizing its land, but now the human greed, short sightedness and inhuman tendency to exploit nature . . . has determined to poison her nectar like water. Will the people of India, who hail Ganga remain the silent spectators of this sinful act like helpless and impotent persons?[64]

While such rhetoric could be read by many as little more than an appeal to sentiment, it is evident that for the local people of the Himalayas it has more than sentimental meaning. Like the desolating sacrilege in the book of Daniel in the Old Testament, the dam is an invasion and a desecration.

Bahuguna's rhetoric is also sometimes developed in explicitly religious terms. He points out that, during the long course of history, religion played a vital role in the regulation of society, especially in the use of the resources of the earth. The observation of these rules of conduct was regarded not just as prudent but as pious. The violation of such codes of behavior was seen as something sinful. Indian culture has traditionally revered the hermit in a loin cloth more than the crowned king sitting on his throne. Gradually, Bahuguna says, "the real face of religion (Dharma) disappeared and it came to be covered with rituals." He associates this change with the industrial revolution that began to see nature as a commodity, and human society as the only interest it should serve. When the idea of development came to be identified with economic growth, the stage was set for the emergence of what he calls a new religion:

> When the objective of development became affluence and prosperity man became the butcher of Nature. He misused the power of science and technology for maximum exploitation of the resources of the planet. We became the rich children of a poor mother.[65]

One of the most visible features of the role of religion in the challenge to the Tehri dam is the standing of Bahuguna and others as prophetic voices against compromise with this new religion:

> A new religion has taken birth in the development era. That is the religion of economic growth. The market is its temple, technocrats and experts its priests, and the Dollar is the new God. Our political leaders are impatient to possess this God. They are prepared to make the highest sacrifice to bring it home to their respective countries.[66]

While the promises of this new religion are enticing, its demands fall heavily upon the poor. In the words of Raturi:

> Dams are eulogized as the symbol of development. What Development? The development which robs the resources of the poor people to fatten the rich and prosperous. Electricity, irrigation to the affluent and

> displacement and disaster to the poor. We in Himalaya experience acute water scarcity because the rivers flow into deep gorges; the water can not be lifted a few meters for the people living on the banks of the river, but it is taken away, 350 Km. away, for the swimming pools of the five star Hotels of Delhi. . . . Electricity generated will be taken to a national grid, 250 Km away, whereas the villages around the dam are and will remain virtually in darkness.[67]

The religious character of this protest can be seen again in the idiom of resistance for which Bahuguna has been known since the early days of the Chipko struggle. When questioned as to whether the methods of *satyāgraha* can be effective today against the construction of the Tehri dam, Bahuguna replied that there are three methods available for solving problems. One is the method of the establishment, through the machinery of law and order, the legislature and judiciary. This is unacceptable, he says, because the machinery of government is breaking down. Representatives no longer represent. The system has failed to see the plight of people most affected by its actions. The second method is that of terrorism. This is ineffective because the power of individual terrorism is no match for the collective terrorism of the state. Just as they claim innocent human lives, both of these methods undermine human values. The only method that remains, he says, is the method of *satyāgraha,* the method of standing for the truth, through peaceful protest, through foot marches, and through fasting. He holds that even if this method is not immediately effective, at least it does no damage to human values. If it does do harm, it will only harm one person; if it does do harm to that one person, it only harms the body.[68]

In discussing the fasts which he had repeatedly undertaken both in his involvement with the Chipko movement and in the protest against the Tehri dam, Sunderlal Bahuguna points out that the motivation for a fast must not be that of anger or manipulation. A fast is not the same thing as a hunger strike. It must be an act of devotion to God. When the devotee feels that all his worldly efforts have failed, he finally leaves the affairs of the world where they stand and puts himself under the protection of God. His fast is the expression of his faith.[69] When nobody listens to you, Bahuguna said to me, then you can sit in prayer so that the almighty will listen to you. Here there is no longer any anger. Rather, one is fully satisfied. One has completely surrendered to the pleasure of the almighty. "If God so chooses, the body can be taken away."[70] Reflecting on one of his most famous fasts, he states: "It was the most pleasant time for me. My undivided attention was towards God, and I had full faith in him. I had surrendered myself to the supreme power and when you totally surrender yourself to anyone, then where is the worry, as you are His responsibility?"[71]

In the course of his agitation against the Tehri dam, Bahuguna has undertaken several fasts that have received much media attention. These have ranged from eleven days to seventy-four days, and it has repeatedly been feared that he has been very close to death. One of the best known of his fasts concluded after forty-five days, on 27 June 1995, on the promise of then prime minister Narasimha Rao to undertake a scrupulous review of all aspects of the Tehri project. But this review was not forthcoming. In 1996 Bahuguna undertook a *prayaścit vrata,* or discipline of self-purification, for letting down those poor people who were looking to his long fast as the act that would finally bring them justice and safety. His penance of atonement lasted seventy-four days and ended at the Rajghat in New Delhi on 25 June 1996, with the further assurance by the prime minister to investigate the objections to the project.

When I visited him in the summer of 1998, the construction of the dam was evidently in progress again. As he has for a number of years, Bahuguna now lives in a small *kuti,* or hut, on the banks of the Bhagīrathi River, not one hundred yards from the site of the proposed dam. If the waters rise, Bahuguna's *kuti* will be the first human dwelling to be submerged. Every morning he makes his way down the steep rocks for his ritual bath in the cold waters of the Bhagīrathi River that gives him, he believes, his strength. Here, he lives a simple life of protest: he meditates, prays, and grants interviews to visiting scholars. In the evening he conducts services of hymn-singing to Mother Gaṅgā, the divine gift whose course has been diverted by human ambition. With his long white beard, his frail body, his simple *khadi* apparel, with a large scarf tied around his head, he looks the part, as one journalist expressed it, of a prophet warning of disaster.[72] We have falsely identified progress with affluence, he told me. This dam is a project to realize a false hope. His duty, he says, is to warn the people that the promise is false, that it is based on a mistaken view of reality. The vision it offers is like a magician's show. We should not kill our sacred river for the promises of a false vision of reality. His work here, he explained to me, is the continuation of his efforts against the exploitation of nature expressed in the earlier phases of the Chipko protest movement.[73]

Conclusions

The question whether Hinduism is capable of supporting an environmental ethic has been the subject of several studies of the Indian religious and philosophical sources. For some, Hinduism's concern for the transcendence of the spiritual world over the natural world precludes without further attention the possibility of any positive contribution toward an ecological ethic. For others, such as Callicott, Hindu insights into the valuation of nature are ambiguous at best. My conclusion here is that Chipko is unquestionably a movement for the negation of the world. The world it negates, however, is not the world of nature, which for the Chipko activist is sacred. The world it negates is the world of scientific forestry and of politicians, technicians, and contractors within whose knowledge nature is reduced to a commodity in a system of economic exchange that leaves the people destitute and dispossessed, that discounts their material needs and the religious life that supports them. The asceticism of Chipko is a renunciation of this world and its promises. It is also certainly correct that Hinduism inspired and grounded the Chipko movement. But the Hinduism of the Chipko activist differs widely from Callicott's characterization of Hinduism as a religion that views the empirical world as morally negligible and judges it as contemptible, because it deludes the soul into crediting appearances and pursuing false ends. For the Chipko movement, the false ends are the ends of scientific forestry: resin, timber, and foreign exchange; those of the Chipko agitation are soil, water, and pure air. The contractors, corporations, and politicians that support the Tehri dam offer the ambivalent promise of development. The Hinduism of Chipko hears the claims of this world and, like the *jīvanmukta,* knows that they are false.

The evident strength of the Chipko movement was its constant ability to achieve the moral high ground in its confrontation with material and political power. That moral high ground is reflective of a deep recognition of temporal Indian values. In the religious traditions of the Himalayan peasant, these values are supported and sustained.

In the Chipko movement peasants found a crusade for the preservation of the earth that supported truthfulness in the face of political doubletalk, that relied on nonviolence against the force of contractors and government officials, and that supported the conscientious care of nature as opposed to its mindless exploitation. In Hindu religion they found authentic insight and genuine support for the maintenance of the earth.

Notes

1. Aldo Leopold, *The Sand County Almanac and Sketches Here and There* (New York: Oxford University Press, 1949), 201ff.

2. Cited in John Passmore, *Man's Responsibility for Nature,* 2d ed. (London: Duckworth, 1980), 4. Editorial, "Towards an Ecological Ethic," *New Scientist* 48, no. 732 (1970): 575

3. Lynn White, Jr., "The Historical Roots of Our Ecologic Crisis," *Science* 155 (10 March 1967): 1203–7.

4. Roderick Nash, *Wilderness and the American Mind* (New Haven: Yale University Press, 1967), 20–21.

5. Ian L. McHarg, *Design with Nature* (Garden City: Doubleday, 1969), 26.

6. Passmore, *Man's Responsibility for Nature,* ix, 3.

7. Ibid., 121–6.

8. J. Baird Callicott, *Earth's Insights: A Survey of Ecological Ethics from the Mediterranean Basin to the Australian Outback* (Berkeley and Los Angeles: University of California Press 1994), xiv–xv.

9. Ibid., 12.

10. Ibid., 48.

11. Lance E. Nelson, "The Dualism of Nondualism: Advaita Vedanta and the Irrelevance of Nature," in *Purifying the Earthly Body of God: Religion and Ecology in Hindu India,* ed. Lance E. Nelson (Albany: State University of New York Press, 1998), 63.

12. Ibid., 79.

13. Ibid., 80.

14. Callicott, *Earth's Insights,* 223

15. Aristotle *Nicomachean Ethics* 1:2.

16. Shanti Nath Gupta, *The Indian Concept of Values* (New Delhi: Manohar, 1977), ix, 1, 31ff, 146f.

17. Ibid., 48; *Arthaśāstra,* 180.1.1.

18. Ibid., 49; *Arthaśāstra,* 1.2.9–12.

19. Ibid., 48–51; *Arthaśāstra,* 17.1; 18.2; 98.3.

20. *Arthaśāstra,* 2.145. Cited in O. P. Dwivedi and B. N. Tiwari, "Environmental Protection in the Hindu Religion," in *Ethical Perspectives on Environmental Ethics in India,* ed. George A. James (New Delhi: A.P.H. Publishing Corporation, 1999), 179.

21. Gupta, *Indian Concept of Values,* 67; *Kāmasūtra,* 1.2. 49–50.

22. N. K. Devaraja, *The Mind and Spirit of India* (Delhi: Motilal Banarsidass, 1967), 168.

23. Gupta, *Indian Concept of Values,* 84–87; *Manusmṛti,* 2.224.

24. *Manusmṛti,* 4.76 (cf. also 3, 5, 6).

25. *Manusmṛti,* 4.88, 93, 96.

26. Ramachandra Guha, *The Unquiet Woods: Ecological Change and Peasant Resistance in the Himalaya* (New Delhi: Oxford University Press, 1991), 157; Mark Shephard, "Chipko: North India's Tree Huggers," *Co-evolution Quarterly,* fall 1981, 65.

27. Madhav Gadgil, "The Indian Heritage of a Conservation Ethic," in *Ethical Perspectives on Environmental Issues in India,* ed. James, 150.

28. Guha, *The Unquiet Woods,* 154.

29. Chandi Prasad Bhatt, "The Chipko Andolan: Forest Conservation Based on People's Power," in *The Fight for Survival,* ed. Anil Agarwal, Darryl D'Monte, and Ujwala Samarth (New Delhi: Centre for Science and Environment, 1987), 47f; see also Thomas Weber, *Hugging the Trees: The History of the Chipko Movement* (New Delhi: Penguin Books, 1988), 61.

30. Guha, *The Unquiet Woods,* 160.

31. Weber, *Hugging the Trees,* 51–52.

32. Guha, *The Unquiet Woods,* 162; Weber, *Hugging the Trees,* 52; Vandana Shiva, *Staying Alive: Women, Ecology, and Development* (London: Zed Books, 1989), 75.

33. Weber, *Hugging the Trees,* 53.

34. Callicott, *Earth's Insights,* 221.

35. A. L. Basham, *The Wonder That Was India* (New York: Grove Press, 1959), 23–24.

36. Steven G. Darian, *The Ganges in Myth and History* (Honolulu: University Press of Hawaii, 1978), 42–43.

37. Shiva, *Staying Alive,* 75.

38. Guha, *The Unquiet Woods,* 162.

39. Weber, *Hugging the Trees,* 52.

40. P. Thomas, *Hindu Religion, Customs, and Manners* (Bombay: D. B. Taraporevala Sons and Co., 1956), 148.

41. Guha, *The Unquiet Woods,* 221; Jayanta Bandyopadhyay and Vandana Shiva, "Chipko: Rekindling India's Forest Culture," *Ecologist* 17, no. 1 (1987): 30.

42. Sunderlal Bahuguna, interview, 24 June 1998.

43. John Dominic Crossan, *Jesus: A Revolutionary Biography* (San Francisco: Harper Collins, 1994), 55–62.

44. When I asked Bahuguna about his present *āśrama,* or stage in life, he stated that he considers himself a *vānaprasta,* or forest dweller.

45. Guha, *The Unquiet Woods,* 171.

46. Shiva, *Staying Alive,* 38.

47. Ibid., 39.

48. Callicott, *Earth's Insights,* 222f.

49. Shiva, *Staying Alive,* 40.

50. Ibid., 56f.

51. Ibid., 56.

52. Ibid., 6.

53. Ibid.

54. Weber, *Hugging the Trees,* 66–69.

55. Guha, *The Unquiet Woods,* 178f.

56. Sunderlal Bahuguna, "Tehri Dam: A Blueprint for Disaster," in *Fire in the Heart, Firewood on the Back,* ed. Tenzin Rigzin (Silvara, Uttar Pradesh: Parvatya Navjeevan Mandal, 1997), 82; Anil Agarwal, "Human Nature Interactions in a Third World Country," in *Ethical Perspectives on Environmental Issues in India,* ed. James, 71.

57. Bahuguna, "Tehri Dam: A Blueprint for Disaster," 84.

58. Ajit Bhattacharjea, "The Old Man and the Dam," *Outlook,* 26 June 1996; reprinted in *Fire in the Heart, Firewood on the Back,* ed. Rigzin, 21–24.

59. Bahuguna, "Tehri Dam: Blueprint for Disaster," 89.

60. Sunderlal Bahuguna, "Development and Environment," in *Chipko Message: Development, Environment, and Survival* (Tehri: Chipko Information Centre, 1997), 13.

61. Bahuguna, "Tehri Dam: A Blueprint for Disaster," 84–87.

62. J. P. Raturi, "Save the Mothers of Culture: Message from Tehri Struggle," in *Save Ganga,* ed. Inderjit Kaur (Amritsar: All India Pingalwara Society, 1997), 6.

63. Sunderlal Bahuguna, "Make Silent Majority Vocal (Message to the Participants of the International Conference)," in *Save Ganga,* ed. Kaur, 5.

64. Virendra Datt Saklani, introduction to *Save Ganga,* ed. Kaur, 1.

65. Bahuguna, "Development and Environment," 7.

66. Ibid.

67. Raturi, "Save the Mothers of Culture," 9.

68. Sunderlal Bahuguna, "Sunderlal Bahuguna's Crusade," interview by Madhu Kishwar, in *Fire in the Heart, Firewood on the Back,* ed. Rigzin, 61–62. First published in *Manushi*, May-June 1992.

69. Ibid., 62.

70. Sunderlal Bahuguna, interview, 24 June 1998.

71. Bahuguna, "Sunderlal Bahuguna's Crusade," 63.

72. Bhattacharjea, "The Old Man and the Dam," 21.

73. Sunderlal Bahuguna, interview, 24 June 1998.

Appendix One

HARRY BLAIR

Table 1. Hinduism and Ecology: Four Perspectives

	Use	Utility	Romance	Asceticism
Relationship with nature	mastery over nature	nature as supporting humans	nature as including humans	humans freed from nature
Dominant theme	use	sustainability	preservation	abstinence
Philosophy	production	social ecology	deep ecology	non-ecology
Human role	promote development	exercise stewardship	respect divinity of nature	practice withdrawal
Model human being and Hindu analogue	consumer; Durgā/Kālī	sustainer; second *āśrama* of life	preserver; Viṣṇu	renunciant; *sādhu*s
Advocates	Julian Simon	Gifford Pinchot	John Muir	M. K. Gandhi

Table 2. Hinduism and Ecology: Illustrations of the Four Perspectives

Author	Use	Utility	Romance	Asceticism
Philip Lutgendorf	city vs. forest; forest as exploitable	*jāṅgala* as pastoral landscape, savannah	forest as paradise	forest as *āśrama*
Madhu Khanna	nature as brute force (Mahiśā demons)			
Laurie Patton	Vedic ritual violence			
Kelly Alley; David Haberman	separation of cleanliness and purity			
Anil Agarwal	pollution of self vs. that of public space	utilitarian conservationism, sustainable agriculture		Hindu "spiritual individualism," simple living
William Fisher	Narmada as resource		tribal life as paradise	
Frédérique Apffel-Marglin and Pramod Parajuli		alteration of production and rest		
Pratyusha Basu and Jael Silliman		Narmada feminist activists as attuned to river as resource		
Ann Grodzins Gold		need for Leviathan to protect resource	king as protector of trees as ends in themselves	
Mary McGee		*Arthaśāstra* injunctions	sentient plants	

Table 2. Hinduism and Ecology:
Illustrations of the Four Perspectives (continued)

Author	Use	Utility	Romance	Asceticism
T. S. Rukmani			Ayurvedic medical theory	
David Lee			medicinal values of plants	
Vijaya Nagarajan			trees as marriage partners	
Chris Deegan			Narmada as paradise	
O. P. Dwivedi			love for nature	reducing consumption
George James			Sunderlal Bahuguna as activist	Sunderlal Bahuguna as meditator
Vinay Lal			Gandhi as doer	Gandhi as thinker
Larry Shinn				Gandhian control of self, desire
K. L. Seshagiri Rao				reducing consumption
Lance Nelson				*Bhagavadgītā* and escape from nature

Appendix Two: The Population, Natural Resources, and Biodiversity of India

compiled by
DAVID LEE

Population and Health

Table 1. Population

Year	Number	% Annual Increase (years for estimate)
1950	357,561,000	X
1990	850,638,000	2.1 (1985–90)
1998	975,772,000	1.6 (1995–2000)
2025 (est.)	1,330,201,000	1.3 (2005–10)

Average Total Increase, 1995–2000 = 15.5 million
Estimated Population, 11 May 2000 = 1 billion

Table 2. Urbanization

Year	% Urban Population
1980	23
2000	28
2020	39

Number of Cities over 775,000 in 1995 = 34
Population Density, 1996 = 317.7/km^2

Table 3. Life Expectancy

Years	Age (in years)
1975–80	52.9
1995–2000	62.4

Absolute Poverty in 1992 = 52.5 %
Infant Mortality, 1975–80 (number/1000 live births) = 129
Infant Mortality, 1995–2000 = 72

Table 4. Literacy

Year	% Male Literacy	% Female Literacy
1970	48	18
1990	62	34

Land

Table 5. Land Use, 1992–94

Type of Use	Area (km^2)	% Change since 1982–84
Total Land Area	2,973,190	
Cropland	1,695,690	+ 0.5
Permanent Pasture	114,240	- 4.8
Forest and Woodland	681,730	+ 1.2
Other	481,360	- 2.1

Total Percentage of Domesticated Land = 61 %

Table 6. Forest Resources, 1995

Type	Area (km^2)	% Change in 10 Years
Forest and Other Wooded Land	826,480	X
Total Forest	650,050	+ 1.1
Natural Forest	503,850	- 0.5
Plantation	132,300	+ 14.0
Other Wooded Land	176,890	X

Table 7. Deforestation Rates

Source and Years	Total Area/Year (km^2)	% Lost Per Year
FAO (1981–85)	1,470	0.3
Satellite (1975–82)	15,000	4.1

Table 8. Wood Production, 1993–95

Type	Volume (millions of cubic meters)	% Change in 10 Years
Total	294.02	+ 21
Fuel	269.22	+ 22
Industrial	36.26	+ 27
Paper	3.18	+ 647
Imports - Exports	0.27	+ 3,520

Resources

Table 9. Crop Production, 1994–96

Type	Amount (kilograms/hectare)	% Change in 10 Years
Cereals	2,136	+ 26
Total (million tons)	213.3	+ 23
Roots and Tubers	16,936	+ 16

Table 10. Index of Food Production Per Capita (1989–91 baseline)

Year	Index
1984–86	92
1994–96	104

Table 11. Irrigated Cropland

Year	Percentage	Annual Fertilizer Use (kilograms/hectare)
1982–84	24	49
1992–94	29	80

Table 12. Livestock, 1992–94

Type	Number	% Change in 10 Years
Cattle	192,777,000	- 1
Sheep/Goats	162,132,000	+ 11
Pigs	11,630,000	+ 14
Horses	2,672,000	+ 29
Buffaloes/Camels	80,153,000	+ 10
Chickens	437,000,000	+136

Table 13. Fisheries, 1993–95

Type	Annual Catch (10^6 metric tons)	% Change since 1983–85
Marine	1.731	+ 62
Freshwater	1.998	+ 90
Aquaculture	1.521	X
Total Per Capita (kg/person/yr)	4.0	+ 24

Table 14. Energy Production and Consumption, 1995

Type	Amount (10^{15} joules, or petajoules)	% Change since 1985
Production		
Total	9,113	70
Solid	6,663	80
Liquid	1,398	12
Gas	718	391
Electrical	334	40
Nuclear	7	41
Hydro	58	40
Consumption		
Firewood	3,065	29
Commercial	10,513	88

Biodiversity

Table 15. Numbers of Species

Group	All	Endemic	Threatened	#/10,000 km^2
Higher plants	15,000	5000	1256	2216
Mammals	316	44	40	47
Birds	1,219	55	71	180
Reptiles	389	185	21	47
Amphibians	197	120	3	29
Freshwater Fish	X	X	2	X

Table 16. Centers of Biodiversity

Areas Protected for Biodiversity: 143,370 km^2, 4.4% of area
Preserves = 339
World Heritage sites = 5; total area of 2,810 km^2
Major Centers

Agastyamali Hills	2000 km^2, 2000 plant species, 870 km^2 protected; dry to wet forests.
Nallamalais	6840 km^2, 750 plant species, 1950 km^2 protected; deciduous forests.
Namdapha	7000 km^2, 3000 plant species, 3485 km^2 protected; tropical evergreen to alpine.
Nanda Devi	2000 km^2, 800 plant species, 1430 km^2 protected (partly as World Heritage Site); coniferous forests and alpine.
Nilgiri Hills	5520 km^2, 3240 plant species, 3240 km^2 (including Silent Valley National Park, 90 km^2); deciduous and evergreen forests.

Note

Data for these tables are taken from World Conservation Monitoring Center, *Global Biodiversity: Status of the Earth's Living Resources* (London: Chapman and Hall, 1992); and *World Resources 1998–99* (New York: Oxford University Press, 1998). X denotes data missing from a table.

Glossary

adharma lack of virture
Ādiśakti primordial power
Aditi Vedic goddess; the primordial vastness
ādivāsi aboriginal or tribal people of India
Advaita Vedānta monistic school of Indian philosophy
Agastya sage (ca. 10,000 B.C.E.) credited with introducing Vedic religion to South India
Agni Vedic god of fire
Agniṣṭoma rite in Vedic sacrifice
ahiṃsā nonviolence
ākāśa space
Akka Mahādevī twelfth-century saint
amṛta nectar of immortality
ānanda bliss
ap water
aparigraha nonpossession
apsaras sprite
araṇya forest
āratī lamps used in worship
Arjuna hero of the *Mahābhārata,* one of the Pāṇḍava brothers
artha material wealth, political and economic value
asakta detached
Aśoka Buddhist emperor of India, ca. 300 B.C.E.
āśrama hermitage, retreat; also stages of life
aśuddha impure
asura demon
Aśvamedha sacrifice of the horse
ātman highest self

Aurobindo twentieth-century philosopher-saint
avatāra incarnation of Viṣṇu
Ayodhyā city ruled by Rāma

Bhagavadgītā;* the *Gītā portion of the *Mahābhārata;* discourse between Arjuna and Lord Kṛṣṇa
Bhīma one of the Pāṇḍava brothers in the *Mahābhārata*
bhū earth
Bhūdevī Earth goddess
bhūtaśuddhi purification of the elements
Brahmā creator deity
brahmacarya celibacy
Brahman the ultimate reality, supreme being without form
Brāhmaṇas texts that follow the Vedas and include:
the *Kauṣītaki Brāhmaṇa*
the *Śatapatha Brāhmaṇa*
Braj region sacred to Kṛṣṇa

cakravartin wise political leader or teacher
catuṣpuruṣārtha principal ends of human life: wealth, pleasure, renunciation, liberation
chela student

dāna donations
Daṇḍaka forest in the *Rāmāyaṇa*
darśan seeing representation of the divine
Daśaratha father of Rāma, hero of the *Rāmāyaṇa*
***deva*s** Vedic gods
Devī Māhātmya important text of the goddess tradition; a section of the *Mārkaṇḍeya Purāṇa*
dharma virtue, law, moral order
dhyāna meditation
Diwali fall festival of lights
Draupadī wife of the five Pāṇḍava brothers
Durgā beautiful and wrathful goddess of Hinduism
Durgā Pūjā October worship of the goddess

Gaṇeśa elephant-headed god, son of Śiva and Pārvatī
Gaṅgā River the Ganges River, known also by:
Dakṣin Gaṅgā
Kāśī Nāḍī
Maheśvari Gaṅgā

Mother Gaṅgā
Pūrva Gaṅgā
gaṅgājala Ganges water
gau cow
gauchar nomadic herder
gauchara grazing land
gāyatri mantra sacred Vedic prayer recited by most Hindus
gopīs cowherding female companions of the young Kṛṣṇa
goseva service of the cow
guṇas strands or constituents of reality

Hanūmān monkey deity who helped Rāma rescue Sītā
havis gift of oblation
hiṃsā violence
homa rite of oblations into fire

Indra Vedic god of war
itihāsa lore

jal* or *jala water
jāti birth niche
jīvanmukta "liberated" saint
jñāna knowledge, wisdom
Jñāneśvarī medieval commentary of the *Bhagavadgītā*

Kālī fierce Hindu goddess, slayer of demons
kāma pleasure
Kāmasūtra a Hindu erotic text
Kaṃsa evil king
Kapila Sāṃkhya philosopher
karma action and its residues
karma-yoga action without attachment to the outcome
kartṛ doer, agent
Kauṭilya author of the Arthaśāstra
khadi simple cotton cloth
Khuṃbha Melā religious gathering held every twelve years on the banks of the Gaṅgā River
Kṛṣṇa incarnation (*avatāra*) of Viṣṇu
kṣatriya dharma warrior duty
kuṅḍ reservoir
Kuru dynasty rulers in the *Mahābhārata* epic
kuti hut

Lakṣmaṇa brother of Rāma
Lakṣmī goddess of wealth
līlā illusion, play
liṅgam representation of Śiva
loka world
loka-saṃgraha world maintenance

Mahābhārata great epic tale of the battle between the Pāṇḍava brothers and the sons of the blind king Dhṛtarāṣṭra
mahābhūta great element
mahāvākyas great sayings from the Upaniṣads
māhāyajña great ceremony
Mahiśāsura the demon who battles Durgā
makara alligator or large aquatic animal; the vehicle of Varuṇa
maṇḍala sacred design
Manu author of famous law book, the *Manusmṛti*
mātā mother
Mathurā city sacred to Kṛṣṇa
māyā illusion
melā religious gathering
mokṣa liberation
mukti liberation
muni sage
mūrti statues of Hindu gods

Navarātrī festival in honor of the goddess; fall season; nine nights
nīm tree of many uses, sacred in India
nirguṇa without attributes
niyamas rules
niyata ordained

pañcamahābhūta the five great elements
Pāṇḍavas the five putative sons of Pāṇḍu, heroes of the *Mahābhārata*
Pāṇḍu king in the *Mahābhārata*
pradakṣina ritual of the sacrificial fire
Prajāpati Lord of Creatures in the Upaniṣads
Prakṛti feminine principle symbolizing activity
prakṛti cosmic matter
pralaya dissolution of matter
prayaschit vrata discipline of self-purification
prema love
pṛthvī earth

pūjā ceremony, process of worship

Purāṇas lengthy Sanskrit texts that tell stories of various deities and include:

the *Bhāgavata Purāṇa*
the *Mārkaṇḍeya Purāṇa*
the *Narmada Purāṇa*
the *Padma Purāṇa*
the *Skanda Purāṇa*
the *Vāmana Purāṇa*
the *Varāha Purāṇa*
the *Viṣṇu Purāṇa*

puroḍāśa grain offering

Puruṣa masculine principle symbolizing pure consciousness

Rādhā consort of Kṛṣṇa

rāj purohit court

rāja king

rajas passionate activity

***rākhi* (*raksha bandhan*)** amulet; ceremonial string bracelet to signify devotion

rākṣasa demon

Rāma hero of the *Rāmāyaṇa*

Ramakrishna nineteenth-century philosopher-saint from Calcutta

Rāmāyaṇa great epic story of Rāma, Lakṣmaṇa, and Sītā, written by Vālmīki, includes the following sections:

Araṇyakāṇḍa
Ayodhyākāṇḍa
Bālakāṇḍa
Uttarakkāṇḍa

rasa flavor

Rāvaṇa demon of the *Rāmāyaṇa*

ṛṣi seer, sage

ṛta harmony

ṛtu flow

sadhu renouncer, holy person

Śaivites followers of Śiva

śakti power

Śakti creative principle; the goddess

samādhi meditative absorption

samatva equanimity

Sāṃkhya philosophy of ancient India, includes psychology and cosmology

saṃnyāsin renouncer, holy person
saṃskāra residue of past action or *karma*
samvega aesthetic power
sanātan dharma eternal faith
saṅgam gathering
Śaṅkara eighth-century philosopher of Advaita Vedānta
śānta peace
Sarasvatī goddess of wisdom
sarvodaya uplift of everyone
śāstra normative instructional religious text for Hindus, including:
the *Arthaśāstra* of Kauṭilya
the *dharmaśāstra* literature
the *Manusmṛti*
the *Viṣṇusmṛti*
sattva lightness, purity
satya truth
satyāgraha holding to truth, employed by Gandhi
siddhas accomplished sages
siddhis powers
Sītā wife of Rāma
Śiva god of destruction
snāna ablutions
soma intoxicating drink mentioned in the Vedas
stotram praise hymn
Sūrdās medieval poet-saint
Sūrya sun, one of the chief Vedic deities
svabhāva inherent natures
svarāj self-rule
svarga heaven
swami learned religious figure

tamas heaviness, lethargy
tanmatras subtle elements
Tantra medieval school of Hinduism; generally Śaivite
tapas self-purification
tīrtha pilgrimage
tīrthasthan place of pilgrimage
tulsi basil

Upanayana ceremony initiation into adulthood
Upaniṣads philosophical discourses, dating from approximately 800 B.C.E., including:

the *Bṛhadāraṇyaka Upaniṣad*
the *Chāndogya Upaniṣad*
the *Kaṭha Upaniṣad*
the *Maitrī Upaniṣad*
the *Nārāyaṇa Upaniṣad*
the *Taittirīya Upaniṣad*

Uṣas dawn, the daughter of heaven

Vāc goddess of speech
vahana vehicle
Vaiṣṇavites followers of Viṣṇu
Vallabhācāryā sixteenth-century saint
Vālmīki author of the *Rāmāyaṇa*
vana forest
Vāna devatās deities of the forest
Varanasi (Banaras) sacred city of India
varṇa caste category
Varuṇa Vedic god of order
Vasiṣṭha Vedic sage
Vāsuki chief of the serpents
Vasus a class of gods
Vātsyāyaṇa author of the *Kāmasūtra*
Vāyu wind, personified in the Vedas
Vedānta school of philosophy that follows the Vedas and Upaniṣads
Vedas the oldest Sanskrit religious literature of India, including:
the *Atharvaveda Saṃhitā*
the *Ṛgveda Saṃhitā*
the *Sāmaveda Saṃhitā*
the *Yajurveda Saṃhitā*
vibhūti power
vidyā wisdom
viniyoga application
vīrya strength
Viṣṇu god of preservation
viśva-rūpa-darśana vision of all form
Vivekananda disciple of Ramakrishna, lecturer, and author of the Hindu revival
Vṛndāvana city sacred to Kṛṣṇa
Vyāsa compiler of the Vedas, *Mahābhārata*

yajña sacrifice
yakṣa forest god

yakṣī forest goddess
Yama god of death
yama injunction
Yamunā daughter of Yama
yoga spiritual practice
yogi meditator
yogin practitioner of yoga
Yudhiṣṭhira hero of the *Mahābhārata,* one of the Pāṇḍava brothers

zamindars, ***zimmedārī*** landlords

Select Bibliography

Action Research in Community Health and Development–Vahini. "Sardar Sarovar Project: An Intellectual Fashion." Pamphlet. Mongrol, Gujarat, 1993.

Agarwal, Anil. "Human-Nature Interactions in a Third World Country." In James, ed., *Ethical Perspectives on Environmental Issues in India,* 31–72.

Akula, Vikram K. "Grassroots Environmental Resistance in India." In *Ecological Resistance Movements: The Global Emergence of Radical and Popular Environmentalism,* ed. Bron Raymond Taylor, 127–45. Albany: State University of New York Press, 1995.

Alagh, Y. K., and D. T. Buch. "The Sardar Sarovar Project and Sustainable Development." In Fisher, ed., *Toward Sustainable Development.*

Alagh, Y. K., R. D. Desai, G. S. Guha, and S. P. Kashyap. *Economic Dimensions of the Sardar Sarovar Project*. New Delhi: Har-anand Publications, 1995.

Allchin, Bridget, "Early Man and Environment in South Asia 10,000 BC–AD 500." In Grove, Damodaran, and Sangwan, eds., *Nature and the Orient.*

Alley, Kelly D. "Ganga and Gandagi: Interpretations of Pollution and Waste in Banaras." *Ethology* 33 (spring 1994): 127–45.

———. "Idioms of Degeneracy: Assessing Ganga's Purity and Pollution." In Nelson, ed. *Purifying the Earthly Body of God.*

———. "On the Banks of the Ganga." *Annals of Tourism Research* 19 (winter 1992): 125–27.

Altvater, Ilmar. "Ecological and Economic Modalities of Time and Space." In *Is Capitalism Sustainable? Political Economy and the Politics of Ecology,* ed. Martin O'Conner, 76-90. New York: Guilford Press, 1994.

Ambasta, S. P., ed. *The Useful Plants of India*. New Delhi: Publications and Information Directorate, Council of Scientific and Industrial Research, 1986.

Apffel-Marglin, Frédérique. "Gender and the Unitary Self: Looking for the Subaltern in Coastal Orissa." *South Asian Research*, 1995.

———. "Introduction: Rationality and the World." In *Decolonizing Knowledge: From Development to Dialogue,* ed. Frédérique Apffel-Marglin and Stephen A. Marglin, 1–39. Oxford: Clarendon Press; New York: Oxford University Press, 1996.

———. "Of Pirs and Pandits: Tradition of Hindu-Muslim Cultural Commonalities in Orissa." *Manushi* (Delhi) 91 (1995): 17–26.

———. "Rationality, the Body and the World: From Production to Regeneration." In *Decolonizing Knowledge: From Development to Dialogue,* ed. Frédérique Apffel-Marglin and Stephen A. Marglin, 142–81. Oxford: Clarendon Press; New York: Oxford University Press, 1996.

———. "Sacred Groves: Regenerating the Body, the Land, the Community." In Sachs, ed., *Global Ecology,* 197–207.

Appadurai, Arjun. "Comments on 'The Jungle and the Aroma of Meats: An Ecological Theme in Hindu Medicine.'" *Social Science and Medicine* 27, no. 3 (1988): 206–7.

———. "How Moral Is South Asia's Economy? A Review Article." *Journal of Asian Studies* 43, no. 3 (1984): 481–97.

———. "Introduction: Commodities and the Politics of Value." In *The Social Life of Things: Commodities in Cultural Perspective,* ed. Arjun Appadurai, 3–63. Cambridge: Cambridge University Press, 1986.

Ardener, Shirley, ed. *Women and Space: Ground Rules and Social Maps*. New York: St. Martin's Press, 1981. Reprint, Oxford: Berg Publishers, 1993.

Arnold, David, and Ramachandra Guha, eds. *Nature, Culture, Imperialism: Essays on the Environmental History of South Asia*. Delhi: Oxford University Press, 1995.

Asia Watch. "Before the Deluge: Human Rights Abuses at India's Narmada Dam." 1992.

Balapure, K. N., J. K. Maheshware, and R. K. Tandon. "Plants of the Ramayana." *Ancient Science of Life* 7 (1987): 76–84.

Banerjee, Suresh Chandra. *Flora and Fauna in Sanskrit Literature*. Calcutta: Naya Prokash, 1980.

Banuri, Tariq, and Frédérique Apffel-Marglin, eds. *Who Will Save the Forests? Knowledge, Power, and Environmental Destruction.* London: Zed Books, 1993.

Banwari. *Pancavati: Indian Approach to Environment*. Trans. Asha Vohra. Delhi: Shri Vinayaka Publications, 1992.

Bartolomé, Leopoldo J. "Forced Resettlement and the Survival Systems of the Urban Poor." *Ethnology* 23, no. 3 (1984): 177–92.

Basu, Helene. *Habshi-Sklaven, Sidi-Fakire: Muslimische Heiligenverehrung im westlichen Indien*. Berlin: Das Arabische Buch Verlag, 1995.

Bauer, Stefan. *Angepasste Technologie: Augewählte ethnologische Fallbeispiele aus Indien unter besonderer Berücksichtigung der Abfallverwertung*. Vienna: Magistergrades der Philosophie, Universität Wien, 1994.

———. "Bambus: Angepasste Anwendugen in der Architektur tropischer und subtropischer Länder." *Archiv für Völkerkunde* 47 (1993): 161–89.

Baviskar, Amita. *In the Belly of the River: Tribal Conflicts over Development in the Narmada Valley*. Delhi: Oxford University Press, 1995.

Benthall, Jonathan. "The Greening of the Purple." *Anthropology Today* 11 (June 1995):18–20.

Benveniste, Emile. *Indo-European Language and Society*. Trans. Elizabeth Palmer. Coral Gables, Fla.: University of Miami Press, 1973.

Berleant, Arnold. *The Aesthetics of Environment*. Philadelphia: Temple University Press, 1992.

Berman, Morris. *The Reenchantment of the World*. Ithaca, N.Y.: Cornell University Press, 1981.

Bhabha, Homi K. *The Location of Culture*. London and New York: Routledge, 1994.

Bharara, L. P. "Notes on the Experience of Drought: Perception, Recollection, and Prediction." In *Desertification and Development: Dryland Ecology in Social Perspective,* ed. Brian Spooner and H. S. Mann, 351–61. London and New York: Academic Press, 1982.

Bhattacharji, Sukumari. *Fatalism in Ancient India*. Calcutta: Baulmon Prakashan, 1995.

Bhide, A. D., and B. B. Sundaresan. "Street Cleansing and Waste Storage and Collection in India." In *Managing Solid Wastes in Developing Countries,* ed. John R. Holmes, 139–49. New York: John Wiley and Sons, 1984.

Biardeau, Madeleine. *Hinduism: The Anthropology of a Civilization*. Trans. Richard Nice. Delhi and New York: Oxford University Press, 1989.

Bishnoi, Shri Krishna. *Bishnoi Dharm-Sanskar*. Bikaner: Dhok Dhora Prakashan, 1991.

Blincow, Malcolm. "Scavengers and Recycling: A Neglected Domain of Production." *Labour, Capital and Society* 19, no. 1 (1986): 94–115.

Blinkhorn, Thomas A., and William T. Smith. "India's Narmada: River of Hope." In Fisher, ed., *Toward Sustainable Development?*

Bormen, Jan. "Labor Relations in the 'Formal' and 'Informal' Sectors: Report of a Case Study in South Gujarat, India." *Journal of Peasant Studies* 4, no. 3 (1977): 171–205; 4, no. 4 (1977): 337–59.

Braidotti, Rosi, Ewa Charkiewicz, Sabine Hausler, and Sakia Wiernga. *Women, the Environment, and Sustainable Development: Towards a Theoretical Synthesis*. London: Zed Books, 1994.

Brooks, Charles R. *The Hare Krishnas in India*. Princeton: Princeton University Press, 1989.

———. "A Unique Conjuncture: The Incorporation of ISKON in Vrindaban." In *Krishna Consciousness in the West,* ed. David G. Bromley and Larry D. Shinn, 165–87. Lewisburg, Pa.: Bucknell University Press: 1989.

Broder, Jonathan. "Pollution Threatens the Ancient Purity of the Ganges." *San Francisco Examiner*, 31 March, 1997, 8–9 (Travel).

Buckley, Thomas, and Alma Gottlieb, eds. *Blood Magic: The Anthropology of Menstruation*. Berkeley and Los Angeles: University of California Press, 1988.

Buttel, Frederick H. "Environmentalization: Origins, Processes, and Implications for Rural Social Change." *Rural Sociology* 57 (spring 1992): 1–27.

———. "New Directions in Environmental Sociology." *Annual Review of Sociology* 13 (1987): 465–88.

Buttel, Frederick, and Peter Taylor. "Environmental Sociology and Global Environmental Change: A Critical Assessment." *Society and Natural Resources* 5 (July-September 1992): 211–30.

Callicott, J. Baird. *Earth's Insights: A Survey of Ecological Ethics from the Mediterranean Basin to the Australian Outback*. Berkeley and Los Angeles: University of California Press, 1994.

Callicott, J. Baird, and Roger T. Ames, eds. *Nature in Asian Traditions of Thought: Essays in Environmental Philosophy*. Albany: State University of New York Press, 1989.

Capra, Fritjof, David Steindl-Rast, and Thomas Matus. *Belonging to the Universe: Explorations on the Frontiers of Science and Spirituality*. San Francisco: HarperSanFrancisco, 1991.

Carman, John B., and Frédérique Apffel-Marglin. *Purity and Auspiciousness in Indian Society*. Leiden: E. J. Brill, 1985.

Cenkner, William. *A Tradition of Teachers: Sankara and the Jagadgurus Today*. Delhi: Motilal Banarsidass, 1983.

Chakrabarty, Dipesh. "Open Space/Public Place: Garbage, Modernity, and India." *South Asia* 14, no. 1 (1991): 15–31.

Champion, Harry. G., and S. K. Seth. *A Revised Survey of the Forest Types of India*. Delhi: Manager of Publications, 1968.

Chandrasekhar, Sripati. "The Hindu Understanding of Population and Population Control." In James, ed. *Ethical Perspectives,* 189–216.

Chapple, Christopher Key, ed. *Ecological Prospects: Scientific, Religious, and Aesthetic Perspectives*. Albany: State University of New York Press, 1994.

———. "Hinduism and Ecology." In Tucker and Grim, eds., *Worldviews and Ecology*.

———. "India's Earth Consciousness." In Tobias and Cowan, eds., *The Soul of Nature*, 145–51.

———. *Nonviolence to Animals, Earth, and Self in Asian Traditions*. Albany: State University of New York Press, 1993.

———. "Toward an Indigenous Indian Environmentalism." In Nelson, ed., *Purifying the Earthly Body of God*.

Childs, John Brown. "Rooted Cosmopolitanism: The Transnational Character of Indigenous Particularity." Stevenson Programme on Global Security Colloquium, University of Santa Cruz, 1992.

Chopra, R. N., I. C. Chopra, K. L. Handa, and L. D. Kapur. *Indigenous Drugs of India*. 2d ed. Calcutta, New York, and Delhi: Academic Publishers, 1958.

Claiborne, William. "Devout Hindus Resist Efforts to Clean Up the Sacred Ganges." *The Washington Post*, 8 May 1983, pp. 18–19 (A).

Clarke, J. J. *Oriental Enlightenment: The Encounter between Asian and Western Thought*. London and New York: Routledge, 1997.

Cobb, John B., Jr. *Sustainability: Economics, Ecology, and Justice*. Maryknoll, N.Y.: Orbis Books, 1992.

Corcoran, Maura. *Vrndavana in Vaisnava Literature: History, Mythology, Symbolism*. Vrindaban: Vrindaban Research Institute; and New Delhi: D. K. Printworld, 1995.

Covarrubias, Miguel. *Island of Bali*. Kuala Lumpur: Oxford University Press, 1974.

Coward, Harold. "The Ecological Implications of Karma Theory." In Nelson, ed., *Purifying the Earthly Body of God*.

———, ed. *Population, Consumption, and the Environment: Religious and Secular Responses*. Albany: State University of New York Press, 1995.

———. *Visions of a New Earth: Religious Perspectives on Population, Consumption, and Ecology*. Albany: State University of New York Press, 2000.

Crawford, S. Cromwell. *The Evolution of Hindu Ethical Ideals*. Delhi: Arnold-Heinemann, 1984.

———. "Hindu Ethics for Modern Life." In *World Religions and Global Ethics,* ed. S. Cromwell Crawford. New York: Paragon House Publishers, 1989.

Cremo, Michael A., and Mukunda Goswami. *Divine Nature: A Spiritual Perspective on the Environmental Crisis*. Los Angeles: Bhaktivedanta Institute, 1995.

Croll, Elisabeth, and David Parkin, eds. *Bush Base, Forest Farm: Culture, Environment, and Development*. London and New York: Routledge, 1992.

Darian, Steven G. *The Ganges in Myth and History*. Honolulu: University Press of Hawaii, 1978.

Das, Veena. *Structure and Cognition: Aspects of Hindu Caste and Ritual*. 2d ed. Delhi: Oxford University Press, 1982.

Dasa, Rancho [Ranchor Prime]. "Reviving the Forests of Vrndavana." *Back to Godhead: The Magazine of the Hare Krishna Movement* 26 (September-October 1992): 24–39.

Das Gupta, S. *Alpana*. Delhi: Publications Division, Ministry of Information and Broadcasting, Government of India, 1960.

Dasgupta, Shashi Bushan. "Evolution of Mother Worship in India." In *Great Women of India,* ed. Swami Madhavananda and Ramesh Chandra Majumdar, 2d. ed. Mayavati Pithoragarh, Himalayas: Advaita Ashrama, 1982.

Deegan, Chris. "The Narmada in Myth and History." In Fisher, ed., *Toward Sustainable Development?*

Denslow, Julie Sloan, and Christine Padoch, eds. *People of the Tropical Rain Forest*. Berkeley and Los Angeles: University of California Press; Washington, D.C.: Smithsonian Institution Traveling Exhibition Service, 1988.

Deutsch, Eliot. "A Metaphysical Grounding for Natural Reverence: East-West." In Callicott and Ames, eds., *Nature in Asian Traditions of Thought*.

———. "Vedanta and Ecology." *Indian Philosophical Annual* (Madras) 7 (1970): 79–88.

Dharmadhikary, Shripad. "Hydropower at Sardar Sarovar: Is It Necessary, Justified, and Affordable?" In Fisher, ed., *Toward Sustainable Development?*

Douglas, Mary, and Baron Isherwood. *The World of Goods: Towards an Anthropology of Consumption*. London: Allen Lane, 1979.

Dove, Michael. "The Dialectical History of 'Jungle' in Pakistan: An Examination of the Relationship between Nature and Culture." *Journal of Anthropological Research* 48 (1992): 231–53.

Dumont, L., and D. F. Pocock. "Pure and Impure." *Contributions to Indian Sociology* 3 (1959): 9–34.

Drèze, Jean, Meera Samson, and Satyajit Singh, eds. *The Dam and the Nation: Displacement and Resettlement in the Narmada Valley*. Delhi: Oxford University Press, 1997.

Dwivedi, O. P. *Darshan: Nature and the Face of God*. Vol. 36. New York: S.Y.D.A. Foundation, 1990.

———. "Environmental Protection in the Hindu Tradition." In James, ed., *Ethical Perspectives,* 161–88.

———. *India's Environmental Policies, Programmes, and Stewardship*. London: Macmillan; New York: St. Martin's Press, 1997.

———. "Satyagraha for Conservation: Awakening the Spirit of Hinduism." In *This Sacred Earth: Religion, Nature, Environment,* ed. Roger S. Gottlieb. New York: Routledge, 1996.

Dwivedi, O. P., and B. N. Tiwari. *Environmental Crisis and Hindu Religion*. New Delhi: Gitanjali Publishing House, 1987.

Eck, Diana L. *Banaras: City of Light*. New York: Knopf, 1982.

———. "Ganga: The Goddess in Hindu Sacred Geography." In Hawley and Wulff, eds., *The Divine Consort,* 166–83.

———. "India's Tirthas: 'Crossings' in Sacred Geography." *History of Religions* 20, no. 4 (1981): 323–24.

———. "A Survey of Sanskrit Sources for the Study of Varanasi." In *Banaras (Varanasi): Cosmic Order, Sacred City, Hindu Tradtions: Festschrift to Professor R. L. Singh,* ed. Rana P. B. Singh, 9–19. Varanasi: Tara Book Agency for the Varanasi Studies Foundation, 1993.

Emett, Carolyn. "The Tree Man." *Resurgence: An International Forum for Ecological and Spiritual Thinking* 183 (July-August 1997): 42.

Erndl, Kathleen M. *Victory to the Mother: The Hindu Goddess of Northwest India in Myth, Ritual, and Symbol*. New York: Oxford University Press, 1993.

Eschmann, Anncharlott, Hermann Kulke, and Gaya Charan Tripathi, eds. *The Cult of Jagannath and the Regional Tradition of Orissa*. South Asian Studies, no. 8. New Delhi: Manohar, 1978.

———. "Prototypes of the Navakelevara Rituals and Their Relation to the Jagannath Cult." In Eschmann et al., eds., *The Cult of Jagannath*.

Erdosy, George. "Deforestation in Pre- and Proto-Historic South Asia." In Grove, Damodaran, and Sagwan, eds., *Nature and the Orient,* 51–69.

Feldhaus, Anne. *Water and Womanhood: Religious Meanings of Rivers in Maharashtra*. New York: Oxford University Press, 1995.

First International Meeting of People Affected by Large Dams, Declaration of Curitiba, Brazil, 14 March 1997.

Fisher, William F. "Development and Resistance in the Narmada Valley." In Fisher, ed., *Toward Sustainable Development?*

———. "Doing Good? The Politics and Anti-Politics of NGO Practices." *Annual Review of Anthropology* 26 (1997): 439–64.

———. "Full of Sound and Fury? Struggling towards Sustainable Development." In Fisher, ed., *Toward Sustainable Development?*

———, ed. *Toward Sustainable Development? Struggling over India's Narmada River*. Armonk, N.Y.: M. E. Sharpe, 1995.

Fleet, John F. *Inscriptions of the Early Gupta Kings and Their Successors*. Varanasi: Indological Book House, 1997.

Flinders, Carol Lee. *At the Root of This Longing: Reconciling a Spiritual Hunger and a Feminist Thirst*. San Francisco: HarperSanFrancisco, 1998.

Frater, Alexander. *Chasing the Monsoon*. Calcutta: Penguin Books, 1991.

Friends of Vrindavan. "Protecting Sacred Forests: Linking Leicester's Community with the Sacred Forests of India." 1993.

Fuller, C. J. "Gods, Priests, and Purity: On the Relation between Hinduism and the Caste System." *Man,* n.s., 14 (September 1979): 459–76.

Furedy, Christine. "Survival Strategies of the Urban Poor—Scavenging and Recuperation in Calcutta." *Geo-Journal* 8, no. 2 (1984): 129–36.

Fürer-Haimendorf, Christoph von. *Tribal Populations and Cultures of the Indian Subcontinent*. Leiden: E. J. Brill, 1985.

———. *Tribes of India: The Struggle for Survival*. Berkeley and Los Angeles: University of California Press, 1982.

Gaard, Greta, ed. *Ecofeminism: Women, Animals, Nature*. Philadelphia: Temple University Press, 1993.

Gadgil, Madhav. "The Indian Heritage of a Conservation Ethic." In James, ed., *Ethical Perspectives,* 141–60.

Gadgil, Madhav, and Subash Chandran. "Sacred Groves." *Indian International Center Quarterly* 19, no. 1-2 (1992): 183–88.

Gadgil, Madhav, and Ramachandra Guha. *Ecology and Equity: The Use and Abuse of Nature in Contemporary India*. London and New York: Routledge, 1995.

———. *This Fissured Land: An Ecological History of India*. Delhi and Melbourne: Oxford University Press, 1992.

Gallagher, Rob. *The Rickshaws of Bangladesh*. Dhaka: The University Press, 1992.

Gandhi, Mohandas. *Collected Works of Mahatma Gandhi*. Delhi: Publications Division, Ministry of Information and Broadcasting, Government of India, 1958–.

Ghosh, Arun. "Ecology and Environment." In *Nature, Man, and the Indian Economy,* ed. Tapas Majumdar. Delhi and New York: Oxford University Press, 1993.

Giddens, Anthony. *The Consequences of Modernity*. Stanford, Calif.: Stanford University Press, 1990.

Gold, Ann Grodzins. "Abandoned Rituals: Knowledge, Time, and Rhetorics of Modernity in Rural India." Paper presented at the annual meeting of the American Anthopological Association, Washington, D.C., 1995.

———. *Fruitful Journeys: The Ways of Rajasthani Pilgrims*. Berkeley and Los Angeles: University of California Press, 1990.

———. “Of Gods, Trees, and Boundaries: Divine Conservation in Rajasthan.” In *Folk, Faith, and Feudalism: Rajasthan Studies,* ed. N. K. Singhi and Rajendra Joshi, 33–54. Jaipur: Rawat Publications, 1995.

———. “Sin and Rain: Moral Ecology in Rural North India.” In Nelson, ed., *Purifying the Earthly Body of God.*

———. “Wild Pigs and Kings: Remembered Landscapes in Rajasthan.” *American Anthropologist* 99, no. 1 (1997): 70–84.

Gold, Ann Grodzins, and Bhoju Ram Gujar, “Drawing Pictures in the Dust: Rajasthani Children’s Landscapes.” *Childhood* 2 (1994):73–91 (special issue, *Children and Environment: Local Worlds and Global Connections,* ed. Sharon Stephens).

Goldsmith, Edward, and Nicholas Hildyard. *The Social and Environmental Effects of Large Dams*. Camelford, United Kingdom: Wadebridge Ecological Centre; San Francisco: Sierra Club Books, 1984.

Government of Gujarat. “Comment on the Report of the Independent Review Mission on Sardar Sarovar Project.” Draft. Gandinagar, 1992.

Greenberg, Brian. “Sustainable Futures and Romantic Pasts: Political Ecology and Environmental History in North India.” Paper presented in SAME Workshop, University of Chicago.

Grothues, Jürgen. *Aladins Neue Lampe: Recycling in der Dritten Welt.* Munich: Trickster Verlag, 1988.

———. “Recycling als Handwerk.” *Archiv für Völkerkunde* 38 (1984): 103–31.

Grove, Richard H. *Green Imperialism: Colonial Expansion, Tropical Island Edens, and the Origins of Environmentalism, 1600–1860*. Cambridge: Cambride University Press, 1995.

Grove, Richard H., Vinita Damodaran, and Satpal Sangwan, eds. *Nature and the Orient: The Environmental History of South and Southeast Asia*. Delhi and New York: Oxford University Press, 1998.

Grualski, Bart. “The Chipko Movement: A Gandhian Approach to Ecologial Sustainability and Liberation from Economic Colonization.” In *Ethical and Political Dilemmas of Modern India,* ed. Ninian Smart and Shivesh Thakur, 100–125. New York: St. Martin’s Press, 1993.

Guha, Ramachandra. *Environmentalism: A Global History*. New York: Longman, 2000.

———. “The Environmentalism of the Poor.” In *Between Resistance and Revolution: Cultural Politics and Social Protest,* ed. Richard G. Fox and Orin Starn. New Brunswick, N.J.: Rutgers University Press, 1997.

———. “Ideological Trends in Indian Environmentalism.” *Economic and Political Weekly* 23 (1988): 29.

———. *Mahatma Gandhi and the Environmental Movement*. Pune: Parisar Annual Lecture, 1993.

———. "Radical American Environmentalism: A Third World Critique." *Environmental Ethics* 11, no. 1 (1989): 71–83. Reprinted in James, ed., *Ethical Perspectives,* 115–30.

———. "Two Phases of American Environmentalism: A Critical History." In *Decolonizing Knowledge: From Development to Dialogue,* ed. Frédérique Apffel-Marglin and Stephen A. Marglin. Oxford: Clarendon Press; New York: Oxford University Press, 1996.

Guha, Sumit. "Kings, Commoners, and the Commons: People and Environments in Western India, 1600–1900." Paper presented at Cornell University, Ithaca, New York.

Guibbert, Jean-Jacques. "Ecologie populaire urbaine et assainissement environnemental dans le Tiers Monde." *Environnement Africain* 8, no. 1-2 (1990): 21–50.

Gupta, Lina. "Ganga: Purity, Pollution, and Hinduism." In *Ecofeminism and the Sacred,* ed. Carol J. Adams, 99–116. New York: Continuum,1994.

Haberman, David. *Acting as a Way of Salvation: A Study of Raganuga Bhakti Sadhana*. New York: Oxford University Press, 1988.

———. *Journey through the Twelve Forests: An Encounter with Krishna*. New York: Oxford University Press, 1994.

Harman, William P. *The Sacred Marriage of a Hindu Goddess*. Bloomington: Indiana University Press, 1989.

Havell, E. B. *Benares: The Sacred City*. 1905. Reprint, Varanasi: Vishwavidyalaya Prakashan, 1990.

Hawley, John Stratton, and Donna M. Wulff, eds. *The Divine Consort: Rādhā and the Goddesses of India*. Berkeley: Religious Studies Series; New Delhi: Motilal Banarsidass, 1982. Reprint, Boston: Beacon Press, 1986.

Haynes, Edward. "Land Use and Land-Use Ethic in Rajasthan, 1850–1980." Paper presented at the Conference on South Asia, Madison, Wisconsin.

Herring, Ronald J. "Rethinking the Commons." *Agriculture and Human Values* 7, no. 2 (1990): 88–104.

Hiltebeitel, Alf. "Draupadi's Garment." *Indo-Iranian Journal* 22 (1980): 98–112.

Hopkins, E. Washburn. "Mythological Aspects of Trees and Mountains in the Great Epic." *Journal of the American Oriental Society* 30 (1910): 347–74.

Hull, R. B., and G. R. G. Revell. "Cross-Cultural Comparison of Landscape Scenic Beauty Evaluations: A Case Study of Bali." *Journal of Environmental Psychology* 9 (1989): 177–91.

Illich, Ivan. *H_2O and the Waters of Forgetfulness: Reflections on the Historicity of "Stuff."* Dallas: Dallas Institute of Humanities and Culture, 1985.

Inden, Ronald. *Imagining India*. Oxford and Cambridge, Mass.: Basil Blackwell, 1990.

Indian National Trust for Art and Cultural Heritage (INTACH). *Bhagirathi Ki Pukar* 4 (1993).

———. *Deforestation, Drought, and Desertification: Perceptions on a Growing Ecological Crisis*. New Delhi: Indian National Trust for Art and Cultural Heritage, 1989.

International Commission on International Humanitarian Issues. *Indigenous Peoples: A Global Quest for Justice*. London and Atlantic Highlands, N.J.: Zed Books, 1987.

International Rivers Network. "Independent Commission to Review the World's Dams." *World Rivers Review* 12, no. 3 (June 1997).

———. "World Commission on Dams Launched." Press Release, 16 February 1998.

Jaini, Padmanabh S. "Indian Perspectives on the Spirituality of Animals." In *Buddhist Philosophy and Culture: Essays in Honour of N. A. Jayawickrema,* ed. David J. Kalupahana and W. G. Weeraratne, 169–78. Colombo: N. A. Jayawickrema Felicitation Volumé Committee, 1987.

James, George A., ed. *Ethical Perspectives on Environmental Issues in India*. New Delhi: A.P.H. Publishing Corporation, 1999.

———. "The Significance of Indian Traditions for Environmental Ethics." In James, ed., *Ethical Perspectives,* 3–30.

Jodha, N. S. "Population Growth and the Decline of Common Property Resources in Rajasthan, India." *Population and Development Review* 11, no. 2 (1985): 247–64.

———. "Rural Common Property Resources: Contributions and Crisis." *Economic and Political Weekly,* 30 June 1990, A65–A78.

Joshi, Vidyut. *Rehabilitation, A Promise to Keep: A Case of the SSP*. Ahmedabad: The Tax Publications, 1991.

Kalpagam, U. "Coping with Urban Poverty in India." *Bulletin of Concerned Asian Scholars* 17, no. 1 (1985): 2–18.

Kalpavriksh. *Narmada: A Campaign Newsletter* (New Delhi), nos. 5–14 (1990–94).

Khilnani, Sunil. *The Idea of India*. New York: Farrar, Straus, Giroux, 1998.

Khoshoo, T. N. "Gandhian Environmentalism." In James, ed., *Ethical Perspectives,* 241–82.

Kilambi, Jyotsna. "Muggu: Threshold Art in South India." *Res: The Journal of Aesthetics and Anthropology* 10 (1986): 71–102.

Kinsley, David R. *The Divine Player: A Study of Krsna Lila*. Delhi: Motilal Banarsidass, 1979.

———. *Ecology and Religion: Ecological Spirituality in Cross-Cultural Perspective*. Englewood Cliffs, N.J.: Prentice-Hall, 1995.

———. *Hindu Goddesses: Visions of the Divine Feminine in the Hindu Religious Tradition*. Berkeley and Los Angeles: University of California Press, 1986; Delhi: Motilal Banarsidass, 1987.

———. "Learning the Story of the Land: Reflections on the Liberating Power of Geography and Pilgrimage in the Hindu Tradition." In Nelson, ed., *Purifying the Earthly Body of God*.

Kipling, John Lockwood. *Beast and Man in India: A Popular Sketch of Indian Animals in Their Relations with the People*. London: Macmillan, 1904.

Klostermaier, Klaus. "Bhakti, Ahimsa, and Ecology." *Journal of Dharma* 16 (July-September 1991): 246–54.

———. "The Body of God." In *The Charles Strong Lectures, 1972–1984*, ed. Robert B. Crotty, 103–20. Leiden: E. J. Brill, 1987.

Korom, Frank J. "On the Ethics and Aesthetics of Recycling in India." In Nelson, ed., *Purifying the Earthly Body of God*.

———. "Recycling." In *South Asian Folklore: An Encyclopedia,* ed. Margaret Mills and Peter J. Claus. New York: Garland Press, forthcoming.

———. "Recycling in India: Status and Economic Realities." In *Recycled Reseen: Folk Art from the Global Scrap Heap,* ed. Charlene Cerny and Suzanne Seriff, 118–29, 190–92. New York: Harry N. Abrams, 1996.

Kothari, Smitu, and Pramod Parajuli. "Nature without Social Justice: A Plea for Cultural and Ecological Pluralism in India." In Sachs, ed., *Global Ecology,* 224–41.

Krishnan, M. *The Hand Book of India's Wildlife*. Madras: Maps and Agencies, 1982.

Kulke, Hermann. "Tribal Deities at Princely Courts: The Feudatory Rajas of Central Orissa and Their Tutelary Deities [Ishtadevata]." In *The Realm of the Sacred: Verbal Symbolism and Ritual Structures,* ed. Sitakant Mahapatra. Calcutta and New York: Oxford University Press, 1992.

Kurin, Richard. "Cultural Conservation through Representation: Festival of India Folklife Exhibitions at the Smithsonian Institution." In *Exhibiting Cultures: The Poetics and Politics of Museum Display,* ed. Ivan Karp and Steven D. Levine. Washington, D.C.: Smithsonian Institution Press, 1991.

Lalas, Sitaram. *Rajasthani Sabad Kos*. 9 vols. Jodhpur: Rajasthani Sodh Sansthan, 1962–78.

Lansing, J. Stephen. *Priests and Programmers: Technologies of Power in the Engineered Landscape of Bali*. Princeton: Princeton University Press, 1991.

Larson, Gerald. "The Aesthetic (*rasasuvada*) and the Religious (*brahmasuvada*) in Abhinavagupta's Kashmir Saivism." *Philosophy East and West* 26 (1976):371–87.

———. "'Conceptual Resources' in South Asia for 'Environmental Ethics.'" In Callicott and Ames, eds., *Nature in Asian Traditions of Thought*.

Lawyers Committee for Human Rights. "Unacceptable Means: India's Sardar Sarovar Project and Violations of Human Rights." October 1992 through February 1993.

Leopold, Aldo. *A Sand County Almanac*. New York: Oxford University Press, 1949.

Leslie, Julia, ed. *Sri and Jyestha: Ambivalent Role Models for Women*. New Delhi: Motilal Banarsidass, 1992.

Ludden, David. "Productive Power in Agriculture: A Survey of Work on the Local History of British India." In *Agrarian Power and Agricultural Productivity in South Asia,* ed. Meghnad Desai, Susanne Hoeber Rudolph, and Ashok Rudra, 51–99. Delhi: Oxford University Press, 1984.

Lynch, Owen. "Pilgrimage with Krishna, Sovereign of Emotions." *Contributions to India Sociology* 22, no. 2 (1988): 171–94.

Madan, T. N. "Concerning the Categories Subha and Suddha in Hindu Culture: An Exploratory Essay." In Carman and Apffel-Marglin, eds., *Purity and Auspiciousness in Indian Society*.

———. *Non-Renunciation: Themes and Interpretations of Hindu Culture*. Delhi and New York: Oxford University Press, 1987.

Majumdar, R. C., ed. *The History and Culture of the Indian People*. Vol. 3, *The Classical Age*. Delhi: Bharatiya Vidya Bhavan, 1988.

Mann, R. S., ed. *Nature-Man-Spirit Complex in Tribal India*. New Delhi: Concept Publishing Company, 1981.

Mathur, Jivanlal. *Braj-Bhavani*. Sawar: Mani Raj Singh, 1977.

Mathur, Sri Rakesh. "Can India's Timeworn Dharma Help Renew a Careworn World?" *Hinduism Today,* July 1995, 1, 9.

Maw, Geoffrey Waring. *Narmada: The Life of a River*. Ed. Marjorie Sykes. Hoshangabad: Distributed by Friends Rural Centre, 1991.

McCully, Patrick. *Silenced Rivers: The Ecology and Politics of Large Dams*. London: Zed Books, 1996.

McKibben, Bill. *Hope, Human and Wild: True Stories of Living Lightly on the Earth*. Boston: Little, Brown, and Co., 1995. Reprinted, St Paul: Hungry Mind Press, 1997.

Mies, Maria, and Vandana Shiva. *Ecofeminism*. London: Zed Books, 1993.

Motichandra, D. *Kashi Ka Itihas*. 2d ed. Varanasi: Vishwavidyalaya Prakashan, 1985.

Mumme, Patricia Y. "Models and Images for a Vaisnava Environmental Theology: The Potential Contribution of Srivaisnavism." In Nelson, ed., *Purifying the Earthly Body of God*.

Murti, C. R. K., K. S. Bilgrami, T. M. Das, and R. P. Mathur, eds. *The Ganga: A Scientific Study*. New Delhi: Ganga Project Directorate, 1991.

Naess, Arne "Self-Realization: An Ecological Approach to Being in the World." In *Thinking Like a Mountain: Towards a Council of All Beings,* ed. John Seed, Joanna Macy, and Arne Naess, 19–30. Philadelphia: New Society Publishers, 1988.

Nagar, Shanti Lal. *Varaha in Indian Art, Culture, and Literature*. New Delhi: Aryan Books International, 1993.

Nagarajan, Vijaya Rettakudi. "The Earth as Goddess Bhu Devi: Toward a Theory of 'Embedded Ecologies' in Folk Hinduism." In Nelson, ed., *Purifying the Earthly Body of God.*

———. "Hosting the Divine: The Kolam in Tamil Nadu." In *Mud, Mirror, and Thread: Folk Traditions of Rural India,* ed. Nora Fisher. Ahmedabad: Mapin Publishing; Middletown, N.J.: Grantha Corp.; Santa Fe: Museum of New Mexico Press, 1993.

Nanavati, R. I., ed. *Purana-Itihasa-Vimarsah.* Delhi: Bharatiya Vidya Prakashan, 1998.

Narayan, Vasudha. "One Tree Is Equal to Ten Sons: Hindu Responses to the Problems of Ecology, Population, and Consumption." *Journal of the American Academy of Religion* 65, no. 2 (January 1997): 291–332.

Nath, K. J. "Metropolitan Solid Waste Management in India." In *Managing Solid Wastes in Developing Countries,* ed. John R. Holmes, 47–69. New York: John Wiley and Sons, 1984.

Nelson, Lance E. "The Dualism of Nondualism: Advaita Vedanta and the Irrelevance of Nature." In Nelson, ed., *Purifying the Earthly Body of God.*

———, ed. *Purifying the Earthly Body of God: Religion and Ecology in Hindu India.* Albany: State University of New York Press, 1998.

———. "Reverence for Nature or the Irrelevance of Nature? Advaita Vedanta and Ecological Concern." *Journal of Dharma* 16 (July-September 1991): 282–301.

———. "Theism for the Masses, Non-dualism for the Monastic Elite: A Fresh Look at Samkara's Trans-theistic Spirituality." In *The Struggle over the Past: Fundamentalism in the Modern World,* ed. William M. Shea, 61–77. Lanham, Md.: University Press of America, 1993.

Omvedt, Gail. *Reinventing Revolution: New Social Movements and the Socialist Tradition in India.* Armonk, N.Y.: M. E. Sharpe, 1993.

Orenstein, Henry. "The Structure of Hindu Caste Values: A Preliminary Study of Hierarchy and Ritual Defilement." *Ethnology* 4 , no. 1 (1995): 1–15.

———. "Toward a Grammar of Defilement in Hindu Sacred Law." In *Structure and Change in Indian Society,* ed. Milton Singer and Bernard S. Cohn. New York: Wenner-Gren Foundation for Anthropological Research, 1968.

Panwalkar, Pratima. "Processus de Recyclage du Plastique à Bombay." *Environnement Africain* 8, no. 1-2 (1990): 153–57.

Parajuli, Pramod. "Tortured Bodies and Altered Earth: Ecological Ethnicities in the Regime of Globalization." Unpublished manuscript.

Paranjpye, Vijay. *High Dams on the Narmada.* Studies in Ecology and Sustainable Development, no. 3. New Delhi: Indian National Trust for Art and Cultural Heritage, 1990.

Parisura, Manimekalai. *Vagai Vagaiyana Kolangal*. Madras: Manimekalai Pirusuram, 1980.

Parkhill, Thomas. *The Forest Setting in Hindu Epics: Princes, Sages, Demons*. Lewiston, N.Y.: Mellen University Press, 1995.

Patel, Anil. "What Do the Narmada Valley Tribals Want?" in Fisher, ed., *Toward Sustainable Development?*

Patel, C. C. "The Sardar Sarovar Project: A Victim of Time." In Fisher, ed., *Toward Sustainable Development?*

Patkar, Medha (in conversation with Dunu Roy and Geeti Sen). "The Strength of a People's Movement." In Sen, ed., *Indigenous Vision*.

——— (in conversation with Smitu Kothari). "The Struggle for Participation and Justice: A Historical Narrative." In Fisher, ed., *Toward Sustainable Development?*

Peritore, N. Patrick. "Environmental Attitudes of Indian Elites: Challenging Western Postmodernist Models." *Asian Survey* 33 (1993): 804–18.

Pillai, Moozhikkulam Chandrasekharam. *Mannarassala: The Serpent Temple*. Trans. Ayyappa Panikker. Harippad: Manasa Publications, 1991.

Powell, Robert, ed. *The Ultimate Medicine: As Prescribed by Sri Nisargadatta Maharaj*. San Diego: Blue Dove Press, 1994.

Prime, Ranchor. *Hinduism and Ecology: Seeds of Truth*. London: Cassell, 1992.

Puri, G. S., V. M. Meher-Homji, R. K. Gupta, and S. Puri. *Forest Ecology*. Vol. 1, *Phytogeography and Forest Conservation*. New Delhi: Oxford and IBH Publishing Company, 1984.

Raheja, Gloria Goodwin. *The Poison in the Gift: Ritual, Prestation, and the Dominant Caste in a North Indian Village*. Chicago: University of Chicago Press, 1988.

Raj, A. R. Victor. *The Hindu Connection: Roots of the New Age*. St. Louis: Concordia Publishing House, 1995.

Ram, Rahul N. "Benefits of the Sardar Sarovar Project: Are the Claims Reliable?" in Fisher, ed., *Toward Sustainable Development?*

Rambachan, Anatanand. "The Value of the World as the Mystery of God in Advaita Vedanta." *Journal of Dharma* 14 (July-September 1989): 287–97.

Ramanujan, A. K., "Is There an Indian Way of Thinking? An Informal Essay." *India through Hindu Categories,* ed. McKim Marriott, 41–58. New Delhi and Newbury Park, Calif.: Sage Publications, 1990.

Rangarajan, Mahesh. *Fencing the Forest: Conservation and Ecological Change in India's Central Provinces, 1860–1914*. Delhi and New York: Oxford University Press, 1991.

Ray, Amit. "Rabindranath Tagore's Vision of Ecological Harmony." In James, ed., *Ethical Perspectives,* 217–40.

Richards, John F., Edward S. Haynes, and James R. Hagen. "Changes in the Land and Human Productivity in Northern India, 1870–1970." *Agricultural History* 59, no. 4 (1985): 523–48.

Redford, Kent. "The Ecologically Noble Savage." *Orion* 9 (1990): 24–29.

Rolston, Holmes, III. "Can the East Help the West to Value Nature?" *Philosophy East and West* 37 (April 1987): 172–90.

Rukmani, T. S., "Environmental Ethics as Enshrined in Sanskrit Sources." *Nidān* (Journal of the Department of Hindu Studies and Indian Philosophy, Durban) 7 (1995).

Sachs, Wolfgang, ed. *Global Ecology: A New Arena of Political Conflict*. London: Zed Books, 1993.

Saint, Kishore. "Aravalli Abhiyan—Save Aravalli Campaign." In *Deforestation, Drought, and Desertification: Perceptions on a Growing Ecological Crisis*. New Delhi: Indian National Trust for Art and Cultural Heritage, 1989.

Sampat, Payal. "What Does India Want?" *World Watch* 11, no. 4 (1998): 30–38.

Sankat Mochan Foundation. *Proposal for GAP Phase II at Varanasi*. Varanasi: Swatcha Ganga Campaign, 1994.

———. *A Seminar on Pollution Control of River Cities in India: A Case Study of Varanasi*. Varanasi: Swatcha Ganga Campaign, 1992.

———. *Swatcha Ganga Campaign Annual Report, 1988–1990*. Varanasi: Swatcha Ganga Campaign, 1990.

Saraswathi, R., and L. Vijayalakshmi. *Kothaiyar idum kolaangal*. Madras: Lakkumi Nilayam, 1991.

Savyasaachi. "An Alternative System of Knowledge: Fields and Forests in Abujhmarh." In Banuri and Apffel-Marglin, eds., *Who Will Save the Forests?*

Sax, William S. *Mountain Goddess: Gender and Politics in a Himalayan Pilgrimage*. New York: Oxford University Press, 1991.

Schiff, Bennett. "A Fantasy Garden by Nek Chand Flourishes in India." *Smithsonian* 15, no. 3 (June 1984): 126–36.

Seed, John. "Spirit of the Earth: A Battle-Weary Rainforest Activist Journeys to India to Renew His Soul." *Yoga Journal* 138 (January-February 1998): 69–71, 132–36.

Sen, Geeti, ed. *Indigenous Vision: Peoples of India, Attitudes to the Environment*. New Delhi: Sage Publications, 1992.

Seshadri, B. *India's Wildlife and Wildlife Reserves*. New Delhi: Sterling Publishers, 1986.

Shah, Ashvin A. "A Technical Overview of the Flawed Sardar Sarovar Project and a Proposal for a Sustainable Alternative." In Fisher, ed., *Toward Sustainable Development?*

Sharma, Arvind. "Attitudes to Nature in the Early Upanishads." In Nelson, ed., *Purifying the Earthly Body of God*.

Sharma, B. D. *Globalization: The Tribal Encounter*. New Delhi: Arand Publications, 1996.

Sharma, B. K. "No Bhagirath Came" (*Koi Bhagirath nahi aya*). *India Today*, 15 July 1987, 80.

Sharma, R. K., ed. *Jnanabhaissjyamanjari*. Delhi: Nag Publishers, 1998.

Sherma, Rita DasGupta. "Sacred Immanence: Reflections of Ecofeminism in Hindu Tantra." In Nelson, ed., *Purifying the Earthly Body of God*.

Sheth, D. L. "Politics and Social Transformation: Grassroots Movements in India." In *The Constitutional Foundations of World Peace,* ed. Richard A. Falk, Robert C. Johansen, and Samuel S. Kim, 275–87. Albany: State University of New York Press, 1993.

Shiva, Vandana, ed. *Close to Home: Women Reconnect Ecology, Health, and Development Worldwide*. Phildadelphia: New Society Publishers, 1994.

———. "Ecology and Indian Myth." In Sen, ed., *Indigenous Vision*.

———. *Staying Alive: Women, Ecology, and Development*. London: Zed Books, 1988.

———. "Women in the Forest." In James, ed., *Ethical Perspectives,* 73–114.

———. "Women's Indigenous Knowledge and Biodiversity Convention." In Sen, ed., *Indigenous Vision*.

Shiva, Vandana, and J. Bandyopadhyay. "The Chipko Movement." In *Deforestation: Social Dynamics in Watersheds and Moutain Ecosystems,* ed. J. Ives and D. C. Pitt, 224–41. London: Routledge, 1988.

Sicular, Daniel T. *Scavengers, Recyclers, and Solutions for Solid Waste Management in Indonesia*. Berkeley: Center for Southeast Asia Studies, University of California at Berkeley, 1992.

Singh, R. P. B., ed. *Banaras (Varanasi): Cosmic Order, Sacred City, Hindu Traditions*. Varanasi: Tara Book Agency, 1993.

———. *Environmental Ethics*. Varanasi: National Geographic Society, 1993.

Singh, Shekar. "Sovereignty, Equality, and the Global Environment." In James, ed., *Ethical Perspectives,* 131–40.

Sivaramakrishnan. "Colonialism and Forestry in India: Imagining the Past in Present Politics." *Comparative Study of Society and History* 37, no. 1 (1995): 3–40.

Sivaramamurti, C. *Ganga*. Delhi: Orient Longman, 1976.

Sochaczewski, Paul Spencer. "The Saga of Krishna's Gardens: Can Love and Faith Heal Environmental Sacrilege?" *International Herald Tribune,* 18 October 1994.

Sohoni, S. Shrinivas, trans. *Prthvisukta*. New Delhi: Sterling Publishers, 1991.

Spanel, Ann. "Interview with Vandana Shiva." *Women of Power* 9 (1988): 27–31.

Strong, Maurice F. Foreword to *Traditional and Modern Approaches to the Environment on the Pacific Rim: Tensions and Values,* ed. Harold Coward. Albany: State University of New York Press, 1998.

Sullivan, Bruce M. "Theology and Ecology at the Birthplace of Krsna." In Nelson, ed., *Purifying the Earthly Body of God.*

Swami, Praveen. "Narmada Home-Truths: A Movement Makes Some Headway." Electronic release to Narmada Action Committee, 16 February 1995.

Tobias, Michael, and Georgianne Cowan, eds., *The Soul of Nature: Visions of a Living Earth.* New York: Continuum, 1994.

Tucker, Mary Evelyn, and John A. Grim, eds., *Worldviews and Ecology: Religion, Philosophy, and the Environment.* Maryknoll, N.Y.: Orbis Books, 1994.

Udall, Lori. "The International Campaign: A Case of Sustained Advocacy." In Fisher, ed., *Toward Sustainable Development?*

Vannucci, Marta. *Ecological Readings in the Veda: Matter, Energy, Life.* New Delhi: D. K. Printworld, 1994.

Vatsyayan, Kapila. "Ecology and Indian Myth." In Sen, ed., *Indigenous Vision.*

———, ed. *Prakrti: The Integral Vision.* 5 vols. New Delhi: Indira Gandhi National Centre for the Arts; D. K. Printworld, 1995.

Veer, Peter van der. *Gods on Earth: The Management of Religious Experience and Identity in a North Indian Pilgrimage Center.* London and Atlantic Highlands, N.J.: Athlone Press, 1988.

Vidyarthi, Lalita Prasad, Makhan Jha, and Baidyanath N. Saraswati. *The Sacred Complex of Kashi: A Microcosm of Indian Civilization.* Delhi: Concept Publishing, 1979.

Visvanathan, Shiv. *A Carnival of Science.* Delhi: Oxford University Press, 1997.

Vogler, J. A. "Waste Recycling in Developing Countries: A Review of the Social, Technological, and Market Forces." In *Managing Solid Wastes in Developing Countries,* ed. John R. Holmes, 241–66. New York: John Wiley and Sons, 1984.

Wadley, Susan S., and Bruce W. Derr. "Eating Sins in Karimpur." In *India through Hindu Categories,* ed. McKim Marriott. New Delhi: Sage Publications, 1990.

Ward, Geoffrey. "Benares, India's Most Holy City, Faces an Unholy Problem." *Smithsonian* 16, no. 6 (September 1985): 83–85.

Wilmer, Frank. *The Indigenous Voice in World Politics: Since Time Immemorial.* London: Sage, 1993.

World Bank. Bank Management Response to the Findings of the Independent Review. Washington, D.C., 1992.

———. India Irrigation Sector Review, volumes 1 and 2, Report No. 9518-IN, 20 December 1991.

———. "Sardar Sarovar Projects: Review of Current Status and Next Steps." 1992.

———. Staff Appraisal Report: India, Narmada River Development—Gujarat, Supplementary Data Volume, 1985.

"World's Toxic Trash Bin: India Accepting Toxic Products from Industrialized Nations." *India Currents Magazine,* 13 July 1995.

World Wide Fund for Nature. "Reviving the Sacred Forests of Vrindavana." World Wide Fund for Nature, India Technical Report, July 1993 to June 1994.

———. "Vrindavan Conservation Project." World Wide Fund for Nature, India brochure, 1995.

———. "Vrindavan Forest Revival Project." New Delhi: World Wide Fund for Nature—India, 1993.

Zimmermann, Francis. *The Jungle and the Aroma of Meats: An Ecological Theme in Hindu Medicine*. Berkeley and Los Angeles: University of California Press, 1987.

Notes on Contributors

Kelly D. Alley is Associate Professor of Anthropology in the Department of Sociology, Anthropology and Social Work at Auburn University. She has studied cultural interpretations of the intersections of wastewater and the river Gaṅgā and has co-directed a citizens' exchange project on environmental law in India. She has published articles in *Ethnology*, *Modern Asian Studies, City and Society,* and *Urban Anthropology,* among others, and chapters in edited volumes. She has just completed a book titled *On the Banks of the Ganga: Sacred Ecology and Waste*.

Anil Agarwal is the founder and director of the Centre for Science and Environment, a leading environmental public interest research institution in India with a deep interest in participatory natural resource management and pollution-related issues. The centre regularly publishes the *Citizens' Reports on the State of India's Environment*. Anil Agarwal has a deep interest in the material culture of India, especially as it relates to natural resource management.

Frédérique Apffel-Marglin is Professor of Anthropology at Smith College where she also codirects the Center for Mutual Learning. She has published the following books based on research in Orissa, India: *Wives of the God-King: The Rituals of the Devadasis of Puri* (Oxford University Press, 1985); *Purity and Auspiciousness in Indian Society,* edited with John Carman (E. J. Brill, 1985); *Dominating Knowledge: Development, Culture, and Resistance,* edited with Stephen A. Marglin (Oxford University Press, Clarendon Press, 1990); and *Decolonizing Knowledge: From Development to Dialogue*, edited with Stephen A. Marglin (Oxford University Press, Clarendon Press, 1996). She now collaborates with several NGOs in Peru and Bolivia and from that work has published *The Spirit of Regeneration: Andean Culture Confronting*

Western Notions of Development, edited with PRATEC (Zed Books and St. Martins Press, 1998).

Pratyusha Basu is a doctoral candidate in the Department of Geography, University of Iowa. Her research focuses on the interactions between gender, development, and resistance and is informed by postcolonial and feminist theoretical frameworks. Her dissertation will analyze the participation of women in dairy development in western India. She is currently engaged in fieldwork, with support from the Social Science Research Council and the National Science Foundation.

Harry W. Blair is Professor of Political Science at Bucknell University. A specialist in South Asia, he has worked and published extensively on that region, mainly on politics and development. He has served as a consultant to the U.S. Agency for International Development on these issues, as well as in natural resources management. During 1998–2000, while on leave from Bucknell, he served as a democracy specialist with USAID. In the natural resources management field, he is coauthor, with Porus D. Olpadwala, of *Forestry in Development Planning: Lessons from the Rural Experience* (Westview Press, 1988) as well as author of "Uses of Political Science in Agroforestry Interventions," in *Social Science Concepts in Agroforestry,* edited byWilliam Burch and Kathy Parker (Oxford/IBH Publishing, 1992), and "Democracy, Equity and Common Property Resource Management in the Indian Subcontinent," *Development and Change* 27, no. 3 (July 1996).

Christopher Key Chapple is Professor of Theological Studies and Director of Asian and Pacific Studies at Loyola Marymount University in Los Angeles. He is the author of *Karma and Creativity* (State University of New York Press, 1986) and *Nonviolence to Animals, Earth, and Self in Asian Traditions* (State University of New York Press, 1993); co-translator of the *Yoga Sutras* (Sri Satguru Publications, 1990); and editor of several books, including *Ecological Prospects: Scientific, Religious, and Aesthetic Perspectives* (State University of New York Press, 1994).

Chris Deegan first went to India in 1970 to work with rural development projects in Madhya Pradesh. He subsequently lived for over fourteen years in a village on the southern bank of the Narmada River and worked as a development consultant and cultural researcher. His doctoral fieldwork in South Asian history focused on circumambulation of the Narmada and secondary pilgrimage patterns in the Narmada valley. In addition to his ongoing work on the Narmada, his current research centers on the development of critical literacies to enhance cross-cultural interpretation. For the past ten years, he has also been the regional director for Asia Programs at the School for International Training.

O. P. Dwivedi, Ph.D., LL.D. (Hon.), Fellow of the Royal Society of Canada, teaches environmental policy and law and public administration. He has published twenty-six books and many articles and chapters in books. From 1986 to 1989 he was a member of a judicial tribunal, the Environmental Assessment Board of Ontario, Canada. He is a past president of the Canadian Political Science Association; a former vice president of the International Association of Schools and Institutes of Administration, Brussels, Belgium; and chair of the Research Committee on Technology and Development, the International Political Science Association.

William F. Fisher, Associate Professor of Anthropology and Social Studies at Harvard University, teaches courses on contemporary social movements, religion, identity and violence, and the politics of development. His recent publications include *Fluid Boundaries: Forming and Transforming Identity in Central Nepal* (Columbia University Press, forthcoming); "Doing Good? The Politics and Anti-Politics of NGO Practices," in the *Annual Review of Anthropology* (1997); and, as editor, *Toward Sustainable Development? Struggling over India's Narmada River* (M. E. Sharpe, 1995).

Ann Grodzins Gold holds a Ph.D. in anthropology from the University of Chicago, and is Professor in the Department of Religion at Syracuse University. Her work in the North Indian state of Rajasthan has included studies of pilgrimage, performance, world renunciation, women's expressive traditions, environmental change, and the transmission of environmental knowledge. Her publications include three books (all published by the University of California Press): *Fruitful Journeys: The Ways of Rajasthani Pilgrims* (1988); *A Carnival of Parting: The Tales of King Bharthari and King Gopi Chand* (1992); and *Listen to the Heron's Words: Reimagining Gender and Kinship in North India,* coauthored with Gloria Raheja (1994).

David L. Haberman (Ph.D., University of Chicago) is Professor in the Department of Religious Studies, Indiana University. His publications include *Journey Through the Twelve Forests: An Encounter with Krishna* (Oxford University Press, 1994) and *Acting as a Way of Salvation: A Study of Raganuga Bhakti Sadhana* (Oxford University Press, 1988). He recently finished a translation of a sixteenth-century Sanskrit text, published as *The Bhaktirasamrtasindhu of Rupa Gosvamin* (Indira Gandhi National Centre for the Arts, 2000). His current research involves a study of the theology and ecology of the Yamuna River of northern India.

George A. James received his Ph.D in religion from Columbia University in 1983. Over the past several years he has been conducting research on India's Environmental Philosophy. He recently edited a volume of essays entitled *Ethical Perspectives on Environmental Problems in India* (New Delhi, 1999). He teaches in the graduate program in environmental philosophy at the University of North Texas.

Madhu Khanna (Ph.D., Oxford University) is Associate Professor at the Indira Gandhi National Centre for the Arts, New Delhi. She is the author of two popular books, *The Tantric Way* (co-author; reprint, 1996) and *Yantra: The Tantric Symbol of Cosmic Unity* (reprint, 1997), and her seminal work *Sricakra of the Cult of Goddess Tripurasundari* is forthcoming. She was awarded the Homi Bhabha Fellowship (1991–1993) for her project on goddess ecology in India. She is the founding member of the Tantra Foundation, New Delhi.

Vinay Lal received his Ph.D. in South Asian studies from the University of Chicago in 1992. In recent years he has held fellowships from the National Endowment for the Humanities, the American Institute of Indian Studies, and the National Museum of Ethnology (Japan). He has been teaching history at the University of California, Los Angeles, since 1993. His papers on Gandhi and on an array of other subjects, including modern Indian history, the Indian diaspora, the politics of knowledge, popular culture and sexuality in India, and American cultural politics, have appeared in three dozen journals.

David Lee is Professor in the Department of Biological Sciences, Florida International University, Miami. A native of eastern Washington State, he received a Ph.D. from Rutgers University in 1970. Most of his research in the past thirty years has been on the functional ecology of plants, particularly in tropical Asia where he has lived over seven years. This research has led to some sixty scientific publications, numerous more general articles, and two books on environmental issues in the Asian tropics. He presently is studying the seedling shade responses of Indian forest trees, the function of anthocyanins in leaves, and structural coloration in plants.

Philip Lutgendorf received a Ph.D. in South Asian languages and civilizations from the University of Chicago in 1987. His research focuses on popular culture and oral performance traditions in the context of medieval and modern South Asia. His topics range from literary epics and rural folklore to twentieth-century films, television serials, and mass socioreligious movements. His book *The Life of a Text: Performing the* Rāmcaritmānas *of Tulsidas* (University of California Press, 1991) received the Ananda K. Coomaraswamy

Prize of the Association for Asian Studies in 1993. He is currently working on a study of the cult of the monkey-god Hanuman.

Mary McGee is Associate Professor of Classical Hinduism at Columbia University. As Director of Dharam Hinduja Indic Research Center at Columbia University from 1995–1999, she inaugurated an initiative on Indic Traditions of Environmental Consciousness, which included support of the 1998 Hinduism and Ecology conference at the Center for the Study of World Religions, Harvard University. Her research and publications focus on ethical, legal, and ritual concerns in classical Hindu texts and contemporary practice. She is currently completing a translation of the Yājñvalkyasmṛti, one of the foremost treatises on Hindu law.

Vijaya Nagarajan (Ph.D., University of California, Berkeley) is Assistant Professor of Theology and Religious Studies at the University of San Francisco. Her research interests include the fields of cultural studies and ecology, Hinduism, folklore, gender, art history, and South Asian literature. Her essay "Hosting the Divine: The Kolam in Tamil Nadu" appeared in *Mud, Mirror, and Thread: Folk Traditions of Rural India* (ed. Nora Fisher; Museum of New Mexico Press, 1993). She is the founder and co-director of the Institute for the Study of Natural and Cultural Resources.

Lance E. Nelson (Ph.D., McMaster University) is Associate Professor of Theology and Religious Studies at the University of San Diego. His writings on Advaita Vedanta and other aspects of South Asian religion have appeared in books and scholarly journals in the United States and India. Most recently, he edited *Purifying the Earthly Body of God: Religion and Ecology in Hindu India* (State University of New York Press, 1998).

Pramod Parajuli teaches anthropology, ecology, and social movements at Syracuse University. His research interests are in analyzing the intersection of social movements, ecology, and traditions of knowledge among ecological ethnicities—peasants, indigenous peoples, rural peasants, fisherfolk. Recently, he completed a book manuscript entitled *Tortured Bodies and Altered Earth: Ecological Ethnicities in the Regime of Globalization.* He is actively involved in various ethno-ecological movements and movements for sustainable livelihoods in his home country, Nepal, and in India.

Laurie L. Patton is Associate Professor of Early Indian Religion at Emory University. She is the author of *Myth as Argument: The Bṛhaddevatā as Canonical Commentary* (de Gruyter, 1996) and the editor of *Authority, Anxiety, and Canon: Essays in Vedic Interpretation* (State University Press of New

York, 1994); *Myth and Method* (University of Virginia Press, 1996); and the forthcoming *Jewels of Authority: Women and Text in the Hindu Tradition* (Oxford University Press, 2000). She is completing a book on the uses of poetry in Vedic ritual. She serves as chair of the Department of Religion at Emory.

K. L. Seshagiri Rao is professor emeritus in the Department of Religious Studies at the University of Virginia. He worked there as full professor from 1971–1995. Before joining the University of Virginia, he was professor and chair of the Guru Gobind Singh Department of Religious Studies, Punjab University, Patiala, India. He is the author or editor of several books and numerous articles. He is the chief editor of the forthcoming eighteen-volume *Encyclopedia of Hinduism* and coeditor of *World Faith Encounter,* a journal of the World Congress of Faiths, London. His areas of study and research are Indic religions, Gandhian studies, and interreligious dialogue.

T. S. Rukmani holds Ph.D. (1958) and D.Litt. (1991) degrees from the University of Delhi. She now holds the Chair in Hindu Studies at Concordia University, Montreal, Canada. She also held the first Chair in Hindu Studies and Indian Philosophy at the University of Durban-Westville, South Africa, from 1993 to 1995. She was the recipient of the Ida Smedley International Fellowship of the International Federation of University Women, 1972–1973 and received the Delhi Sanskrit Academy Award in 1993. She is the author of ten books, including the four-volume Yogavārttika of Vijñānabhikṣu, and numerous research papers. Her main work is in Indian philosophy, especially Advaita Vedānta, Sāṅkhya, and Yoga.

Larry D. Shinn is currently President of Berea College, a liberal arts college in Berea, Kentucky, dedicated to serving the Appalachian region. He received his Ph.D. in the history of religions from Princeton University, taught at Oberlin College for fourteen years, and served as Dean and Vice President at Bucknell University for ten years. He has authored two books, *Two Sacred Worlds: Experience and Structure in the World's Religions* (Abingdon Press, 1977) and *The Dark Lord: Cult Images and the Hare Krishnas in America* (Westminster Press, 1987), and has coauthored and edited four others.

Jael Silliman is Assistant Professor in the Women's Studies Program at the University of Iowa. She has an Ed.D. from the Graduate School of Education, Columbia University. She works on gender and development issues and transnational movements for social change. She has recently coedited *Dangerous Intersections: Feminist Perspectives on Population, Environment, and Development* (Southend Press and Zed Press, 1999). She has also written

"Making the Connections: Environmental Justice and the Women's Health Movement," in *Journal of Race, Class and Gender: Issue on Environmental Justice*. She is currently a Fellow of the Open Society Institute and is working on a project entitled "The Politics of Inclusion: Women of Color and the Reproductive Rights Movement."

Mary Evelyn Tucker is Professor of Religion at Bucknell University in Lewisburg, Pennsylvania. She received her Ph.D. from Columbia University in the history of religions, specializing in Confucianism in Japan. She has published *Moral and Spiritual Cultivation in Japanese Neo-Confucianism* (State University of New York Press, 1989) and is coeditor, with John Grim, of *Worldviews and Ecology* (Bucknell University Press, 1993); with Duncan Williams, of *Buddhism and Ecology* (Harvard University Center for the Study of World Religions, 1997); and, with John Berthrong, of *Confucianism and Ecology* (Harvard University Center for the Study of World Religions, 1998).

Index